ADVANCED MODEL PREDICTIVE CONTROL

Edited by **Tao ZHENG**

Advanced Model Predictive Control
Edited by Tao ZHENG

Published by InTech

Janeza Trdine 9, 51000 Rijeka, Croatia

Publishing Process Manager Romina Krebel

Technical Editor Teodora Smiljanic

Cover Designer InTech Design Team

Additional hard copies can be obtained from orders@intechopen.com

Advanced Model Predictive Control, Edited by Tao ZHENG
p. cm.
ISBN 978-953-307-298-2

For EU product safety concerns:
IN TECH d.o.o., Prolaz Marije Krucifikse Kozulić 3, 51000 Rijeka, Croatia,
info@intechopen.com or visit our website at intechopen.com.

Meet the editor

Dr. Tao ZHENG was born in Chongqing, China, in1981. In 2003 and 2008, he received his B.E. degree and Ph. D. degree from University of Science and Technology of China, both in Automatic Control. He was awarded the President's Excellent Prize of Chinese Academy of Sciences in 2008. He is currently at the Department of Automation, Hefei University of Technology, China. Dr. ZHENG's recent interests are in Advanced Model Predictive Control (including Nonlinear Model Predictive Control and Multi-objective Model Predictive Control), Multi-objective Optimization and Multi-objective Genetic Algorithm. He is also interested in the application of optimization theory in economic field, such as the optimization problem in currency-circulation system.

Contents

Preface

Since the earliest algorithm of Model Predictive Control was proposed by French engineer Richalet and his colleagues in 1978, the explicit background of industrial application has made MPC develop rapidly. Different from most other control algorithms, the research trajectory of MPC is originated from engineering application and then expanded to theoretical field, while ordinary control algorithms often have applications after sufficient theoretical work.

Nowadays, MPC is not just the name of one or some specific computer control algorithms, but the name of a specific controller design thought, which can derive many kinds of MPC controllers for almost all kinds of systems, linear or nonlinear, continuous or discrete, integrated or distributed. However, the basic characters of MPC can be simply summarized as a model used for prediction, online optimization based on prediction and feedback compensation, while there is no special demand on the form of the system model, the computational tool for online optimization and the form of feedback compensation.

The linear MPC theory is now comparatively mature, so its applications can be found in almost every domain in modern engineering. But robust MPC and nonlinear MPC (NMPC) are still problems for us. Though there are some constructive results because many efforts have been made on them in these years, they will remain the focus of MPC research for a long period in the future.

In the first part of this book, to present recent theoretical developments of MPC, Chapter 1 to Chapter 3 introduce three kinds of Fast Model Predictive Control, and Chapter 4 presents Model Predictive Control for distributed systems. Model Predictive Control for nonlinear systems, multi-variable systems and other special model are proposed in Chapters 5 through 10.

To give the readers successful examples of MPC's recent applications, in the second part of the book, Chapters 11 through 18 introduce some of them, from sugar crystallization process to paper-making system, from linear system to nonlinear system. They can, not only help the readers understand the characteristics of MPC more clearly, but also give them guidance how to use MPC to solve practical problems.

Authors of this book truly want it to be helpful for researchers and students who are concerned about MPC, and further discussions on the contents of this book are warmly welcome.

Finally, thanks to InTech and its officers for their efforts in the process of edition and publication, and thanks to all the people who have made contributes to this book, including our dear family members.

ZHENG Tao
Hefei University of Technology,
China

Part 1

New Theory of Model Predictive Control

1

Fast Model Predictive Control and its Application to Energy Management of Hybrid Electric Vehicles

Sajjad Fekri and Francis Assadian
Automotive Mechatronics Centre, Department of Automotive Engineering
School of Engineering, Cranfield University
UK

1. Introduction

Modern day automotive engineers are required, among other objectives, to maximize fuel economy and to sustain a reasonably responsive car (i.e. maintain driveability) while still meeting increasingly stringent emission constraints mandated by the government. Towards this end, Hybrid Electric Vehicles (HEVs) have been introduced which typically combine two different sources of power, the traditional internal combustion engine (ICE) with one (or more) electric motors, mainly for optimising fuel efficiency and reducing Carbon Dioxide (CO_2) and greenhouse gases (GHG) (Fuhs, 2008).

Compared to the vehicles with conventional ICE, hybrid propulsion systems are potentially capable of improving fuel efficiency for a number of reasons: they are able to recover some portion of vehicle kinetic energy during braking and use this energy for charging the battery and hence, utilise the electric motor at a later point in time as required. Also, if the torque request (demanded by driver) is below a threshold torque, the ICE can be switched off as well as during vehicle stop for avoiding engine idling. These are in fact merely few representative advantages of the hybrid vehicles compared to those of conventional vehicles. There are also other benefits hybrid electric vehicles could offer in general, e.g. engine downsizing and utilising the electric motor/motors to make up for the lost torque. It turns out that the internal combustion engine of the hybrid electric vehicle can be potentially designed with a smaller size and weight which results in higher fuel efficiency and lower emissions (Steinmaurer & Del Re, 2005).

Hybrid electric vehicles have been received with great enthusiasm and attention in recent years (Anderson & Anderson, 2009). On the other hand, complexity of hybrid powertrain systems have been increased to meet end-user demands and to provide enhancements to fuel efficiency as well as meeting new emission standards (Husain, 2003).

The concept of sharing the requested power between the internal combustion engine and electric motor for traction during vehicle operation is referred to as "vehicle supervisory control" or "vehicle energy management" (Hofman & Druten, 2004). The latter term, employed throughout this chapter, is particularly referred to as a control allocation for delivering the required wheel torque to maximize the average fuel economy and sustain the battery state of charge (SoC) within a desired charging range (Fekri & Assadian, 2011).

The vehicle energy management development is a challenging practical control problem and a significant amount of research has been devoted to this field for full HEVs and Electric Vehicles (EVs) in the last decade (Cundev, 2010). To tackle this challenging problem, there are currently extensive academic and industrial research interests ongoing in the area of hybrid electric vehicles as these vehicles are expected to make considerable contributions to the environmentally conscious requirements in the production vehicle sector in the future – see (Baumann et al., 2000) and other references therein.

In this regard, we shall analysis and extend the study done by (Sciarretta & Guzzella, 2007) on the number of IEEE publications published between 1985 and 2010. Figure 1 depicts the number of publications recorded at the IEEE database[1] whose abstract contains at least one of the strings "hybrid vehicle" or "hybrid vehicles".

From Figure 1, it is obvious that the number of publications in the area of hybrid electric vehicles (HEVs) has been drastically increased during this period, from only 2 papers in 1985 to 552 papers in 2010. Recall that these are only publications of the IEEE database - there are many other publications than those of the IEEE including books, articles, conference papers, theses, filed patents, and technical reports which have not been taken into account in this study. Besides, a linear regression analysis of the IEEE publications shown in Figure 1 indicates that research in the field of hybrid vehicles has been accelerated remarkably since 2003. One may also predict that the number of publications in this area could be increased up to about 1000 articles in 2015, that is nearly twice as many as in 2010 - this is a clear evidence to acknowledge that HEVs research and development is expected to make considerable contributions to both academia and industry of production automotive sector in the future.

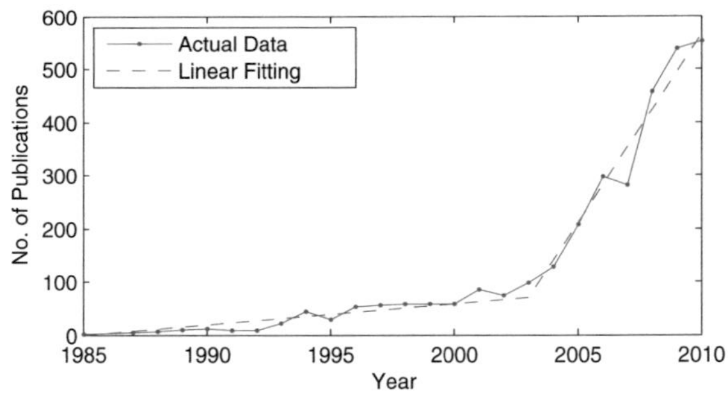

Fig. 1. Hybrid vehicle research trend based on the number of publications of the IEEE over the period 1985 to 2010.

Here are the facts and regulations which must be taken into consideration by automotive engineers:

- Due to the ever increasing stringent regulations on fuel consumption and emissions, there are tremendous mandates on Original Equipment Manufacturers (OEMs) to deliver fuel-efficient less-polluting vehicles at lower costs. Hence, the impact of advanced controls for the application of the hybrid vehicle powertrain controls has become extremely important (Fekri & Assadian, 2011).

[1] See http://ieeexplore.ieee.org for more information.

- It is essential to meet end-user demands for increasingly complex new vehicles towards improving vehicle performance and driveability (Cacciatori et al., 2006), while continuing to reduce costs and meeting new emission standards.
- There is a continuous increase in the gap between the theoretical control advancement and the control strategies being applied to the existing production vehicles. This gap is resulting on significant missed opportunities in addressing some fundamental functionalities, e.g. fuel economy, emissions, driveability, unification of control architecture and integration of the Automotive Mechatronics units on-board vehicle. It seems remarkably vital to address how to bridge this gap.
- Combined with ever-increasing computational power, fast online optimisation algorithms are now more affordable to be developed, tested and implemented in the future production vehicles.

There are a number of energy management methods proposed in the literature of hybrid vehicles to minimize fuel consumption and to reduce CO_2 emissions (Johnson et al., 2000). Among these energy management strategies, a number of heuristics techniques, say e.g. using rule-based or Fuzzy logic, have attempted to offer some improvements in the HEV energy efficiency (Cikanek & Bailey, 2002; Schouten et al., 2002) where the optimisation objective is, in a heuristic manner, a function of weighted fuel economy and driveability variables integrated with a performance index, to obtain a desired closed-loop system response. However, such heuristics based energy management approaches suffer from the fact that they guarantee neither an optimal result in real vehicle operational conditions nor a robust performance if system parameters deviate from their nominal operating points. Consequently, other strategies have emerged that are based on optimisation techniques to search for sub-optimal solutions. Most of these control techniques are based on programming concepts (such as linear programming, quadratic programming and dynamic programming) and optimal control concepts, to name but a few (Ramsbottom & Assadian, 2006; Ripaccioli et al., 2009; Sciarretta & Guzzella, 2007). Loosely speaking, these techniques do not offer a feasible casual solution, as the future driving cycle is assumed to be entirely known. Moreover, the required burdensome calculations of these approaches put a high demand on computational resources which prevent them to be implemented on-line in a straightforward manner. Nevertheless, their results could be used as a benchmark for the performance of other strategies, or to derive rules for rule-based strategies for heuristic based energy management of HEVs (Khayyam et al., 2010).

Two new HEV energy management concepts have been recently introduced in the literature. In the first approach, instead of considering one specific driving cycle for calculating an optimal control law, a set of driving cycles is considered resulting in the stochastic optimisation approach. A solution to this approach is calculated off-line and stored in a state-dependent lookup table. Similar approach in this course employs Explicit Model Predictive Control (Beccuti et al., 2007; Pena et al., 2006). In this design methodology, the entire control law is computed offline, where the online controller will be implemented as a lookup table, similar to the stochastic optimisation approach. The lookup table provides a quasi-static control law which is directly applicable to the on-line vehicle implementation. While this method has potential to perform well for systems with fewer states, inputs, constraints, and "sufficiently short" time-horizons (Wang & Boyd, 2008), it cannot be utilised in a wide variety of applications whose dynamics, cost function and/or constraints are time-varying due to e.g.

parametric uncertainties and/or unmeasurable exogenous disturbances. In other words, any lookup table based optimisation approach may end up with severe difficulties in covering a real-world driving situation with a set of individual driving cycle. A recent approach has endeavored to decouple the optimal solution from a driving cycle in a game-theoretic (GT) framework (Dextreit et al., 2008). In this approach, the effect of the time-varying parameters (namely drive cycle) is represented by the actions of the first player while the effect of the operating strategy (energy management) is modeled by the actions of the second player. The first player (drive cycle) wishes to maximize the performance index which reflects the optimisation objectives, say e.g. to minimise emission constraints and fuel consumption, while the second player aims to minimize this performance index. Solutions to these approaches are calculated off-line and stored in a state-dependent lookup tables. These look up tables provide a quasi-static control law which is directly suitable for on-line vehicle implementation. Similar to previous methods, the main drawbacks of the game-theoretic approach are the lack of robustness and due to quasi-static nature of this method, it cannot address vehicle deriveability requirements.

If only the present state of the vehicle is considered, optimisation of the operating points of the individual components can still be beneficial. Typically, the proposed methods define an optimisation criterion to minimise the vehicle fuel consumption and exhaust emissions (Kolmanovsky et al., 2002). A weighting factor can be included to prevent a drift in the battery from its nominal energy level and to guarantee a charge sustaining solution. This approach has been considered in the past, but it is still remained immensely difficult task to select a weighting factor that is mathematically sound (Rousseau et al., 2008). An alternative approach is to extend the objective function with a fuel equivalent term. This term includes the corresponding fuel use for the energy exchange with the battery in the optimisation criterion (Kessels, 2007).

Hybrid modeling tools have been recently developed to analyse and optimise a number of classes of hybrid systems. Among many other modeling tools developed to represent the hybrid systems, we shall refer to Mixed Logical Dynamical (MLD) (Bemporad & Morary, 1999), HYbrid Systems Description Language (HYSDEL) (Torrisi & Bemporad, 2004), and Piecewise Affine (PWA) models (Ripaccioli et al., 2009; Sontag, 1981), to name but a few. In addition, Hybrid Toolbox for MATLAB (Bemporad, 2004) is developed for modeling, simulation, and verifying hybrid dynamical models and also for designing hybrid model predictive controllers. Almost all of these hybrid tools, however, are only suitable for slow applications and can not attack the challenging fast real-time optimisation problems, e.g., for the use of practical hybrid electric vehicle energy management application.

Two fundamental drawbacks of aforementioned strategies are firstly their consideration of driveability being an afterthought and secondly the driveability issue is considered in an ad-hoc fashion as these approaches are not model-based dynamic. Applicable techniques such as game-theoretic based optimisation method utilise quasi-static models which are not sufficient to address driveability requirements (Dextreit et al., 2008).

Towards a feasible and tractable optimisation approach, there are a number of model-based energy management methods such as Model Predictive Controls (MPC). A recently developed package for the hybrid MPC design is referred to as Hybrid and Multi-Parametric Toolboxes (Narciso et al., 2008) which is based on the traditional model predictive control optimisation alternatives using generic optimisers. The main shortcoming of traditional model predictive control methods is that they can only be used in applications with "sufficiently slow" dynamics

(Wang & Boyd, 2008), and hence are not suitable for many practical applications including HEV energy management problem. For this reason the standard MPC algorithms have been retained away from modern production vehicles. In fact, a number of inherent hardware constraints and limitations integrated with the vehicle electronic control unit (ECU), such as processing speed and memory, have made on-line implementations of these traditional predictive algorithms almost impossible. In a number of applications, MPC is currently applied off-line to generate the required maps and then these maps are used on-line. However, generation and utilisation of maps defeat the original purpose of designing a dynamic compensator which maintains driveability. Therefore, there is a vital need of increased processing speed, with an appropriate memory size, so that an online computation of "fast MPC" control law could be implemented in real applications.

In this chapter, we shall describe a method for improving the speed of conventional model predictive control design, using online optimisation. The method proposed would be a complementary for offline methods, which provide a method for fast control computation for the problem of energy management of hybrid electric vehicles. We aim to design and develop a practical fast model predictive feedback controller (FMPC) to replace the current energy management design approaches as well as to address vehicle driveability issues. The proposed FMPC is derived based on the dynamic models of the plant and hence driveability requirements are taken into consideration as part of the controller design. In this development, we shall extend the previous studies carried out by Stephen Boyd and his colleagues at Stanford University, USA, on fast model predictive control algorithms. In this design, we are also able to address customising the robustness analysis in the presence of parametric uncertainties due to, e.g., a change in the dynamics of the plant, or lack of proper estimation of the vehicle load torque (plant disturbance).

In this chapter, we shall also follow and overview some of theoretical and practical aspects of the fast online model predictive control in applying to the practical problem of hybrid electric vehicle energy management along with representing some of simulation results. The novelty of this work is indeed in the design and development of the fast robust model predictive control concept with practical significance of addressing vehicle driveability and automotive actuator control constraints. It is hoped that the results of this work could make automotive engineers more enthusiastic and motivated to keep an eye on the development of state-of-the-art Fast Robust Model Predictive Control (FMPC) and its potential to attack a wide range of applications in the automotive control system designs.

In the remaining of this chapter, we will describe in detail the mathematical description, objectives and constraints along with the optimisation procedure of the proposed fast model predictive control. We shall also provide dynamical model of the hybrid electric vehicle (parallel, with diesel engine) to which the FMPC will be applied. Simulation results of the HEV energy management system will be demonstrated to highlight some of the concepts proposed in this chapter which will offer significant improvements in fuel efficiency over the base system.

2. Fast Model Predictive Control

The Model Predictive Control (MPC), referred also to as Receding Horizon Control (RHC), and its different variants have been successfully implemented in a wide range of practical applications in industry, economics, management and finance, to name a few (Camacho &

Bordons, 2004; Maciejowski, 2002). A main advantage of MPC algorithms, which has made these optimisation-based control system designs attractive to the industry, is their abilities to handle the constraints directly in the design procedure (Kwon & Han, 2005). These constraints may be imposed on any part of the system signals, such as states, outputs, inputs, and most importantly actuator control signals which play a key role in the closed-loop system behaviour (Tate & Boyd, 2001).

Although very efficient algorithms can currently be applied to some classes of practical problems, the computational time required for solving the optimisation problem in real-time is extremely high, in particularly for fast processes, such as energy management of hybrid electric vehicles. One method to implement a fast MPC is to compute the solution of a multiparametric quadratic or linear programming problem explicitly as a function of the initial state which could turn into a relatively easy-to-implement piecewise affine controller (Bemporad et al., 2002; Tondel et al., 2003). However, as the control action implemented online is in the form of a lookup table, it could exponentially grow with the horizon, state and input dimensions. This means that any form of explicit MPC could only be applied to small problems with few state dimensions (Milman & Davidson, 2003). Furthermore, due to there being off-line lookup table, explicit MPC cannot deal with applications whose dynamics, cost function and/or constraints are time-varying (Wang & Boyd, 2008). A non-feasible active set method was proposed in (Milman & Davidson, 2003) for solving the Quadratic Programming (QP) optimisation problem of the MPC. However, to bear further explanation, these studies have not addressed any comparison to the other earlier optimisation methods using primal-dual interior point methods (Bartlett et al., 2000; Rao et al., 1998). Another fast MPC strategy was introduced in (Wang & Boyd, 2010) which has tackled the problem of solving a block tridiagonal system of linear equations by coding a particular structure of the QPs arising in MPC applications (Vandenberghe & Boyd, 2004; Wright, 1997), and by solving the problem approximately. Starting from a given initial state and input trajectory, the fast MPC software package solves the optimization problem fast by exploiting its special structure. Due to using an interior-point search direction calculated at each step, any problem of any size (with any number of state dimension, input dimension, and horizon) could be tackled at every operational time step which in return will require only a limited number of steps. Therefore, the complexity of MPC is significantly reduced compared to the standard MPC algorithms. While this algorithm could be scaled in any problem size in principle, a drawback of this method is that it is a custom hand-coded algorithm, ie. the user should transform their problem into the standard form (Wang & Boyd, 2010; 2008) which might be very time-consuming. Moreover, one may require much optimisation expertise to generate a custom solver code. To overcome this shortcoming, a very recent research (Mattingley & Boyd, 2010a;b; 2009) has studied a development of an optimisation software package, referred to as CVXGEN, based on an earlier work by (Vandenberghe, 2010), which automates the conversion process, allowing practitioners to apply easily many class of convex optimisation problem conversions. CVXGEN is effectively a software tool which helps to specify one's problem in a higher level language, similar to other parser solvers such as SeDuMi or SDPT3 (Ling et al., 2008). The drawback of CVXGEN is that it is limited to optimization problems with up to around 4000 non-zero Karush-Kuhn-Tucker (KKT) matrix entries (Mattingley & Boyd, 2010b). In the next section, we will extend the work done by (Mattingley & Boyd, 2010b) and propose a new fast KKT solving approach, which alleviates the aforementioned limitation to some extent. We will implement our method on a hybrid electric vehicle energy management application in Section 4.

2.1 Quadratic Programming (QPs)

In convex QP problems, we typically minimize a convex quadratic objective function subject to linear (equality and/or inequality) constraints. Let us assume a convex quadratic generalisation of the standard form of the QP problem is

$$\begin{aligned} \min \;\; & (1/2)x^T Qx + c^T x \\ \text{subject to} \;\; & Gx \leq h, \\ & Ax = b. \end{aligned} \tag{1}$$

where $x \in R^n$ is the variable of the QP problem and Q is a symmetric $n \times n$ positive semidefinite matrix.

An interior-point method, in comparison to other methods such as primal barrier method, is particularly appropriate for embedded optimization, since, with proper implementation and tuning, it can reliably solve to high accuracy in 5-25 iterations, without even a "warm start" (Wang & Boyd, 2010).

In order to obtain a cone quadratic program (QP) using the QP optimisation problem of Equation (1), it is expedient for the analysis and implementation of interior-point methods to include a slack variable s and solve the equivalent QP

$$\begin{aligned} \min \;\; & (1/2)x^T Qx + c^T x \\ \text{subject to} \;\; & Gx + s = h, \\ & Ax = b, \\ & s \geq 0. \end{aligned} \tag{2}$$

where $x \in R^n$ and $s \in R^p$ are the variables of the cone QP problem.

The dual problem of Equation (3) can be simply derived by introducing an additional variable ω: (Vandenberghe, 2010)

$$\begin{aligned} \max \;\; & -(1/2)\omega^T Q\omega - h^T z - b^T y \\ \text{subject to} \;\; & G^T z + A^T y + c + Q\omega = 0, \\ & z \geq 0. \end{aligned} \tag{3}$$

where $y \in R^m$ and $z \in R^p$ are the Lagrange multiplier vectors for the equality and the inequality constraints of (1), respectively.

The dual objective of (3) provides a lower bound on the primal objective, while the primal objective of (1) gives an upper bound on the dual (Vandenberghe & Boyd, 2004). The vector $x^* \in R^n$ is an optimal solution of Equation (1) if and only if there exist Lagrange multiplier vectors $z^* \in R^p$ and $y^* \in R^m$ for which the following necessity KKT conditions hold for $(x, y, z) = (x^*, y^*, z^*)$; see (Potra & Wright, 2000) and other references therein for more details.

$$F(x,y,z,s) = \begin{bmatrix} Qx + A^T y + G^T z + c \\ Ax - b \\ Gx + s - h \\ ZS\mathbf{e} \end{bmatrix} = 0, \quad (4)$$
$$(s,z) \geq 0$$

where $S = diag(s_1, s_2, \ldots, s_n)$, $Z = diag(z_1, z_2, \ldots, z_n)$ and $\mathbf{e}$ is the unit column vector of size $n \times 1$.

The primal-dual algorithms are modifications of Newton's method applied to the KKT conditions $F(x,y,z,s) = 0$ for the nonlinear equation of Equation (4). Such modifications lead to appealing global convergence properties and superior practical performance. However, they might interfere with the best-known characteristic of the Newton's method, that is "fast asymptotic convergence" of Newton's method. In any case, it is possible to design algorithms which recover such an important property of fast convergence to some extent, while still maintaining the benefits of the modified algorithm (Wright, 1997). Also, it is worthwhile to emphasise that all primal-dual approaches typically generate the iterates (x_k, y_k, z_k, s_k) while satisfying nonnegativity condition of Equation (4) strictly, i.e. $s_k > 0$ and $z_k > 0$. This particular property is in fact the origin of the generic term "interior-point" (Wright, 1997) which will be briefly discussed next.

2.2 Embedded QP convex optimisation

There are several numerical approaches to solve standard cone QP problems. One alternative which seems suitable to the literature of fast model predictive control is the path-following algorithm – see e.g. (Potra & Wright, 2000; Renegar & Overton, 2001) and other references therein.

In the path-following method, the current iterates are denoted by (x_k, y_k, z_k, s_k) while the algorithm is started at initial values $(x_k, y_k, z_k, s_k) = (x_0, y_0, z_0, s_0)$ where $(s_0, z_0) > 0$. For most problems, however, a strictly feasible starting point might be extremely difficult to find. Although it is straightforward to find a strictly feasible starting point by reformulating the problem – see (Vandenberghe, 2010, §5.3)), such reformulation may introduce distortions that can potentially make the problem harder to solve due to an increased computational time to generate real-time control law which is not desired for a wide range of practical applications, e.g. the HEV energy management problem – see Section 4. In §2.4, we will describe one tractable approach to obtain such feasible starting points.

Similar to many other iterative algorithms in nonlinear programming and optimisation literature, the primal-dual interior-point methods are based on two fundamental concepts: First, they contain a procedure for determining the iteration step and secondly they are required to define a measure of the attraction of each point in the search space. The utilised Newton's method in fact forms a linearised model for $F(x,y,z,s)$ around the current iteration point and obtains the search direction $(\Delta x, \Delta y, \Delta z, \Delta s)$ by solving the following set of linear equations:

$$J(x,y,z,s) \begin{bmatrix} \Delta_x \\ \Delta_y \\ \Delta_z \\ \Delta_k \end{bmatrix} = -F(x,y,z,s) \quad (5)$$

where J is the Jacobian of F at point (x_k, y_k, z_k, s_k).
Let us assume that the current point is strictly feasible. In this case, a Newton "full step" will provide a direction at

$$\begin{bmatrix} Q & A^T & G^T & 0 \\ A & 0 & 0 & 0 \\ G & 0 & 0 & I \\ 0 & 0 & S & Z \end{bmatrix} \begin{bmatrix} \Delta x_k \\ \Delta y_k \\ \Delta z_k \\ \Delta s_k \end{bmatrix} = -F(x_k, y_k, z_k, s_k) \tag{6}$$

and the next starting point for the algorithm will be

$$(x_{k+1}, y_{k+1}, z_{k+1}, s_{k+1}) = (x_k + \Delta x_k, y_k + \Delta y_k, z_k + \Delta z_k, s_k + \Delta s_k)$$

However, the pure Newton's method, i.e. a full step along the above direction, could often violate the condition $(s, z) > 0$ – see (Renegar & Overton, 2001). To resolve this shortcoming, a line search along the Newton direction is in a way that the new iterate will be (Wright, 1997)

$$(x_k, y_k, z_k, s_k) + \alpha_k(\Delta x_k, \Delta y_k, \Delta z_k, \Delta s_k)$$

for some line search parameter $\alpha \in (0, 1]$. If α is to be selected by user, one could only take a small step $(\alpha \ll 1)$ along the direction of Equation (6) before violating the condition $(s, z) > 0$. However, selecting such a small step is not desirable as this may not allow us to make much progress towards a sound solution to a broad range of practical problems which usually are in need of fast convenance by applying "sufficiently large" step sizes.
Following the works (Mattingley & Boyd, 2010b) and (Vandenberghe, 2010), we shall intend to modify the basic Newton's procedure by two scaling directions (i.e. affine scaling and combined centering & correction scaling). Loosely speaking, by using these two scaling directions, it is endeavoured to bias the search direction towards the interior of the nonnegative orthant $(s, z) > 0$ so as to move further along the direction before one of the components of (s, z) becomes negative. In addition, these scaling directions keep the components of (s, z) from moving "too close" to the boundary of the nonnegative orthant $(s, z) > 0$. Search directions computed from points that are close to the boundary tend to be distorted from which an inferior progress could be made along those points – see (Wright, 1997) for more details. Here, we shall list the scaling directions as follows.

2.3 Scaling iterations

We follow the works by (Vandenberghe, 2010, §5.3) and (Mattingley & Boyd, 2010b) with some modifications that reflect our notation and problem format. Starting at initial values $(\hat{x}, \hat{y}, \hat{z}, \hat{s}) = (x_0, y_0, z_0, s_0)$ where $s_0 > 0$, $z_0 > 0$, we consider the scaling iterations as summarised here.

- Step 1. Set $k = 0$.
- Step 2. Start the iteration loop at time step k.
- Step 3. Define the residuals for the three linear equations as:

$$\begin{bmatrix} r_x \\ r_y \\ r_z \end{bmatrix} = \begin{bmatrix} 0 \\ 0 \\ \hat{s} \end{bmatrix} + \begin{bmatrix} Q & A^T & G^T \\ A & 0 & 0 \\ G & 0 & 0 \end{bmatrix} \begin{bmatrix} \hat{x} \\ \hat{y} \\ \hat{z} \end{bmatrix} + \begin{bmatrix} c \\ -b \\ -h \end{bmatrix}$$

- Step 4. Compute the optimality conditions:

$$\begin{bmatrix} 0 \\ 0 \\ s \end{bmatrix} = \begin{bmatrix} Q & A^T & G^T \\ -A & 0 & 0 \\ -G & 0 & 0 \end{bmatrix} \begin{bmatrix} x \\ y \\ z \end{bmatrix} + \begin{bmatrix} c \\ b \\ h \end{bmatrix}, \quad (s,z) \geq 0.$$

- Step 5. If the optimality conditions obtained at Step 4 satisfy $\|(x,y,z,s) - (\hat{x},\hat{y},\hat{z},\hat{s})\|_\infty \leq \epsilon$, for some small positive $\epsilon > 0$, go to Step 13.
- Step 6. Solve the following linear equations to generate the affine direction (Mattingley & Boyd, 2010b):

$$\begin{bmatrix} Q & A^T & G^T & 0 \\ A & 0 & 0 & 0 \\ G & 0 & 0 & I \\ 0 & 0 & S & Z \end{bmatrix} \begin{bmatrix} \Delta x_k^{aff} \\ \Delta y_k^{aff} \\ \Delta z_k^{aff} \\ \Delta s_k^{aff} \end{bmatrix} = -F(x_k, y_k, z_k, s_k) \tag{7}$$

- Step 7. Compute the duality measure μ, step size $\alpha \in (0,1]$, and centering parameter $\sigma \in [0,1]$

$$\mu = \frac{1}{n}\sum_{i=1}^{n} s_i z_i = \frac{z^T s}{n}$$
$$\sigma = \left(\frac{(s+\alpha_c \Delta s^{aff})^T (z+\alpha_c \Delta z^{aff})}{s^T z} \right)^3$$

and

$$\alpha_c = \sup\{\alpha \in [0,1] | (s + \alpha_c \Delta s^{aff}, z + \alpha_c \Delta z^{aff}) \geq 0\}$$

- Step 8. Solve the following linear equations for the combined centering-correction direction[2]:

$$\begin{bmatrix} Q & A^T & G^T & 0 \\ A & 0 & 0 & 0 \\ G & 0 & 0 & I \\ 0 & 0 & S & Z \end{bmatrix} \begin{bmatrix} \Delta x^{cc} \\ \Delta y^{cc} \\ \Delta z^{cc} \\ \Delta s^{cc} \end{bmatrix} = \begin{bmatrix} 0 \\ 0 \\ 0 \\ \sigma\mu\mathbf{e} - \operatorname{diag}(\Delta s^{aff})\Delta z^{aff} \end{bmatrix}$$

  This system is well defined if and only if the Jacobian matrix within is nonsingular (Peng et al., 2002, §6.3.1).
- Step 9. Combine the two affine and combined updates of the required direction as:

$$\Delta x = \Delta x^{aff} + \Delta x^{cc}$$
$$\Delta y = \Delta y^{aff} + \Delta y^{cc}$$
$$\Delta z = \Delta z^{aff} + \Delta z^{cc}$$
$$\Delta s = \Delta s^{aff} + \Delta s^{cc}$$

- Step 10. Find the appropriate step size to retain nonnegative orthant $(s,z) > 0$,

$$\alpha = \min\{1, 0.99 \sup(\alpha \geq 0 | (s + \alpha\Delta s, z + \alpha\Delta z) \geq 0)\}$$

[2] This is another variant of Mehrotra's predictor-corrector algorithm (Mehrotra, 1992), a primal-dual interior-point method, which yields more consistent performance on a wide variety of practical problems.

- Step 11. Update the primal and dual variables using:

$$\begin{bmatrix} x \\ y \\ z \\ s \end{bmatrix} := \begin{bmatrix} x \\ y \\ z \\ s \end{bmatrix} + \alpha \begin{bmatrix} \Delta x \\ \Delta y \\ \Delta z \\ \Delta s \end{bmatrix}$$

- Step 12. Set $(\hat{x}, \hat{y}, \hat{z}, \hat{s}) = (x_k, y_k, z_k, s_k)$ and $k := k+1$; Go to Step 2.
- Step 13. Stop the iteration and return the obtained QP optimal solution (x, y, z, s).

The above iterations will modify the the search direction so that at any step the solutions are moved closer to feasibility as well as to centrality. It is also emphasised that most of the computational efforts required for a QP problem are due to solving the two matrix equalities in steps 6 and 8. Among many limiting factors which may make the above algorithm failed, floating-point division is perhaps the most critical problem of an online optimisation algorithm to be considered (Wang & Boyd, 2008). In words, stability of an optimisation-based control law (such as model predictive control) are significantly dependent on the risk of algorithm failures, and therefore it is vital to develop robust algorithms for solving these linear systems leading towards fast optimisation-based control designs, which is the focal point of our work. We should also stress that robustness of any algorithm must be taken into account at starring point. In particular, many practical problems are prone to make optimisation procedures failed at the startup. For instance, (possibly large) disparity between the initial states of the plant and the feedback controller might lead to large transient control signals which consequently could violate feasibility assumptions of the control law – this in turn may result into an unstable feedback loop. Therefore, the initialisation of the optimisation-based control law is significantly important and must be taken into consideration in advance. The warm start is also an alternative to resolve this shortcoming – see e.g. (Wang & Boyd, 2010). Here, we shall discuss a promising initialisation method for the solution of the linear systems which is integrated within the framework of our fast model predictive control algorithm.

2.4 Initialisation

We shall overview the initialisation procedure addressed in (Vandenberghe, 2010, §5.3) and (Mattingley & Boyd, 2010b). If primal and dual starting points $(\hat{x}, \hat{y}, \hat{z}, \hat{s})$ are not specified by the user, they are chosen as follows:

- Solve the linear equations (see §2.5 for a detailed solution of this linear system.)

$$\begin{bmatrix} Q & A^T & G^T \\ A & 0 & 0 \\ G & 0 & -I \end{bmatrix} \begin{bmatrix} x \\ y \\ z \end{bmatrix} = \begin{bmatrix} -c \\ b \\ h \end{bmatrix} \tag{8}$$

  to obtain optimality conditions for the primal-dual pair problems of

$$\begin{aligned} \min\ & (1/2)x^T Q x + c^T x + (1/2)\|s\|_2^2 \\ \text{subject to}\ & Gx + s = h \\ & Ax = b \end{aligned} \tag{9}$$

and

$$\begin{aligned} \max \quad & -(1/2)\omega^T Q\omega - h^T z - b^T y - (1/2)\|z\|_2^2 \\ \text{subject to} \quad & Q\omega + G^T z + A^T y + c = 0. \end{aligned} \tag{10}$$

- From the above, $\hat{x} = x, \hat{y} = y$ are found as the two initialisation points. The initial value of $\hat{s}$ is calculated from the residual $h - Gx = -z$, as

$$\hat{s} = \begin{cases} -z & \alpha_p < 0 \\ -z + (1+\alpha_p)\mathbf{e} & \text{otherwise} \end{cases}$$

where $\alpha_p = \inf\{\alpha | -z + \alpha\mathbf{e} \geq 0\}$. Also, the initial value of $\hat{z}$ is computed as

$$\hat{z} = \begin{cases} z & \alpha_d < 0 \\ z + (1+\alpha_d)\mathbf{e} & \text{otherwise} \end{cases}$$

where $\alpha_d = \inf\{\alpha | z + \alpha\mathbf{e} \geq 0\}$.

2.5 Fast KKT solution

As explained earlier, the most time-consuming parts of QP optimisation problem is due to solving the linear KKT systems of the format

$$\begin{bmatrix} Q & A^T & G^T & 0 \\ A & 0 & 0 & 0 \\ G & 0 & 0 & I \\ 0 & 0 & S & Z \end{bmatrix} \begin{bmatrix} x \\ y \\ z \\ s \end{bmatrix} = \begin{bmatrix} r_x \\ r_y \\ r_z \\ r_s \end{bmatrix} \tag{11}$$

In Ref. (Mattingley & Boyd, 2010b) a numerical method has been introduced, using the permuted LDL^T factorisation, to solve the KKT linear systems of Equation (11) in the compact form of $KX = R$ to find optimal variables of X. In the so-called "iterative refinement approach", see (Mattingley & Boyd, 2010b, §5.3)), the original KKTs is regularised by choosing a small $\epsilon > 0$ to ensure that such a factorisation exists and that it is numerically stable. However, since the solution of the modified KKT system is an approximation to the original KKT system, it could potentially affect both the affine and combined step sizes, as well as the feasibility conditions and the rate of global convergence. In words, such an approximation could introduce additional "hold-ups" to the QP problem at each time step which is not desirable for the purpose of the fast real-time optimisation applications, such the one considered in Section 4 as a case study.

In order to obtain "fast" and "reliable" solutions of Equation (11) at each iteration, it is significantly important to avoid any sort of calculation of unstructured (possibly sparse) matrix inverse, as well as to reduce the number of the KKT linear systems. Due to the particular structure of the original exact KKT system given in Equation (11), however, we shall employ a more reliable and stable interior-point solver of the convex QP optimisation problem, even for the KKT systems with sparse matrices. To this end, we start by eliminating the variable s among the KKT linear systems. After some algebra, we will have

$$\begin{bmatrix} Q & A^T & G^T \\ A & 0 & 0 \\ G & 0 & -Z^{-1}S \end{bmatrix} \begin{bmatrix} x \\ y \\ z \end{bmatrix} = \begin{bmatrix} r_x \\ r_y \\ r_z - Z^{-1}r_s \end{bmatrix} \tag{12}$$

which reduces the number of original KKT systems solved per iteration by three. To calculate s, we could use $s = -Gx + r_z$.

The Cholesky factorisation method is the preferred KKT equation solver for linear and quadratic programs. However, due to the particular structure of $Z^{-1}S$, being a diagonal matrix, there is no longer a need to carry out the Cholesky factorization of the diagonal matrix of $Z^{-1}S$ given in Equation (12). In fact, $Z^{-1}S = W^TW$ with diagonal matrix $W = W^T$ will lead to $Z^{-1}S = W^2$. We can now obtain a reduced order of KKT system of Equation (12) with only two equations as

$$\begin{bmatrix} Q + G^TW^{-2}G & A^T \\ A & 0 \end{bmatrix} \begin{bmatrix} x \\ y \end{bmatrix} = \begin{bmatrix} r_x + G^TW^{-2}(r_z - Z^{-1}r_s) \\ r_y \end{bmatrix} \tag{13}$$

From x and y, the solution z follows as $Wz = W^{-T}(Gx - r_z + Z^{-1}r_s)$. Recall that matrices S, Z and W are diagonal matrices, and hence calculation of inverses of $W^{-T} = W^{-1} = \text{diag}(1/W(i,i))$ and $Z^{-1} = \text{diag}(1/Z(i,i))$, $i = 1, 2, \ldots, n$ are straightforward and fast, even for large sparse problems.

Using the Cholesky factorization $Q + G^TW^{-2}G = LL^T$, the KKT solutions of Equation (13) are computed as follows – see also (Vandenberghe, 2010).

- Case 1. If $Q + G^TW^{-2}G$ is nonsingular, y and x are computed from the following equations, respectively:

$$\begin{aligned} AL^{-T}L^{-1}A^Ty &= AL^{-T}L^{-1}\left(r_x + G^TW^{-2}(r_z - Z^{-1}r_s)\right) - r_y \\ LL^Tx &= r_x + G^TW^{-2}(r_z - Z^{-1}r_s) - A^Ty \end{aligned} \tag{14}$$

- Case 2. If $Q + G^TW^{-2}G$ is singular, the exact solutions of y and x are obtained respectively as

$$\begin{aligned} AL^{-T}L^{-1}A^Ty &= AL^{-T}L^{-1}\left(r_x + G^TW^{-2}(r_z - Z^{-1}r_s) + A^Tr_y\right) - r_y \\ LL^Tx &= r_x + G^TW^{-2}(r_z - Z^{-1}r_s) + A^T(r_y - y) \end{aligned} \tag{15}$$

The above algorithm will provide the KKT linear systems to be solved reliably, and to precise accuracy, in a limited number of iterations. This, along with previous optimisation requirements addressed earlier, will help develop a fast reliable optimisation algorithm leading towards fast model predictive control which is briefly discussed in the subsequent section.

2.6 Tracking control problem using fast MPC

As we discussed earlier, standard MPC-based algorithms are great tools in the literature of feedback control system designs, mainly due to their abilities in handling constraints, e.g. actuator saturations, which are successfully taken into consideration in the design of an MPC. However, to provide an appropriate input control signal, MPC and its standard variants suffer from a major drawback due to having a desperate need of excessive computational

time for solving the online minimization problems, at each sampling interval, particularly in the presence of large number of horizons, constraints, optimisation parameters or parametric uncertainties. This shortcoming is an outstanding motivation to look for some sort of efficient model predictive algorithms, to solve an integrated optimisation problems "sufficiently fast" in real time.
Any MPC tracking reference problem subject to constraints, being also the focal point of this section, results in the optimization of a quadratic programming (QP) problem as the objective function is quadratic – see §2.1.
The reference tracking control for a discrete-time linear dynamical system is described as the following quadratic optimisation problem:

$$\begin{array}{ll} \min & \sum_{k=1}^{N} (y_k - y_k^{ref})^T Q_y (y_k - y_k^{ref}) + u_k^T Q_u u_k \\ \text{subject to} & \mathrm{x}_{k+1} = Ax_k + Bu_k + w_k \\ & \mathrm{y}_k = Cx_k + v_k \\ & \mathrm{x}_{\min} \leq x_k \leq \mathrm{x}_{\max} \\ & \mathrm{u}_{\min} \leq u_k \leq \mathrm{u}_{\max} \\ & \dot{\mathrm{u}}_{\min} \leq \dot{u}_k \leq \dot{\mathrm{u}}_{\max} \end{array} \quad (16)$$

where Q_y and Q_u are positive semidefinite weighting matrices for penalizing the tracking error and control effort, respectively; $A \in R^{n\times n}$, $B \in R^{n\times m}$, $C \in R^{p\times n}$ and $x_0 \in R^n$ are the discrete-time plant data, w_k and v_k are plant disturbance and measurement noise, respectively; y_k^{ref} is the reference signal to be tracked at the plant output; N is the horizon; The optimisation variables are the system state $x_k(k = 1, 2, \ldots, N)$ and input control signals $u_k(k = 0, 1, \ldots, N - 1)$.
Recall that here we only consider the linear time-invariant (LTI) systems. Nonetheless, the proposed method could be extended to the time-varying and/or nonlinear cases (Del Re et al., 2010). Also, regarding the fact that most of the physical plants are continuous-time, we shall consider a continuous-time linear dynamic system driven by stationary continuous-time white noise. To simulate such a continuous-time dynamics on a digital computer (or microprocessor) using an equivalent discrete-time dynamics, it is required to utilise equated linear discrete-time system and its noise statistics, so that both systems have identical statistical properties at the discrete time instants (Gelb, 1974, pp. 72-75). Moreover, any type of continuous-time plant dynamics could be transformed, with an appropriate sampling time, to the equivalent discrete-model in the format of the one shown in the subjective of Equation (16) – see (Grewal & Andrews, 1993, pp. 88-91).
The output tracking control system design in Equation (16) could be transformed to the standard quadratic programming problem illustrated in Equation (1). Therefore, we could use the fast KKT solutions following the initialisation. In Section 4, the proposed optimisation procedure of the fast model predictive control design approach will be implemented to the case study of the HEV energy management.

3. Hybrid diesel electric vehicle model

In Section 1, we briefly discussed the history of hybrid electric vehicles which, in some extent, could clarify the importance of our work carried out in the field of advanced energy management for the HEV applications. In this section, we will investigate how to model a

simplified hybrid electric vehicle to replace the sophisticated nonlinear dynamic of the diesel internal combustion engine. We shall integrate this simplified HEV model, for the first time, with recent advances on fast model predictive control architecture described in Section 2 based on embedded convex optimisation.
Before describing the structure of our hybrid diesel electric vehicle, let us first overview a generic HEV structure. A representative configuration of an advanced $4x4$ parallel hybrid electric vehicle configuration is shown in Figure 2.

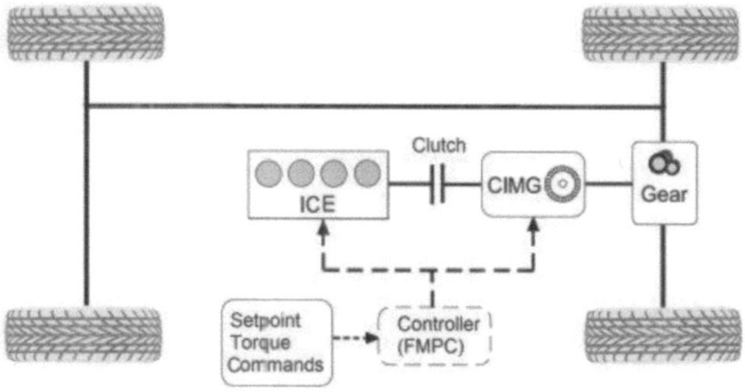

Fig. 2. Schematic structure of a parallel $4x4$ Hybrid Electric Vehicle (HEV). Low-level control components such as high voltage electric battery, electric rear axle drive etc are excluded in this high-level energy management configuration.

The hybrid electric vehicle structure shown in Figure 2 is equipped with a turbocharged diesel engine and a crankshaft integrated motor/generator (CIMG) which is directly mounted on the engine crankshaft. The CIMG is used for starting and assisting the engine in motoring-mode, and also for generating electric energy via charging a high-voltage battery (not shown in the figure). As our intention in this study is to investigate the "full-hybrid" mode, we shall assume that the integrated ICE-CIMG clutch is fully engaged and hence our descriptive HEV dynamical model (see §3.3) excludes a clutch dynamics as it is shown in Figure 2. Likewise, the gearbox is shown in Figure 2 but no gear setting was considered in our simplified HEV demonstration. This is due to the fact that our empirical diesel engine model is derived with engine speed range of $\omega = [1200, 2000]$rpm running at the first gear.
It is also worthwhile to emphasise that our design methodology on the development of the HEV energy management is a high-level design strategy. For this reason, most of the common low-level subsystems, integrated within the typical HEV dynamics, are not considered in the HEV configuration as shown in Figure 2. These low-level subsystems include CIMG low-level motor control, high voltage battery management, low level clutch control, low level transmission control, and electrical distribution including DC-DC converter, to name just a few. Furthermore, the dynamics of the engine model includes the average torque responses of both diesel engine and CIMG over all four cylinders, which are the quantities of interest.
For designing a well balanced feedback control law, control engineers should possess a good comprehension of the physics of the plant under investigation. One of the challenging aspect of any model based engine control development is the derivation of a simplified yet insightful model. For instance, the frequency range of the engine system, the nonlinearities associated with the internal engine processes (i.e. combustion process, heat distribution, air flow), the severe cross coupling, the inherent sampling nature of the four cycle internal

combustion engine, and the limited output sensor dynamic capabilities all contribute to make this modelling step a most arduous task (Lewis, 1980).
There are two main reasons to highlight the importance of simplified HEV dynamical models: First, it is not usually possible to obtain a detailed diesel engine data (or model) from the production vehicle manufacturer (Kaszynski & Sawodny, 2008). Secondly, obtaining a precise mathematical model of a HEV powertrain is a very challenging task particularly due to multi-energetic nature and switching dynamics of a powertrain.
For the above reasons, and for the ease of development of an advanced HEV energy management system, it is essential to obtain a straightforward and realistic model of the propulsion system to which an efficient control strategy, such as our proposed fast MPC design methodology, could be applied. Generally speaking, this model shall be used for the simulation of the overall vehicle motion (at longitudinal direction). Therefore, we do not intend to utilise any detailed model of the internal engine processes, but rather a high-level torque manager model that will generate control efforts based on a given set-point torque commands. Recall that this torque management structure could be easily adopted to other engine configurations in a straightforward manner.
As stated earlier, our developments towards a simplified hybrid model are based on a high fidelity simulation model of the overall diesel hybrid electric vehicle. This HEV dynamical model is modeled using two subsystems, a diesel internal combustion engine (ICE) and an armature-controller DC electric motor. The mathematical modeling of these two subsystems will be discussed in the remainder of this section.

3.1 Simplified diesel engine model

In this section, we shall present a simplified dynamical model of a turbo-charged diesel engine. This simplified model is based upon the nonlinear diesel engine dynamics and the fact that it must capture both the transient and steady-state dominant modes of the diesel engine during operational conditions.
The engine indicated torque T_{ind} is assumed to be mapped from the delayed fueling input proportionally, and has limited bandwidth due to internal combustion dynamic effects, arising e.g. due to combustion and turbo lag. In a mathematical representation, we will have

$$T_{ind}(t) = \frac{1}{\tau(\omega)s+1} T_B^{dem}(t - t_d(\omega)) \tag{17}$$

where τ is the speed-dependant time constant due to combustion lag, t_d is the speed-dependant time-delay due to fueling course and T_B^{dem} is the mapped fueling input representing the required ICE crankshaft (brake) torque.
Our simplified diesel engine model is empirically derived using a turbo-charged diesel engine at speed range of $\omega = [1200, 2000]$rpm with operational brake torque acting at $T_B = [50, 100]$NM. The speed-dependant fueling delay (t_d) and combustion lag (τ) are given in Table 1.
Our diesel model also contains a speed-dependant torque loss T_{Loss} arising due to friction torque, ancillary torque and pumping loss. Such a total torque loss is typically a nonlinear function of the engine speed. However, at the studied operating range of engine speed and brake torque, namely $\omega = [1200, 2000]$rpm and $T_B = [50, 100]$NM, the total engine torque loss is a linear function of ω modeled as $T_{Loss} = m\omega$ where $m = 0.12$ with ω's dimension in [rad/sec].

ω	T_B	t_d	τ
1200	50	100	144
1200	100	140	142
1600	50	84	140
1600	100	96	137
2000	50	80	140
2000	100	72	134

Table 1. Experimental results of fueling delay, t_d [msecs], and combustion lag, τ [msecs], as functions of diesel engine speed [rpm] and brake torque [NM]. These results are captured by measuring the step response of the engine to a step change in the engine brake torque.

For the purpose of this study, we shall employ a 1-st order Pade approximation to model the fueling time-delay by a rational 1st-order LTI model of

$$e^{-t_d s} \cong \frac{-s + 2/t_d}{s + 2/t_d} \tag{18}$$

The simplified diesel engine model can now be described as the following state-space equations:

$$\begin{aligned} \dot{x}_1 &= -\tfrac{2}{t_d}x_1 + T_B^{dem} \\ \dot{x}_2 &= \tfrac{4}{t_d}x_1 - \tfrac{1}{\tau}x_2 - T_B^{dem} \\ T_{Loss} &= m\omega \\ T_B &= \tfrac{1}{\tau}x_2 - T_{Loss} \end{aligned} \tag{19}$$

where x_1 and x_2 are the states associated with the Pade approximation, and combustion lag dynamics, respectively.
The diesel dynamic shown in Equation (19) will be used in the overall configuration of the HEV dynamics.

3.2 Simplified CIMG Model

Assuming that the hybrid electric drivetrain includes an armature-controlled CIMG (DC motor), the applied voltage v_a controls the motor torque (T_M) as well as the angular velocity ω of the shaft.
The mathematical dynamics of the CIMG could be represented as follows.

$$\begin{aligned} I_a &= \frac{1}{L_a s + R_a}(v_a^{dem} - v_{emf}) \\ v_{emf} &= k_b\omega \\ T_M &= k_m I_a \end{aligned} \tag{20}$$

where k_m and k_b are torque and back emf constants, v_a^{dem} is control effort as of armature voltage, v_{emf} is the back emf voltage, I_a is armature current, L_a and R_a are inductance and resistance of the armature, respectively.
Regarding the fact that the engine speed is synchronised with that of the CIMG in full-hybrid mode, the rotational dynamics of the driveline (of joint crankshaft and motor) is given as follows:

$$J\dot{\omega} + b\omega = T_B + T_M - T_L \tag{21}$$

where ω is the driveline speed, J is the effective combined moment of rotational inertia of both engine crankshaft and motor rotor, b is the effective joint damping coefficient, and T_L is the vehicle load torque, which is representing the plant disturbance.
The armature-controlled CIMG model in Equation (20) along with the rotational dynamics of Equations (20) and (21) could be integrated within the following state-space modelling:

$$\begin{aligned} \dot{x}_3 &= v_a^{dem} - \tfrac{R_a}{L_a} x_3 - \tfrac{K_b}{J} x_4 \\ \dot{x}_4 &= \tfrac{1}{\tau} x_2 + \tfrac{K_b}{L_a} x_3 - \tfrac{b}{J} x_4 - T_{Loss} - T_L \\ \omega &= \tfrac{1}{J} x_4 \\ T_M &= \tfrac{K_b}{L_a} x_3 \end{aligned} \tag{22}$$

where x_3 and x_4 are the states associated with the armature circuit, and driveline rotational dynamics, respectively.
A simplified but realistic simulation model with detailed component representations of diesel engine and DC electric motor (CIMG) will be used as a basis for deriving the hybrid model as discussed in the subsequent section.

3.3 Simplified hybrid diesel electric vehicle model

Based on the state-space representation of both the diesel ICE and electric CIMG, given in Equation (19) and Equation (22), respectively, we can now build our simplified 4-state HEV model to demonstrate our proposed approach.
A schematic representation of the simplified parallel hybrid diesel electric vehicle model is shown in Figure 3.

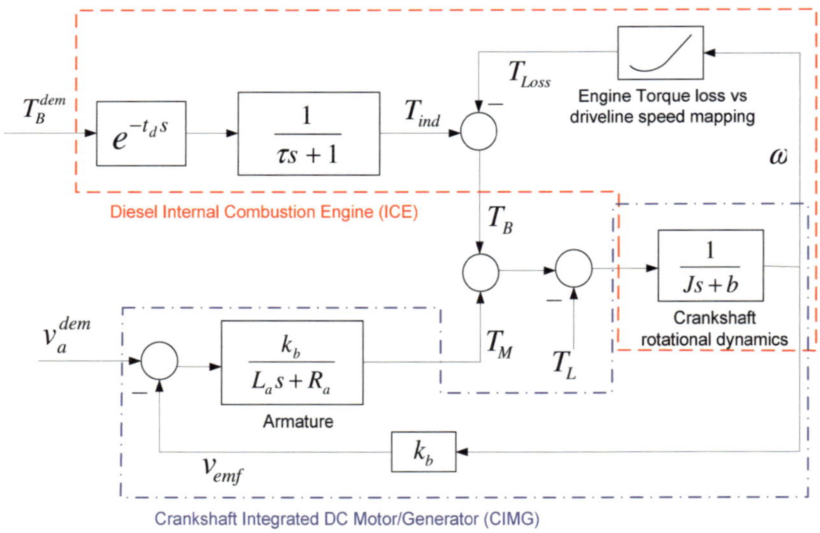

Fig. 3. Simplified model of the parallel Hybrid Diesel Electric Vehicle.

Recall that, as illustrated in Figure 2, the setpoint torque commands (indicated by T_B^{req} and T_M^{req}) are provided to the controller by a high-level static optimisation algorithm, not discussed in this study – see (Dextreit et al., 2008) for more details. Also, in this figure the engine

brake torque and the CIMG torque are estimated feedback signals. However, the details of the estimation approach are not included here. For the sake of simplicity, in this work we shall assume that both engine and CIMG output torques are available to measure.
In addition, due to there being in "full hybrid" mode, it is assumed that the ICE-CIMG clutch is fully engaged and hence the clutch model is excluded from the main HEV dynamics - it was previously shown in Figure 2. Also, the gear setting is disregarded at this simplified model, as discussed earlier. Furthermore, the look-up mapping table of CIMG torque request vs armature voltage request (v_a^{dem}) is not shown in this model for the sake of simplicity.
The overall state-space equations of the simplified HEV model is represented by

$$
\begin{aligned}
\dot{x} &= \begin{bmatrix} -\frac{2}{t_d} & 0 & 0 & 0 \\ \frac{4}{t_d} & -\frac{1}{\tau} & 0 & 0 \\ 0 & 0 & -\frac{R_a}{L_a} & -\frac{K_b}{J} \\ 0 & \frac{1}{\tau} & \frac{K_b}{L_a} & -\frac{m+b}{J} \end{bmatrix} x + \begin{bmatrix} 1 & 0 \\ -1 & 0 \\ 0 & 1 \\ 0 & 0 \end{bmatrix} u + \begin{bmatrix} 0 \\ 0 \\ 0 \\ -1 \end{bmatrix} T_L \\
y &= \begin{bmatrix} 0 & \frac{1}{\tau} & 0 & -\frac{m}{J} \\ 0 & 0 & \frac{K_b}{L_a} & 0 \end{bmatrix} x
\end{aligned}
\tag{23}
$$

where $x \in R^4$ is the state of the system obtained from Equations (19) and (22), $u = [T_B^{dem}\ v_a^{dem}]^T$ and $y = [T_B\ T_M]^T$ are control signals and HEV torque outputs, respectively.
The state-space equations of Equation (23) will be used in designing the proposed fast model predictive control described in Section 2. Some representative simulation results of HEV energy management case study will be shown in the next section to highlight some advances of our proposed embedded predictive control system.

4. Simulation results

In this section, we shall present our proposed Fast MPC algorithm described in Section 2 for the application of the simplified HEV energy management system discussed in Section 3. The problem addressed in the next subsection is to discuss required setpoint torque tracking problem with appropriate optimisation objective leading towards applying our fast MPC design to the HEV energy management problem as illustrated by some of our simulation results.

4.1 HEV energy management optimisation objective and control strategy

For the HEV energy management application subject to the objective function and constraints, HEV demanded torques are found at each time step by solving the optimisation problem of Equation (16) with the following data:

$$
\begin{aligned}
x_{min} &= [0,\ -56,\ -300,\ 0]^T \\
x_{max} &= [18,\ 56,\ 300,\ 360]^T \\
u_{min} &= [0,\ -380]^T \\
u_{max} &= [400,\ 380]^T \\
\dot{u}_{max} &= -\dot{u}_{min} = [0.5,\ 4]^T
\end{aligned}
\tag{24}
$$

For our HEV setpoint tracking problem, based on Equation (16), $y_k = [T_B \;\; T_M]^T$ is the HEV torque outputs (ICE torque and CIMG torque, respectively), $y_k^{req} = [T_B^{req} \;\; T_M^{req}]^T$ is the tracking setpoint torques commands , $w_k \in R^4$ is the discretised vehicle load torque, $u_k = [T_B^{dem} \;\; v_a^{dem}]^T$ is the demanded HEV torques (control efforts) generated in real-time by the controller.
An equated LTI discrete-time system of the continuous-time state-space dynamics described in Equation (23) is obtained using a sampling interval t_s (see Table 2). The plant initial condition $x_0 \in R^4$ is assumed zero in our simulations.
The parameters used in the proposed Fast MPC design together with other physical constants of the simplified HEV model are provided in Table 2.

Parameter	Value	Unit
Sampling time (t_s)	8	msecs
ICE fueling delay (t_d)	90	msecs
ICE combustion lag (τ)	140	msecs
Motor armature resistance (R_a)	1	Ohms
Motor armature inductance (L_a)	0.3	Henrys
Motor torque constant (k_m)	0.25	NM.Amp^{-1}
Motor back emf constant (k_b)	0.25	Volts.secs.rad^{-1}
Effective hybrid rotational inertia (J)	0.6	kg.m^2/s^2
Effective hybrid rotational damping (b)	0.125	Nms
FMPC horizon (N)	20	-
Output penalising matrix (Q_y)	diag(400,200)	-
Control penalising matrix (Q_u)	diag(0.01,0.01)	-

Table 2. Physical constants and FMPC design parameters in regard to the HEV model case study.

In the next subsection, the closed-loop behavior of the HEV energy management problem with our FMPC controller placed in the feedback loop has been evaluated based on the high-fidelity simplified model of the HEV described in Section 3.

4.2 Simulation results

Our simulations have been carried out in Simulink and implemented in discrete-time using a zero-order hold with a sampling time of $t_s = 8$ msecs – see Table 2.
We shall emphasis that optimization based model predictive control (MPC) techniques, including the proposed fast MPC design methodology, require knowledge about future horizon (driving conditions in this case study). These future driving conditions in our case study include setpoint torque commands (requested by driver) and vehicle load torque. This fact will make implementation of all sort of optimisation based predictive control algorithms even more arduous to be applied in real time.
For the purpose of simulations, assuming that the future driving cycle (i.e. torque references and vehicle load) are entirely known could be perhaps an acceptable assumption. In our simulations, the future driving cycle is unknown whilst retaining constant for the whole horizon of N samples. However, if the future driving cycle could be entirely known, the performance of the proposed FMPC would be superior than those shown here.

Figure 4 shows a typical simulation results for the period of 20 secs in tracking requested setpoint HEV torques. During this simulation period, the system is in hybrid mode as both ICE torque and CIMG torque are requested.

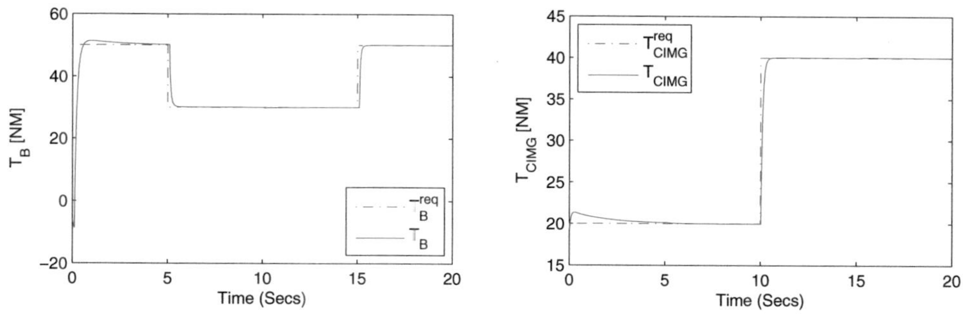

(a) Engine Brake Torque.

(b) CIMG Torque.

Fig. 4. Simulation results of the HEV torque setpoints and outputs using the proposed FMPC algorithm.

As shown in Figure 4, despite the fact that the HEV energy management is a coupled Two-Input Two-Output (TITO) dynamical system, both the diesel ICE and the DC electric motor have successfully tracked the requested torque setpoints. At times $t = 5$ secs and $t = 15$ secs , the TITO controller is requested for an increased and decreased ICE torques, respectively to which the fast MPC algorithm could precisely follow those commands, as illustrated in Figure 4(a). Similarly, there was an increased request for the CIMG torque (from 20 NM to 40 NM) at time $t = 10$ secs, and the controller has successfully delivered this torque request, as depicted in Figure 4(b).

This is noted that our torque manager structure, as stated earlier, assumes that setpoint torque commands are provided by some sort of static optimisation algorithms. The designed FMPC is then enquired to optimise control efforts so as to track the requested torque references.

Figure 5 shows the load torque transient used in our simulations (being modeled as a plant disturbance), ICE torque loss and control efforts generated by the FMPC. We have assumed that plant disturbance (vehicle load) is known and available to controller. In reality, this might be an infeasible assumption where an estimation algorithm is required to estimate the vehicle load torque w_k over the prediction horizon. Also, as mentioned earlier, the estimation of future driving conditions must be made online. Due to lack of space, however, we shall preclude addressing a detailed discussion in this course.

Figure 5(c) shows that the FMPC fully satisfies the required optimisation constraints as of Equation (24).

Figure 6 shows simulation results in regard to driveline speed and vehicle speed. It is worthwhile to point out that as illustrated in Figure 6(a), by requesting large torque commands, we have in fact violated our empirical HEV modeling assumption in that driveline speed must be limited to $\omega = [1200, 2000]$rpm. However, it can be seen that the FMPC can still successfully control the HEV energy endamagement dynamics in real-time. The vehicle speed shown in Figure 6(b) has been calculated using a dynamic model of the vehicle as a function of the driveline speed which is not discussed here.

It is also important to mention that fueling delay and combustion lag are functions of engine speed and brake torque – see Table 1. However, in designing our fast MPC algorithm we

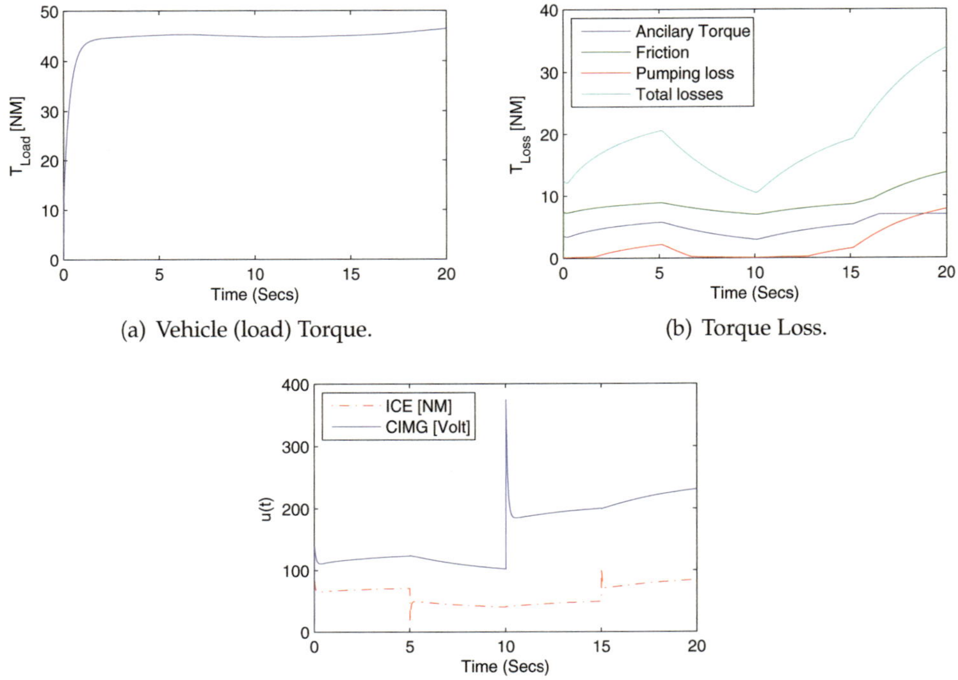

(a) Vehicle (load) Torque.

(b) Torque Loss.

(c) Control signals.

Fig. 5. Simulation results of vehicle load, Torque loss, and Control efforts.

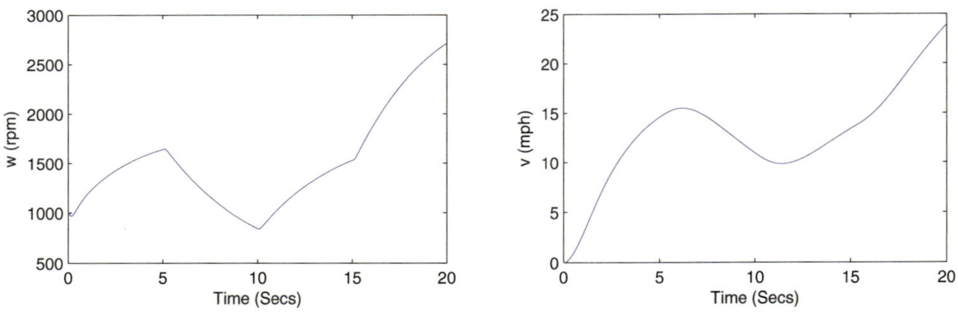

(a) Driveline Speed.

(b) Vehicle Speed.

Fig. 6. Simulation results of parallel diesel HEV driveline speed and vehicle speed.

require to utilise an LTI model of the HEV energy management plant. Towards this end, we use the numerical values of $\tau = 140$ msecs and $t_d = 90$ msecs, in our design to capture worst case of the ICE speed-dependant parameters. However, the simulation results are based on the actual time-varying speed-dependant parameters of the ICE, namely τ and t_d.

Regarding the real-time simulations in Simulink (fixed-step) using our Matlab custom S-function codes with a sampling time of t_s, the simulation time required for a single run of 20 secs was approximately 500 times faster than real-time running a Toshiba Portege laptop with an Intel(R) Core(TM) i5 processor, at 2.4GHz under Windows 7 Pro platform.

Without doubt, this shows a significant improvement on the computational capability of the control action that could potentially permit any sort of fast MPC algorithms to be run using inexpensive low-speed CPUs under possibly kilo Hertz control rates.

5. Conclusions

The aim of this chapter was to present a new Fast Model Predictive Control (FMPC) algorithm with an application for the energy management of hybrid electric vehicles (HEVs). The main goal of energy management in hybrid electric vehicles is to reduce the CO_2 emissions with enhanced fuel consumption for a hybrid powertrain control system. The applicability of conventional MPC in the energy management setting, however, has shown a main drawback of these algorithms where they currently cannot be implemented on-line due to the burdensome real-time numerical optimisation, arising due to e.g. hardware constraints and limitation of online calculations. The proposed FMPC design architecture could resolve such shortcomings of the standard MPC algorithms. In fact, such a custom method, is able to speed up the control action, by exploiting particular structure of the MPC problem, much faster than that of the conventional MPC methods. Moreover, our proposed FMPC design methodology does not explicitly utilise any knowledge in regard to the future driving cycle. Simulation results illustrated that FMPC could be a very promising on-line control design algorithm and could play a key role in a wide variety of challenging complex automotive applications in the future.

6. Acknowledgment

This work was supported by EPSRC, UK, under framework "Low Carbon Vehicles Integrated Delivery Programme", Ref. EP/H050337/1.
We would like to thank Dr. Jacob Mattingley and Yang Wang, from Stanford University, for their valuable comments and discussions which helped us in preparation of an earlier version of our simulation results.

7. References

Anderson, C.D. & Anderson, J. (2010). *Electric and Hybrid Cars: A History*, 2nd Edition, McFarland & Co Inc.

Bartlett, R.A.; Wachter, A. & Biegler, L.T. (2000). Active set vs. interior point strategies for model predictive control, *Proc. of the American Control Conf.*, Vol. 6, pp. 4229-4233.

Baumann, B.M.; Washington, G.; Glenn, B.C. & Rizzoni, G. (2000). Mechatronic design and control of hybrid electric vehicles, *IEEE/ASME Trans. On Mechatronics*, 5(1): 58-72.

Beccuti, A.G.; Papafotiou, G.; Frasca, R. & Morari, M.; . (2007). Explicit Hybrid Model Predictive Control of the dc-dc Boost Converter, *IEEE Power Electronics Specialists Conference, PESC 2007*, Orlando, Florida, USA, pp. 2503-2509

Bemporad, A. (2004). *Hybrid Toolbox - User's Guide.* URL: *http://www.ing.unitn.it/bemporad/hybrid/toolbox*

Bemporad, A.; Morari, M; Dua, V. & Pistikopoulos, E.N. (2002). The explicit linear quadratic regulator for constrained systems, *Automatica* 38: 3-20.

Bemporad, A. & Morary, M. (1999). Control of systems integrating logic, dynamics, and constraints, *Automatica* 35: 407-427.

Cacciatori, E.; Vaughan, N.D. & Marco, J. (2006). Evaluating the Impact of Driveability Requirements on the Performance of an Energy Management Control Architecture for a Hybrid Electric Vehicle, *The 2nd International IEE Conference on Automotive Electronics*, IEE Savoy Place, UK.

Camacho, E.F. & Bordons C. (2004). *Model Predictive Control*, Springer-Verlag.

Cikanek, S.R. & Bailey, K.E. (2002). Regenerative braking system for a hybrid electric vehicle, *In Proc. of the American Control Conf.*, Anchorage, AK, Vol. 4, pp. 3129-3134.

Cundev, D. (2010). *Hybrid-electric Cars and Control Systems: Control strategy of car hybrid system and its experimental confirmation*, LAP LAMBERT Academic Publishing.

De la Pena, D.M.; Bemporad, A. & Filippi, C. (2006). Robust explicit MPC based on approximate multiparametric convex programming, *IEEE Transactions on Automatic Control* 51: 1399-1403.

Del Re, L.; Glielmo, L.; Guardiola, C. & Kolmanovsky, I. (Eds.) (2010). *Automotive Model Predictive Control: Models, Methods and Applications*, 1st edn, Springer.

Dextreit, C.; Assadian, F.; Kolmanovsky, I.; Mahtani, J. & Burnham, K. (2008). Hybrid electric vehicle energy management using game theory, *SAE World Congress & Exhibition*, SAE Paper No. 2008-01-1317.

Fekri, S. & Assadian, F. (2011). Application of Robust Multivariable Control on Energy Management of Hybrid Electric Vehicles, *International Journal of Vehicle Design (IJVD), Special Issue on "Automotive Mechatronics: Innovations in Design, Analysis and Implementation"*, March 2011, under review.

Fuhs, A. (2008). *Hybrid Vehicles*, 1st edn, CRC Press.

Gelb, A. (1974). *Applied Optimal Estimation*, MIT Press, MA, USA.

Grewal, M.S. & Andrews, A.P. (1993). *Kalman Filtering: Theory and Practice*, Prentice-Hall.

Hofman, T. & Druten, R.V. (2004). Energy analysis of hybrid vehicle powertrains, *In Proc. Of the IEEE Int. Symp. On Vehicular Power and Propulsion*, Paris, France.

Husain, I. (2003). *Electric and Hybrid Vehicles: Design Fundamentals*, 1st edn, CRC Press.

Johnson, V.H; Wipke, K.B. & Rausen, D.J. (2000). HEV control strategy for real-time optimization of fuel economy and emissions, *In Proc. of the Future Car Congress*, Washington DC, USA.

Kaszynski, M. & Sawodny, O. (2008). Modeling and identification of a built-in turbocharged diesel engine using standardized on-board measurement signals, *Proc. of IEEE International Conference on Control Applications (CCA 2008)*, pp. 7-12.

Kessels, J. (2007). *Energy Management for Automotive Power Nets*, PhD thesis, Faculteit Electrical Engineering, Technische Universiteit Eindhoven, The Netherlands.

Khayyam, H.; Kouzani, A.; Nahavandi, S.; Marano, V. & Rizzoni, G. (2010). *Energy Management*, InTech, chapter Intelligent Energy Management in Hybrid Electric Vehicles, pp. 147-175.

Kolmanovsky, I.; Siverguina, I. & Lygoe, B. (2002b). Optimization of powertrain operating policy for feasibility assessment and calibration: Stochastic dynamic programming approach, *In Proc. of the American Control Conf.*, Anchorage, AK, pp. 1425-1430.

Kwon, W.H. & Han, S. (2005). *Receding Horizon Control*, Springer-Verlag.

Lewis, J.B. (1980). *Automotive engine control: A linear quadratic approach*, PhD thesis, Dept. of Electrical Engineering and Computer Science, Massachusetts Institute of Technology, MA, USA.

Ling, K.V.; Wu, B.F. & Maciejowski, J.M. (2008). Embedded model predictive control (MPC) using a FPGA, *Proc. 17th IFAC World Congress*, pp. 15250-15255.

Maciejowski, J.M. (2002). *Predictive Control with Constraints*, Prentice-Hal.

Mattingley, J. & Boyd, S. (2010). Real-time convex optimization in signal processing, *IEEE Signal Processing Magazine* 27: 50-61.

Mattingley, J. & Boyd, S. (2010). CVXGEN: A code generator for embedded convex optimization, *working manuscript*.

Mattingley & Boyd (2009). *Automatic Code Generation for Real-Time Convex Optimization*, Cambridge University, chapter Convex Optimization in Signal Processing and Communications.

Mehrotra, S. (1992). On the implementation of a primal-dual interior point method, *SIAM Journal on Optimization* 2: 575-601.

Milman, R. & Davidson, E.J. (2008). A fast MPC algorithm using nonfeasible active set methods, *Journal of Optimization Theory and Applications* 139: 591-616.

Milman, R. & Davidson, E.J. (2003). Fast computation of the quadratic programming subproblem in MPC, *Proc. of the American Control Conf.*, pp. 4723-4729.

Narciso, D.A.C.; Faisca, N.P. & Pistikopoulos, E.N. (2008). A framework for multi-parametric programming and control Ű an overview, *Proc. of IEEE International Engineering Management Conference, IEMC Europe 2008*, pp. 1-5.

Peng, J.; Roos, C. & Terlaky, T. (2002). *Self-Regularity: A New Paradigm for Primal-Dual Interior-Point Algorithms*, Princeton University Press.

Potra P.A. & Wright S.J. (2000). Interior-point methods, *Journal of Computational and Applied Mathematics* 124: 281-302.

Ramsbottom, M. & Assadian, F. (2006). Use of Approximate Dynamic Programming for the Control of a Mild Hybrid , *Proceedings of WMG Hybrid Conference*, University of Warwick, Warwick, United Kingdom, December 12-13.

Rao, C.V.; Wright, S.J. & Rawlings, J.B. (1998). Applications of interior-point methods to model predictive control, *Journal of Optimization, Theory and Applications* 99: 723-757.

Renegar, J. & Overton, M. (2001). *A Mathematical View of Interior-point Methods in Convex Optimization*, Society for Industrial & Applied Mathematics, USA.

Ripaccioli, G.; Bemporad, A.; Assadian, F.; Dextreit, C.; Di Cairano, S. & Kolmanovsky I.V. (2009). chapter Hybrid Modeling, Identification, and Predictive Control: An Application to Hybrid Electric Vehicle Energy Management, R. Majumdar and P. Tabuada (Eds.), *Proc. of the 12th International Conference on Hybrid Systems Computation and Control (HSCC 2009)*, Springer-Verlag Berlin Heidelberg, pp. 321Ű335.

Rousseau, G.; Sinoquet, D.; Sciarretta, A & Yohan Milhau. (2008). Design Optimisation and Optimal Control for Hybrid Vehicles, *In Proc. of the International Conference on Engineering Optimization*, EngOpt 2008, Rio de Janeiro, Brazil, June 2008.

Schouten, N.J.; Salman, M.A. & Kheir, N.A. (2002). Fuzzy logic control for parallel hybrid vehicles, *IEEE Trans. on Control Systems Technology* 10(3): 460-468.

Sciarretta, A. & Guzzella, L. (2007). Control of hybrid electric vehicles, *IEEE Control Systems Magazine* pp. 60-70.

Sontag, E. (1981). Nonlinear regulation: The piecewise linear approach, *IEEE Trans. on Automatic Control* 26(2): 346-358.

Steinmaurer, G. & Del Re, L. (2005). Optimal energy management for mild hybrid operation of vehicles with an integrated starter generator, *In Proc. of the SAE World Congress*, Detroit, Michigan.

Tate, E. & Boyd, S. (2001). Towards finding the ultimate limits of performance for hybrid electric vehicles, *In Proc. of the SAE Future Transportation Technology Conf.*, Costa Mesa, CA.

Tondel, P.; Johansen, T.A. & Bemporad, A. (2003). An algorithm for multi-parametric quadratic programming and explicit MPC solutions, *Automatica*, 39(3): 489-497.

Torrisi, F.D. & Bemporad, A. (2004). Hysdel - a tool for generating computational hybrid models, *IEEE Trans. on Control Systems Technology* 12(2): 235-249.

Vandenberghe, L. (2010). The CVXOPT linear and quadratic cone program solvers. *working manuscript*.

Vandenberghe, L. & Boyd S. (2004). *Convex Optimization*, Cambridge University Press.

Wang, Y. & Boyd, S. (2010). Fast model predictive control using online optimization, *J. Optim. Theory Appl* 18: 267-278.

Wang, Y. & Boyd, S. (2008). Fast model predictive control using online optimization, *Proceedings of the 17th IFAC World Congress*, Seoul, Korea, pp. 6974-6979.

Wright, S.J. (1997). *Primal-dual interior-point methods*, SIAM.

2

Fast Nonlinear Model Predictive Control using Second Order Volterra Models Based Multi-agent Approach

Bennasr Hichem and M'Sahli Faouzi
Institut Supérieur Des Etudes Technologiques de SFAX
Ecole Nationale d'ingénieur de Monastir
Tunisia

1. Introduction

Model predictive control (MPC) refers to a class of computer control algorithms that utilize a process model to predict the future response of a plant. During the past twenty years, a great progress has been made in the industrial MPC field. Today, MPC has become the most widely implemented process control technology. One of the main reasons for its application in the industry is that it can take account of physical and operational constraints. In classical model predictive control (MPC), the control action at each time step is obtained by solving an online optimization problem. If it is possible, MPC algorithms based on linear models should be used because of low computational complexity [Maciejowski J,2002]. Since properties of many technological processes are nonlinear, different nonlinear MPC techniques have been developed [Qin, S. J et al, 2003]. The structure of the nonlinear model and the way it is used on-line affect the accuracy, the computational burden and the reliability of nonlinear MPC. Several different attempts to reduce computational complexity have been released during the last thirty years. The simplest way to reduce on-line computation is to transform the NMPC problem into LMPC. The nonlinear system is transformed into a linear system using a feedback-linearizing law, the input constraints are mapped into constraints on the manipulated input of the transformed system and the obtained constrained linear system is controlled using LMPC [Kurtz M.J et al,1997]. An interesting strategy is presented in [Arahal M.R et al.,1998], when the linear model is used to predict future process behavior and the nonlinear model is used to compute the effect of the past input moves. The most straightforward technique used to implement fuzzy models [Fischer M et al.,1998] is based on a linearization method. The accuracy of the linear model can be improved by relinearizing the model equations several times over a sampling period or by linearizing the model along the computed trajectory [Mollov S.,et al.,2004].Another approach has been used by a number of researchers such as in [Brooms A et al., 2000], where the NMPC problem is reduced to an LMPC problem at each time step using a successive linearization. The structure of certain nonlinear empirical models allows the NMPC optimization problem to be solved more efficiently than is possible with other forms. Such an approach will be followed in [Abonyi, J et al,2000]. An algorithm for controller reconfiguration for non-linear systems based on a combination of a multiple model estimator and a generalized predictive controller is presented in [Kanev, S et al., 2000], in which a set of models are constructed. Each corresponding to a different operating condition of the system and an interacting multiple model estimators is utilized to yield a reconstruction of the state of the non-linear system. For unconstrained control based on linear process models and a quadratic cost function, the control sequence can be analytically calculated. When linear constraints are taken into account, the solution can be found using quadratic programming techniques. With the introduction of a nonlinear model into MPC scheme, a nonlinear programming technique (NLP) has to be solved at each sampling time to compute the future manipulated variables in on-line optimization that is generally non-convex which make their implementation difficult for real time control. During the past decade significant theoretical results as well as advances in the implementation strategies of NMPC have been obtained and NMPC has been successfully applied in practice to relatively slow plants, mainly in the process industry. However, the application of such techniques for fast nonlinear systems remains a widely opened problem due to the computation burden associated with solving an open loop optimal control problem. Most of the research has focused on computations carried out by one agent. In [Negenborn R et al., 2004], a survey how a distributed multi-agent MPC setting can reduce the computations of a single MPC agent. Moreover, researchers have investigated feedback linearization model predictive control (FLC-MPC) schemes for their ability to handle constraints on input and output [Soest Van W.R et al., 2005]. These approaches reduce the on-line computation by transforming the NLMPC problem into a LMPC and quadratic programming can be used to handle constraints. When sampling times become so short, the computation times for QP solution can no longer be neglected [Jaochim H et al., 2006]. In [Didier G ,2006], a distributed model predictive control is considered and the proposed strategy allows dramatic reduction of the computational requirement for solving large-scale nonlinear MPC problem due to computation parallelism. However, recent advancements in MPC allow for a faster online solution by shifting some of the computational burden off-line. We can notice that many optimization algorithm solutions for NMPC have been investigated lately; however, an analytical solution in NMPC approach is usually impossible to find. One possible way to address computational complexity is to decentralize the optimization tasks. Attention has been focused on multi-agent model predictive control approach [H.Ben Nasr et al.,2008a,b,c,d,e]. There are multiple agents in multi-agent model predictive control. Each uses a model of its sub-system to determine which action to take. Decentralized agent architecture and decentralized model decomposition are then chosen, in which there are numerous agents that do not have any interaction among one another. A methodology based multiagent has been investigated in the implementation of a given predictive control law for nonlinear systems. Such procedure relies on the decomposition of the overall system into subsystems and a multiple agents each uses a model of its sub-system to determine which action to take.

In this chapter book, new NMPC scheme based MAMPC (Multiagent model predictive control) is implemented to reduce the computational effort. The performance of the proposed controllers is evaluated by applying to single input-single output (SISO) control of non linear system. Moreover, in general, the optimization problem is nonconvex and leads to many difficulties impacting on implementation of MPC. These difficulties are related to feasibility and optimality, computation and stability aspects. In order to avoid solving nonconvex optimization problem, MAMPC (Multiagent model predictive control) optimization procedure, a method for convex NMPC was also developed in this chapter book. Theoretical analysis and simulation results demonstrate better performance of the MAMPC over a conventional NMPC based on sequential quadratic programming (SQP) in tracking the set point changes as well as stabilizing the operation in the presence of input disturbances. In this work, our main objective has been to illustrate the potential advantage of nonlinear predictive control based multiagent when applied to nonlinear systems. The suggested approach was to identify a new control algorithm that in essence is a bridge between linear and nonlinear control. This resulted in the development of the MAMPC approach. Through simulation-based comparisons, it is shown that a MAMPC control algorithm is capable of delivering significantly improved control performance in comparison to a conventional NMPC, so that the difficulty of minimizing the performance function for nonlinear predictive control is avoided, which is usually carried by the use of NLP solved at each sampling time that generally is non-convex. In this chapter book we describe algorithm that find the solution of a non-convex programming and also demonstrated that global nonlinear requirements can effectively be resolved by considering smaller regimes. The simulation example shows that the multi-agent compares favorably with respect to a numerical optimization routine. Moreover, the MAMPC reduces the online computational burden and hence has the potential to be applied to the system with faster time constants.

2. Statement of the problem

2.1 Process model

A broad class of physical systems can be represented using the Volterra model. Particularly, it was shown that a truncated Volterra model could represent any non-linear system, time-invariant with fading memory. This model is thus particularly attractive for non-linear systems modeling and identification purpose. One of the main advantages of the Volterra model is its linearity-in-parameters, i.e. the kernel coefficients. This property allows the extension of some results established for linear model identification to this model. In this work, we consider the control of a class of single-input single output non-linear system described by the following non-linear discrete-time parametric second-order Volterra model (Haber *et al.* 1999a,b):

$$y(k) = y_0 + \sum_{i=1}^{n_y} a_i y(k-i) + \sum_{i=1}^{n_u} b_i u(k-i) + \sum_{i=1}^{n_u}\sum_{j=1}^{i} b_{ij} u(k-i)u(k-j) + \varepsilon(k) \tag{1}$$

Where y_0 is a bias term, $y(k)$ is the output, $u(k)$ is the input, a_i, b_i and b_{ij} are the parameters of the parametric Volterra model, n_u and n_y are the number of lags on the input and the output, respectively. $\varepsilon(k)$ Contains all terms up to second-order. One advantage of using the parametric Volterra model is that the one-ahead prediction problem can be formulated as a linear regression, which simplifies the identification of the parameters from input-output data. Therefore, the model given by "Equation (1)" can be written as:

$$y(k) = \theta^T \phi(k) + \varepsilon(k). \tag{2}$$

With:

$$\theta^T = \left[y_0, a_1, a_2, \cdots, a_{n_y}, b_1, b_2, \cdots, b_{n_u}, b_{1,1}, \cdots, b_{n_u,n_u} \right] \tag{3}$$

$$\phi^T(k) = \left[1, y(k-1), \cdots, y(k-n_y), u(k-1), \cdots, u^2(k-1), \cdots, u^2(k-n_u)\right] \quad . \tag{4}$$

Where $\phi(k)$ and θ are the regressor and the parameter vectors, respectively. The model "Equation (3)" is linear in parameters, and its regressors and parameters may be identified from input output information. Moreover, from identification point of view, parametric Volterra models are superior to Volterra series models in the sense that the number of parameters needed to approximate a process are generally much less with parametric Volterra models. This is due to the fact that Volterra models only include previous inputs, while the model (1) includes previous outputs as well as previous inputs.

2.2 Optimization criteria

The purpose of the control strategy is to compute future control moves which will minimize some performance function based on the desired output trajectory over a prediction horizon, subject to constraints on input and output signals [D.W. Clarke et al.,1987]. The most common objective cost function, also used here, is:

$$J(N_1, N_2, N_{u,}\delta) = \sum_{j=N_1}^{N_2} (w(k+j) - \hat{y}(k+j/k))^2 + \sum_{j=1}^{N_u} (\lambda_j \Delta u(k+j-1))^2 \tag{5}$$

Subject to

$$\begin{aligned} \Delta u_{low} \le \Delta u(j+j-1) \le \Delta u_{high} \quad & for \quad 1 \le j \le N_u \\ u_{low}(k) \le u(k+j-1) \le u_{high}(k) \quad & for \quad 1 \le j \le N_u \end{aligned} \tag{6}$$

Where N_1 is the minimum prediction horizon, N_2 is the maximum prediction horizon, $\hat{y}(k+j/k)$ is an optimum j-step ahead prediction of the system output on data up to time k, $w(k+j)$ is a sequence of future set points, $N_u \le N_2$ is the control horizon, and $(\lambda_j)_{j=1 \cdots N_u} = (\lambda_1, \cdots, \lambda_{N_u})$ are control-weighting factors usually assumed to be equal to each other used to penalize the control increments. $\Delta u(k+j-1), j \in [1, N_u]$, is a sequence of future control increments computed by the optimization problem at time k; $\Delta u(k+j-1) = 0$ for $j > N_u$. For the constraints $u_{low}, u_{high}, \Delta u_{low}, \Delta u_{high}$, are respectively the lower limit, upper limit, lower derivative limit and higher derivative limit of the control input. Using the quadratic prediction equation of the model, the cost function becomes fourth degree equation in the control increments. Th objective finction never exeeds fourth order, regardless of the value of the prediction horizon. (Haber, 1999a, 1999b)

2.3 Nonlinear Predictive Control

Despite of the wide exposure of and the intensive research efforts attracted over the past few decades on Nonlinear model predictive control (NLMPC), this control strategy is still being perceived as an academic concept rather than a practicable control technique. However, nonlinear model predictive control is gaining popularity in the industrial community. The formulations for these controllers vary widely, and almost the only common principle is to retain nonlinearities in the process model [Matthew et al.,2002]. In nonlinear control, a receding horizon approach is typically used, which can be summarized in the following steps:

1. At time k, solve, on-line, an open-loop optimal control problem over some future interval, taking into account the current and future constraints.
2. Apply the first step in the optimal control sequence.
3. Receding strategy so that at each instant the horizon is displaced towards the future, which involves the application of the first control signal of the sequence calculated at each step.

The process to control is assumed to be represented by a mono-variable second order parametric Volterra model. The model given by (1) can be expressed as:

$$A(q^{-1})y(k) = y_0 + B_1(q^{-1})u(k) + B_2(q_1^{-1},q_2^{-1})u^2(k) + \frac{\varepsilon(k)}{\Delta(q^{-1})} \tag{7}$$

Where are two polynomials of the backward shifting operator q^{-1} given by :

$$\begin{aligned} A(q^{-1}) &= 1 + a_1 q^{-1} + \cdots + a_{na} q^{-na} \\ B_1(q^{-1}) &= 1 + b_{11} q^{-1} + \cdots + b_{1nb} q^{-nb} \end{aligned} \tag{8}$$

$B_2(q_1^{-1},q_2^{-1})$ represents the quadratic term of the Volterra model, this quantity is defined by:

$$B_2(q_1^{-1},q_2^{-1})u^2(k) = \sum_{n=0}^{nb}\sum_{m=n}^{nb} b_{2nm} u(k-n)u(k-m) \tag{9}$$

The incremental predictive form of the parametric Volterra model can be expressed as a function of the current and future control increments :

$$\hat{y}(k+j) = v_0^j + v_1^j(q^{-1})\Delta u(k+j) + v_2^j(q_1^{-1},q_2^{-1})\Delta u^2(k+j) \tag{10}$$

With

$$\begin{aligned} v_0^j &= y_o + G_j y(k) + \sum_{i=j+1}^{nb+j-1}\left[\delta_{1i} + \sum_{m=i}^{nb+j-1}\delta_{2im}\Delta u^*(k+j-m)\right]\Delta u^*(k+j-i) \\ v_{1i}^j &= v_{1i} + \sum_{m=j+1}^{nb1+j-1}\delta_{2im}\Delta u^*(k+j-m) \qquad i = 1,2,\cdots,j \\ v_{2im}^j &= \delta_{2im} \qquad i = 1,2,\cdots,j \;\; and \;\; m = 1,2,\cdots,j \end{aligned} \tag{11}$$

The effect of selecting the parameters and the coefficient of the predictive control are not investigated here, for more detail see (Haber et al., 1999a) . Replacing the incremental output by his expression, the cost function (5) can be written as follows:

$$J = (v_0 - w + v_1\tilde{u} + v_2\tilde{u}^2)^T(v_0 - w + v_1\tilde{u} + v_2\tilde{u}^2) + \lambda\tilde{u}^T\tilde{u} \tag{12}$$

With constraints, the cost function can be minimized numerically by a one-dimensional search algorithm (dynamic algorithm programming). Without constraints the solution leads to a third-degree one-dimensional equation [F.J.Doyle et al.,1995].

3. Multi-agent Model Predictive Control

3.1 Control and design

The main idea of the proposed concept model predictive control is to transform the nonlinear optimization procedure used in a standard way into sub-problems, in which the global task can be resolved. The objective of this approach is to regulate the nonlinear system output to the expected values and satisfying the above constraints. This can be done as follows. The global system can first be decomposed on sub-systems independent of one another, for each sub-system an MPC unit sub-system is made constituting the agent controller *i*. Based on an analytical solution, which corresponds to the solution of the local receding horizon sub-problems, a logic unit switching tries to find the best sequence of actions sent to the nonlinear system and gives the desired trajectory. Sequences of actions that bring the global system in a desired trajectory are made and avoid any violated constraints on actions. The multi-agent controller consists of synchronizing the output of the true system at every decision step k with the reference trajectory. In fact, at every decision step the right action is the one that will cause the agent to be the most successful. The parallel controller structure is based on the fact that a neural network can be used to learn from the feedback error controller non linear system. A neural network controller is also made on, in objective to take handle the results of the actions on the global system and monitor the closed-loop system. Figure 1, shows the architecture of the multi-agent controller. In the multi-agent context, the agents are the controllers and the non linear system is the environment.

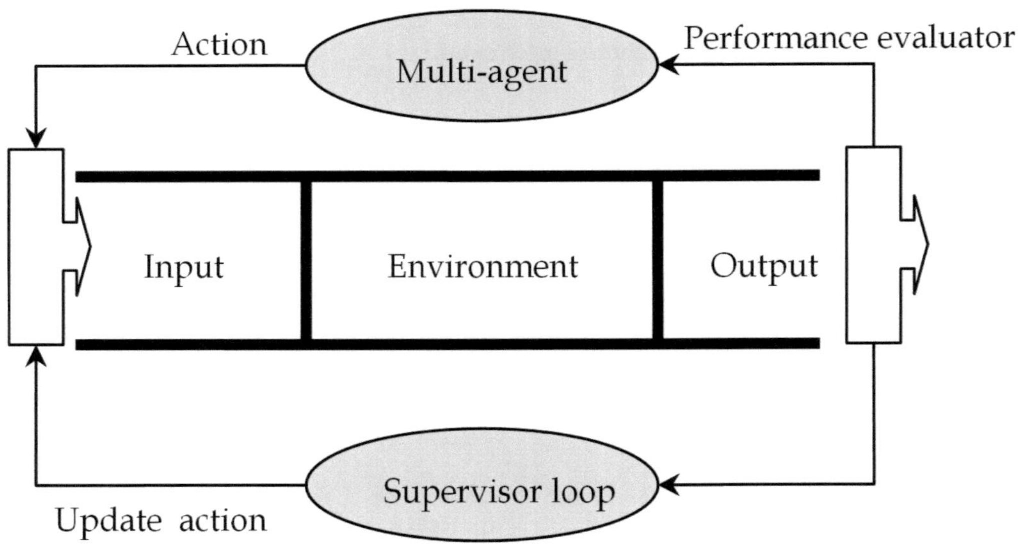

Fig. 1. Architecture of Multi-agent Controller

The basic structure of the control strategy proposed is shown in figure 2. The control problem to solve should be decomposed into supposedly independent subproblems. Each subproblem is solved by designing a controller-agent. The controller-agent is realized by some control algorithm that is operational only under particular operating conditions of the plant being controlled. Moreover, the controller-agent's action consist of the analytical optimal control sequence elaborated in each sub-system after having learned the trajectory of the control to follow and by minimizing a local cost function. The individual solutions or controller-agents are combined into one overall solution. This implies addressing the global problems by selecting an appropriate coordination mechanism. The conceptual design consists of the following three stages:

Structuring: The control problem to solve should be decomposed into supposedly independent subproblems. The global system can first be decomposed on sub-systems independent of one another.

Solving individual subproblems: Each subproblem is solved by designing a controller-agent. An MPC unit sub-system is made constituting the controller agent. A supervisor based on performance measure J_k is used. By means of the output errors ε_k for each agent's action, the supervisor decides then what action should be applied to the plant during each sampling interval k. The performance measure is given by:

$$J_k = \varepsilon_k - \varepsilon_{k-1}e^{-\lambda}, \lambda > 0 \tag{13}$$

Where, ε_k is the error for the agent I defined by:

$$\varepsilon_k = setpoint - y_a \tag{14}$$

And y_a is the plant output after agent's action.

Combining individual solutions The individual solutions or controller-agents are combined into one overall solution. The parallel controller structure is based on the fact that a neural network can be used to learn from the feedback error controller nonlinear system., to take handle the results of the actions on the global system and monitor the closed-loop system.

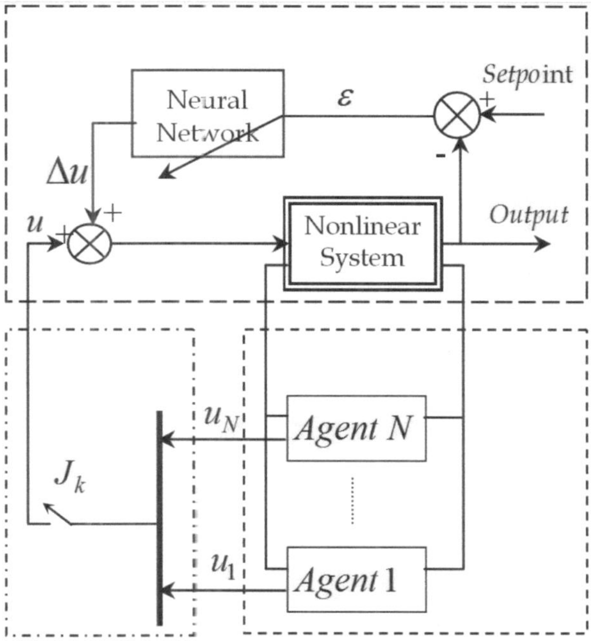

Fig. 2. Architecture of Multi-agent Controller

3.2 Control problem decomposition

The extension of MPC for the use of nonlinear process models is one of the most interesting research topics. These algorithms generally lead to the use of computationally intensive nonlinear techniques that make application almost impossible. In order to avoid this problem, the proposed concept algorithm utilizes a linear model extracted from the nonlinear model. A decentralized model and decentralized goals are then considered. A decentralized problem model consists of multiple smaller, independent subsystems in witch subsystem in an overall nonlinear system have his own independent goals and represented by a discrete model of the form:

$$\begin{cases} x_l(k+1) = A_l x_l(k) + B_l u_l(k) \\ y_l(k) = C_l x_l(k) \end{cases} \tag{15}$$

Where $x_l \in \Re^{n_x}$ is the local state space; $y_l \in \Re^{n_y}$ is the measurement output of each subsystem; $u_l \in \Re^{n_u}$ is the local control input. Therefore the overall nonlinear system can be seen as a collection of smaller subsystems that are completely independent from one another witch is referred as a decentralized model. The variable control of every agent sent to the nonlinear system consists of its agent's optimal input control given by minimizing local standard MPC cost function:

$$J_l = \sum_{N_1}^{N_2} \left\| y_l(k+j) - Setpoint(k+j) \right\|_{Q_l}^2 + \sum_{j=1}^{N_u} \left\| \Delta u_l(k+j|k) \right\|_{R_l}^2 \tag{16}$$

Where Q_l, R_l are suitable weighting matrixes.

One of the advantages of the state-space representation is that it simplifies the prediction; the prediction for this model is given:

$$\hat{y}_l(k+i|k) = C_l(A_l^{\,i}\,\hat{x}_l(k|k) + \sum_{j=1}^{i} A_l^{\,i-1} B_l u_l(k+i-j|k) \tag{17}$$

For local suitable matrix $\Psi_l, \Gamma_l, \Theta_l$ and Λ_l, we can rewrite the local predictive model output for future time instants as:

$$Y_l(k) = \Psi_l x_l(k) + \Gamma_l u_l(k-1) + \Theta_l \Delta u_l(k) \tag{18}$$

Where

$$\Psi = \begin{bmatrix} C_l A_l \\ \vdots \\ C_l A_l^{N_u} \\ \vdots \\ C_l A_l^{N_u+1} \\ \vdots \\ C_l A_l^{N_2} \end{bmatrix} \quad \Theta_l = \begin{bmatrix} C_l B_l \\ \vdots \\ C_l \sum_{j=0}^{N_u-1} A_l^j B_l \\ \vdots \\ C_l \sum_{j=0}^{N_2-1} A_l^j B_l \end{bmatrix} \tag{19}$$

$$\Gamma_l = \begin{bmatrix} C_l B_l \cdots & \cdots & \cdots 0 \\ C_l(A_l B_l + B_l) & \cdots & \cdots 0 \\ C_l \sum_{j=0}^{N_u-1} A_l^j B_l & \cdots & C_l B_l \\ \cdots & \cdots & \cdots \\ C_l \sum_{j=0}^{N_2-1} A_l B_l & \cdots & C_l \sum_{j=0}^{N_2-N_u} A_l^j B_l \end{bmatrix}$$

The cost function (16) can be rewritten as:

$$J_l = \varepsilon_l(k)^T Q_l \varepsilon_l(k) - \Delta u_l(k)^T G_l + \Delta u_l(k)^T H_l \Delta u_l(k) \tag{20}$$

Where:

$$\begin{aligned} &\varepsilon_l(k) = Setpoint(k) - \Psi_l x_l(k) - \Gamma_l u_l(k-1) - \Lambda_l \\ &G_l = 2\Theta_l{}^T Q_l \varepsilon_l(k) \\ &H_l = \Theta_l{}^T Q_l \Theta_l + R_l \end{aligned} \tag{21}$$

Therefore the control law that minimizes the local cost function (16) is given by:

$$\Delta u_l(k) = \frac{1}{2} H_l^{-1} G_l \tag{22}$$

In order to take into account constraints on the manipulated variables, a transformation method for each action is made. The control action based on (22) is transformed into new action with the following transformation [R. Fletcher, 1997].

$$\begin{cases} u_l(k) = f_{moy} + f_{amp} \tanh(\dfrac{u_l(k) - f_{moy}}{f_{amp}}) \\ f_{moy} = \dfrac{f_{\max} - f_{\min}}{2} \\ f_{amp} = \dfrac{f_{\max} + f_{\min}}{2} \\ f_{\max} = \min(u_{l\max}, u_l(k-1) + \Delta u_{l\max}) \\ f_{\min} = \max(u_{l\min}, u_l(k-1) + \Delta u_{l\min}) \end{cases} \tag{23}$$

The optimum control law (22) for each agent does not guarantee the global optimum. Accordingly to that, nonlinear system requires coordination among the control agent's action. The required coordination is done by a logic switch added to supervisory loop based neural networks which compute the global optimum control subject to constraints.

3.3 The supervisor loop

A neural network is used with the proper control architecture by changing the results of switched input u_i of each agent's action through a stable online NN weights which can guarantee the tracking performance of the overall closed-loop nonlinear system. Moreover, the neural network should reduce the deleterious effect of constraints attached with the different actions [Wenzhi, G et al., 2006]. In this work the neural network is represented by feed-forward single-input single output. The neural network tries to optimize the control action Δu.

$$\Delta u = f_{NN}(\sum_{i=1}^{n_u} b_i u(k-i) + \sum_{i=1}^{n_u}\sum_{j=1}^{i} b_{ij} u(k-i)u(k-j)) \tag{24}$$

The method of Levenberg Marquardt was designed for the optimization due to its properties of fast convergence and robustness. The main incentive of the choice of the algorithm of Levenberg Marquardt rests on the fast guarantee of the convergence toward a minimum.

4. Simulation results

The chosen example used in aim to valid the theory exposed above is given [B.Laroche et al.,2000]. A continuous state space representation of this example is as follow:

$$\begin{cases} \dot{x}_1 = x_3 - x_2 u \\ \dot{x}_2 = -x_2 + u \\ \dot{x}_3 = x_2 - x_1 + 2x_2(u - x_2) \end{cases} \tag{25}$$

The system model is implanted in the Matlab-simulink environment of which the goal is to get the input/output vector for the identification phase. Matlab® discrete these equations by the 4th order Runge-Kutta method. The vector characterizing the Volterra model that linking the output x_3 with the input u is given by:

$$A = [1 \quad -1.9897 \quad .9997]^T, B_1 = [-0.0318 \quad -0.0096]$$
$$B_2 = \begin{pmatrix} 0.0396 & 0.0656 & 0 \\ 0 & 0.0388 & 0 \\ 0 & 0 & 0 \end{pmatrix} \tag{26}$$

Moreover, the Chiu procedure is developed to divide the nonlinear system into independent subsystem [Chiu S.L, 1994]. The modeling of the dynamic system, led to the localization of two centers with respective values $c_1 = 0.0483, c_2 = 0.5480$. The classification parameters adopted for the algorithm are as follows : $r_a = .6;\ r_b = 1.25 r_a;\ \varepsilon_1 = .5;\ \varepsilon_2 = .1$. The procedure of identification and modeling has been applied to the whole measures input/output come out of the global system, driving to the different following subsystem models:

$$A_1 = \begin{pmatrix} 0.552 & 1 \\ 0.2155 & 0 \end{pmatrix}, B_1 = \begin{bmatrix} 0.0496 \\ 0.1419 \end{bmatrix}$$
$$A_2 = \begin{pmatrix} 0.7962 & 1 \\ 0.0481 & 0 \end{pmatrix}, B_2 = \begin{bmatrix} 0.0239 \\ -.0088 \end{bmatrix} \quad (27)$$

The result of modelisation is reported in figure 3. These results showed the application Chiu algorithm for the classification which has a better quality of local approximation of the system.

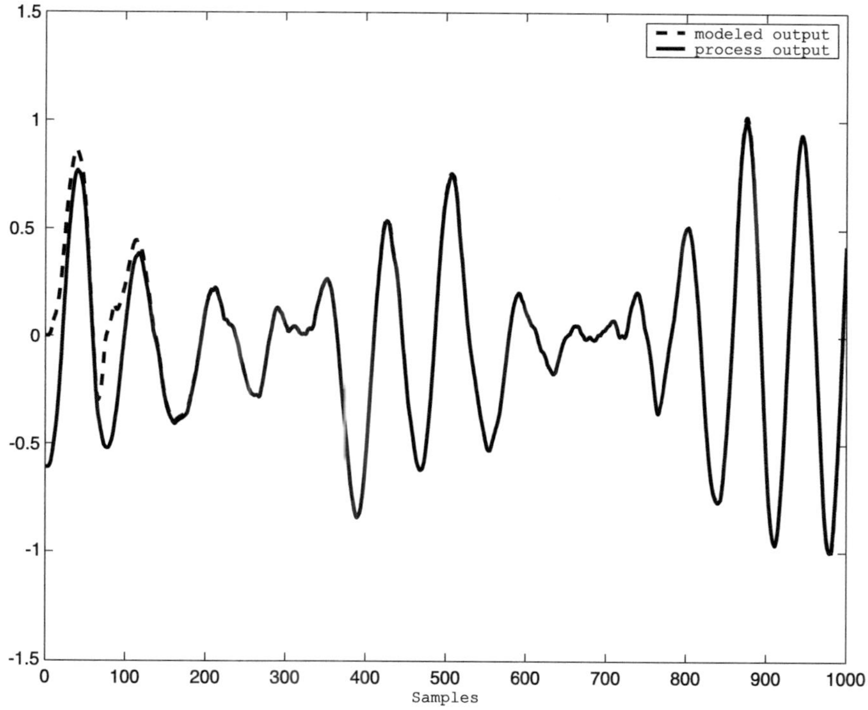

Fig. 3. Validation of the obtained model

4.1 Set point tracking

The proposed concept as seen in section 3 is used, to control the nonlinear system. The tuning parameters of the multi-agent consists of the parameters values of each agent given by: $N_1 = 1$;$N_2 = 5$;$N_u = 1$;$R_1 = R_2 = 4$;. Assuming for the sake of simplicity but without loss of generality, the prediction and control horizons are the same for each agent. The tuning parameters for the NMPC are: $N_1 = 1$;$N_2 = 5$;$N_u = 1$;$\delta = .001$. The gradient of the control $\Delta u_{\min}$ and $\Delta u_{\max}$ are taken, respectively, equal to −0.2 and 0.1 and the control is limited between 0 and 1. In this application example, the neural network was a feedforward network and it consisted of three hidden layer nodes with tangent sigmoid transfer functions and one output layer node with linear transfer function .In this section, we present a comparative study between the proposed method and the NMPC procedure. The results shown in Fig. 4 and Fig. 5 are obtained in the constrained case

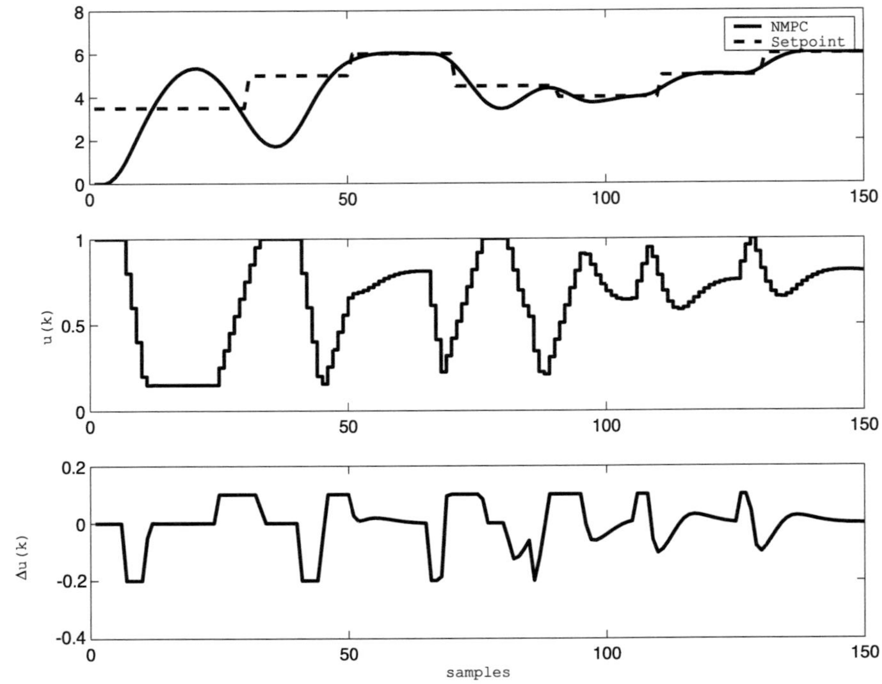

Fig. 4. Evolution of the set point, the output and the control (NMPC): constrained case

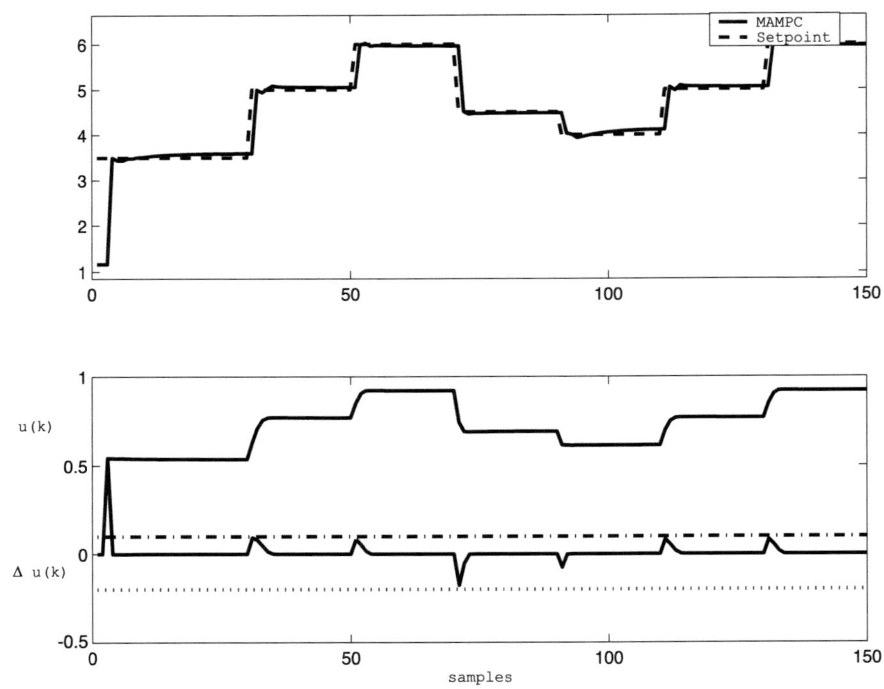

Fig. 5. Evolution of the set point, the output and the control (MAMPC): constrained case

It is clear from this figures that the new strategy of control leads to satisfactory results with respect to set-point changes. Indeed, the tracking error is reduced and with a smooth control action. It is shown that NMPC also gives consistently a good performance for the range examined. The two controllers are remarkably similar, which indicates that the MAMPC controller is close to optimal for this control problem. Moreover, the new controller meets all the required performance specifications within given input constraints and the results show a significant improvement in the system performance compared with the results obtained when only nonlinear programming model is used and the multi-agent compares favorably with respect to a numerical optimization routine as shown in Figure 6, the final control law to the nonlinear system obeying the specified constraints and with the proposed concept the constrained input and rate of change inputs cannot violate the specified range premise.

4.2 Effect of load disruptions and noise

In order to test the effect of load disruptions, we have added to the system output a constant equals to 0.02 from iteration 100 to iteration 125 and from iteration 200 to iteration 225. And in the case of noise, we have added to the output of the process an uncertain pseudo-noise of maximal amplitude equal to 0.025. Figs. 6 and 7 present the evolutions of the set point, the outputs obtained, respectively, with the presence of load disruption and noise. Fig. 6 shows the evolutions of the set point, the outputs signals obtained with both NMPC and MAMPC control strategy. It is clear from this figure that the presence of load disruptions, from iteration70 to iteration 90 and from iteration 120 to iteration 140, does not lead to a correct pursuit. Thus, the presence of load disruptions has more effect on NMPC control than the MAMPC strategy. Fig. 8 shows the evolutions of the set point, the outputs obtained with NMPC and MAMPC strategy. According to the obtained results, we notice that the MAMPC controller is capable to deliver a less fluctuate output than that obtained with NMPC approach.

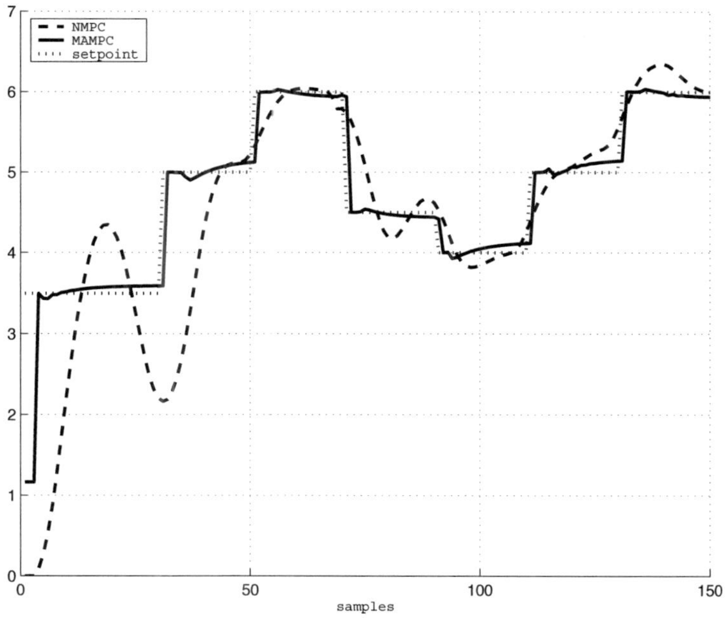

Fig. 6. Evolution of the set point, the output NMPC and MAMPC control in the case of load disruptions.

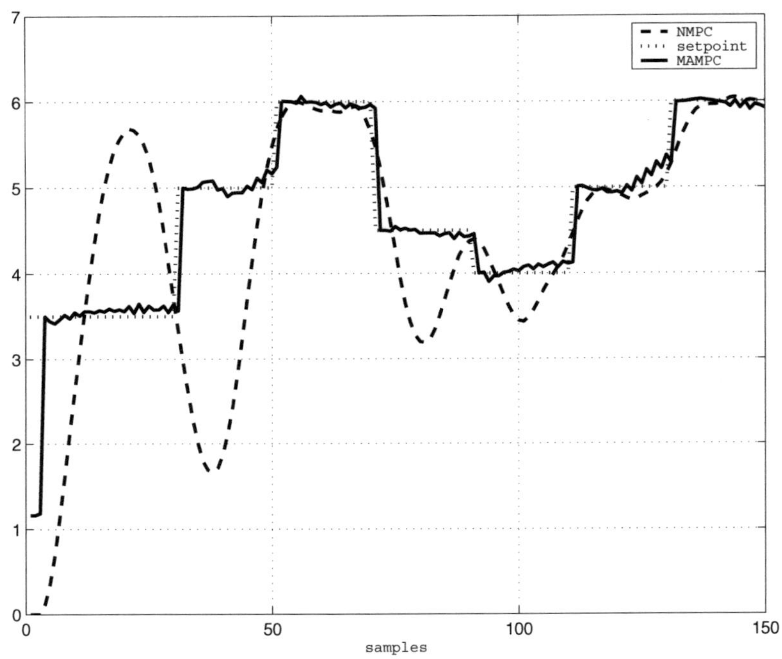

Fig. 7. Evolution of the set point, the outputs NMPC and MAMPC control in the case of the effect of the noise.

4.3 Convex optimization approach

In order to avoid solving nonconvex optimization problem, MAMPC optimization procedure, a method for convex NMPC was also developed in this chapter book. The performance of the proposed controllers is evaluated by applying to the same process and the attention has been focused on multi-agent model predictive control approach as a possible way to resolve non-convex optimization tasks. We have shown in Figure 8, a new constraint where the control is limited between 0 and .5. The nonlinear programming algorithm (NLP) cannot find a solution for the optimization problem. So because of the use of a nonlinear model, the NMPC calculation usually involves a non-convex nonlinear program, for which the numerical solution is very challenging. Therefore, finding a global optimum can be a difficult and computationally very demanding task, if possible at all. In other words, non-convexity makes the solution of the NLP uncertain. The proposed approach describe algorithm that find the solution of a non-convex programming.

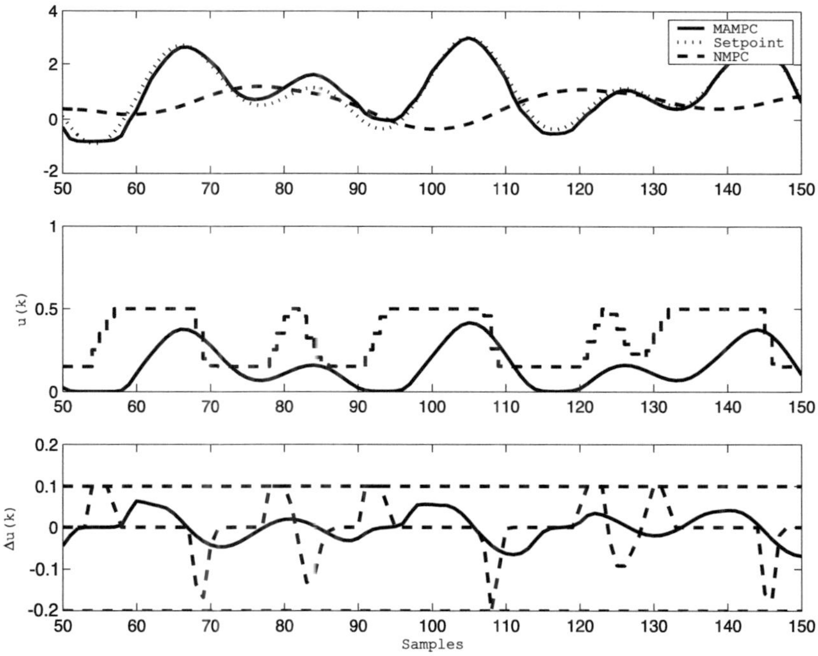

Fig. 8. Evolution of the set point, the outputs NMPC and MAMPC control: Restriction applicability of the NMPC

4.4 Computational time study

The load computational time constitutes a gate in the scheme of the predictive control indeed (Leonidas et al, 2005). The performances of the computational load established with the proposed concept are compared to a nonlinear programming In Figure 9 the time required to compute the control input at each time step k for the two approaches is plotted. We also reported in Table I, the mean and the maximum value of the implementation time required for the control law for the two cases. In Figure 10, the CPU time required to compute the control input at each time step k for the two approaches is plotted. It is very easy to see, from figure 9, 10 and table 1, that the NLMPC controller is too CPU time consuming and the computation for optimization in the new design procedure is simpler, faster and has good response curve and control performance because it uses a simple analytical solution to the minimization of the performance objective. On average, the NMPC method was about ten times slower than the novel approach and the control input in the MAMPC procedure require a twenty time smaller in the operating action.

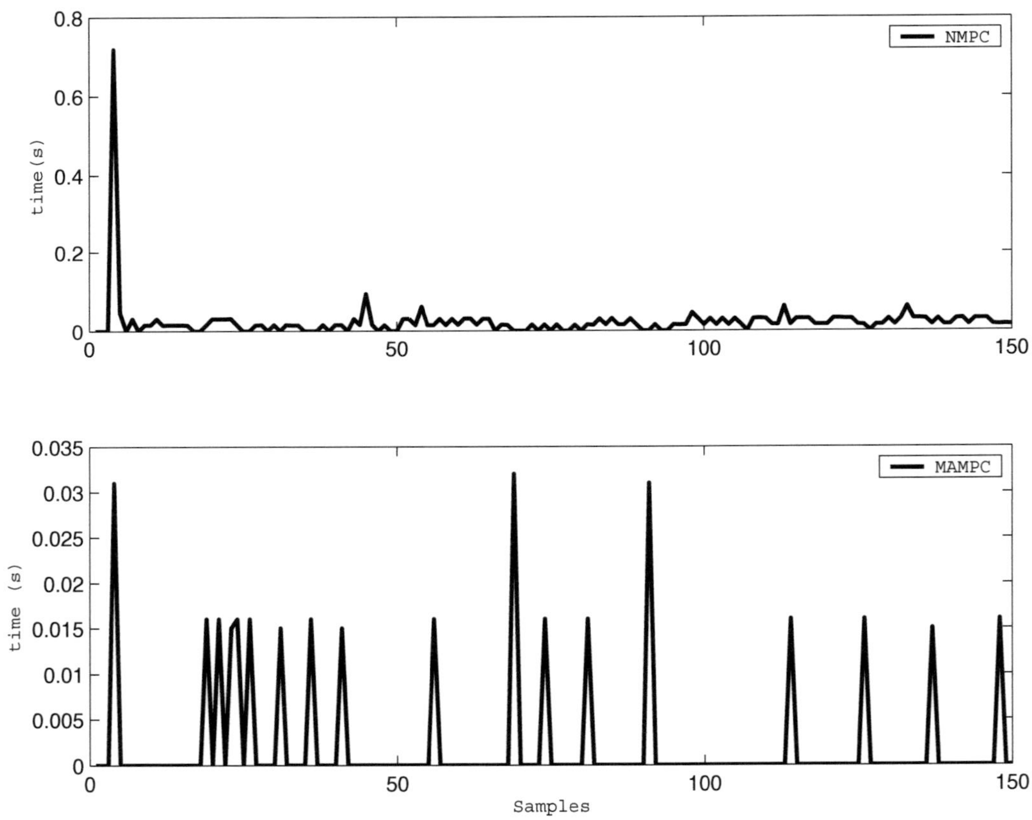

Fig. 9. Computational time requirement

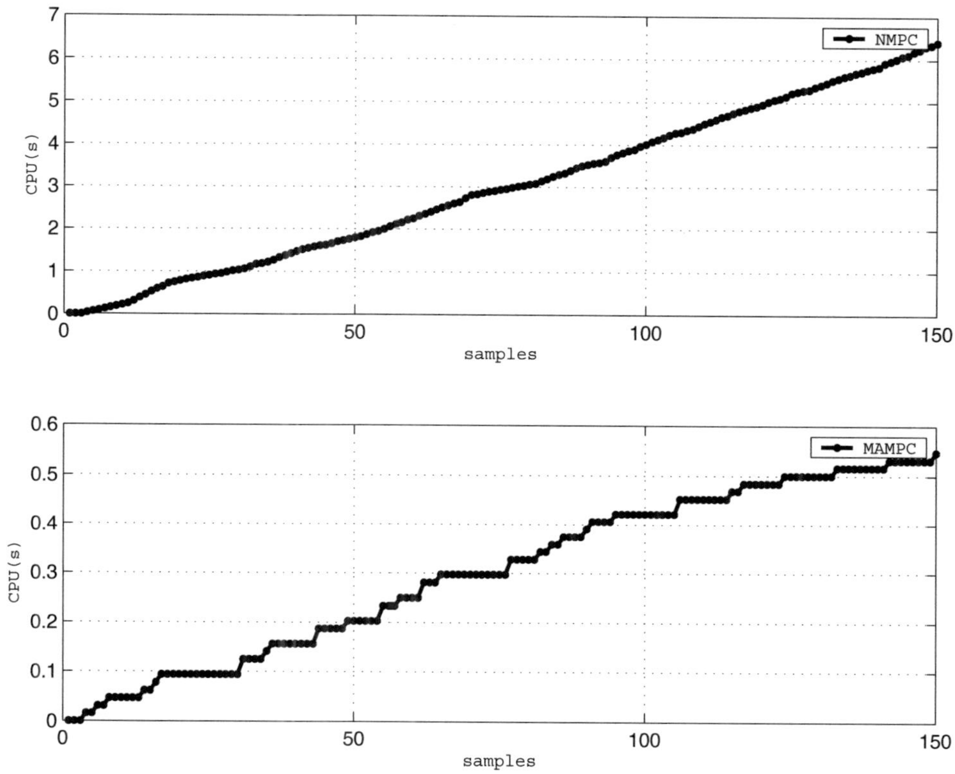

Fig. 10. CPU time comparison

	Mean	Max
NMPC	0.0224	0.7190
MAMPC	9.6875e-004	0.032

Table 1. Comparison of operating time

4.5 Controller performance comparison

Through simulation-based comparisons, it is shown that a MAMPC control system is capable of delivering significantly improved control performance in comparison to a conventional NMPC, so that the difficulty of minimizing the performance function for nonlinear predictive control is avoided, which is usually carried by the use of NLP solved at each sampling time that generally is non-convex. Moreover, the nonlinear controls based on MAMPC approach provide excellent performance, both in terms of disturbance rejection, noise suppression and set point tracking. The NMPC controller is also good for disturbance rejection and noise suppression, but the set point tracking is not succeeded. In order to make a comparison of the novel concept to the NLMPC controller, the performance of the controller was measured by the following performance indices in unconstrained and constrained cases given by [Abonyi J,2003]:

$$\begin{aligned} SSE &= \sum_{k=1}^{N} (Setpoint - y(k))^2. \\ SSU &= \sum_{k=1}^{N} (u(k) - u(k-1))^2. \\ SS\Delta U &= \sum_{k=1}^{N} \Delta u^2(k). \end{aligned} \tag{28}$$

Where SSE denotes the sum of the square error, SSU the sum of the square of the control signal, SSΔU the sum of the square of the change of the control signal and N is the number of samples. The values are summarized in Table1, shows that the MAMPC achieving control performance improves more with the use of the NLMPC controller. Moreover the MAMPC produces the best tracking performance and the smallest energy consumption.

	Constrained case		
	SSE	SSU e-011	SSΔU e-013
NMPC	0.0135	5.5415	16754
MAMPC	0.0042	1.2464	3.8622

Table 2. Control performance comparison

5. Conclusions

One of the main drawbacks of NMPC schemes is the enormous computational effort these controllers require. On the other hand, linear MPC methods can be implemented solving just Quadratic Programming (QP) or Linear Programming problems (LP).The main focus of this chapter is to develop a new control algorithm that in essence is a bridge between linear and nonlinear control. This resulted in the development of the MAMPC (Multiagent model predictive control) approach. The new NMPC scheme based MAMPC is implemented to reduce the computational effort. The control performance of MAMPC algorithm is evaluated by illustrative comparison with general NMPC. All the results prove that MAMPC approach is a fairly promising algorithm by delivering significantly improved control. The performance of the proposed controllers is evaluated by applying to single input-single output control of non linear system. Theoretical analysis and simulation results demonstrate better performance of the MAMPC over a conventional NMPC based on sequential quadratic programming in tracking the setpoint changes as well as stabilizing the operation in the presence of input disturbances.

6. References

Maciejowski J (2002), Predictive Control with constraints, Prentice Hall, London.

Qin, S. J and Badgwell T.A A survey of industrial model predictive control technology, Control Engineering Practice 11 733-764.(2003)

Kurtz M.J and.Henson M.A Input-output linearizing control of constrained nonlinear process, Journal of Process Control, 7(1) 3-17(1997).

Arahal M.R and Berenguel M and Camacho E.F , Neural identification applied to predictive control of a solar plant, Control Engineering Practice 6 334- 344,(1998)

Fischer M and.Nelles O , Predictive control based on local linear fuzzy models, International Journal of system Science, 29 679-697(1998).

Mollov S., Babuska R., Abonyi J., and Henk B.. Effective Optimization for Fuzzy Model Predictive Control . IEEE Transactions on fuzzy systems, Vol. 12, No. 5,661-675 (2004)

Brooms A and Kouvaritakis B, Successive constrained optimization and interpolation on non-Linear model based predictive control, International journal of control, 68 (3) 599- 623(2000)

Abonyi, J and Babuska.R and Abotto M and Szeifert F and Nagy L. Identification and control of nonlinear systems using fuzzy Hammerstein models, Ind.Eng.Chem.Res, 39 4302-4314.(2000)

Kanev, S and Vergaegen, M. "Controller e-configuration for non-linear systems", Control Engineering Practice, Vol. 8 No. 11,(2000) pp. 1223-35.

Negenborn R.R, De Schutter B. and Hellendroom J Multi-agent model predictive control A survey, Technical report 04-010, Delf Center for systems and control.(2004)

Soest Van W.R. and Q.P. Chu Q.P, and Mulder J.ACombined Feedback Linearization and Constrained Model Predictive Control for Entry Flight Journal of Guidance, Control, and Dynamics, 29(2) 427- 434 (2005)

Jaochim H, Bock H.G and Diehl MAn online active set strategy for fast parametric quadratic programming in mpc applications, Proceeding IFAC Workshop on Nonlinear Model Predictive Control for fast Systems NMPC- FS'06, France,pp 13-22.(2006),

Didier G., Distributed model predictive control via decomposition- coordination techniques and the use of an augmented lagrangian. IFAC. Workshop on NMPC, France, 111-116(,2006)

H.Ben Nasr and F.M'Sahli, A Computational Time Requirement Comparison Between Two Approaches In MBPC, International Journal of soft computing, 3 (2), pp:147-154 (2008a)

H.Ben Nasr and F.M'Sahli, Une supervision d'action pour la commande prédictive multi-agent, 1st International Workshop on Systems Engineering Design & Applications,,SENDA (2008b).

H.Ben Nasr and F.M'Sahli, Une nouvelle stratégie de commande prédictive des systèmes non linéaires à dynamique rapides. Conférence Internationale Francophone d'Automatique Bucarest, Roumanie, 3-5 septembre CIFA(2008c)

H.Ben Nasr, M'Sahli F, A Multi-agent Approach To TS-Fuzzy Modeling and Predictive Control of Constrained Nonlinear System", International Journal of Intelligent Computing and Cybernetics. Vol,3,pp:398-424 (2008d)

H.Ben Nasr, M'Sahli F, A Multi-agent predictive control approach based on fuzzy supervisory loop for fast dynamic systems. The fifth International Multi-Conference on Systems, Signal and Devices, SSD (2008e).

D.W. Clarke, C.Mohtadi and P.S.Tuffs, Generalized predictive control. I: The basic algorithm II: Extensions and interpretations, Automatica, 23, pp. 137-160, 1987.

Haber R and Bars R.and Engel L.(1999a) Three Extended Horizon Adaptive Nonlinear predictive Control Schemes on the Parametric Volterra Model, ECC, Proceedings of the European Control Conference, Karlshruhe, Germany

Haber R and Bars R. and Engel L(1999b) Sub-optimal Nonlinear Predictive and Adaptive Control based on the Parametric Volterra Model, Applied Mathematics and Computer Science 70 161-173

Matthew J. Tenny_and James B. Rawlings TWMCC (Texas-Wisconsin Modeling and Control Consortium) Technical report number 2002-04

F.J.Doyle, B.A.Ogunnike and R.K.Pearson, "Non-linear model-based control using seconds order Volterra models". Automatica.1995,Vol5, N31.pp.697-714.

B.Laroche, M. Philippe and P. Rochon. "Motion planning for the heat equation", international journal of robotic and nonlinear control. pp. 629-643. 2000.

R. Fletcher. Pratical Methods of Optimization. JohnWiley and Sons, 1987.

Wenzhi, G. and Rastko, R. , "Neural network control of a class of nonlinear systems with actuator saturation", IEEE Trans. On Neural Networks, 17 (1) (2006) 147-56.

Chiu S.L., "Fuzzy Model identification based on cluster estimation", Journal of Intelligent Fuzzy Systems 2 (1994) 267-278.

Leonidas, G.B. and Mayuresh, V.K. (2005), "Real-time implementation of model predictive control", American Control Conference, Portland, OR, pp. 4166-71.

Abonyi J. (2003) Fuzzy Model identification for control, Birkhauser Boston.

3

Improved Nonlinear Model Predictive Control Based on Genetic Algorithm

Wei CHEN, Tao ZHENG, Mei CHEN and Xin LI
Department of automation, Hefei University of Technology, Hefei, China

1. Introduction

Model predictive control (MPC) has made a significant impact on control engineering. It has been applied in almost all of industrial fields such as petrochemical, biotechnical, electrical and mechanical processes. MPC is one of the most applicable control algorithms which refer to a class of control algorithms in which a dynamic process model is used to predict and optimize process performance. Linear model predictive control (LMPC) has been successfully used for years in numerous advanced industrial applications. It is mainly because they can handle multivariable control problems with inequality constraints both on process inputs and outputs.

Because properties of many processes are nonlinear and linear models are often inadequate to describe highly nonlinear processes and moderately nonlinear processes which have large operating regimes, different nonlinear model predictive control (NMPC) approaches have been developed and attracted increasing attention over the past decade [1-5].

On the other hand, since the incorporation of nonlinear dynamic model into the MPC formulation, a non-convex nonlinear optimal control problem (NOCP) with the initial state must be solved at each sampling instant. At the result only the first element of the control policy is usually applied to the process. Then the NOCP is solved again with a new initial value coming from the process. Due the demand of an on-line solution of the NOCP, the computation time is a bottleneck of its application to large-scale complex processes and NMPC has been applied almost only to slow systems. For fast systems where the sampling time is considerably small, the existing NMPC algorithms cannot be used. Therefore, solving such a nonlinear optimization problem efficiently and fast has attracted strong research interest in recent years [6-11].

To solve NOCP, the control sequence will be parameterized, while the state sequence can be handled with two approaches: sequential or simultaneous approach. In the sequential approach, the state vector is handled implicitly with the control vector and initial value vector. Thus the degree of freedom of the NLP problem is only composed of the control parameters. The direct single shooting method is an example of the sequential method. In the simultaneous approach, state trajectories are treated as optimization variable. Equality constraints are added to the NLP and the degree of freedom of the NLP problem is composed of both the control and state parameters. The most well-known simultaneous method is based on collocation on finite elements and multiple shooting.

Both single shooting method and multiple shooting based optimization approaches can then be solved by a nonlinear programming (NLP) solver. The conventional iterative optimization method , such as sequential quadratic programming (SQP) has been applied to NMPC. As a form of the gradient-based optimization method, SQP performs well in local search problems. But it cannot assure that the calculated control values are global optimal because of its relatively weak global search ability. Moreover, the performance of SQP greatly depends on the choice of some initialization values. Improper initial values will lead to local optima or even infeasible solutions.

Genetic Algorithms (GAs) are a stochastic search technique that applies the concept of process of the biological evolution to find an optimal solution in a search space. The conceptual development of the technique is inspired by the ability of natural systems for adaptation. The increasing application of the algorithm has been proved to be efficient in solving complicated nonlinear optimization problems, because of their ability to search efficiently in complicated nonlinear constrained and non-convex optimization problem, which makes them more robust with respect to the complexity of the optimization problem compared to the more conventional optimization techniques.

Compared with SQP, GAs can reduce the dimension of search space efficiently. Indeed, in SQP the state sequence is treated as additional optimization variables; as such, the number of decision variables is the sum of the lengths of both the state sequence and the control sequence. In contrast, in GAs, state equations can be included in the objective function, thus the number of decision variables is only the length of control sequence. Furthermore, the search range of the input variable constraints can be the search space of GA during optimization, which makes it easier to handle the input constraint problem than other descent-based methods.

However, a few applications of GAs to nonlinear MPC [12][13] can partially be explained by the numerical complexity of the GAs, which make the suitable only for processes with slow dynamic. Moreover, the computational burden is much heavier and increases exponentially when the horizon length of NMPC increases. As a result, the implementation of NMPC tends to be difficult and even impossible.

In this paper an improved NMPC algorithm based on GA is proposed to reduce the severe computational burden of conventional GA-based NMPC algorithms. A conventional NMPC algorithm seeks the exact global solution of nonlinear programming, which requires the global solution be implemented online at every sampling time. Unfortunately, finding the global solution of nonlinear programming is in general computationally impossible, not mention under the stringent real-time constraint. We propose to solve a suboptimal descent control sequence which satisfies the control, state and stability constraints in the paper. The solution does not need to minimize the objective function either globally or locally, but only needs to decrease the cost function in an effective manner. The suboptimal method has relatively less computational demands without deteriorating much to the control performance.

The rest of the paper is organized as follows. Section 2 briefly reviews nonlinear model predictive control. Section 3 describes the basics of GAs, followed by a new GA-based computationally efficient NMPC algorithm. Section 4 analyses the stability property of our nonlinear model predictive control scheme for closed-loop systems. Section 5 demonstrates examples of the proposed control approach applied to a coupled-tank system and CSTR. Finally we draw conclusions and give some directions for future research.

2. Nonlinear model predictive control

2.1 System

Consider the following time-invariant, discrete-time system with integer k representing the current discrete time event:

$$x(k+1) = f[x(k),u(k)] \tag{1}$$

In the above, $x(k) \in X \subseteq R^{nx}$ is the system state variables; $u(k) \in U \subseteq R^{nu}$ is the system input variables; the mapping $f: R^{nx} \times R^{nu} \to R^{nx}$ is twice continuously differentiable and $f(0,0)=0$.

2.2 Objective function

The objective function in the NMPC is a sum over all stage costs plus an additional final state penalty term [14], and has the form:

$$J(k) = F(x(k+P \mid k)) + \sum_{j=0}^{P-1} l(x(k+j \mid k), u(k+j \mid k)) \tag{2}$$

where $x(k+j \mid k)$ and $u(k+j \mid k)$ are predicted values at time k of $x(k+j),u(k+j)$. P is the prediction horizon. In general, $F(x) = x^T \bar{Q} x$ and $l(x,u) = x^T Q x + u^T R u$. For simplicity, $\bar{Q} > 0$ defines a suitable terminal weighting matrix and $Q \geq 0, R > 0$.

2.3 General form of NMPC

The general form of NMPC law corresponding to (1) and (2) is then defined by the solution at each sampling instant of the following problem:

$$\min_{u(k|k),u(k+1|k),\ldots,u(k+P-1|k)} J(k) \tag{3a}$$

$$s.t.\, x(k+i+1 \mid k) = f(x(k+i \mid k), u(k+i \mid k)) \tag{3b}$$

$$x(k+i \mid k) \in X, u(k+i \mid k) \in U, i = 0,1,\ldots,P-1 \tag{3c}$$

$$x(k+P \mid k) \in X_F \tag{3d}$$

where X_F is a terminal stability constraint, and $\mathbf{u}(k)=[u(k \mid k), \ldots, u(k+P-1 \mid k)]$ is the control sequence to be optimized over.

The following assumptions A1 - A4 are made:

A1: $X_F \subset X$, X_F closed, $0 \in X_F$

A2: the local controller $\kappa_F(x) \in U$, $\forall x \in X_F$

A3: $f(x, \kappa_F(x)) \in X_F$, $\forall x \in X_F$

A4: $F(f(x, \kappa_F(x))) - F(x) + l(x, \kappa_F(x)) \leq 0$, $\forall x \in X_F$

Based on the formulation in (3), model predictive control is generally carried out by solving online a finite horizon open-loop optimal control problem, subject to system dynamics and constraints involving states and controls. At the sampling time k, a NMPC algorithm attempts to calculate the control sequence $\mathbf{u}(k)$ by optimizing the performance index (3a) under constraints (3b) (3c) and terminal stability constraint (3d). The first input $u(k|k)$ is then sent into the plant, and the entire calculation is repeated at the subsequent control interval k+1.

3. NMPC algorithm based on genetic algorithm

3.1 Handling constraints

An important characteristic of process control problems is the presence of constraints on input and state variables. Input constraints arise due to actuator limitations such as saturation and rate-of-change restrictions. Such constraints take the form:

$$\mathbf{u}_{\min} \leq \mathbf{u}(k) \leq \mathbf{u}_{\max} \tag{4a}$$

$$\Delta\mathbf{u}_{\min} \leq \Delta\mathbf{u}(k) \leq \Delta\mathbf{u}_{\max} \tag{4b}$$

State constraints usually are associated with operational limitations such as equipment specifications and safety considerations. System state constraints are defined as follows:

$$\mathbf{x}_{\min} \leq \mathbf{x}(k) \leq \mathbf{x}_{\max} \tag{4c}$$

where $\Delta\mathbf{u}(k)=[u(k|k)-u_k-1,\ldots, u(k+P-1|k)-u(k+P-2|k)]$, $x(k)=[x(k+1|k),\ldots,x(k+P|k)]$.
The constraints (4a) and (4b) can be written as an equivalent inequality:

$$\begin{bmatrix} I \\ -I \\ I \\ -I \end{bmatrix} \mathrm{u}(k) \leq \begin{bmatrix} \mathrm{u}_{\max} \\ -\mathrm{u}_{\min} \\ S\Delta\mathrm{u}_{\max} + cu_{k-1} \\ -S\Delta\mathrm{u}_{\min} - cu_{k-1} \end{bmatrix} \tag{5}$$

where

$$S = \begin{bmatrix} 1 & 0 & 0 & 0 \\ 1 & 1 & 0 & \ldots \\ \ldots & \ldots & \ldots & 0 \\ 1 & \ldots & \ldots & 1 \end{bmatrix}, \quad c = [I, I, \ldots, I]^T .$$

3.2 Genetic algorithm

GA is known to have more chances of finding a global optimal solution than descent-based nonlinear programming methods and the operation of the GA used in the paper is explained as follows.

3.2.1 Coding

Select the elements in the control sequence $\mathbf{u}(k)$ as decision variables. Each decision variable is coded in real value and n_u*P decision variables construct the n_u*P -dimensional search space.

3.2.2 Initial population

Generate initial control value in the constraint space described in (5). Calculate the corresponding state value sequence $\mathbf{x}(k)$ from (3b). If the individual (composing control value and state value) satisfies the state constraints (4c) and terminal constraints (3d), select it into the initial population. Repeat the steps above until PopNum individuals are selected.

3.2.3 Fitness value

Set the fitness value of each individual as $1/(J+1)$.

3.2.4 Genetic operators

Use roulette method to select individuals into the crossover and mutation operator to produce the children. Punish the children which disobey the state constraints (4c) and terminal constraints (3d) with death penalty. Select the best PopNum individuals from the current parent and children as the next generation.

3.2.5 Termination condition

Repeat the above step under certain termination condition is satisfied, such as evolution time or convergence accuracy.

3.3 Improved NMPC algorithm based on GA

In recent years, the genetic algorithms have been successfully applied in a variety of fields where optimization in the presence of complicated objective functions and constraints. The reasons of widely used GAs are its global search ability and independence of initial value. In this paper GAs are adopted in NMPC applications to calculate the control sequence. If the computation time is adequate, GAs can obtain the global optimal solution. However, it needs to solve on line a non-convex optimization problem involving a total number of n_u*P decision variables at each sampling time. To obtain adequate performance, the prediction horizon should be chosen to be reasonably large, which results in a large search space and an exponentially-growing computational demand. Consequently, when a control system requires fast sampling or a large prediction horizon for accurate performance, it becomes computationally infeasible to obtain the optimal control sequence via the conventional GA approach. There is thus a strong need for fast algorithms that reduce the computational demand of GA.

The traditional MPC approach requires the global solution of a nonlinear optimization problem. This is in practice not achievable within finite computing time. An improved NMPC algorithm based on GA does not necessarily depend on a global or even local minimum. The optimizer provides a feasible decedent solution, instead of finding a global or local minimum. The feasible solution decreases the cost function instead of minimizing the cost function. Judicious selection of the termination criterions of GA is the key factor in reducing the computation burden in the design of the suboptimal NMPC algorithm. To this end, the following two strategies at the $(k+1)$-th step are proposed.

- The control sequence output at the k-th control interval in the genetic algorithm is always selected as one of the initial populations at the $(k+1)$-th control interval. Furthermore, some of the best individuals at the k-th control interval are also selected into the current initial population. Most of all, the elite-preservation strategy is adopted. Figure 1 shows one of the choices of the initial population per iteration. This strategy guarantees the quality of current population and the stability of the NMPC algorithm.

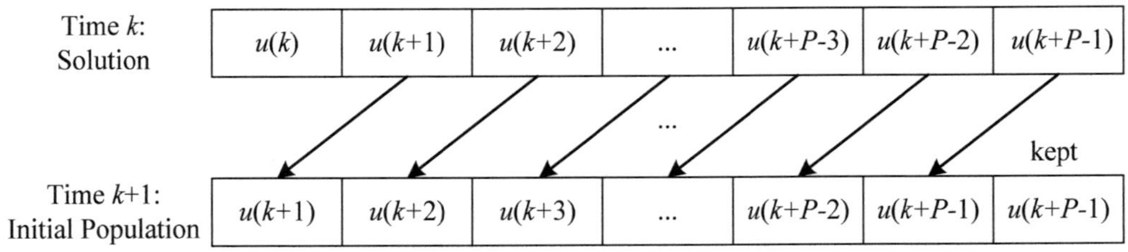

Fig. 1. The choice of initial population per iteration

- Stopping criterions of GA are the key factor of decreasing the computation burden. GA is used to compute the control sequence. The objective value $J(k+1)$ at the $(k+1)$-th control interval is computed and compared with $J(k)$ that stored at the k-th control interval. If $J(k+1)$ is smaller than $J(k)$, that at the k-th control interval, then the control sequence $\mathbf{u}(k+1)$ is retained as a good feasible solution, and its first element $u(k+1\,|\,k+1)$ is sent to the plant. Otherwise, if there does not exist a feasible value for $\mathbf{u}(k+1)$ to yield $J(k+1) < J(k)$, then the best $\mathbf{u}(k+1)$ is chosen to decrease the objective function the most.

The traditional MPC approach requires the global solution of a nonlinear optimization problem. This is in practice not achievable within finite computing time. An improved NMPC algorithm based on GA does not necessarily depend on a global or even local minimum. The optimizer provides a feasible decedent solution, instead of finding a global or local minimum. The feasible solution decreases the cost function instead of minimizing the cost function. Judicious selection of the termination criterions of GA is the key factor in reducing the computation burden in the design of the suboptimal NMPC algorithm. To this end, the following two strategies at the $(k+1)$-th step are proposed.

With the above two strategies, the computational complexity of the control calculation is substantially reduced. Summarizing, our proposed improved NMPC algorithm performs the following iterative steps:

Step 1. [Initialization]:

choose parameters P, X_F, Q, R, Q' and model $x(k+1) = f(x(k),u(k))$; initialize the state and control variables at $k = 0$; compute and store $J(0)$.

Step 2. [modified Iteration]:

- at the k-th control interval, determine the control sequence $\mathbf{u}(k)$ using GA satisfies constraints (3b) (3c), terminal stability constraint (3d) and $J(k) < J(k\text{-}1)$. The first input $u(k\,|\,k)$ is then sent into the plant.
- store $J(k)$ and set $k = k+1$;
- if there does not exist a feasible value for $\mathbf{u}(k)$ to yield $J(k) < J(k\text{-}1)$, then the best $\mathbf{u}(k)$ is chosen to decrease the objective function the most.

Step 3. [Termination]

The entire calculation is repeated at subsequent control interval k+1 and goes to Step 2.

Though the proposed method does not seek a globally or locally optimal solution within each iteration step, it may cause little performance degradation to the original GA due to its iterative nature, which is known to be capable of improving suboptimal solution step by step until reaching near-optimal performance at the final stage. Besides its near-optimal performance, the proposed algorithm possesses salient feature; it guarantees overall system stability and, most of all, leads to considerable reduction in the online computation burden. Finding a control sequence that satisfies a set of constraints is significantly easier than solving a global optimization problem. Here it is possible to obtain the suboptimal control sequence via GA for practical systems with very demanding computation load, that is, systems with a small sampling time or a large prediction horizon.

4. Stability of nonlinear model predictive control system

The closed-loop system controlled by the improved NMPC based on GA is proved to be stable.
Theorem 1: For a system expressed in (1) and satisfying the assumption A1-A4, the closed-loop system is stable under the improved NMPC framework.
Proof: Suppose there are an admissible control sequence **u**(k) and a state sequence **x**(k) that satisfy the input, state and terminal stability constraints at the sampling time k.
At the sampling time k, the performance index, which is related to **u**(k) and **x**(k), is described as

$$J(k) = J^*(k;\mathrm{u}(k),\mathrm{x}(k)) \tag{6}$$

In the closed-loop system controlled by the improved NMPC, define the feasible input and state sequences for the successive state are $x^+=f(x, u(k\,|\,k))$.

$$\begin{aligned} \mathrm{u}^+(k+1) &= [u^T(k+1\,|\,k),\ldots,u^T(k+P-1\,|\,k), K_F(x(k+P\,|\,k))^T]^T \\ \mathrm{x}^+(k+1) &= [x^T(k+1\,|\,k),\ldots,x^T(k+P\,|\,k), f(x(k+P\,|\,k), K_F(x(k+P\,|\,k)))^T]^T \end{aligned} \tag{7}$$

The resulting objective function of **u**(k+1) and **x**(k+1) at the (k+1)-th sampling time is

$$J^+(k+1) = J(k+1;\mathrm{u}(k+1),\mathrm{x}(k+1)) \tag{8}$$

If the optimal solution is found, it follows that

$$\begin{aligned} &J^*(k+1) - J^*(k) \le J^+(k+1) - J^*(k) \\ &= -l(x,K_P(x)) + F(f(x,K_F(x))) - F(x) + l(x,K_F(x)) \end{aligned} \tag{9}$$

From A4, the following inequality holds

$$J^*(k+1) - J^*(k) \leq= -l(x,K_P(x)) \tag{10}$$

Or using the improved NMPC algorithm, it follows that

$$J^*(k+1) - J^*(k) \le J^+(k+1) - J^*(k) \le 0 \tag{11}$$

Thus the sequence $J^*(k+i)$ over P time indices decreases. As such and given the fact that the cost function $l(x,u)$ is lower-bounded by zero, it is evident that $J^*(k+i)$ converges. Taking the sum, we obtain

$$J^*(k+P) - J^*(k) \le \sum_{i=1}^{P}[-l(x(k+i),u(k+i))] \tag{12}$$

Also, because the sequence $J^*(k+i)$ is decreasing, then as $N \to \infty$, we have $l(x(k+i),u(k+i)) \to 0$ and $x \to 0$. Hence, the closed-loop system is stable.

5. Simulation and experiment results

5.1 Simulation results to a continuous stirred tank reactor plant

5.1.1 Model of continuous stirred tank reactor plant

Consider the highly nonlinear model of a chemical plant (continuous stirred tank reactor-CSTR). Assuming a constant liquid volume, the CSTR for an exothermic, irreversible reaction, A→B, is described by

$$\begin{aligned} \dot{C}_A &= \frac{q}{V}(C_{Af} - C_A) - k_0 e^{-E/(RT)} C_A \\ \dot{T} &= \frac{q}{V}(T_f - T) + \frac{-\Delta H}{\rho C_P} k_0 e^{-E/(RT)} C_A + \frac{UA}{V \rho C_P}(T_C - T) \end{aligned} \tag{13}$$

where C_A is the concentration of A in the reactor, T is the reactor temperature and T_c is the temperature of the coolant stream. The parameters are listed in the Table 1.

Variables	Meaning	Value	Unit
q	the inlet flow	100	l/min
V	the reactor liquid volume	100	l
C_{Af}	the concentration of inlet flow	1	mol/l
k_0	reaction frequency factor	$7.2*10^{10}$	min-1
E/R		8750	K
E	activation energy		
R	gas constant	$8.3196*10^{3}$	J/(mol K)
T_f	the temperature of inlet flow	350	K
ΔH	the heat of reaction	$-5*10^{4}$	J/mol
ρ	the density	1000	g/l
C_P	the specific heat capacity of the fluid	0.239	J/(g K)
UA		$5*10^{4}$	J/(min K)
U	the overall heat transfer coefficient		
A	the heat transfer area		

Table 1. List of the model parameters

5.1.2 Simulation results

The paper present CSTR simulated examples to confirm the main ideas of the paper. The nominal conditions, C_A = 0.5mol/l, T = 350K, T_c = 300K, correspond to an unstable operating point. The manipulated input and controlled output are the coolant temperature (T_c) and reactor temperature (T). And the following state and input constraints must be enforced:

$$\begin{aligned} \mathsf{X} &= \left\{ \begin{bmatrix} C_A \\ T \end{bmatrix} \in \mathsf{R}^2 \mid 0 \le C_A \le 1, 280 \le T \le 370 \right\} \\ \mathsf{U} &= \left\{ T_C \in \mathsf{R} \mid 280 \le T_C \le 370 \right\} \end{aligned} \tag{14}$$

The simulation platform is MATLAB and simulation time is 120 sampling time. The sampling time is Ts = 0.05s, mutation probability is P_c =0.1, population size is 100, maximum generation is 100, and the fitness value is 1/(J+1).

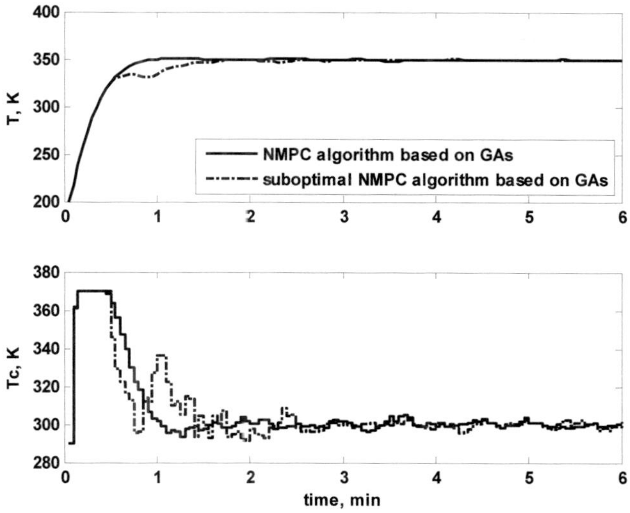

Fig. 2. Comparative simulation between the conventional NMPC algorithm and the suboptimal NMPC algorithm

method	settling time, min	percent overshoot,%
NMPC algorithm based on GA	0.75	1
Suboptimal NMPC algorithm based on GA	2.5	0

Table 2. Performance comparisons of simulation results

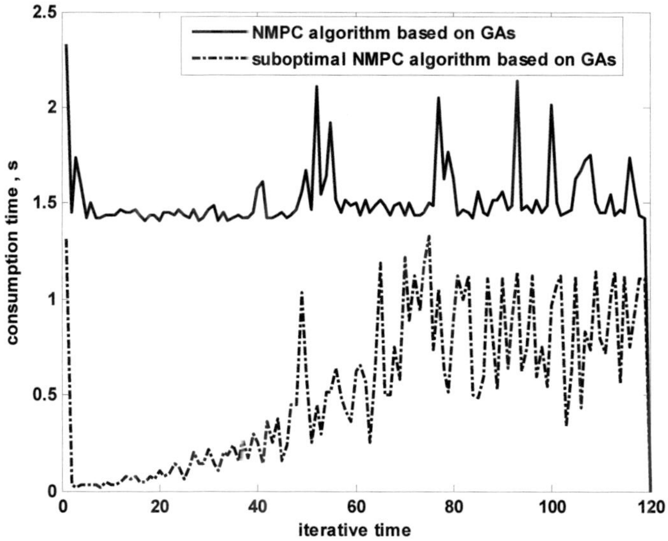

Fig. 3. The time consumptions of the two methods

For comparison, the same simulation setup is used to test both the conventional NMPC algorithm based GA and the suboptimal NMPC algorithm. The resulting control values are depicted in Fig 2. Table 2 compares the performance of the two algorithms using the metrics settling time and percent of overshoot. The conventional NMPC algorithm has a faster transient phase and a smaller percentage of overshoots.
When the population sizes or the maximum generation is relatively large, the time consumption of the two methods is compared in Fig 3.
From Fig 2 and Table 2, it is apparent that the control performance of the two methods is almost same. But from Fig 3, it is evident that the suboptimal NMPC algorithm based on GA has a considerably reduced demand on computational complexity.

5.2 Simulation results to a coupled-tank system

5.2.1 Model of coupled-tank system

The apparatus [15], see Fig.4, consists of two tanks T_1 and T_2, a reservoir, a baffle valve V_1 and an outlet valve V_2. T_1 has an inlet commanded through a variable pump based on *PMW* and T_2 has an outlet that can be adjusted through a manually controlled valve only. The outlets communicate to a reservoir from which the pumps extract the water to deliver it to the tank. The two tanks are connected through the baffle valve, which again can only be adjusted manually. The objective of the control problem is to adjust the inlet flow so as to maintain the water level of the second tank close to a desired setpoint.
The water levels h_1 and h_2, which are translated through the pressure transducer into a *DC* voltage ranging from $0V$ to $5V$, are sent to *PC* port via *A/D* transition. The tank pump control, which is computed by the controller in *PC* with the information of the water level h_1 and h_2, is a current level in the range $4mA$ to $20mA$, where these correspond to the pump not operating at all, and full power respectively.

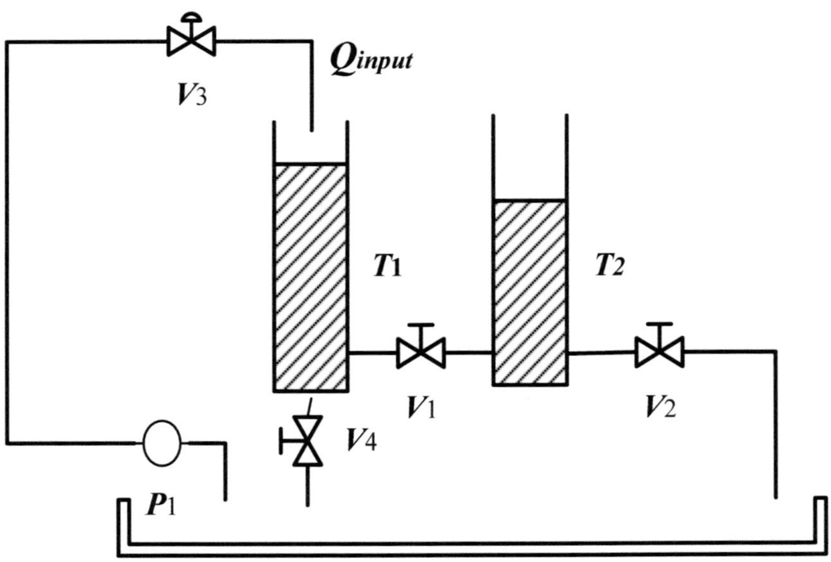

Fig. 4. Coupled-tank apparatus

The dynamics of the system are modeled by the state-space model equations:

$$A\frac{dh_1}{dt} = Q_{input} - Q_{12}$$
$$A\frac{dh_2}{dt} = Q_{12} - Q_{20} \tag{15}$$

where the flows obey Bernoulli's equation[16], i.e.

$$Q_{12} = \mu_1 S\,\text{sgn}(h_1 - h_2)(2g|h_1 - h_2|)^{1/2} \tag{16}$$

$$Q_{20} = \mu_2 S(2gh_2)^{1/2} \tag{17}$$

and $\text{sgn}(z) = \begin{cases} 1, z \geq 0 \\ -1, z < 0 \end{cases}$ is the symbol function of parameter z.

The output equation for the system is

$$y = h_2 \tag{18}$$

The cross-section areas, i.e. A and S, are determined from the diameter of the tanks and pipes. The flow coefficients, μ_1 and μ_2, have experimentally (from steady-state measurements) been determined. Table 3 is the meanings and values of all the parameters in Eqn.15

Signal	Physics Meaning	Value
A	Cross-section area of tank	$6.3585\times10^{-3}m^2$
S	Cross-section area of pipe	$6.3585\times10^{-5}m^2$
g	acceleration of gravity	$9.806m/s^2$
μ_1	flow coefficient 1	0.3343
μ_2	flow coefficient 2	0.2751

Table 3. Meanings and value of all the parameters

Several constraints have to be considered. Limited pump capacity implies that values of Q_{input} range from 0 to $50cm^3/s$. The limits for the two tank levels, h_1 and h_2, are from 0 to $50cm$.

5.2.2 Simulation results

The goal of the couple-tank system is to control the level of Tank 2 to setpoint. The initial levels of the two tanks, h_1, h_2, are $0cm$. The objectives and limits of the tank system: Input constraint is $0 \leq u \leq 100\%$; State objectives are $0 \leq h_1, h_2 \leq 0.5m$, and the setpoint of Tank 2 is 0.1m. The simulation platform is MATLAB and simulation time is 80 sample time. In NMPC, select prediction horizon P=10, weighting parameters $\bar{Q} = Q = 8$, R=1, sample time Ts=5s, mutation probability Pc=0.1, population size is 200, maximum generation is 100, the fitness value is $1/(J+1)$.

For the purpose of comparison the same simulation is carried out with the conventional NMPC algorithm based GA and the fast NMPC algorithm. The result is shown in Fig 5. The performance indexes of the two algorithms are shown in Table 4.

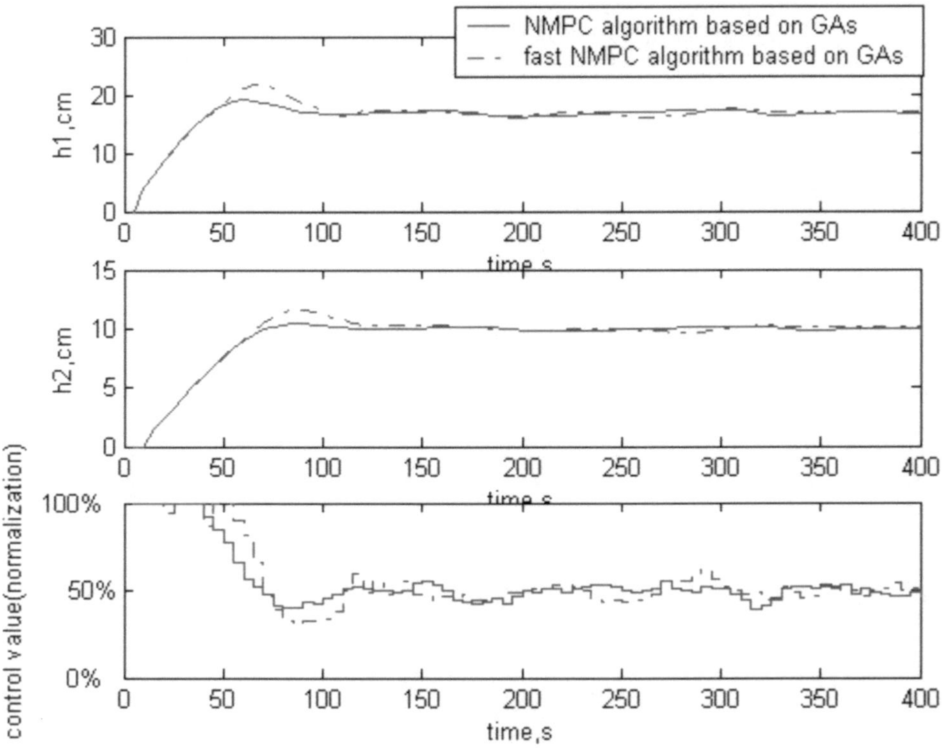

Fig. 5. Compared simulation results based on conventional NMPC and fast NMPC algorithm

method	Settling time, s	percent overshoot, %
NMPC algorithm based on GA	70	3
Fast NMPC algorithm based on GA	100	7

Table 4. Performance index of simulation results

When the population sizes or the maximum generation is relatively larger, the time consumptions of the two method is shown in Fig 6.

From Fig 5 and Table 4, it is apparent that the control performance of the two methods is almost same. But from Fig 6, the computation demand reduces significant when the fast NMPC algorithm based on GA is brought into the system.

5.2.3 Experiment results

The objectives and limits of the system: Input constraint is $0 \le u \le 100\%$; State objectives are $0 \le h_1, h_2 \le 0.5m$, and the setpoint of Tank 2 is 0.1m. Select prediction horizon P=10, weighting parameters $\bar{Q} = Q = 8$, R=1, sample time Ts=5s, mutation probability Pc=0.1, population size is 200, maximum generation is 100.

The tank apparatus is controlled with the NMPC algorithm based on conventional GA, the experimental curve is shown in Fig 7 and performance index is shown in Table 5.

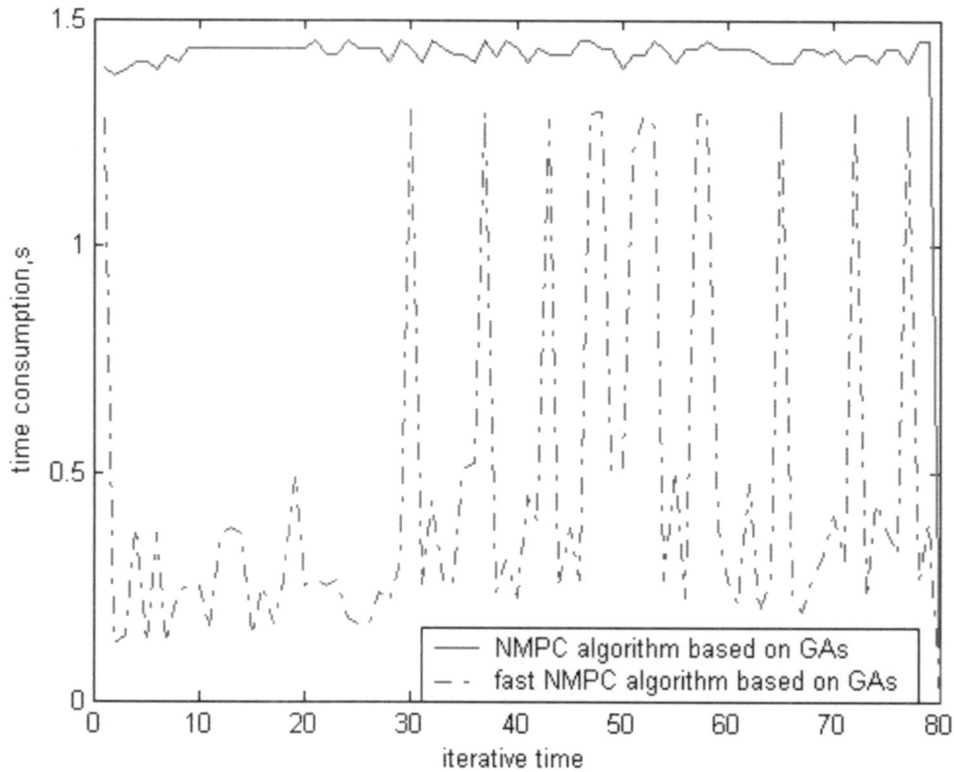

Fig. 6. Time consumptions for two methods

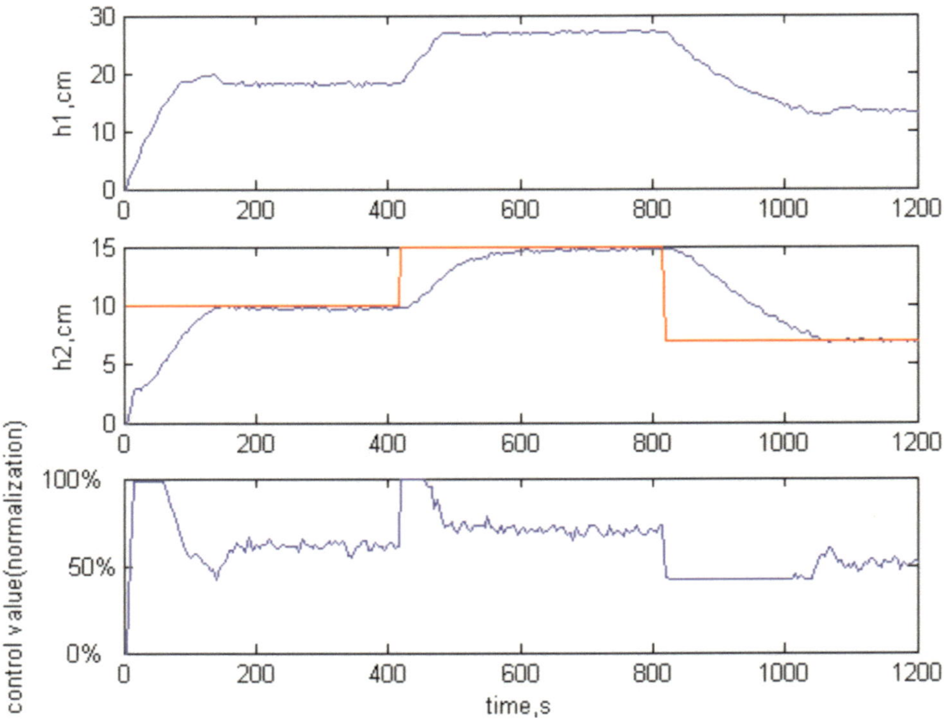

Fig. 7. Experimental curve of NMPC based on GA

Time, s	0-400	401-800	801-1200
Setpoint, m	0.1	0.15	0.07
Settling time, s	159	190	210
Percent overshoot	None	None	None

Table 5. Performance index of experimental result with conventional NMPC

The same experiment is carried out with fast NMPC algorithm based on GA. The result is shown in Fig 8 and performance index is shown in Table 5.

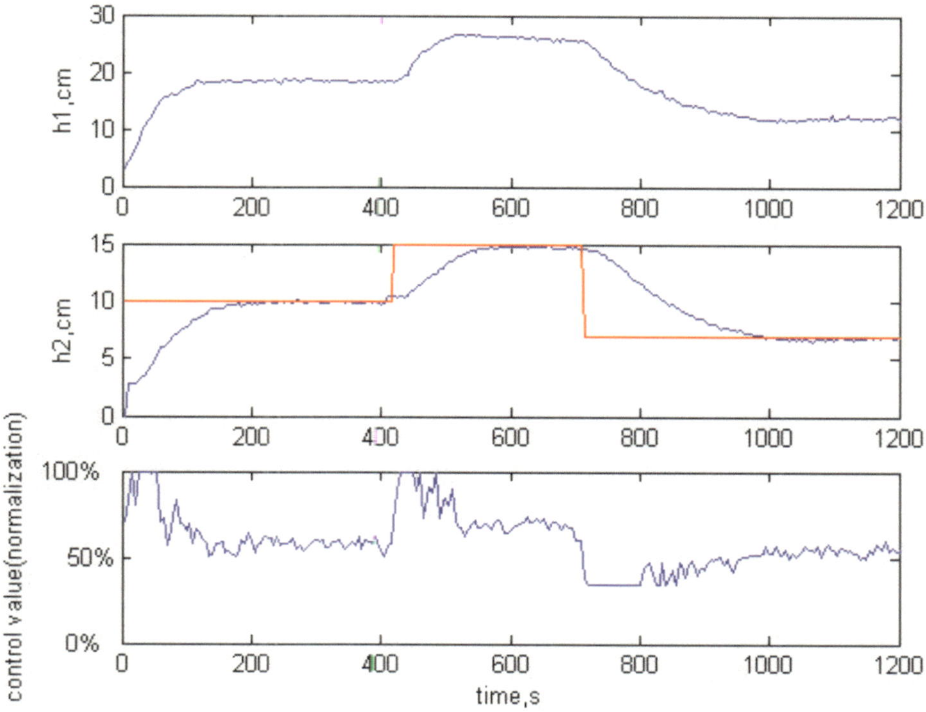

Fig. 8. Experimental curve of fast NMPC based on GA

Time, s	0-400	401-800	801-1200
Setpoint, m	0.1	0.15	0.07
Settling time, s	190	190	260
Percent overshoot	None	None	None

Table 6. Performance index of experimental result with fast NMPC

6. Conclusions

In this paper an improved NMPC algorithm based on GA has been proposed. The aim is to reduce the computational burden without much deterioration to the control performance. Compared with traditional NMPC controller, our approach has much lower computational burden, which makes it practical to operate in systems with a small sampling time or a large prediction horizon.

The proposed approach has been tested in CSTR and a real-time tank system. Both computer simulations and experimental testing confirm that the suboptimal NMPC based on GA resulted in a controller with less computation time.

7. Acknowledgment

This work is supported by National Natural Science Foundation of China (Youth Foundation, No. 61004082) and Special Foundation for Ph. D. of Hefei University of Technology (No. 2010HGBZ0616).

8. References

Michael A.Henson, Nonlinear model predictive control: current status and future directions [J], Computers and Chemical Engineering 23(1998):187-202

D.Q.Mayne, J.B.Rawlings, C.V.Rao and P.O.M.Scokaert, Constrained model predictive control: Stability and optimality [J], Automatica 36(2000): 789-814

S.Joe Qin and Thomas A.Badgwell, a survey of industrial model predictive control technology [J], Control Engineering Practice 11(2003): 733-764

Mark Cannon, Efficient nonlinear model predictive control algorithms [J], Annual Reviews in Control 28(2004): 229-237

Basil Kouvaritakis and Mark Cannon, Nonlinear predictive control: Theory and practice [M], the institution of electrical engineers, 2001

Frode Martinse, Lorenz T.Biegler and Bjarne A.Foss, A new optimization algorithm with application to nonlinear MPC[J], Journal of Process Control, 2004:853-865

Moritz Diehl, Rolf Findeisen, Frank Allgower, Hans Georg Bock and Johannes Schloder, stability of nonlinear model predictive control in the presence of errors due to numerical online optimization, Proceedings of the 42nd IEEE Conference Decision and Control, December 2003: 1419-1424

Rolf Findeisen, Moritz Diehl, Ilknur Disli-Uslu and Stefan Schwarzkopf, Computation and Performance Assessment of Nonlinear Model Predictive Control[J]. Proc. Of the 41st IEEE Conference on Decision and Control December 2002:4613-4618

Moritz Diehl, Rolf Findeisen, Stefan Schwarzkopf, An efficient algorithm for nonlinear predictive control of large-scale systems, Automatic Technology, 2002:557-567

L.T.Biegler, Advaneces in nonlinear programming concepts for process control, Proc. Cont. Vol.8, Nos.5-6, pp.301-311, 1998

Alex Zheng and Frank Allgower, towards a practical nonlinear predictive control algorithm with guaranteed stability for large-scale systems, Proceedings of the American Control Conference, 1998:2534-2538

Jianjun Yang, Min Liu, Cheng Wu. Genetic Algorithm Based on Nonlinear Model Predictive Control Method [J]. Control and Decision. 2003,2(18):141-144.

Fuzhen Xue, Yuyu Tang, Jie Bai. An Algorithm of Nonlinear Model Predictive Control Based on BP Network [J]. Journal of University of Science and Technology of China. 2004, 5(34): 593-598.

H.Michalska and D.Q.Mayne, robust receding horizon control of constrained nonlinear systems, IEEE Transactions on Automatic Control, Vol.38, No.11,1993:1623-1633

Wei Chen, Gang Wu. Nonlinear Modeling of Two-Tank System and Its Nonlinear Model Predictive Control [A]. Proceedings of the 24th Chinese Control Conference. 2005, 7:396-301

E.John Finnemore and Joseph B.Franzini, Fluid Mechanics with Engineering Applications(Tenth Edition)[M], March 2003

N.K.Poulsen, B.Kouvaritakis and M.Cannon, Nonlinear constrained predictive control applied to a coupled-tank apparatus[J], IEE Proc.-Control Theory Appl., Vol.148, No 1, January 2001:17-24

Rolf Findeisen, Moritz Diehl, Ilknur Disli-Uslu and Stefan Schwarzkopf, Computation and Performance Assessment of Nonlinear Model Predictive Control[J]. Proc. Of the 41st IEEE Conference on Decision and Control December 2002:4613-4618

4

Distributed Model Predictive Control Based on Dynamic Games

Guido Sanchez[1], Leonardo Giovanini[2], Marina Murillo[3] and Alejandro Limache[4]

[1,2]*Research Center for Signals, Systems and Computational Intelligence*
Faculty of Engineering and Water Sciences, Universidad Nacional del Litoral
[3,4]*International Center for Computer Methods in Engineering*
Faculty of Engineering and Water Sciences, Universidad Nacional del Litoral
Argentina

1. Introduction

Model predictive control (*MPC*) is widely recognized as a high performance, yet practical, control technology. This model-based control strategy solves at each sample a discrete-time optimal control problem over a finite horizon, producing a control input sequence. An attractive attribute of *MPC* technology is its ability to systematically account for system constraints. The theory of *MPC* for linear systems is well developed; all aspects such as stability, robustness,feasibility and optimality have been extensively discussed in the literature (see, e.g., (Bemporad & Morari, 1999; Kouvaritakis & Cannon, 2001; Maciejowski, 2002; Mayne et al., 2000)). The effectiveness of *MPC* depends on model accuracy and the availability of fast computational resources. These requirements limit the application base for *MPC*. Even though, applications abound in process industries (Camacho & Bordons, 2004), manufacturing (Braun et al., 2003), supply chains (Perea-Lopez et al., 2003), among others, are becoming more widespread.

Two common paradigms for solving system-wide *MPC* calculations are centralised and decentralised strategies. Centralised strategies may arise from the desire to operate the system in an optimal fashion, whereas decentralised *MPC* control structures can result from the incremental roll-out of the system development. An effective centralised *MPC* can be difficult, if not impossible to implement in large-scale systems (Kumar & Daoutidis, 2002; Lu, 2003). In decentralised strategies, the system-wide *MPC* problem is decomposed into subproblems by taking advantage of the system structure, and then, these subproblems are solved independently. In general, decentralised schemes approximate the interactions between subsystems and treat inputs in other subsystems as external disturbances. This assumption leads to a poor system performance (Sandell Jr et al., 1978; Šiljak, 1996). Therefore, there is a need for a cross-functional integration between the decentralised controllers, in which a coordination level performs steady-state target calculation for decentralised controller (Aguilera & Marchetti, 1998; Aske et al., 2008; Cheng et al., 2007; 2008; Zhu & Henson, 2002).

Several distributed *MPC* formulations are available in the literature. A distributed *MPC* framework was proposed by Dumbar and Murray (Dunbar & Murray, 2006) for the class

of systems that have independent subsystem dynamic but link through their cost functions and constraints. Then, Dumbar (Dunbar, 2007) proposed an extension of this framework that handles systems with weakly interacting dynamics. Stability is guaranteed through the use of a consistency constraint that forces the predicted and assumed input trajectories to be close to each other. The resulting performance is different from centralised implementations in most of cases. Distributed *MPC* algorithms for unconstrained and *LTI* systems were proposed in (Camponogara et al., 2002; Jia & Krogh, 2001; Vaccarini et al., 2009; Zhang & Li, 2007). In (Jia & Krogh, 2001) and (Camponogara et al., 2002) the evolution of the states of each subsystem is assumed to be only influenced by the states of interacting subsystems and local inputs, while these restrictions were removed in (Jia & Krogh, 2002; Vaccarini et al., 2009; Zhang & Li, 2007). This choice of modelling restricts the system where the algorithm can be applied, because in many cases the evolution of states is also influenced by the inputs of interconnected subsystems. More critically for these frameworks is the fact that subsystems-based *MPC*s only know the cost functions and constraints of their subsystem. However, stability and optimality as well as the effect of communication failures has not been established.
The distributed model predictive control problem from a game theory perspective for *LTI* systems with general dynamical couplings, and the presence of convex coupled constraints is addressed. The original centralised optimisation problem is transformed in a dynamic game of a number of local optimisation problems, which are solved using the relevant decision variables of each subsystem and exchanging information in order to coordinate their decisions. The relevance of proposed distributed control scheme is to reduce the computational burden and avoid the organizational obstacles associated with centralised implementations, while retains its properties (stability, optimality, feasibility). In this context, the type of coordination that can be achieved is determined by the connectivity and capacity of the communication network as well as the information available of system's cost function and constraints. In this work we will assume that the connectivity of the communication network is sufficient for the subsystems to obtain information of all variables that appear in their local problems. We will show that when system's cost function and constraints are known by all distributed controllers, the solution of the iterative process converge to the centralised MPC solution. This means that properties (stability, optimality, feasibility) of the solution obtained using the distributed implementation are the same ones of the solution obtained using the centralised implementation. Finally, the effects of communication failures on the system's properties (convergence, stability and performance) are studied. We will show the effect of the system partition and communication on convergence and stability, and we will find a upper bound of the system performance.

2. Distributed Model Predictive Control

2.1 Model Predictive Control

MPC is formulated as solving an on-line open loop optimal control problem in a receding horizon style. Using the current state $x(k)$, an input sequence $U(k)$ is calculated to minimize a performance index $J(x(k), U(k))$ while satisfying some specified constraints. The first element of the sequence $u(k,k)$ is taken as controller output, then the control and the prediction horizons recede ahead by one step at next sampling time. The new measurements are taken to compensate for unmeasured disturbances, which cause the system output to be different from its prediction. At instant k, the controller solves the optimisation problem

$$\begin{aligned} &\min_{U(k)} J\left(x(k), U(k)\right) \\ &st. \\ X(k+1) &= \Gamma x(k) + \mathcal{H} U(k) \\ U(k) &\in \mathcal{U} \end{aligned} \tag{1}$$

where Γ and $\mathcal{H}$ are the observability and Haenkel matrices of the system (Maciejowski, 2002) and the states and input trajectories at time k are given by

$$X(k) = \left[x(k,k) \quad \cdots \quad x(k+V,k)\right]^T \quad V > M,$$
$$U(k) = \left[u(k,k) \quad \cdots \quad u(k+M,k)\right]^T.$$

The integers V and M denote the prediction and control horizon. The variables $x(k+i,k)$ and $u(k+i,k)$ are the predicted state and input at time $k+i$ based on the information at time k and system model

$$x(k+1) = Ax(k) + Bu(k), \tag{2}$$

where $x(k) \in R^{n_x}$ and $u(k) \in \mathcal{U} \subseteq R^{n_u}$. The set of global admissible controls $\mathcal{U} = \{u \in R^{n_u} \, | Du \leq d, d > 0\}$ is assumed to be non-empty, compact and convex set containing the origin in its interior.

Remark 1. *The centralised model defined in (2) is more general than the so-called composite model employed in (Venkat et al., 2008), which requires the states of subsystems to be decoupled and allows only couplings in inputs. In this approach, the centralised model can represent both couplings in states and inputs.*

In the optimisation problem (1), the performance index $J\left(x(k), U(k)\right)$ measures of the difference between the predicted and the desired future behaviours. Generally, the quadratic index

$$J\left(x(k), U(k)\right) = \sum_{i=0}^{V} x^T(k+i,k) Q_i x(k+i,k) + \sum_{i=0}^{M} u^T(k+i,k) R_i u(k+i,k) \tag{3}$$

is commonly employed in the literature. To guarantee the closed–loop stability, the weighting matrices satisfy $Q_i = Q > 0$, $R_i = R > 0 \quad \forall i \leq M$ and $Q_i = \bar{Q} \quad \forall i > M$, where $\bar{Q}$ is given by $A^T \bar{Q} A - \bar{Q} = -Q$ (Maciejowski, 2002). For this choice of the weighting matrices, the index (3) is equivalent to a performance index with an infinite horizon.

$$J_\infty\left(x(k), U(k)\right) = \sum_{i=0}^{\infty} x^T(k+i,k) Q x(k+i,k) + u^T(k+i,k) R u(k+i,k).$$

In many formulations an extra constraint or extra control modes are included into (1) to ensure the stability of the closed-loop system (Maciejowski, 2002; Rossiter, 2003).

2.2 Distributed MPC framework

Large-scale systems are generally composed of several interacting subsystems. The interactions can either be: *a*) dynamic, in the sense that the states and inputs of each subsystem influence the states of the ones to which it is connected, *b*) due to the fact that the subsystems share a common goal and constraint, or *c*) both. Systems of this type admit a decomposition into m subsystems represented by

$$x_l(k+1) = \sum_{p=1}^{m} A_{l\,p} x_p(k) + B_{l\,j\in\mathcal{N}_p} u_{j\in\mathcal{N}_p}(k) \quad l = 1, \ldots, m \tag{4}$$

where $x_l \in R^{n_{x_l}} \subseteq R^{n_x}$ and $u_l \in \mathcal{U}_l \subseteq R^{n_{u_l}} \subset R^{n_u}$ are the local state and input respectively. The set of control inputs indices of subsystem l is denoted $\mathcal{N}_l$, and the set $\mathcal{I}$ denotes all control input indices such that $u(k) = u_{j\in\mathcal{I}}(k)$.

Remark 2. *This is a very general model class for describing dynamical coupling between subsystems and includes as a special case the combination of decentralised models and interaction models in (Venkat et al., 2008). The subsystems can share input variables such that*

$$\sum_{l=1}^{m} n_{u_l} \geq n_u. \tag{5}$$

Each subsystem is assumed to have local convex independent and coupled constraints, which involve only a small number of the others subsystems. The set of local admissible controls $\mathcal{U}_l = \{u_l \in R^{n_{u_l}} \mid D_l u_l \leq d_l, d_l > 0\}$ is also assumed to be non-empty, compact, convex set containing the origin in their interior.

The proposed control framework is based on a set of m independent agents implementing a small-scale optimizations for the subsystems, connected through a communication network such that they can share the common resources and coordinate each other in order to accomplish the control objectives.

Assumption 1. *The local states of each subsystem $x_l(k)$ are accessible.*

Assumption 2. *The communication between the control agents is synchronous.*

Assumption 3. *Control agents communicates several times within a sampling time interval.*

This set of assumption is not restrictive. In fact, if the local states are not accessible they can be estimated from local outputs $y_l(k)$ and control inputs using a Kalman filter, therefore **Assumption 1** is reasonable. As well, **Assumptions 2** and **3** are not so strong because in process control the sampling time interval is longer with respect the computational and the communication times.

Under these assumptions and the decomposition, the cost function (3) can be written as follows

$$J(x(k), U(k), A) = \sum_{l=1}^{m} \alpha_l J_l \left(x(k), U_{j\in\mathcal{N}_l}(k), U_{j\in\mathcal{I}-\mathcal{N}_l}(k) \right), \tag{6}$$

where $A = [\alpha_l]$, $\alpha_l \geq 0$, $\sum_{l=1}^{m} \alpha_l = 1$, $U_j(k)$ is the j-th system input trajectory. This decomposition of the cost function and input variable leads to a decomposition (1) into m coupled optimisation problems

$$\min_{U_{j\in\mathcal{N}_l}(k)} J(x(k), U(k), A)$$

$$\begin{aligned} &st. \\ X(k+1) &= \Gamma x(k) + \mathcal{H}U(k) \\ U_{j\in\mathcal{N}_l}(k) &\in \mathcal{U}_{j\in\mathcal{N}_l} \\ U_{j\in\mathcal{I}-\mathcal{N}_l}(k) &\in \mathbf{U}_{j\in\mathcal{I}-\mathcal{N}_l} \end{aligned} \tag{7}$$

where $\mathbf{U}_{j\in\mathcal{I}-\mathcal{N}_l}$ denotes the assumed inputs of others agents. The goal of the decomposition is to reduce the complexity of the optimisation problem (1) by ensuring that subproblems (7) are smaller than the original problem (fewer decision variables and constraints), while they retain the properties of the original problem. The price paid to simplify the optimisation problem (1) is the needs of coordination between the subproblems (7) during their solution. In this way, the optimisation problem (1) has been transformed into a *dynamic game* of m agents where each one searches for their optimal decisions through a sequence of *strategic games*, in response to decisions of other agents.

Definition 1. *A dynamic game $\langle m, \mathcal{U}, J_l\left(x(k), U^q(k), A\right), \mathcal{D}(q,k)\rangle$ models the interaction of m agents over iterations q and is composed of: i) $m \in \mathbb{N}$ agents; ii) a non empty set $\mathcal{U}$ that corresponds to the available decisions $U_l^q(k)$ for each agent; iii) an utility function $J_l\left(x(k), U^q(k), A\right) : x(k) \times U^q(k) \to \mathbb{R}^+$ for each agent; iv) an strategic game $\mathcal{G}(q,k)$ that models the interactions between agents at iteration q and time k; v) a dynamic process of decision adjustment $\mathcal{D}(q,k) : (U^q(k), \mathcal{G}(q,k), q) \to U^{q+1}(k)$.*

At each stage of the dynamic game, the joint decision of all agents will determine the outcome of *the strategic game* $\mathcal{G}(q,k)$ and each agent has some preference $U^q_{j\in\mathcal{N}_l}(k)$ over the set of possible outcomes $\mathcal{U}$. Based on these outcomes and the adjustment process $\mathcal{D}(q,k)$, which in this framework depends on the cost function $J_l(\cdot)$ and constraints, the agents reconcile their decisions. More formally, a strategic game is defined as follows (Osborne & Rubinstein, 1994)

Definition 2. *A finite strategic game $\mathcal{G}(q,k) = \langle m, \mathcal{U}_l, J_l\left(x(k), U^q(k), A\right)\rangle$ models the interactions between m agents and is composed of: i) a non empty finite set $\mathcal{U}_l \subseteq \mathcal{U}$ that corresponds the set of available decisions for each agent; ii) an utility function $J_l\left(x(k), U^q(k), A\right) : x(k) \times U^q(k) \to \mathbb{R}$ $U^q(k) \in \mathcal{U}$ for each agent.*

In general, one is interested in determining the choices that agents will make when faced with a particular game, which is sometimes referred to as the *solution of the game*. We will adopt the most common solution concept, known as *Nash equilibrium* (Nash, 1951): a set of choices where no individual agent can improve his utility by unilaterally changing his choice. More formally, we have:

Definition 3. *A group of control decisions $U(k)$ is said to be Nash optimal if*

$$J_l\left(x(k), U^q_{j\in\mathcal{N}_l}(k), U^{q-1}_{j\in\mathcal{I}-\mathcal{N}_l}(k)\right) \leq J_l\left(x(k), U^{q-1}_{j\in\mathcal{N}_l}(k), U^{q-1}_{j\in\mathcal{I}-\mathcal{N}_l}(k)\right)$$

where $q > 0$ is the number of iterations elapsed during the iterative process.

If Nash optimal solution is achieved, each subproblem does not change its decision $U^q_{j\in\mathcal{N}_l}(k)$ because it has achieved an equilibrium point of the coupling decision process; otherwise the local performance index J_l will degrade. Each subsystem optimizes its objective function using its own control decision $U^q_{j\in\mathcal{N}_l}(k)$ assuming that other subsystems' solutions $U^q_{j\in\mathcal{I}-\mathcal{N}_l}(k)$ are known. Since the mutual communication and the information exchange are adequately taken into account, each subsystem solves its local optimisation problem provided that the other subsystems' solutions are known. Then, each agent compares the new solution with that obtained in the previous iteration and checks the stopping condition

$$\left\| U^q_{j\in\mathcal{N}_l}(k) - U^{q-1}_{j\in\mathcal{N}_l}(k) \right\|_\infty \leq \varepsilon_l \quad l = 1,\ldots,m. \tag{8}$$

If the algorithm is convergent, condition (8) will be satisfied by all agents, and the whole system will arrive to an equilibrium point. The subproblems m (7) can be solved using the following iterative algorithm

Algorithm 1

Given $Q_l,\ R_l, 0 < q_{max} < \infty, \varepsilon_l > 0 \quad \forall l = 1,\cdots,m$
For each agent $l \quad l = 1,\cdots,m$
Step 1 Initialize agent l
 1.a Measure the local state $x_l(k), q = 1$,
  $\rho_l = \phi\varepsilon_l \quad \phi \gg 1$
 1.b $U^0(k) = [u(k,k-1) \ \cdots \ u(k+M,k-1) \ 0\,]$
Step 2 **while** $\rho_l > \varepsilon_l$ and $q < q_{max}$
 2.a Solve problem (7) to obtain $\tilde{U}^q_{j\in\mathcal{N}_l}(k)$
 2.b **for** $p = 1,\cdots,m$ and $p \neq l$
  Communicate $\tilde{U}^q_{j\in\mathcal{N}_l}(k)$ to agent p
 end
 2.c Update the solution iterate q $\forall j \in \mathcal{N}_l$
$$U^q_j(k) = \Sigma^m_{p=1}\alpha_p\tilde{U}^q_{i\in\mathcal{N}_p\cap j}(k) + \left(1 - \Sigma^m_{p=1}\alpha_p\, card\,(\,j\cap\mathcal{N}_l)\right) U^{q-1}_j(k)$$
 2.d $\rho_l = \left\| U^q_{j\in\mathcal{N}_l}(k) - U^{q-1}_{j\in\mathcal{N}_l}(k) \right\|_\infty$
  $q = q + 1$
 end
Step 3 Apply $u_l(k,k)$
Step 4 $k = k + 1$ and goto Step 1

At each k, q_{max} represents a design limit on the number of iterates q and ε_l represents the stopping criteria of the iterative process. The user may choose to terminate Algorithm 1 prior to these limits.

3. Properties of the framework

3.1 Performance

Given the distributed scheme proposed in the previous Section, three fundamental questions naturally arise: a) the behavior of agent's iterates during the negotiation process, b) the

location and number of equilibrium points of the distributed problem and *c*) the feasibility of the solutions. One of the key factors in these questions is the effect of the cost function and constraints employed by the distributed problems. Therefore, in a first stage we will explore the effect of the performance index in the number and position of the equilibrium points.
Firstly, the optimality conditions for the centralised problem (1) are derived in order to have a benchmark measure of distributed control schemes performance. In order to make easy the comparison, the performance index (3) is decomposed into m components related with the subsystems, like in the distributed problems (7), as follows

$$J\left(x(k), U(k), \Theta\right) = \sum_{l=1}^{m} \theta_l J_l\left(x(k), U(k)\right), \theta_l \geq 0, \sum_{l=1}^{m} \theta_l = 1. \tag{9}$$

This way writing the performance index corresponds to multiobjective characterization of the optimisation problem (1). Applying the first–order optimality conditions we obtain

$$\sum_{l=1}^{m} \theta_l \frac{\partial J_l\left(x(k), U(k)\right)}{\partial U_{j\in\mathcal{N}_p}(k)} + \lambda^T D^{j\in\mathcal{N}_p} = 0 \qquad p = 1, \ldots, m, \tag{10a}$$

$$\lambda^T \left(D^{j\in\mathcal{N}_p} U_{j\in\mathcal{N}_p}(k) - b\right) = 0, \tag{10b}$$

where D^j is the j–th column vector of D. The solution of this set of equations $U^*(k)$ is the optimal solution of the optimisation problem (1) and belongs to *Pareto set*, which is defined as (Haimes & Chankong, 1983).

Definition 4. *A solution* $U^*(k) \in \mathcal{U}$ *is said to be Pareto optimal of the optimisation problem (1) if there exists no other feasible solution* $\forall U(k) \in \mathcal{U}$ *such that* $J_l\left(x(k), U(k)\right) \leq J_l\left(x(k), U^*(k)\right) \quad \forall l = 1, \ldots, m$.

In distributed control the agents coordinate their decisions, through a negotiation process. Applying the first–order optimality conditions to decentralised cost (6) we obtain

$$\sum_{l=1}^{m} \alpha_l \frac{\partial J_l\left(x(k), U(k)\right)}{\partial U_{j\in\mathcal{N}_p}(k)} + \lambda^T D^{j\in\mathcal{N}_p} = 0 \qquad p = 1, \ldots, m, \tag{11a}$$

$$\lambda^T \left(D^{j\in\mathcal{N}_p} U_{j\in\mathcal{N}_p}(k) - b\right) = 0. \tag{11b}$$

By simple inspection of (10) and (11) we can see that these equations have the same structure, they only differ on the weights. Therefore, the location of the distributed schemes equilibrium will depend on the selection of $\alpha_l \quad l = 1, \ldots, m$. There are two options:

- If $\alpha_l = 1, \alpha_{p \neq l} = 0$ the optimality condition (11) becomes

$$\frac{\partial J_l\left(x(k), U(k)\right)}{\partial U_{j\in\mathcal{N}_l}(k)} + \lambda^T D^{j\in\mathcal{N}_l} = 0 \qquad l = 1, \ldots, m, \tag{12a}$$

$$\lambda^T \left(D^{j\in\mathcal{N}_l} U_{j\in\mathcal{N}_l}(k) - b\right) = 0. \tag{12b}$$

  This condition only evaluates the effect of $U_{j\in\mathcal{N}_l}$, given $U_{j\in\mathcal{I}-\mathcal{N}_l}$, in subsystem l without taking into account its effects in the remaining agents (**selfish behavior**). This configuration of the distributed problem leads to an **incomplete and perfect information game** that can achieve Nash optimal solutions for a pure strategy (**Cournot equilibrium**)

(Osborne & Rubinstein, 1994). By simple comparison of (10) and (12) we can conclude that the solution of this equations lies outside of the Pareto set (Dubey & Rogawski, 1990; Neck & Dockner, 1987). The reason of Nash equilibrium inefficiency lies in the fact that the information of each agent decision variable effects' on the remaining agents is neglected ($\alpha_{p\neq l} = 0$ incomplete information game). Therefore, each agent minimizes their performance index, accommodating the effects of other agents' decisions, without taking in account its effects on the rest of the system. Besides the lack of optimality, the number of equilibrium points generated by the optimality condition (12) can grow with the number of agents (Bade et al., 2007).

- If $\alpha_l > 0$ the optimality condition (11) becomes

$$\sum_{l=1}^{m} \alpha_l \frac{\partial J_l\left(x(k), U(k)\right)}{\partial U_{j\in\mathcal{N}_p}(k)} + \lambda^T D^{j\in\mathcal{N}_p} = 0 \qquad p = 1, \ldots, m \tag{13a}$$

$$\lambda^T \left(D^{j\in\mathcal{N}_p} U_{j\in\mathcal{N}_p}(k) - b \right) = 0. \tag{13b}$$

This condition evaluates the effect of $U_{j\in\mathcal{N}_l}$, given $U_{j\in\mathcal{I}-\mathcal{N}_l}$, in the entire system, taking in account the effect of interactions between the subsystems (**cooperative behavior**), leading to a **complete and perfect information game**. By simple comparison of (10) and (13) it is easy to see that these two equations have a similar structure, therefore we can conclude that their solutions lie in the Pareto set. The position of distributed *MPC* solutions will depend on the values of α_l. In the particular case of $\alpha_l = \theta_l \quad l = 1, \ldots, m$ the solution of the centralised and distributed schemes are the same.

The value of weights $\alpha_l \quad l = 1, \ldots, m$ depends on the information structure; that is the information of the cost function and constraints available in each agent. If the cost function and constraints of each agent are known by all the others, for example a retailer company, α_l can be chosen like the second distributed scheme ($\alpha_l > 0 \quad \forall l = 1, \ldots, m$). In this case the centralised optimisation problem is distributed between m independent agents that coordinate their solutions in order to solve the optimisation problem in a distributed way. For this reason we call this control scheme **distributed MPC**. On the other case, when the local cost function and constraints are only known by the agents, for example a power network where several companies compete, the weights α_l should be chosen like the first scheme ($\alpha_l = 1, \alpha_{p\neq l} = 0 \quad \forall l, p = 1, \ldots, m$). In this case the centralised optimisation problem is decentralised into m independent agents that only coordinate the effects of their decisions to minimize the effect of interactions. For this reason we call this control scheme **coordinated decentralised MPC**.

Remark 3. *The fact that agents individually achieve Nash optimality does not imply the global optimality of the solution. This relationship will depend on the structure of agents' cost function and constraints, which depends on the value of weights α_l, and the number of iterations allowed.*

The structure of $U_{j\in\mathcal{N}_l}$ determine the structure of constraints that can be handled by the distributed schemes. If the subproblems share the input variables involved in the coupled constraints ($\mathcal{N}_l \cap \mathcal{N}_{p\neq l} \neq \emptyset$), the distributed MPC schemes can solve optimisation problems with coupled constraints. On the other hand, when subproblems do not include the input variables of coupled constraints ($\mathcal{N}_l \cap \mathcal{N}_{p\neq l} = \emptyset$), the distributed MPC schemes can only solves optimisation problems with independent constraints (Dunbar, 2007; Jia & Krogh, 2001; Venkat et al., 2008). These facts become apparent from optimality conditions (12) and (13).

3.2 Convergence

During the operation of the system, the subproblems (7) can compete or cooperate in the solution of the global problem. The behavior of each agent will depend on the existence, or not, of conflictive goals that can emerge from the characteristics of the interactions, the control goals and constraints. The way how the system is decomposed is one of the factors that defines the behavior of the distributed problem during the iterations, since it defines how the interactions will be addressed by distributed schemes.
The global system can be partitioned according to either the physical system structure or on the basis of an analysis of the mathematical model, or a combination of both. Heuristic procedures for the partitioning the system based on input–output analysis (see (Goodwin et al., 2005; Henten & Bontsema, 2009; Hovd & Skogestad, 1994)), an state–space analysis based (see (Salgado & Conley, 2004; Wittenmark & Salgado, 2002) or on performance metric for optimal partitioning of distributed and hierarchical control systems (see (Jamoom et al., 2002; Motee & Sayyar-Rodsari, 2003)) have been proposed. In all these approaches the objective is to simplify the control design by reducing the dynamic couplings, such that the computational requirements are evenly distributed to avoid excessive communication load. It is important to note that the partitioning of a state–space model can lead to overlapping states both due to coupled dynamics in the actual continuous system and due to discrete-time sampling, which can change the sparsity structure in the model.

Assumption 4. *The model employed by the distributed MPC algorithms are partitioned following the procedures described in (Motee & Sayyar-Rodsari, 2003).*

To analysed the effect of the system decomposition on the distributed constrained scheme, firstly we will analysed its effects on unconstrained problem. Solving the optimality condition (11) for an unconstrained system leads to

$$U^q(k) = \mathcal{K}_C U^{q-1}(k) + \mathcal{K}_1 x(k) \quad \forall q > 0, \tag{14}$$

which models the behavior of the distributed problem during the iterative process. Its stability induces the convergence of the iterative process and it is given by

$$|\lambda(\mathcal{K}_0)| < 1. \tag{15}$$

The gain $\mathcal{K}_1$ is the decentralised controller that computes the contribution of $x(k)$ to $U(k)$ and has only non–zero elements on its main diagonal $\mathcal{K}_1 = [\mathcal{K}_{ll}] \quad l = 1, \ldots, m$. On other hand, $\mathcal{K}_0$ models the interaction between subsystems during the iterative process, determining its stability, and has non zero elements on its off diagonal elements

$$\mathcal{K}_0 = \begin{bmatrix} 0 & \mathcal{K}_{12} & \cdots & \mathcal{K}_{1m} \\ \mathcal{K}_{21} & 0 & & \mathcal{K}_{2m} \\ \vdots & & \ddots & \vdots \\ \mathcal{K}_{m1} & \cdots & \mathcal{K}_{mm-1} & 0 \end{bmatrix}. \tag{16}$$

The structure of the components of $\mathcal{K}_0$ and $\mathcal{K}_1$ depends on the value of the weights α_l:

- If the *coordinated decentralised* MPC is adopted ($\alpha_l = 1, \alpha_{p \neq l} = 0$) the elements of $\mathcal{K}_0$ are given by given by

$$\mathcal{K}_{lp} = -\mathcal{K}_{ll}\mathcal{H}_{lp} \quad l, p = 1, \ldots, m \tag{17}$$

where $K_{ll} = \left(\mathcal{H}_{ll}^T Q_{ll} \mathcal{H}_{ll} + R_{ll}\right)^{-1} \mathcal{H}_{ll}^T Q_{ll}$. Therefore, the way in which the global problem (1) was partitioned and how the controllers' parameters were tuned defines the convergence of the *coordinated decentralised MPC*. Under **Assumption 2** the convergence of the algorithm can be guaranteed for those systems that exhibit weak interactions.

- On the other hand, when the *distributed MPC* is adopted ($\alpha_l > 0$) the gain $\mathcal{K}_0$ is given by

$$\mathcal{K}_{lp} = -K_{lp}\mathcal{H}_{lp} \quad l,p = 1,\ldots,m \tag{18}$$

where the controller gains are given by $K_{lp} = \left(\mathcal{H}_{lp}^T Q_{lp} \mathcal{H}_{lp} + R_{lp}\right)^{-1} \mathcal{H}_{lp}^T Q_{lp}$. Since the *distributed MPC* is designed to guarantee the stability of the entire system, its convergence is guaranteed independently of the way of partitioning the system.

Now, we will consider constrained systems. In this case, under Assumption 2, the convergence for constrained systems can be analysed using Lyapunov arguments. The key idea is to show the contractivity of the sequence of global cost functions $J\left(x(k, U^q(k)), A\right)$ generated by **Algorithm 1** along the iterative process.

Lemma 1. *Let's assume that the system has been partitioned following a decentralised design procedure and the distributed MPC problems (7) $\forall l = 1,\ldots,m$ are feasible, then the sequence of cost functions $J\left(x(k, U^q(k)), A\right)$ generated by **Algorithm 1** during the iterative process is non increasing $\forall q > 0$ at any time k.*

Proof. See appendix 8.A. □

3.3 Feasibility

Although in current literature it is typically assumed that an initial centralised feasible solution exist and is available, in this Section we will provide a simple and implementable way of constructing it in a distributed way assuming that the global initial state is available in advanced.

An initial feasible solution input $U^0_{j\in\mathcal{N}_l}(k)$ at $k = 0$ can be computed locally by using an inner approximation of the global feasible set $\mathcal{U}$ based on all the constraints appearing in (1) and the global initial state $x(0)$, which is assumed to be available. Consider an inner-hyperbox approximation Ω of $\mathcal{U}$, which then takes the form of a Cartesian product

$$\Omega = \Omega_1 \times \cdots \Omega_m \subset \mathcal{U}. \tag{19}$$

This approximation essentially decomposes and decouples the constraints among subsystems by performing constraint tightening. Each subsystem l will thus have to include Ω_l in their local problem setup. Since the Cartesian product of these local constraint sets are included in the globally feasible set $\mathcal{U}$, any combination of local solutions within Ω_l will be globally feasible as well. The local constraint sets that arise from this inner-hyperbox approximation will be in general quite conservative, but at the same time will allow the construction of a feasible solution locally to initialize **Algorithm 1**.

Calculation of the inner-hyperbox approximation can be performed a priori and the local Ω_l constraints distributed to each subsystem. A polynomial-time procedure to compute a maximum volume inner box of could follow the procedure described in (Bemporad et al., 2004). Obtaining the local component-wise constraints Ω_l is then straightforward. For time steps $k > 0$, we construct a feasible solution by performing **Step 1** of **Algorithm 1**

$$U^0_{j\in\mathcal{N}_l}(k) = \left[u_{j\in\mathcal{N}_l}(k, k-1) \ \cdots \ u_{j\in\mathcal{N}_l}(k+M, k-1)\ 0\right]$$

The feasibility throughout the iterations is maintained because in **step 2.**a m feasible solutions $\tilde{U}^q_{j\in\mathcal{N}_l}(k)$ are obtained. Then, in **step 2.**c, the new control profile $U^q_{j\in\mathcal{N}_l}(k)$ is built as a convex combination of these solutions. Since problem (7) is a convex constrained QP, any convex combination of $U^q_{j\in\mathcal{N}_l}(k)$ also satisfies the convex constraint set. Therefore $U^q(k)$ is a feasible solution of optimisation problem (7) for all l.

3.4 Stability

Showing nominal stability of the resulting closed-loop system follows standard arguments for the most part (Mayne et al., 2000). The proof in this section is closely related to the stability proof of the FC-MPC method in (Venkat et al., 2008) with the addition of Assumption 2. The key idea is to show the contractivity of the sequence of global cost functions J generated by **Algorithm 1** along the system operation and the stability of the origin.

Theorem 1. *Let us assume that the system has been partitioned following a decentralised design procedure and the optimisation problem (7) solved using Algorithm 1 is feasible, then the origin is an exponentially stable equilibrium point.*

Proof. See appendix 8.B. □

4. System behavior under communication failures

In the proposed framework agents coordinate their actions by exchanging information through the communication network. Since the agents extensively use the communication network some questions related to the system behavior arise if communications fail: Which are the conditions for the convergence of the iterative process? How closed-loop stability is affected? How does system performance change?

In a first stage, the failures in the communication system are modeled introducing three matrices: *i*) the *connection matrix* $\mathcal{C}$ which represents the communication structure, *ii*) the *transmission failure matrix* $\mathcal{T}$ which models the transmission failures and *iii*) *reception failure matrices* $\mathcal{R}$ that models the reception failures in the system. The matrix $\mathcal{C}$ is defined as

$$\mathcal{C} = \left[c_{lp}\right],\ c_{lp} = \begin{cases} 0 & l = p, \\ 1 \text{ or } 0 & l \neq p, \end{cases} \tag{20}$$

where $c_{lp} = 1$ indicates the connection between agents l and p, while $c_{lp} = 0$ shows no connection between these agents. Then, the failures in the communication system can be modeled combining the connection matrix with the others matrices that models the reception ($\mathcal{R}$) and transmission ($\mathcal{T}$) failures, $\mathcal{RCT}$, which are given by

$$\mathcal{R} = \left[r_{lp}\right],\ r_{lp} = \begin{cases} 1 & l = p, \\ 0 & l \neq p, \end{cases} \tag{21a}$$

$$\mathcal{T} = \left[t_{lp}\right],\ t_{lp} = \begin{cases} 1 & l = p, \\ 0 & l \neq p. \end{cases} \tag{21b}$$

An element $t_{ll} = 1$ ($r_{ll} = 1$) corresponds to a perfect transmission (reception) of agent l, while $t_{ll} = 0$ ($r_{ll} = 0$) corresponds to a transmission (reception) failure of agent l. A failure between agents l and p is represented with the transition from $1 \to 0$ of the corresponding elements of $\mathcal{R}$ and $\mathcal{T}$.

Following the same procedure like in Section 3.2, the solution for the distributed problem at each iteration is

$$U^q(k) = \mathcal{K}_0 \mathcal{RCT} U^{q-1}(k) + \mathcal{K}_1 \mathcal{RCT} x(k) \quad q \geq 1$$

Its behavior is related with its stability, which is given by

$$\|\lambda(\mathcal{K}_0 \mathcal{RCT})\|_1 < 1 \tag{22}$$

Under communication failure each agent cannot exchange information properly, which will modify the iterative process driving it to another solution. In this case, the agent with communication failure will become a decentralised controller that will receive information about the decision of the others agents through the system. This will deteriorate the stability margins and performance due to the presence of the interactions not accounted during the iterative process, which will act like non measurable disturbances. In the extreme case $\mathcal{K}_0 \mathcal{RCT} = 0$, the control structure will correspond to the *fully decentralised control architecture*, and the stability will depend on the way that the system was partitioned. If the controllers are designed following a decentralised design procedure (Wittenmark & Salgado, 2002), the stability of the closed-loop system can be guaranteed.
Once the convergence of the iterates can be guaranteed, the next issue to be addressed is the effect of the communication failures on the closed-loop stability. In order to establish their on the closed-loop behavior, the control action

$$\tilde{U}(k) = (I - \mathcal{K}_0 \mathcal{RCT})^{-1} \mathcal{K}_1 \mathcal{RCT} \Gamma x(k)$$

is replaced in the open-loop model of the system, leading to the closed-loop system

$$x(k+1) = \left(A - B\mathcal{I}(I - \mathcal{K}_0 \mathcal{RCT})^{-1} \mathcal{K}_1 \mathcal{RCT}\Gamma\right) x(k).$$

Then, the stability of the closed–loop system under communication failures is determined by

$$\left|\lambda\left(A - B\mathcal{I}(I - \mathcal{K}_0 \mathcal{RCT})^{-1} \mathcal{K}_1 \mathcal{RCT}\Gamma\right)\right| < 1. \tag{23}$$

Under the communication failure, each agent can not exchange information properly therefore the stability of the closed–loop system will depend on the dynamic characteristics of the interactions between subsystems. In the extreme case $\mathcal{RCT} = 0$, the stability condition is always satisfied corresponding to the full decentralised architecture. The interactions act like non measurable disturbances for the controllers, reducing the stability margins and degrading the system performance.

Theorem 2. *Let us assume that the system has been partitioned in a way that the* convergence condition *(15) is satisfied, its performance at time instant k under the local communication failure is*

$$\tilde{J}(k) \leq \left(1 + \frac{\|\mathcal{W}(k)\|}{\lambda_{min}(\mathcal{F})}\right) J^*, \tag{24}$$

where the performance degradation is bounded by

$$\frac{\tilde{J}(k) - J^*}{J^*} \leq \frac{\|\mathcal{W}_{max}\|}{\lambda_{min}(\mathcal{F})} \tag{25}$$

where $\lambda_{min}(\mathcal{F})$ *denotes the minimal eigenvalue of*

$$\mathcal{F} = \left(\mathcal{K}_1^{-1}(I - \mathcal{K}_0) - \mathcal{H}\right)^T \mathcal{Q}\left(\mathcal{K}_1^{-1}(I - \mathcal{K}_0) - \mathcal{H}\right) + \mathcal{R},$$
$$\mathcal{W}_{\max} = \mathcal{S}_{\max}^T \left(\mathcal{H}^T \mathcal{Q}\mathcal{H} + \mathcal{R}\right)\mathcal{S}_{\max},$$
$$\mathcal{S}_{\max} = 2I - \left[I + (I - \mathcal{K}_0)^{-1}(I + \mathcal{K}_0)\right]^{-1}.$$

Proof. See appendix 8.C. □

5. Simulations

In this Section, we will illustrate the applicability and limitations of the theoretical results presented in this paper through two problems: *i)* a *LTI MIMO* system and *ii)* the operation of a heat–exchanger network (HEN). In the first case we analyse and evaluate the ideas discussed in previous sections through the control of a strongly coupled MIMO LTI system. In the second problem we will evaluate the applicability and limitations of the proposed framework to system with complex dynamic.

5.1 LTI System

To explore the ideas discussed in previous Sections, let's consider the following *MIMO* linear system

$$\begin{bmatrix} y_1(s) \\ y_2(s) \end{bmatrix} = \begin{bmatrix} \frac{1}{10s+1} & \frac{-1}{7.5s+1} \\ \frac{1.5}{11s+1} & \frac{1.5}{8s+1} \end{bmatrix} \begin{bmatrix} u_1(s) \\ u_2(s) \end{bmatrix} \tag{26}$$

This system shows a strong interaction with a non cooperative behavior between both subsystem due to the difference in the sign of the gain. Besides, the interaction between y_1 and u_2 is faster than the dynamic between y_1 and u_1. The models for the distributed and coordinated decentralized *MPC* algorithms were obtained by dividing the system in two agents

$$\textbf{Agent 1}: y_1(s) = \frac{1}{10s+1}u_1(s) - \frac{1}{7.5s+1}u_2(s), \tag{27a}$$
$$\textbf{Agent 2}: y_2(s) = \frac{1.5}{11s+1}u_1(s) + \frac{1.5}{8s+1}u_2(s). \tag{27b}$$

Agent 1 solves the optimization problem using u_1 as decision variable, while **agent 2** solves its optimization problem using u_2. The parameters of the predictive control algorithms are $M = 5$, $V = 20$, $Q_i = I_{2\times2}$, $R_i = 2I_{2\times2}$ $i = 0, \cdots, M-1$, and the stopping condition for the decentralized and distributed *MPC* algorithms was fixed to $\varepsilon_1 = \varepsilon_2 = 0.005$ and $q_{max} = 30$. Figure 1 shows the closed–loop responses for different *MPC* schemes. In this figure we can see that the performance of *centralized* and *distributed MPC* are similar, while the performance of the *coordinated decentralized MPC* is worst than the others control schemes. In general, the response obtained by the *coordinated decentralized MPC* shows stronger interactions that deteriorate the overall system performance. This phenomenon is due to the fact that the *coordinated decentralized MPC* does not optimize the effect of the agents decision variable on the performance of the other agent. Figure 1.*b* shows the resulting manipulated variables where we can see the behavior of the three MPC schemes. Even though all algorithms achieve

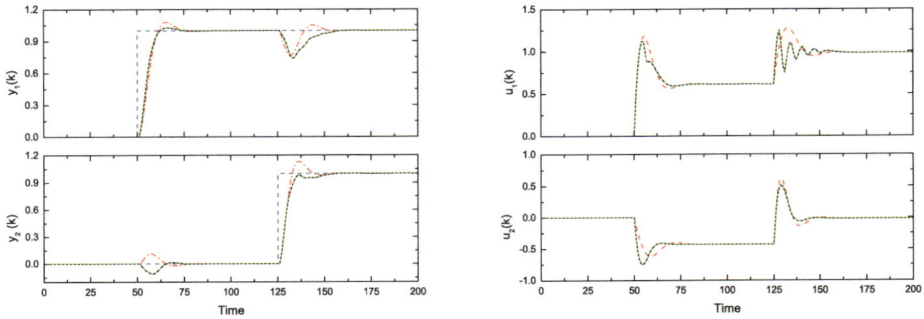

(a) System output (b) Manipulated variables

Fig. 1. System responses to different *MPC* schemes (- - Centralized MPC, — Distributed MPC and -.- Coordinated decentralized MPC).

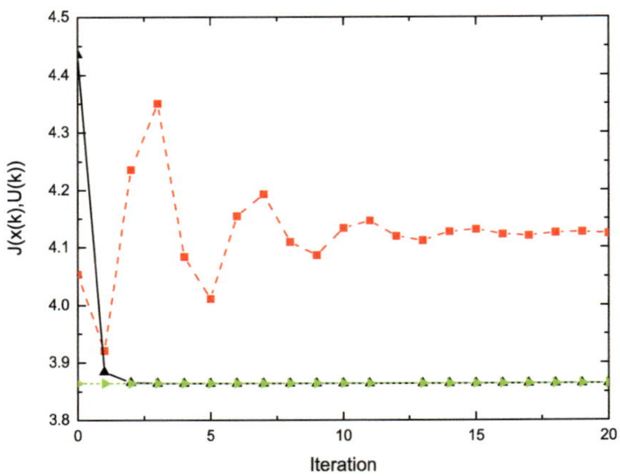

Fig. 2. Behavior of the controllers' cost functions during the iterative procedure (- - Centralized MPC, — Distributed MPC and -.- Coordinated decentralized MPC).

the same steady–state conditions, the inputs / outputs trajectories followed by the *coordinated decentralized MPC* are different from the *centralized* and *distributed MPC*.

Figure 2 shows the behavior of *MPC* controllers cost functions during the iterative procedure for the first set point change. The first thing to see is the oscillatory behavior of the iterative process of the *coordinated decentralized MPC*, in contrast with the monotonous behavior of the *distributed MPC*. This behavior is due to the nature interactions and the characteristics Balderud et al. (2008). The cost function of the *distributed MPC* converges to the solution of the *centralized MPC*, which is globally optimal, in few iterations. If the iterative process is stopped before ($q < 4$), the resulting solution will be suboptimal however it will lead to a stable closed–loop system. The earlier the iterative process is stopped, the bigger the difference to the centralized solution.

5.2 Heat-exchanger network

The heat-exchanger network (*HEN*) system studied here is represented schematically in Figure 3. It is a system with only three recovery exchangers (I_1, I_2 and I_3) and three service (S_1, S_2 and S_3) units. Two hot process streams (h_1 and h_2) and two cold process streams (c_1 and c_2) take part of the heat exchange process. There are also three utility streams (s_1, s_2 and s_3) that can be used to help reaching the desired outlet temperatures.

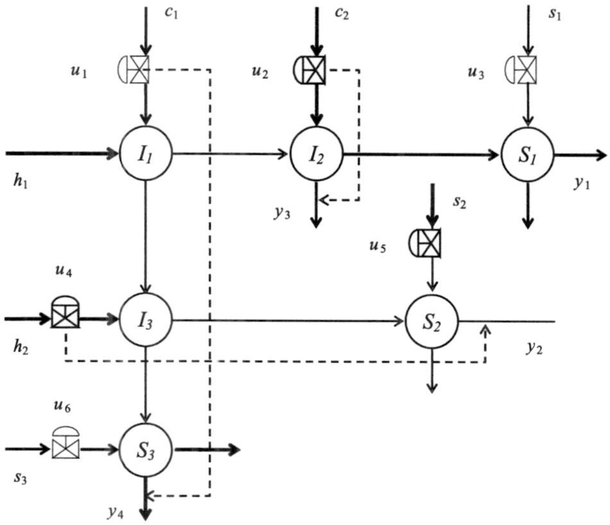

Fig. 3. Schematic representation of the HEN system.

The main purpose of a *HEN* is to recover as much energy as necessary to achieve the system requirements from high–temperature process streams (h_1 and h_2) and to transfer this energy to cold–process streams (c_1 and c_2). The benefits are savings in fuels needed to produce utility streams s_1, s_2 and s_3. However, the *HEN* has to also provide the proper thermal conditioning of some of the process streams involved in the heat transfer network. This means that a control system must *i*) drive the exit process–stream temperatures (y_1, y_2, y_3 and y_4) to the desired values in presence of external disturbances and input constraints while *ii*) minimizes the amount of utility energy.

The usual manipulated variables of a *HEN* are the flow rates at bypasses around heat exchangers (u_1, u_2 and u_4) and the flow rates of utility streams in service units (u_3, u_5 and u_6), which are constrained

$$0 \leq u_j(k) \leq 1.0 \qquad j = 1, \ldots, 6.$$

A fraction $0 < u_j < 1$ of bypass j means a fraction u_j of corresponding stream goes through the bypass and a fraction $1 - u_j$ goes through the exchangers, exchanging energy with other streams. If $u_j = 0$ the bypass is *completely closed* and the whole stream goes through the exchangers, maximizing the energy recovery. On the other hand, a value of $u_j = 1$ the bypass is *completely open* and the whole stream goes through the bypass, minimizing the energy recovery.

The *HEN* studied in this work has more control inputs than outlet temperatures to be controlled and so, the set of input values satisfying the output targets is not unique. The

possible operation points may result in different levels of heat integration and utilities consumption. Under nominal conditions only one utility stream is required (s_1 or s_3) for the operation of the *HEN*, the others are used to expand the operational region of the *HEN*.
The inclusion of the control system provides new ways to use the extra utility services (s_2 and s_3) to achieve control objectives by introducing new interactions that allow the redirection of the energy through the *HEN* by manipulating the flow rates. For example, any change in the utility stream s_3 (u_6) has a direct effect on output temperature of c_1 (y_4), however the control system will redirect this change (through the modification of u_1) to the output temperature of h_1 (y_1), h_2 (y_2), and c_2 (y_3). In this way, the *HEN* has energy recycles that induces feedback interaction, whose strength depends on the operational conditions, and leads to a complex dynamic: *i*) small energy recycles induce weak couplings among subsystems, whereas *ii*) large energy recycles induce a time scale separation, with the dynamics of individual subsystems evolving in a fast time scale with weak interactions, and the dynamics of the overall system evolving in a slow time scale with strong interactions Kumar & Daoutidis (2002).
A complete definition of this problem can be found in Aguilera & Marchetti (1998). The controllers were developed using the following linear model

$$Y = A(s) * U,$$

where

$$A(s) = \begin{bmatrix} \frac{20.6\,e^{-61.3s}}{38.8s+1} & \frac{19.9\,e^{-28.9s}}{25.4s+1} & \frac{17.3\,e^{-4.8s}}{23.8s+1} & 0 & 0 & 0 \\ \frac{4.6\,e^{-50.4s}}{48.4s+1} & 0 & 0 & 79.1\frac{31.4s+0.8}{31.4s+1.0} & \frac{20.1\,e^{-4.1s}}{25.6s+1.0} & 0 \\ \frac{16.9\,e^{-24.7s}}{39.5s+1} & -39.2\frac{22.8s+0.8}{22.8s+1.0} & 0 & 0 & 0 & 0 \\ 24.4\frac{48.2s^2+4.0s+0.05}{48.2s^2+3.9s+0.06} & 0 & 0 & -8.4\frac{e^{-18.8s}}{27.9s+1} & 0 & \frac{16.3\,e^{-3.5s}}{20.1s+1.0} \end{bmatrix}$$

and

$$U = [\, u_1\, u_2\, u_3\, u_4\, u_5\, u_6 \,]^T,$$
$$Y = [\, y_1\, y_2\, y_3\, y_4 \,]^T.$$

The first issue that we need to address in the development of the distributed controllers is selection of the input and output variables associated to each agent. The decomposition was carried after consideration of the multi-loop rules (Wittenmark & Salgado, 2002). The resulting decomposition is given in Table 1: **Agent 1** corresponds to the first and third rows of $A(s)$, while **agents 2** and **3** correspond to the second and fourth rows of $A(s)$ respectively. Agents 1 and 2 will mainly interact between them through the process stream c_1.
For a *HEN* not only the dynamic performance of the control system is important but also the cost associated with the resulting operating condition must be taken into account. Thus, the performance index (3) is augmented by including an economic term J_U, such that the global cost is given by $J + J_U$, defined as follows

$$J_U = u_{SS}^T R_U u_{SS}. \tag{28}$$

where $u_{SS} = [u_3(k+M,k)\; u_5(k+M,k)\; u_6(k+M,k)]$ for the *centralized MPC*. In the case of the *distributed* and *coordinated decentralized MPC*, u_{SS} is decomposed among the agents of the control schemes ($u_{SS} = u_3(k+M,k)$ for Agent 1, $u_{SS} = u_5(k+M,k)$ for Agent 2 and $u_{SS} = u_6(k+M,k)$ for Agent 3). Finally, the tuning parameters of the *MPC* controllers are: $t_s =$

0.2 min; $V_l = 50$; $M_l = 5$; $\varepsilon_l = 0.01$; $q_{max} = 10$ $\quad l = 1,2,3$, the cost functions matrices are given in Table 2.

MATLAB based simulation results are carried out to evaluate the proposed *MPC* algorithms (*coordinated decentralized* and *distributed MPC*) through performance comparison with a *centralized* and *decentralized MPC*. The *MPC* algorithms used the same routines during the simulations, which were run in a computer with an Intel Quad-core Q9300 CPU under Linux operating system. One of the processors was used to execute the *HEN* simulator, while the others were used to execute the *MPC* controllers. Only one processor was used to run the *centralized MPC* controller. In the case of the distributed algorithms, the controllers were distributed among the other processors. These configurations were adopted in order to make a fair comparison of the computational time employed for each controller.

We consider the responses obtained for disturbance rejection. A sequence of changes is introduced into the system: after stabilizing at nominal conditions, the inlet temperature of h_1 ($T_{h_1}^{in}$) changes from 90°C to 80°C; 10 min later the inlet temperature of h_2 ($T_{h_2}^{in}$) goes from 130°C to 140°C and after another 10 min the inlet temperature of c_1 ($T_{c_1}^{in}$) changes from 30°C to 40°C.

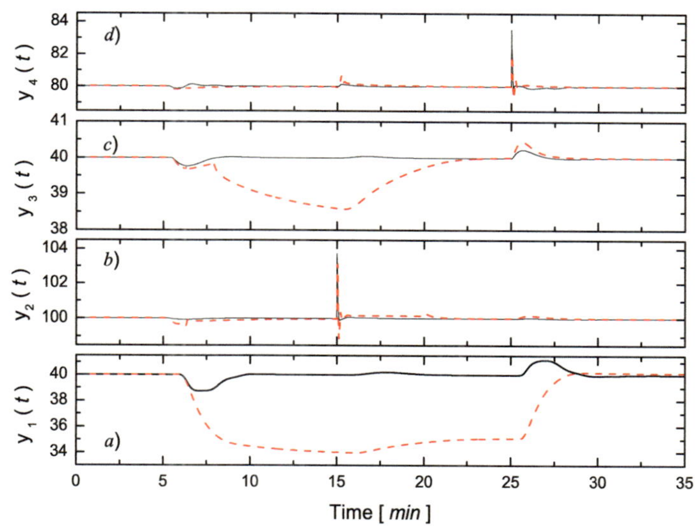

Fig. 4. Controlled outputs of the HEN system using (—) distributed MPC and (-.-) coordinated decentralized MPC.

Figures 4 and 5 show the dynamic responses of the *HEN* operating with a *distributed MPC* and a *coordinated decentralized MPC*. The worse performance is observed during the first and second load changes, most notably on y_1 and y_3. The reasons for this behavior can be found by observing the manipulated variables. The first fact to be noted is that under nominal steady-state conditions, u_4 is completely closed and y_2 is controlled by u_5 (see Figures 5.*b*), achieving the maximum energy recovery. Observe also that u_6 is inactive since no heating service is necessary at this point. After the first load change occurs, both control variables u_2 and u_3 fall rapidly (see Figures 5.*a*). Under this conditions, the system activates the heater flow rate u_6 (see Figures 5.*b*). The dynamic reaction of the heater to the cool disturbance is also stimulated by u_2, while u_6 takes complete control of y_1, achieving the maximum energy recovery. After the initial effect is compensated, y_3 is controlled through u_2 –which never saturates–, while u_6 takes complete control of y_1. Furthermore, Figure 5.*b* show that the cool perturbation also affects y_2, where u_5 is effectively taken out of operation by u_4. The ensuing pair of load changes are heat perturbations featuring manipulated movements in the opposite sense to those indicated above. Though the input change in h_2 allows returning the control of y_1 from u_6 to u_3 (see Figures 5.*a*).

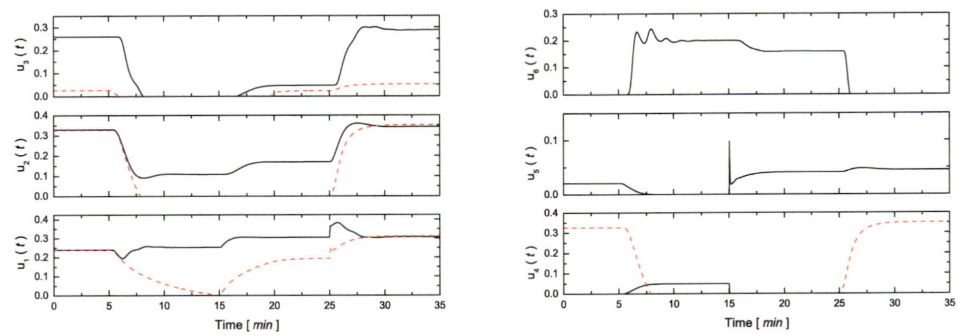

(a) $u_1(t), u_2(t), u_3(t)$ (b) $u_4(t), u_5(t), u_6(t)$

Fig. 5. Manipulated inputs of the HEN system using (—) distributed MPC and (-.-) coordinated decentralized MPC.

In these figures we can also see that the *coordinated decentralized MPC* fails to reject the first and second disturbances on y_1 and y_3 (see Figures 4.*a* and *c*) because it is not able to properly coordinate the use of utility service u_6 to compensate the effects of active constraints on u_2 and u_3. This happens because the *coordinated decentralized MPC* is only able to address the effect of interactions between agents but it can not coordinate the use of utility streams s_2 and s_3 to avoid the *output-unreachability under input constraint* problem. The origin of the problem lies in the cost function employed by the *coordinated decentralized MPC*, which does not include the effect of the local decision variables on the other agents. This fact leads to different steady–state values in the manipulated variables to those ones obtained by the *distributed MPC* along the simulation.

Figure 6 shows the steady–state value of the recovered energy and utility services used by the system for the distributed *MPC* schemes. As mentioned earlier, the *centralized* and *distributed MPC* algorithms have similar steady–state conditions. These solutions are *Pareto optimal*, hence they achieve the best plant wide performance for the combined performance index. On the other hand, the *coordinated decentralized MPC* exhibited a good performance in energy terms, since it employs less service energy, however it is not able of achieving the control objectives, because it is not able of properly coordinate the use of utility flows u_5 and u_6. As it was pointed out in previous Sections, the fact that the agents achieve the Nash equilibrium does not implies the optimality of the solution.

Figure 7 shows the CPU time employed for each *MPC* algorithm during the simulations. As it was expected, the *centralized MPC* is the algorithm that used more intensively the CPU. Its CPU time is always larger than the others along the simulation. This fact is originated on the size of the optimization problem and the dynamic of the system, which forces the

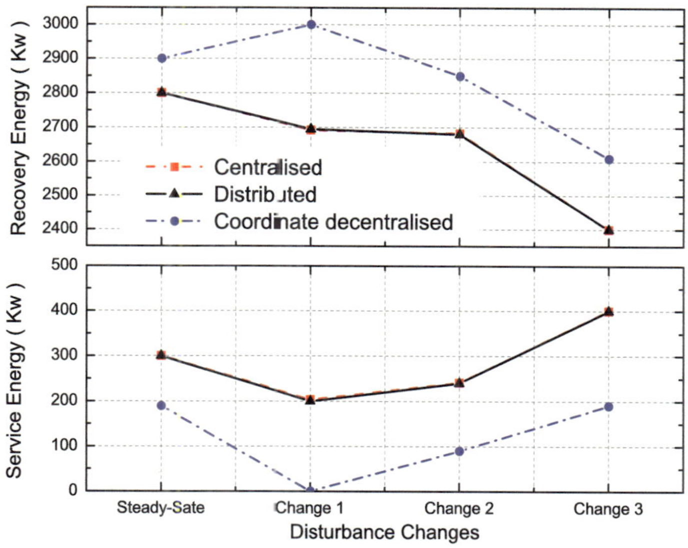

Fig. 6. Steady-state conditions achieved by the HEN system for different MPC schemes.

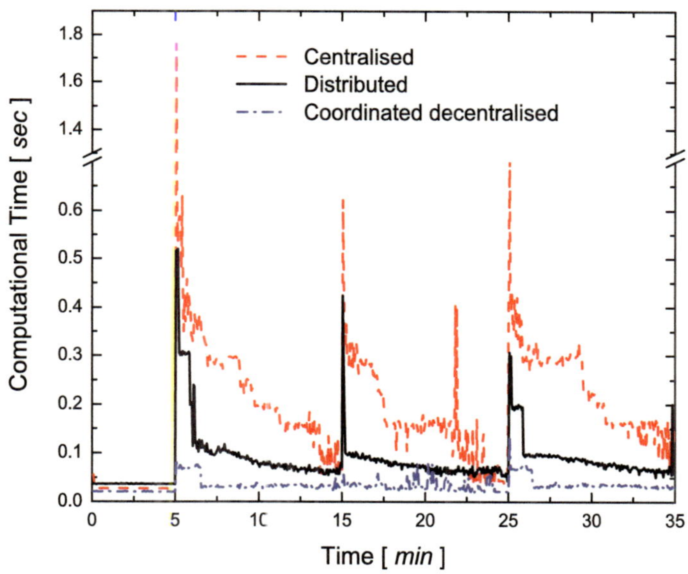

Fig. 7. CPU times for different MPC schemes.

centralized MPC to permanently correct the manipulated variable along the simulation due to the system interactions. On the other hand, the *coordinated decentralized MPC* used the CPU less intensively than the others algorithms, because of the size of the optimization problem. However, its CPU time remains almost constant during the entire simulation since it needs to compensate the interactions that had not been taken into account during the computation. In general, all algorithms show larger CPU times after the load changes because of the recalculation of the control law. However, we have to point out that the value of these peak are smaller than sampling time.

6. Conclusions

In this work a distributed model predictive control framework based on dynamic games is presented. The *MPC* is implemented in distributed way with the inexpensive agents within the network environment. These agents can cooperate and communicate each other to achieve the objective of the whole system. Coupling effects among the agents are taken into account in this scheme, which is superior to other traditional decentralized control methods. The main advantage of this scheme is that the on-line optimization can be converted to that of several small-scale systems, thus can significantly reduce the computational complexity while keeping satisfactory performance. Furthermore, the design parameters for each agent such as prediction horizon, control horizon, weighting matrix and sample time, etc. can all be designed and tuned separately, which provides more flexibility for the analysis and applications. The second part of this study is to investigate the convergence, stability, feasibility and performance of the distributed control scheme. These will provide users better understanding to the developed algorithm and sensible guidance in applications.

7. Acknowledgements

The authors wishes to thank: the *Agencia Nacional de Promoción Científica y Tecnológica*, the *Universidad Nacional de Litoral* and the *Consejo Nacional de Investigaciones Científicas y Técnicas* (CONICET) from Argentina, for their support.

8. Appendices

A. Proof of Lemma 1

Proof. From definition of the $J(\cdot)$ we have

$$J\left(x(k), U^q(k), A\right) = J\left(x(k), \left[U^q_{j\in\mathcal{N}_1}(k) \cdots U^q_{j\in\mathcal{N}_m}(k)\right], A\right) \tag{29}$$

From definition of $U^q_{j\in\mathcal{N}_l}$.we have

$$\begin{aligned} J\left(x(k), U^q(k), A\right) =& J\Big(x(k), \Big[\alpha_{j\in\mathcal{N}_1}\tilde{U}^q_1(k) + \left(1-\alpha_{j\in\mathcal{N}_1}\right)U^{q-1}_{j\in\mathcal{N}_1}(k) \cdots \\ & \alpha_{j\in\mathcal{N}_m}\tilde{U}^q_{j\in\mathcal{N}_m}(k) + \left(1-\alpha_{j\in\mathcal{N}_m}\right)U^{q-1}_{j\in\mathcal{N}_m}(k)\Big], A\Big) \\ =& J\Big(x(k), \Big[\alpha_{j\in\mathcal{N}_1}\left[\tilde{U}^q_{j\in\mathcal{N}_1}(k) \cdots U^{q-1}_{j\in\mathcal{N}_m}(k)\right] \cdots \\ & \alpha_{j\in\mathcal{N}_m}\left[U^{q-1}_{j\in\mathcal{N}_1}(k) \cdots \tilde{U}^q_{j\in\mathcal{N}_m}(k)\right]\Big], A\Big) \end{aligned}$$

By convexity of $J(\cdot)$ we have

$$J\left(x(k), U^q(k), A\right) \leq \sum_{l=1}^{m} \alpha_l J\left(x(k), \tilde{U}^q_{j\in\mathcal{N}_l}(k), U^{q-1}_{j\in\mathcal{I}-\mathcal{N}_l}(k), A\right) \tag{30}$$

and from Algorithm 1 we know that

$$J\left(x(k), \tilde{U}^q_{j\in\mathcal{N}_l}(k), U^{q-1}_{j\in\mathcal{I}-\mathcal{N}_l}(k), A\right) \leq J\left(x(k), U^{q-1}(k), A\right),$$

then

$$J\left(x(k), U^q(k), A\right) \leq J\left(x(k), \tilde{U}^q_{j\in\mathcal{N}_l}(k), U^{q-1}_{j\in\mathcal{I}-\mathcal{N}_l}(k), A\right) \leq J\left(x(k), U^{q-1}(k), A\right). \tag{31}$$

Subtracting the cost functions at $q-1$ and q we obtain

$$\Delta J\left(x(k), U^{q-1}(k), A\right) \leq -\Delta U^{q-1}_{j\in\mathcal{N}_l}(k)^T R \,\Delta U^{q-1}_{j\in\mathcal{N}_l}(k).$$

This shows that the sequence of cost $\left\{J^q_l(k)\right\}$ is non-increasing and the cost is bounded below by zero and thus has a non-negative limit. Therefore as $q \to \infty$ the difference of cost $\Delta J^q(k) \to 0$ such that the $J^q(k) \to J^*(k)$. Because $R > 0$, as $\Delta J^q(k) \to 0$ the updates of the inputs $\Delta U^{q-1}(k) \to 0$ as $q \to \infty$, and the solution of the optimisation problem $U^q(k)$ converges to a solution $\bar{U}(k)$. Depending on the cost function employed by the distributed controllers, $\bar{U}(k)$ can converge to $U^*(k)$ (see Section 3.1). □

B. Proof of Theorem 1

Proof. First it is shown that the input and the true plant state converge to the origin, and then it will be shown that the origin is an stable equilibrium point for the closed-loop system. The combination of convergence and stability gives asymptotic stability.
Convergence. Convergence of the state and input to the origin can be established by showing that the sequence of cost values is non-increasing.
Showing stability of the closed-loop system follows standard arguments for the most part Mayne et al. (2000), Primbs & Nevistic (2000). In the following, we describe only the most important part for brevity, which considers the nonincreasing property of the value function. The proof in this section is closely related to the stability proof of the *FC-MPC* method in Venkat et al. (2008).
Let $q(k)$ and $q(k+1)$ stand for iteration number of **Algorithm 1** at time k and $k+1$ respectively. Let $J(k) = J\left(x(k), U(k), A\right)$ and $J(k+1) = J\left(x(k+1), U(k+1), A\right)$ denote the cost value associated with the final combined solution at time k and $k+1$. At time $k+1$, let $J_l(k+1) = J\left(x(k+1), U^q_{j\in\mathcal{N}_l}(k), U^{q-1}_{j\in\mathcal{I}-\mathcal{N}_l}(k), A\right)$ denote the global cost associated with solution of subsystem l at iterate q.
The global cost function $J\left(x(k), U(k)\right)$ can be used as a Lyapunov function of the system, and its non-increasing property can be shown following the chain

$$J\left(x(k+1), U(k+1), A\right) \leq \cdots \leq J\left(x(k+1), U^q(k+1), A\right) \leq \cdots$$
$$\cdots \leq J\left(x(k+1), U^1(k+1), A\right) \leq J\left(x(k), U(k), A\right) - x(k)^T Q x(k) - u(k)^T R u(k)$$

The inequality $J\left(x(k+1), U^{q}(k+1), A\right) \leq J\left(x(k+1), U^{q-1+}(k+1), A\right)$ is consequence of Lemma 1. Using this inequality we can trace back to $q = 1$

$$\begin{aligned} &J\left(x(k+1), U(k+1), A\right) \leq \cdots \leq J\left(x(k+1), U^{q}(k+1), A\right) \leq \cdots \\ &\cdots \leq J\left(x(k+1), U^{1}(k+1), A\right). \end{aligned}$$

At time step $q = 1$, we can recall the initial feasible solution $U^{0}(k+1)$. At this iteration, the distributed MPC optimizes the cost function with respect the local variables starting from $U^{0}(k+1)$, therefore $\forall l = 1, \ldots, m$

$$\begin{aligned} J\left(x(k+1), U^{1}_{j\in\mathcal{N}_l}(k), U^{0}_{j\in\mathcal{I}-\mathcal{N}_l}(k), A\right) &\leq J\left(x(k+1), U^{0}(k), A\right) \\ &\leq \sum_{i=1}^{\infty} x(k+i,k^{T}Qx(k+i,k) + u(k+i,k)^{T}Ru(k+i,k) \\ &\leq J\left(x(k), U(k), A\right) - x(k)^{T}Qx(k) - u(k)^{T}Ru(k) \end{aligned}$$

Due to the convexity of J and the convex combination up date (Step 2.c of Algorithm 1), we obtain

$$J\left(x(k), U^{1}(k), A\right) \leq \sum_{l=1}^{m} \alpha_l J\left(x(k+1), U^{1}_{j\in\mathcal{N}_l}(k), U^{0}_{j\in\mathcal{I}-\mathcal{N}_l}(k), A\right) \tag{32}$$

then,

$$\begin{aligned} J\left(x(k), U^{1}(k), A\right) &\leq \sum_{l=1}^{m} \alpha_l \left[J\left(x(k), U(k), A\right) - x(k)^{T}Qx(k) - u(k)^{T}Ru(k)\right], \\ &\leq J\left(x(k), U(k), A\right) - x(k)^{T}Qx(k) - u(k)^{T}Ru(k). \end{aligned}$$

Subtracting $J^{*}(k)$ from $J^{*}(k+1)$

$$J^{*}(k+1) - J^{*}(k) \leq -x(k)^{T}Qx(k) - u(k)^{T}Ru(k) \quad \forall k. \tag{33}$$

This shows that the sequence of optimal cost values $\{J^{*}(k)\}$ decreases along closed-loop trajectories of the system. The cost is bounded below by zero and thus has a non–negative limit. Therefore as $k \to \infty$ the difference of optimal cost $\Delta J^{*}(k+1) \to 0$. Because Q and R are positive definite, as $\Delta J^{*}(k+1) \to 0$ the states and the inputs must converge to the origin $x(k) \to 0$ and $u(k) \to 0$ as $k \to \infty$.
Stability. Using the QP form of (6), the feasible cost at time $k = 0$ can be written as follows $J(0) = x(0)^{T}\bar{Q}x(0)$, where $\bar{Q}$ is the solution of the Lyapunov function for dynamic matrix $\bar{Q} = A^{T}QA + Q$.
From equation (33) it is clear that the sequence of optimal costs $\{J^{*}(k)\}$ is non-increasing, which implies $J^{*}(k) \leq J^{*}(0) \quad \forall k > 0$. From the definition of the cost function it follows that $x^{T}(k)Qx(k) \leq J^{*}(k) \quad \forall k$, which implies

$$x^{T}(k)Qx(k) \leq x(0)^{T}\bar{Q}x(0) \quad \forall k.$$

Since Q and $\bar{Q}$ are positive definite it follows that

$$\|x(k)\| \leq \gamma \|x(0)\| \quad \forall k > 0$$

where

$$\gamma = \sqrt{\frac{\lambda_{\max}(\bar{Q})}{\lambda_{\min}(Q)}}.$$

Thus, the closed-loop is stable. The combination of convergence and stability implies that the origin is asymptotically stable equilibrium point of the closed-loop system. □

C. Proof of Theorem 2

Proof. The optimal solution of the distributed control system with communications faults is given by

$$\tilde{U}(k) = (I - \mathcal{K}_0 \mathcal{RCT})^{-1} \mathcal{K}_1 \mathcal{RCT} \Gamma x(k). \tag{34}$$

Using the matrix decomposition technique, it gives

$$(I - \mathcal{K}_0 \mathcal{RCT})^{-1} = (I - \mathcal{K}_0)^{-1} \left(2I - \left[I + (I - \mathcal{K}_0)^{-1} \right.\right.$$
$$\left.\left.(I + \mathcal{K}_0 - 2\mathcal{K}_0 \mathcal{RCT})\right]^{-1}\right) + (I - \mathcal{K}_0)^{-1}$$

In general $(I - \mathcal{K}_0)^{-1}$ and $(I + \mathcal{K}_0 - 2\mathcal{K}_0 \mathcal{RCT})^{-1}$ all exist, therefore the above equation holds. Now, from (34) we have $\mathcal{K}_1 \Gamma x(k) = (I - \mathcal{K}_0) U(k)$, then $\tilde{U}(k)$ can be written as a function of the optimal solution $U(k)$ as follows

$$\tilde{U}(k) = (\mathcal{S} + I) U(k)$$

where $\mathcal{S} = 2I - \left[I + (I - \mathcal{K}_0)^{-1} (I + \mathcal{K}_0 - 2\mathcal{K}_0 \mathcal{RCT})\right]^{-1}$.
The cost function of the system free of communication faults J^* can be written as function of $U(k)$ as follows

$$J^* = \left\| \mathcal{K}_1^{-1} (I - \mathcal{K}_0) U(k) - \mathcal{H} U(k) \right\|_Q^2 + \| U(k) \|_R^2 = \| U(k) \|_{\mathcal{F}}^2 \tag{35}$$

where

$$\mathcal{F} = \left(\mathcal{K}_1^{-1} (I - \mathcal{K}_0) - \mathcal{H}\right)^T Q \left(\mathcal{K}_1^{-1} (I - \mathcal{K}_0) - \mathcal{H}\right) + R.$$

In the case of the system with communication failures we have

$$\tilde{J} \leq J^* + \| U(k) \|_{\mathcal{W}}^2 \tag{36}$$

where $\mathcal{W} = \mathcal{S}^T \left(\mathcal{H}^T Q \mathcal{H} + R\right) \mathcal{S}$. Finally, the effect of communication can be related with J^* through

$$\| U(k) \|_{\mathcal{W}}^2 \leq \frac{\| \mathcal{W} \|}{\lambda_{min}(\mathcal{F})} J^*, \tag{37}$$

where λ_{min} denotes the minimal eigenvalue of $\mathcal{F}$. From the above derivations, the relationship between $\tilde{J}$ and J^* is given by

$$\tilde{J} \leq \left(1 + \frac{\| \mathcal{W} \|}{\lambda_{min}(\mathcal{F})}\right) J^*. \tag{38}$$

and the degradation is

$$\frac{\tilde{J} - J^*}{J^*} \leq \frac{\| \mathcal{W} \|}{\lambda_{min}(\mathcal{F})}. \tag{39}$$

Inspection of (36) shows that $\|\mathcal{W}\|$ depends on $\mathcal{R}$ and $\mathcal{T}$. So in case of all communication failures existed, $\|\mathcal{W}\|$ can arrive at the maximal value

$$\mathcal{W}_{\max} = \left(2I - \left[I + (I - \mathcal{K}_0)^{-1}(I + \mathcal{K}_0)\right]^{-1}\right)^T \left(\mathcal{H}^T Q \mathcal{H} + R\right)$$
$$\left(2I - \left[I + (I - \mathcal{K}_0)^{-1}(I + \mathcal{K}_0)\right]^{-1}\right),$$

and the upper bound of performance deviation is

$$\frac{\tilde{J} - J^*}{J^*} \leq \frac{\|\mathcal{W}_{\max}\|}{\lambda_{min}(\mathcal{F})}. \tag{40}$$

□

9. References

Aguilera, N. & Marchetti, J. (1998). Optimizing and controlling the operation of heat exchanger networks, *AIChe Journal* 44(5): 1090–1104.

Aske, E., Strand, S. & Skogestad, S. (2008). Coordinator mpc for maximizing plant throughput, *Computers and Chemical Engineering* 32(1-2): 195–204.

Bade, S., Haeringer, G. & Renou, L. (2007). More strategies, more nash equilibria, *Journal of Economic Theory* 135(1): 551–557.

Balderud, J., Giovanini, L. & Katebi, R. (2008). Distributed control of underwater vehicles, *Proceedings of the Institution of Mechanical Engineers, Part M: Journal of Engineering for the Maritime Environment* 222(2): 95–107.

Bemporad, A., Filippi, C. & Torrisi, F. (2004). Inner and outer approximations of polytopes using boxes, *Computational Geometry: Theory and Applications* 27(2): 151–178.

Bemporad, A. & Morari, M. (1999). Robust model predictive control: A survey, in robustness in identification and control, *Lecture Notes in Control and Information Sciences* 245: 207–226.

Braun, M., Rivera, D., Flores, M., Carlyle, W. & Kempf, K. (2003). A model predictive control framework for robust management of multi-product, multi-echelon demand networks, *Annual Reviews in Control* 27(2): 229–245.

Camacho, E. & Bordons, C. (2004). *Model predictive control*, Springer.

Camponogara, E., Jia, D., Krogh, B. & Talukdar, S. (2002). Distributed model predictive control, *IEEE Control Systems Magazine* 22(1): 44–52.

Cheng, R., Forbes, J. & Yip, W. (2007). Price-driven coordination method for solving plant-wide mpc problems, *Journal of Process Control* 17(5): 429–438.

Cheng, R., Fraser Forbes, J. & Yip, W. (2008). Dantzig–wolfe decomposition and plant-wide mpc coordination, *Computers and Chemical Engineering* 32(7): 1507–1522.

Dubey, P. & Rogawski, J. (1990). Inefficiency of smooth market mechanisms, *Journal of Mathematical Economics* 19(3): 285–304.

Dunbar, W. (2007). Distributed receding horizon control of dynamically coupled nonlinear systems, *IEEE Transactions on Automatic Control* 52(7): 1249–1263.

Dunbar, W. & Murray, R. (2006). Distributed receding horizon control for multi-vehicle formation stabilization, *Automatica* 42(4): 549–558.

Goodwin, G., Salgado, M. & Silva, E. (2005). Time-domain performance limitations arising from decentralized architectures and their relationship to the rga, *International Journal of Control* 78(13): 1045–1062.

Haimes, Y. & Chankong, V. (1983). *Multiobjective decision making: Theory and methodology*, North Holland, New York.

Henten, E. V. & Bontsema, J. (2009). Time-scale decomposition of an optimal control problem in greenhouse climate management, *Control Engineering Practice* 17(1): 88–96.

Hovd, M. & Skogestad, S. (1994). Sequential design of decentralized controllers, *Automatica* 30: 1601–1601.

Jamoom, M., Feron, E. & McConley, M. (2002). Optimal distributed actuator control grouping schemes, *Proceedings of the 37th IEEE Conference on Decision and Control*, Vol. 2, pp. 1900–1905.

Jia, D. & Krogh, B. (2001). Distributed model predictive control, *American Control Conference, 2001. Proceedings of the 2001*, Vol. 4.

Jia, D. & Krogh, B. (2002). Min-max feedback model predictive control for distributed control with communication, *American Control Conference, 2002. Proceedings of the 2002*, Vol. 6.

Kouvaritakis, B. & Cannon, M. (2001). *Nonlinear predictive control: theory and practice*, Iet.

Kumar, A. & Daoutidis, P. (2002). Nonlinear dynamics and control of process systems with recycle, *Journal of Process Control* 12(4): 475–484.

Lu, J. (2003). Challenging control problems and emerging technologies in enterprise optimization, *Control Engineering Practice* 11(8): 847–858.

Maciejowski, J. (2002). *Predictive control: with constraints*, Prentice Hall.

Mayne, D., Rawlings, J., Rao, C. & Scokaert, P. (2000). Constrained model predictive control: Stability and optimality, *Automatica* 36: 789–814.

Motee, N. & Sayyar-Rodsari, B. (2003). Optimal partitioning in distributed model predictive control, *Proceedings of the American Control Conference*, Vol. 6, pp. 5300–5305.

Nash, J. (1951). Non-cooperative games, *Annals of mathematics* pp. 286–295.

Neck, R. & Dockner, E. (1987). Conflict and cooperation in a model of stabilization policies: A differential game approach, *J. Econ. Dyn. Cont* 11: 153–158.

Osborne, M. & Rubinstein, A. (1994). *A course in game theory*, MIT press.

Perea-Lopez, E., Ydstie, B. & Grossmann, I. (2003). A model predictive control strategy for supply chain optimization, *Computers and Chemical Engineering* 27(8-9): 1201–1218.

Primbs, J. & Nevistic, V. (2000). Feasibility and stability of constrained finite receding horizon control, *Automatica* 36(7): 965–971.

Rossiter, J. (2003). *Model-based predictive control: a practical approach*, CRC press.

Salgado, M. & Conley, A. (2004). Mimo interaction measure and controller structure selection, *International Journal of Control* 77(4): 367–383.

Sandell Jr, N., Varaiya, P., Athans, M. & Safonov, M. (1978). Survey of decentralized control methods for large scale systems, *IEEE Transactions on Automatic Control* 23(2): 108–128.

Vaccarini, M., Longhi, S. & Katebi, M. (2009). Unconstrained networked decentralized model predictive control, *Journal of Process Control* 19(2): 328–339.

Venkat, A., Hiskens, I., Rawlings, J. & Wright, S. (2008). Distributed mpc strategies with application to power system automatic generation control, *IEEE Transactions on Control Systems Technology* 16(6): 1192–1206.

Šiljak, D. (1996). Decentralized control and computations: status and prospects, *Annual Reviews in Control* 20: 131–141.

Wittenmark, B. & Salgado, M. (2002). Hankel-norm based interaction measure for input-output pairing, *Proc. of the 2002 IFAC World Congress*.

Zhang, Y. & Li, S. (2007). Networked model predictive control based on neighbourhood optimization for serially connected large-scale processes, *Journal of process control* 17(1): 37–50.

Zhu, G. & Henson, M. (2002). Model predictive control of interconnected linear and nonlinear processes, *Industrial and Engineering Chemistry Research* 41(4): 801–816.

5

Efficient Nonlinear Model Predictive Control for Affine System

Tao ZHENG and Wei CHEN
Hefei University of Technology
China

1. Introduction

Model predictive control (MPC) refers to the class of computer control algorithms in which a dynamic process model is used to predict and optimize process performance. Since its lower request of modeling accuracy and robustness to complicated process plants, MPC for linear systems has been widely accepted in the process industry and many other fields. But for highly nonlinear processes, or for some moderately nonlinear processes with large operating regions, linear MPC is often inefficient. To solve these difficulties, nonlinear model predictive control (NMPC) attracted increasing attention over the past decade (Qin *et al.*, 2003, Cannon, 2004). Nowadays, the research on NMPC mainly focuses on its theoretical characters, such as stability, robustness and so on, while the computational method of NMPC is ignored in some extent. The fact mentioned above is one of the most serious reasons that obstruct the practical implementations of NMPC.

Analyzing the computational problem of NMPC, the direct incorporation of a nonlinear process into the linear MPC formulation structure may result in a non-convex nonlinear programming problem, which needs to be solved under strict sampling time constraints and has been proved as an NP-hard problem (Zheng, 1997). In general, since there is no accurate analytical solution to most kinds of nonlinear programming problem, we usually have to use numerical methods such as Sequential Quadric Programming (SQP) (Ferreau *et al.*, 2006) or Genetic Algorithm (GA) (Yuzgec *et al.*, 2006). Moreover, the computational load of NMPC using numerical methods is also much heavier than that of linear MPC, and it would even increase exponentially when the predictive horizon length increases. All of these facts lead us to develop a novel NMPC with analytical solution and little computational load in this chapter.

Since affine nonlinear system can represents a lot of practical plants in industry control, including the water-tank system that we used to carry out the simulations and experiments, it has been chosen for propose our novel NMPC algorithm. Follow the steps of research work, the chapter is arranged as follows:

In Section 2, analytical one-step NMPC for affine nonlinear system will be introduced at first, then, after description of the control problem of a water-tank system, simulations will be carried out to verify the result of theoretical research. Error analysis and feedback compensation will be discussed with theoretical analysis, simulations and experiment at last.

Then, in Section 3, by substituting reference trajectory for predicted state with stair-like control strategy, and using sequential one-step predictions instead of the multi-step

prediction, the analytical multi-step NMPC for affine nonlinear system will be proposed. Simulative and experimental control results will also indicate the efficiency of it. The feedback compensation mentioned in Section 2 is also used to guarantee the robustness to model mismatch.
Conclusion and further research direction will be given at last in Section 4.

2. One-step NMPC for affine system

2.1 Description of NMPC for affine system

Consider a time-invariant, discrete, affine nonlinear system with integer k representing the current discrete time event:

$$x_{k+1}=f(x_k)+g(x_k)\times u_k+\xi_k \tag{1a}$$

$$\text{s. t. } x_k \in X \subseteq R^n \tag{1b}$$

$$u_k \in U \subseteq R^m \tag{1c}$$

$$\xi_k \in R^n \tag{1d}$$

In the above, u_k, x_k, ξ_k are input, state and disturbance of the system respectively, $f{:}R^n \to R^n$, $g{:}R^n \to R^{n\times m}$ are corresponding nonlinear mapping functions with proper dimension.
Assume $\hat{x}_{k+j|k}$ are predictive values of x_{k+j} at time k, $\Delta u_k=u_k-u_{k-1}$ and $\Delta\hat{u}_{k+j|k}$ are the solutions of future increment of u_{k+j} at time k, then the objective function J_k can be written as follow:

$$J_k=F(\hat{x}_{k+p|k})+\sum_{j=0}^{p-1}G(\hat{x}_{k+j|k},\Delta u_{k+j|k}) \tag{2}$$

The function F (.) and G (. , .) represent the terminal state penalty and the stage cost respectively, where p is the predictive horizon.
In general, J_k usually has a quadratic form. Assume $w_{k+j|k}$ is the reference value of x_{k+j} at time k which is called reference trajectory (the form of $w_{k+j|k}$ will be introduced with detail in Section 2.2 and 3.1 for one-step NMPC and multi-step NMPC respectively), semi-positive definite matrix Q and positive definite matrix R are weighting matrices, (2) now can be written as :

$$J_k=\sum_{j=1}^{p}\left\|\hat{x}_{k+j|k}-w_{k+j|k}\right\|_Q^2+\sum_{j=0}^{p-1}\left\|\Delta u_{k+j|k}\right\|_R^2 \tag{3}$$

Corresponding to (1) and (3), the NMPC for affine system at each sampling time now is formulated as the minimization of J_k, by choosing the increments sequence of future control input $[\Delta u_{k|k} \quad \Delta u_{k+1|k} \quad \cdots \quad \Delta u_{k+p-1|k}]$, under constraints (1b) and (1c).

By the way, for simplicity, In (3), part of J_k is about the system state x_k, if the output of the system $y_k = Cx_k$, which is a linear combination of the state (C is a linear matrix), we can rewrite (3) as follow to make an objective function J_k about system output:

$$J_k = \sum_{j=1}^{p} \left\| C\hat{x}_{k+j|k} - w_{k+j|k} \right\|_Q^2 + \sum_{j=0}^{p-1} \left\| \Delta u_{k+j|k} \right\|_R^2 = \sum_{j=1}^{p} \left\| \hat{y}_{k+j|k} - w_{k+j|k} \right\|_Q^2 + \sum_{j=0}^{p-1} \left\| \Delta u_{k+j|k} \right\|_R^2 \quad (4)$$

And sometimes, $\Delta u_{k+j|k}$ in J_k could also be changed as $u_{k+j|k}$ to meet the need of practical control problems.

2.2 One-step NMPC for affine system

Except for some special model, such as Hammerstein model, analytic solution of multi-step NMPC could not be obtained for most nonlinear systems, including the NMPC for affine system mentioned above in Section 2.1. But if the analytic inverse of system function exists (could be either state-space model or input-state model), the one-step NMPC always has the analytic solution. So all the research in this chapter is not only suitable for affine nonlinear system, but also suitable for other nonlinear systems, that have analytic inverse system function.

Consider system described by (1a-1d) again, the one-step prediction can be deduced directly as follow with only one unknown data $\Delta u_{k|k} = u_{k|k} - u_{k-1}$ at time k:

$$\hat{x}_{k+1|k} = f(x_k) + g(x_k) \cdot u_{k|k} = f(x_k) + g(x_k) \cdot u_{k-1} + g(x_k) \cdot \Delta u_{k|k} = \hat{x}_{k+1|k}^1 + g(x_k) \cdot \Delta u_{k|k} \quad (5)$$

In (5), $\hat{x}_{k+1|k}^1$ means the part which contains only known data (x_k and u_{k-1}) at time k, and $g(x_k) \cdot \Delta u_{k|k}$ is the unknown part of predictive state $\hat{x}_{k+1|k}$.

If there is no model mismatch, the predictive error of (5) will be $\tilde{x}_{k+1|k} = x_{k+1} - \hat{x}_{k+1|k} = \xi_{k+1}$.

Especially, if ξ_k is a stationary stochastic noise with zero mean and variance $E[\xi_k] = \delta^2$, it is easy known that $E[\tilde{x}_{k+1|k}] = 0$, and $E[(\tilde{x}_{k+1|k} - E[\tilde{x}_{k+1|k}])^T \cdot (\tilde{x}_{k+1|k} - E[\tilde{x}_{k+1|k}])] = n\delta^2$, in another word, both the mean and the variance of the predictive error have a minimum value, so the prediction is an optimal prediction here in (5).

Then if the setpoint is x_{sp}, and to soften the future state curve, the expected state value at time k+1 is chosen as $w_{k+1|k} = \alpha x_k + (1-\alpha) x_{sp}$, where $\alpha \in [0,1)$ is called soften factor, thus the objective function of one-step NMPC can be written as follow:

$$J_k = \left\| \hat{x}_{k+1|k} - w_{k+1|k} \right\|_Q^2 + \left\| \Delta u_{k|k} \right\|_R^2 \quad (6)$$

To minimize J_k without constraints (1b) and (1c), we just need to have $\frac{\partial J_k}{\partial \Delta u_{k|k}} = 0$ and $\frac{\partial^2 J_k}{\partial \Delta u_{k|k}^2} > 0$, then:

$$\Delta u_{k|k} = -(g(x_k)^T \cdot Q \cdot g(x_k) + R)^{-1} \cdot (g(x_k) \cdot Q \cdot (\hat{x}_{k+1|k}^1 - w_{k+1|k})) \quad (7)$$

Mark $H = g(x_k)^T \cdot Q \cdot g(x_k) + R$ and $F = g(x_k) \cdot Q \cdot (w_{k+1|k} - \hat{x}^1_{k+1|k})$, so the increment of instant future input is:

$$\Delta u_{k|k} = -H^{-1}F \tag{8}$$

But in practical control problem, limitations on input and output always exist, so the result of (8) is usually not efficient. To satisfy the constraints, we can just put logical limitation on amplitudes of u_k and x_k, or some classical methods such as Lagrange method could be used. For simplicity, we only discuss about the Lagrange method here in this chapter.
First, suppose every constraint in (1b) and (1c) could be rewritten in the form as $a_i^T \Delta u_{k|k} \le b_i$, $i = 1, 2, \cdots q$, then the matrix form of all constraints is:

$$A\Delta u_{k|k} \le B \tag{9}$$

In which, $A = \begin{bmatrix} a_1^T & a_2^T & \ldots & a_q^T \end{bmatrix}^T$ $B = \begin{bmatrix} b_1 & b_2 & \ldots & b_q \end{bmatrix}^T$.

Choose Lagrange function as $L_k(\lambda_i) = J_k + \lambda_i^T (a_i^T \Delta u_{k|k} - b_i)$, $i = 1, 2, \cdots q$, let $\dfrac{\partial L}{\partial \Delta u_{k|k}} = H\Delta u_{k|k} + F + a_i \lambda_i = 0$ and $\dfrac{\partial L}{\partial \lambda_i} = a_i^T \Delta u_{k|k} - b_i = 0$, then:

$$\Delta u_{k|k} = -H^{-1}(F + a_i \lambda_i) \tag{10a}$$

$$\lambda_i = -\frac{a_i^T H^{-1} F + b_i}{a_i^T H^{-1} a_i} \tag{10b}$$

If $\lambda_i \le 0$ in (10b), means that the corresponding constraint has no effect on $\Delta u_{k|k}$, we can choose $\bar{\lambda}_i = 0$, but if $\lambda_i > 0$ in (10b), the corresponding constraint has effect on Δu_k indeed, so we must choose $\bar{\lambda}_i = \lambda_i$, finally, the solution of one-step NMPC with constraints could be:

$$\Delta u_{k|k} = -H^{-1}(F + A^T \bar{\Lambda}) \tag{11}$$

In which, $\bar{\Lambda} = \begin{bmatrix} \bar{\lambda}_1 & \bar{\lambda}_2 & \cdots & \bar{\lambda}_q \end{bmatrix}^T$

2.3 Control problem of the water-tank system

Our plant of simulations and experiments in this chapter is a water-tank control system as that in Fig. 1. and Fig. 2. (We just used one water-tank of this three-tank system). Its affine nonlinear model is achieved by mechanism modeling (Chen *et al.*, 2006), in which the variables are normalized, and the sample time is 1 second here:

$$x_{k+1} = x_k - 0.2021\sqrt{x_k} + 0.01923u_k \tag{12a}$$

$$\text{s. t. } x_k \in [0\%, 100\%] \tag{12b}$$

$$u_k \in [0\%, 100\%] \tag{12c}$$

In (12), x_k is the height of water in the tank, and u_k is the velocity of water flow into the tank, from pump P_1 and valve V_1, while valve V_2 is always open. In the control problem of the water-tank, for convenience, we choose the system state as the output, that means $y_k = x_k$, and the system functions are $f(x_k) = x_k - 0.2021\sqrt{x_k}$ and $g(x_k) = 0.01923$.

Fig. 1. Photo of the water-tank system

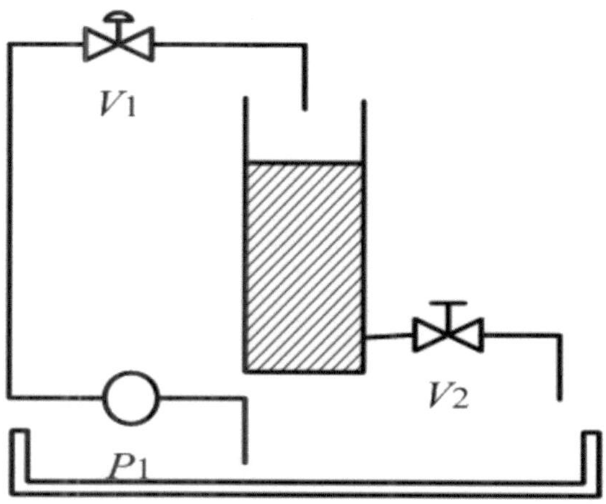

Fig. 2. Structure of the water-tank system

To change the height of the water level, we can change the velocity of input flow, by adjusting control current of valve V_1, and the normalized relation between the control current and the velocity u_k is shown in Fig. 3.

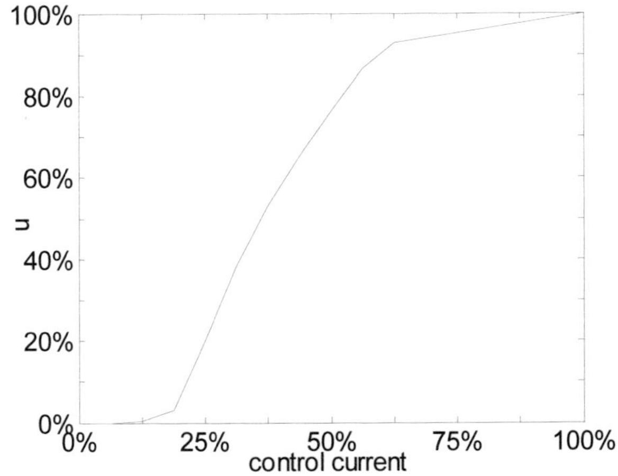

Fig. 3. The relation between control current and input u_k

2.4 One-step NMPC of the water-tank system and its feedback compensation

Choose objective function $J_k = (\hat{x}_{k+1|k} - w_{k+1|k})^2 + 0.001\Delta u_{k|k}^2$, $x_{sp} = 30\%$ and soften factor $\alpha = 0.95$, to carry out all the simulations and the experiment in this section. (except for part of Table 1., where we choose $\alpha = 0.975$)

Suppose there is no model mismatch, the simulative control result of one-step NMPC for water-tank system is obtained as Fig. 4. and it is surely meet the control objective.

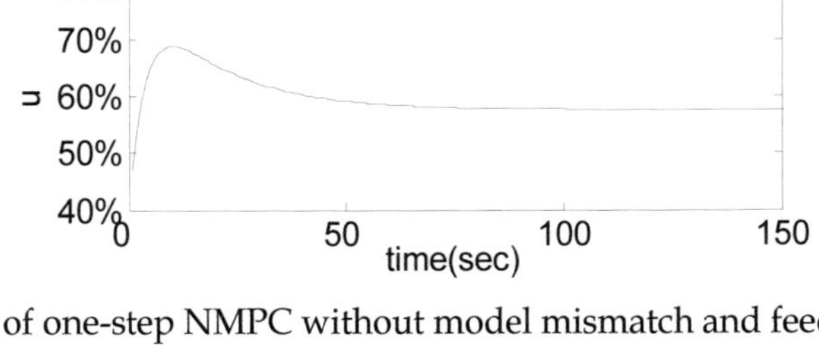

Fig. 4. Simulation of one-step NMPC without model mismatch and feedback compensation

To imitate the model mismatch, we change the simulative model of the plant from $x_{k+1} = x_k - 0.2021\sqrt{x_k} + 0.01923u_k$ to $x_{k+1} = x_k - 110\% \times 0.2021\sqrt{x_k} + 90\% \times 0.01923u_k$, but still use $x_{k+1} = x_k - 0.2021\sqrt{x_k} + 0.01923u_k$ to be the predictive model in one-step NMPC, the result in Fig. 5. now indicates that there is obvious steady-state error.

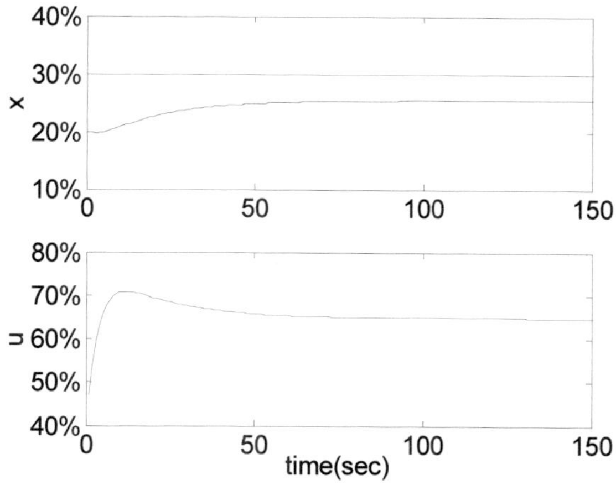

Fig. 5. Simulation of one-step NMPC with model mismatch but without feedback compensation

Proposition 1: For affine nonlinear system $x_{k+1} = f'(x_k) + g'(x_k) \cdot u_k$, if the setpoint is x_{sp}, steady-state is u_s and x_s, and the predictive model is $x_{k+1} = f(x_k) + g(x_k) \cdot u_k$, without consideration of constraints, the steady-state error of one-step NMPC is $e = x_s - x_{sp} = \dfrac{(f'(x_s) - f(x_s)) + (g'(x_s) - g(x_s)) \cdot u_s}{1-\alpha}$, in which α is the soften factor.

Proof: If the system is at the steady-state, then we have $x_{k-1} = x_k = x_s$ and $u_{k-1} = u_k = u_s$. Since $u_{k-1} = u_k = u_s$, so $\Delta u_k = 0$, from (8), we know matrix F=0, or equally $(w_{k+1|k} - \hat{x}^1_{k+1|k}) = 0$.

Update the process of one-step NMPC at time k, we have:

$$w_{k+1|k} = \alpha x_k + (1-\alpha) x_{sp} = \alpha x_s + (1-\alpha) x_{sp} \tag{13}$$

$$\hat{x}^1_{k+1|k} = f(x_k) + g(x_k) \cdot u_{k-1} = f(x_s) + g(x_s) \cdot u_s \tag{14}$$

(13)-(14), and notice that $x_s = f'(x_s) + g'(x_s) \cdot u_s$ for steady-state, we get:

$$0 = \alpha x_s + (1-\alpha) x_{sp} - f(x_s) - g(x_s) \cdot u_s = x_s - f(x_s) - g(x_s) \cdot u_s + (1-\alpha) x_{sp} - (1-\alpha) x_s$$

$$= (f'(x_s) - f(x_s)) + (g'(x_s) - g(x_s)) \cdot u_s + (1-\alpha) \cdot (x_{sp} - x_s)$$

So:

$$e = x_s - x_{sp} = \frac{(f'(x_s) - f(x_s)) + (g'(x_s) - g(x_s)) \cdot u_s}{1-\alpha} \tag{15}$$

Proof end.

Because the soften factor $\alpha \in [0,1)$, thus $1-\alpha \neq 0$ always holds, the necessary condition for $e = 0$ is $(f'(x_s) - f(x_s)) + (g'(x_s) - g(x_s)) \cdot u_s = 0$. When there is model mismatch, there will be steady-state error, while this error is independent of weight matrix Q and dependent of the

soften factor α. For corresponding discussion on steady-state error of one-step NMPC with constraints, the only difference is (11) will take the place of (8) in the proof.
Table 1. is the comparison on $e = x_s - x_{sp}$ between simulation and theoretical analysis, and they have the same result. (simulative model $x_{k+1} = x_k - 110\% \times 0.2021\sqrt{x_k} + 90\% \times 0.01923u_k$, predictive model $x_{k+1} = x_k - 0.2021\sqrt{x_k} + 0.01923u_k$)

α	Q	$e = x_s - x_{sp}$ Simulation(%)	$e = x_s - x_{sp}$ Value of (15)(%)
0.975	0	-8.3489	-8.3489
	0.001	-8.3489	-8.3489
	0.01	-8.3489	-8.3489
0.95	0	-4.5279	-4.5279
	0.001	-4.5279	-4.5279
	0.01	-4.5279	-4.5279

Table 1. Comparison on $e = x_s - x_{sp}$ between simulation and theoretical analysis

From (15) we know, we cannot eliminate this steady-state error by adjusting α, so feedback compensation could be used here, mark the predictive error e_k at time k as follow:

$$e_k = x_k - \hat{x}_{k|k-1} = x_k - (\hat{x}^1_{k|k-1} + g(x_{k-1}) \cdot \Delta u_{k-1}) \tag{16}$$

In which, x_k is obtained by system feedback at time k, and $\hat{x}_{k|k-1}$ is the predictive value of x_k at time k-1.
Then add e_k to the predictive value of x_{k+1} at time k directly, so (5) is rewritten as follow:

$$\hat{x}_{k+1|k} = f(x_k) + g(x_k) \cdot u_{k-1} + g(x_k) \cdot \Delta u_{k|k} + e_k = \hat{x}^1_{k+1|k} + g(x_k) \cdot \Delta u_{k|k} + e_k \tag{17}$$

Use this new predictive value to carry out one-step NMPC, the simulation result in Fig. 6. verify its robustness under model mismatch, since there is no steady-state error with this feedback compensation method.
The direct feedback compensation method above is easy to understand and carry out, but it is very sensitive to noise. Fig. 7. is the simulative result of it when there is noise add to the system state, we can see that the input vibrates so violently, that is not only harmful to the actuator in practical control system, but also harmful to system performance, because the actuator usually cannot always follow the input signal of this kind.
To develop the character of feedback compensation, simply, we can use the weighted average error $\overline{e}_k$ instead of single e_k in (17):

$$\hat{x}_{k+1|k} = \hat{x}^1_{k+1|k} + g(x_k) \cdot \Delta u_{k|k} + \sum_{i=1}^{s} h_i \cdot e_{k+1-i} = \hat{x}^1_{k+1|k} + g(x_k) \cdot \Delta u_{k|k} + \overline{e}_k, \ \sum_{i=1}^{s} h_i = 1 \tag{18}$$

Choose $i = 20$, $h_i = 0.05$, the simulative result is shown in Fig. 8. Compared with Fig. 7. it has almost the same control performance, but the input is much more smooth now. Using the same method and parameters, experiment has been done on the water-tank system, the result in Fig. 9. also verifies the efficiency of the proposed one-step NMPC for affine systems with feedback compensation.

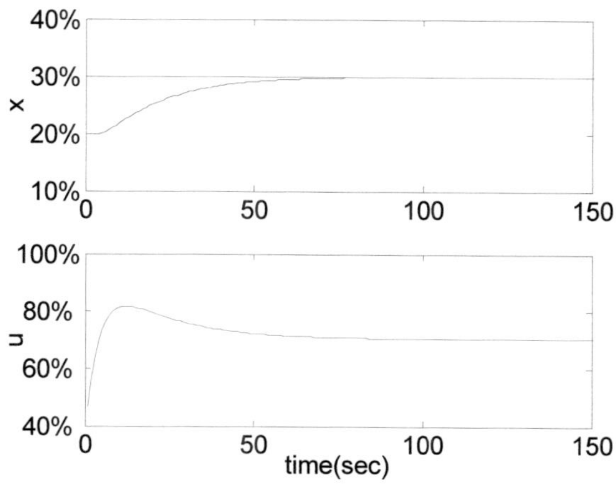

Fig. 6. Simulation of one-step NMPC with model mismatch and direct feedback compensation

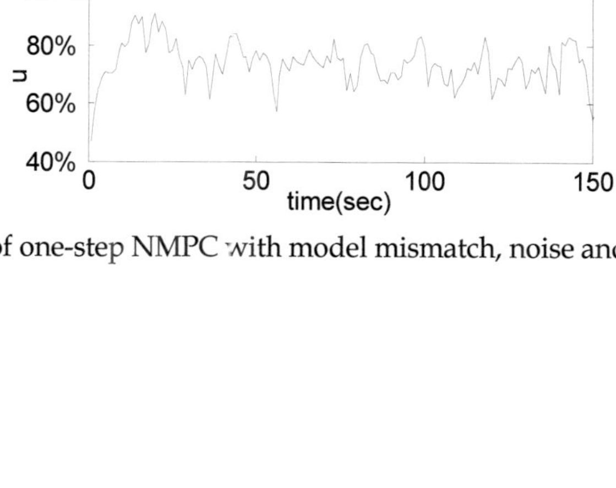

Fig. 7. Simulation of one-step NMPC with model mismatch, noise and direct feedback compensation

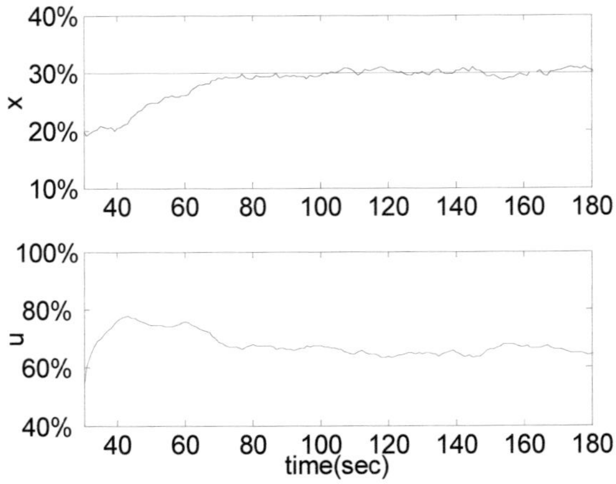

Fig. 8. Simulation of one-step NMPC with model mismatch, noise and smoothed feedback compensation

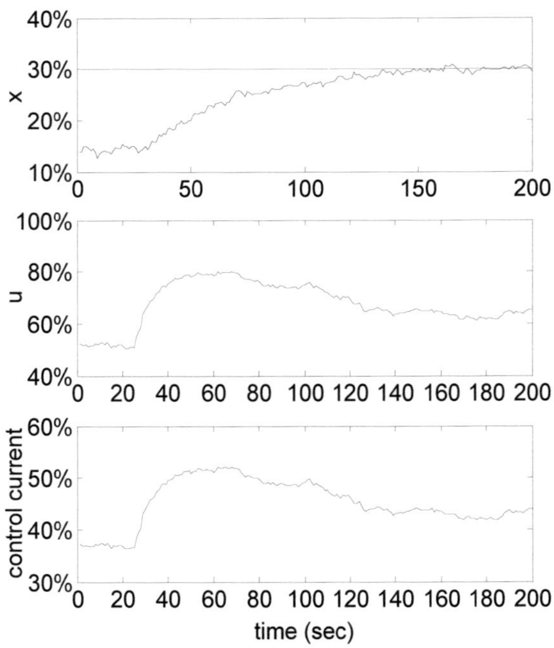

Fig. 9. Experiment of one-step NMPC with setpoint $x_{sp} = 30\%$

3. Efficient multi-step NMPC for affine system

Since reference trajectory and stair-like control strategy will be used to establish efficient multi-step NMPC for affine system in this chapter, we will introduce them in Section 3.1 and 3.2 at first, and then, the multi-step NMPC algorithm will be discussed with theoretical research, simulations and experiments.

3.1 Reference trajectory for future state

In process control, the state usually meets the objective in the form of setpoint along a softer trajectory, rather than reach the setpoint immediately in only one sample time. This may because of the limit on control input, but a softer change of state is often more beneficial to actuators, even the whole process in practice. This trajectory, usually called reference trajectory, often can be defined as a first order exponential curve:

$$w_{k+j|k} = \alpha w_{k+j-1|k} + (1-\alpha)x_{sp}, j = 1,2,\cdots,p-1 \tag{19}$$

In which, x_{sp} still denotes the setpoint, $\alpha \in [0,1)$ is the soften factor, and the initial value of the trajectory is $w_{k|k} = x_k$ .The value of α determines the speed of dynamic response and the curvature of the trajectory, the larger it is, the softer the curve is. Fig. 10. shows different trajectory with different α . Generally speaking, suitable α could be chosen based on the expected setting time in different practical cases.

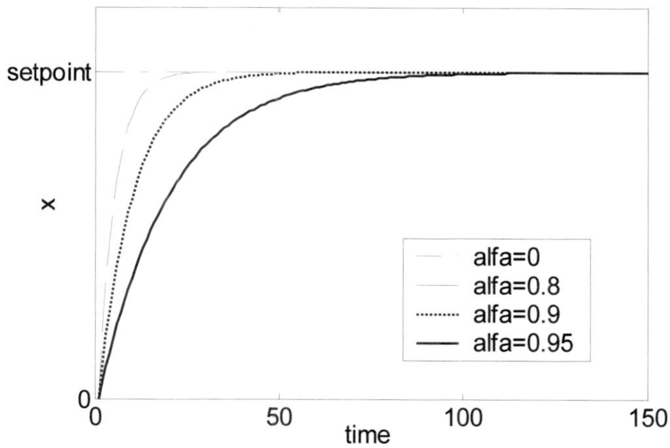

Fig. 10. Reference trajectory with different soften factor α

3.2 Stair-like control strategy

To lighten the computational load of nonlinear optimization, which is one of the biggest obstacles in NMPC's application, stair-like control strategy is introduced here. Suppose the first unknown control input's increment $\Delta u_k = u_k - u_{k-1} = \Delta$, and the stair coefficient β is a positive real number, then the future control input's increment can be decided by the following expression:

$$\Delta u_{k+j} = \beta \cdot \Delta u_{k+j-1} = \beta^j \cdot \Delta u_k = \beta^j \cdot \Delta, j = 1,2,\cdots,p-1 \tag{20}$$

Instead of the full future sequence of control input's increment: $[\Delta u_k \quad \Delta u_{k+1} \quad \cdots \quad \Delta u_{k+p-1}]$, which has p independent variables. Using this strategy, in multi-step NMPC, it now need only compute Δu_k. The computational load now is independent of the length of predictive horizon, which is very convenient for us to choose long predictive horizon in NMPC to obtain a better control performance (Zheng *et al.*, 2007).

Since the dynamic optimization process will be repeated at every sample time, and only instant input $u_k = u_{k-1} + \Delta u_k$ will be carried out actually in NMPC, this strategy is efficient here. In the strategy, it supposes the future increase of control input will be in a same direction, which is the same as the experience in control practice of the human beings, and prevents the frequent oscillation of the input, which is very harmful to the actuators in real control plants. Fig. 11. shows the input sequences with different β.

3.3 Multi-step NMPC for affine system

The one-step NMPC in Section 2 is simple and fast, but it also has one fatal disadvantage. Its predictive horizon is only one step, while long predictive horizon is usually needed for better performance in MPC algorithms. One-step prediction may lead overshooting or other bad influence on system's behaviour. So we will try to establish a novel efficient multi-step NMPC based on proposed one-step NMPC in this section.

In this multi-step NMPC algorithm, the first step prediction is the same as (5), then follows the prediction of $\hat{x}_{k+1|k}$ in (5), the one-step prediction of $\hat{x}_{k+j|k}, j = 2,3,\cdots,p$ could be obtained directly:

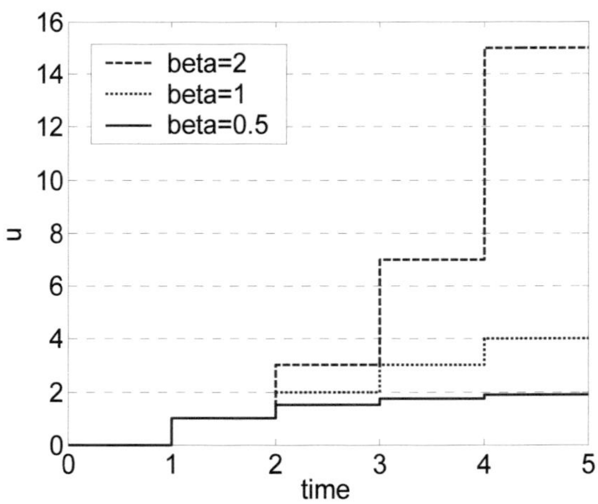

Fig. 11. Stair-like control strategy

$$\hat{x}_{k+j|k} = f(\hat{x}_{k+j-1|k}) + g(\hat{x}_{k+j-1|k}) \cdot u_{k+j-1|k} \tag{21}$$

Since $\hat{x}_{k+j-1|k}$ already contains nonlinear function of former data, one may not obtain the analytic solution of (21) for prediction more than one step. Take the situation of j=2 for example:

$$\hat{x}_{k+2|k} = f(\hat{x}_{k+1|k}) + g(\hat{x}_{k+1|k}) \cdot u_{k+1|k} = f(f(x_k) + g(x_k) \cdot u_{k|k}) + g(f(x_k) + g(x_k) \cdot u_{k|k}) \cdot u_{k+1|k} \tag{22}$$

For most nonlinear f(.) and g(.), the embedding form above makes it impossible to get an analytic solution of u_{k+1} and further future input. So, using reference trajectory, we modified the one-step predictions when $j \geq 2$ as follow:

$$\hat{x}_{k+j|k} = f(w_{k+j-1|k}) + g(w_{k+j-1|k}) \cdot u_{k+j-1} = f(w_{k+j-1|k}) + g(w_{k+j-1|k}) \cdot (u_{k-1} + \sum_{i=0}^{j-1} \Delta u_{k+i|k}) \tag{23}$$

Using the stair-like control strategy, mark $\Delta u_{k|k} = \Delta$, (23) can be transformed as:

$$\hat{x}_{k+j|k} = f(w_{k+j-1|k}) + g(w_{k+j-1|k}) \cdot (u_{k-1} + \sum_{i=0}^{j-1} \beta^i \Delta)$$

$$= f(w_{k+j-1|k}) + g(w_{k+j-1|k}) \cdot u_{k-1} + g(w_{k+j-1|k}) \cdot \sum_{i=0}^{j-1} \beta^i \Delta$$

$$= \hat{x}^1_{k+j|k} + g(w_{k+j-1|k}) \cdot \sum_{i=0}^{j-1} \beta^i \Delta \tag{24}$$

Here, $\hat{x}^1_{k+j|k}$ contains only the known data at time k, while the other part is made up by the increment of future input, thus the unknown data are separated linearly by (24), so the analytic solution of Δ can be achieved.

For $j = 1, 2, \cdots, p$, write the predictions in the form of matrix:

$$\hat{X}_k = \begin{bmatrix} \hat{x}_{k+1|k} \\ \hat{x}_{k+2|k} \\ \vdots \\ \hat{x}_{k+p|k} \end{bmatrix} \quad X^1_k = \begin{bmatrix} \hat{x}^1_{k+1|k} \\ \hat{x}^1_{k+2|k} \\ \vdots \\ \hat{x}^1_{k+p|k} \end{bmatrix} \quad W_k = \begin{bmatrix} w_{k+1|k} \\ w_{k+2|k} \\ \vdots \\ w_{k+p|k} \end{bmatrix} \quad \Delta U_k = \begin{bmatrix} \Delta u_{k|k} \\ \Delta u_{k+1|k} \\ \vdots \\ \Delta u_{k+p-1|k} \end{bmatrix} = \begin{bmatrix} \Delta \\ \beta\Delta \\ \vdots \\ \beta^{p-1}\Delta \end{bmatrix}$$

$$S_k = \begin{bmatrix} g(w_{k|k}) = g(x_k) & 0 & \cdots & 0 \\ g(w_{k+1|k}) & g(w_{k+1|k}) & \cdots & 0 \\ \vdots & \vdots & \ddots & \vdots \\ g(w_{k+p-1|k}) & g(w_{k+p-1|k}) & \cdots & g(w_{k+p-1|k}) \end{bmatrix} = \begin{bmatrix} s_1 & 0 & \cdots & 0 \\ s_2 & s_2 & \cdots & 0 \\ \vdots & \vdots & \ddots & \vdots \\ s_p & s_p & \cdots & s_p \end{bmatrix}$$

$$S_k \cdot \Delta U_k = \begin{bmatrix} s_1 \\ s_2(1+\beta) \\ \vdots \\ s_p(1+\beta+\cdots+\beta^{p-1}) \end{bmatrix} \Delta = \bar{S}_k \cdot \Delta \tag{25}$$

Thus $\hat{X}_k = X_k^1 + S_k \cdot \Delta U_k = X_k^1 + \bar{S}_k \cdot \Delta$, for minimization of traditional quadric objective function $\min_{\Delta} J_k = \min_{\Delta}[(\hat{X}_k - W_k)^T Q(\hat{X}_k - W_k) + \Delta U_k^T R \Delta U_k]$, where semi-positive definite matrix Q and positive definite matrix R are weighting matrices, by $\frac{\partial J_k}{\partial \Delta} = 0$ and $\frac{\partial^2 J_k}{\partial \Delta^2} > 0$, the control solution of multi-step prediction is then obtained. Especially for single input problem, with objective function $\min_{\Delta} J_k = \min_{\Delta}[(\hat{X}_k - W_k)^T(\hat{X}_k - W_k) + r\Delta U_k^T \Delta U_k]$, it is easily denoted as follow:

$$\Delta = \frac{\bar{S}_k^T(W_k - X_k^1)}{\bar{S}_k^T \bar{S}_k + r(1+\beta^2+\cdots+\beta^{2(p-1)})} \tag{26}$$

At last, the instant input $u_{k|k} = u_{k-1} + \Delta$ can be carried out actually. As mentioned in Section 2, and if the model mismatch can be seen as time-invariant in p sample time (usually satisfied in the case of steady state in practice), to maintain the robustness, e_k or $\bar{e}_k$ can be also added to every prediction as mentioned in (17) and (18):

$$\hat{x}_{k+1|k} = f(x_k) + g(x_k) \cdot u_{k|k} + \bar{e}_k$$

$$\hat{x}_{k+j|k} = f(w_{k+j-1|k}) + g(w_{k+j-1|k}) \cdot u_{k+j-1|k} + \bar{e}_k, j = 2,3,\cdots,p \tag{27}$$

Though there are approximate predictions in the novel NMPC which may take in some inaccuracy, the feedback compensation mentioned above and the new optimization process at every sample time will eliminate the error before its accumulation, to keep the efficiency of the algorithm. The constraints also could be handled by methods mentioned Section 2 or by other numerical optimizing algorithm, thus we would not discuss about it here again.

3.4 Multi-step NMPC of the water-tank system

Choose $\alpha = 0.975$, $\beta = 0.5$, $r = 0.005$, $x_{sp} = 60\% \text{or} 30\%$ and predictive horizon $p = 10$ to carry out simulations. Still use the different plant model and predictive model as that of Fig. 5. and Fig. 6. to imitate the model mismatch, the result in Fig. 12. and Fig. 13. shows the efficiency and robustness of this efficient multi-objective NMPC.

Choose $\alpha = 0.975$, $\beta = 0.5$, $r = 0.005$, $x_{sp} = 60\%$ to carry out experiments. Comparing control result between one-step NMPC and multi-step NMPC in Fig. 14. and Fig. 15., we can see the obvious developments on both input and output of the water-tank system when longer predictive horizon is used. It also verifies the efficiency of proposed novel multi-step NMPC algorithm. At last, Fig. 16. is the satisfactory performance of the efficient multi-step NMPC under disturbance (we open an additional outlet valve of the tank for 20 seconds).

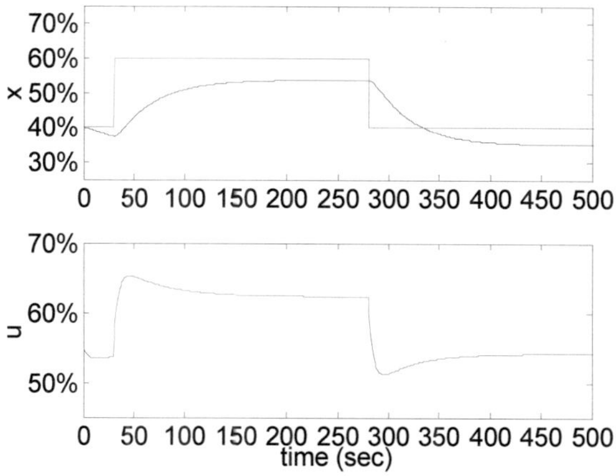

Fig. 12. Simulation of multi-step NMPC with model mismatch but without feedback compensation

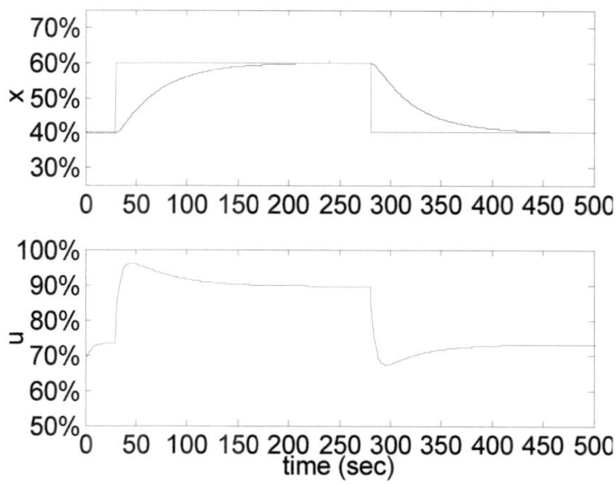

Fig. 13. Simulation of multi-step NMPC with model mismatch and feedback compensation

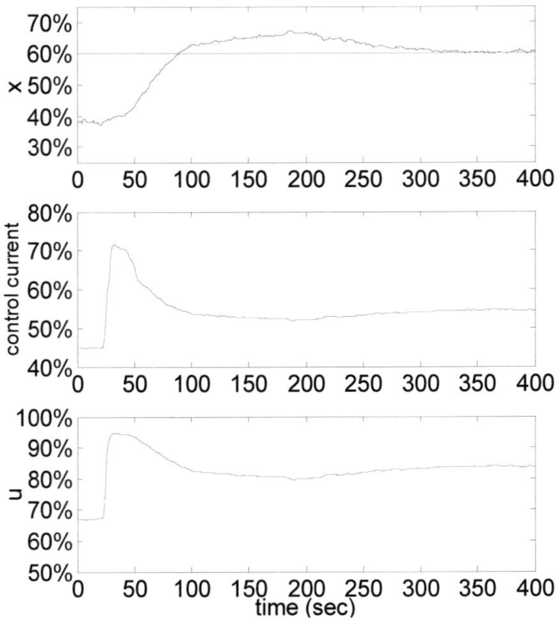

Fig. 14. Experiment of one-step NMPC with setpoint $x_{sp} = 60\%$ (Overshooting exists when setpoint is higher than Fig. 9.)

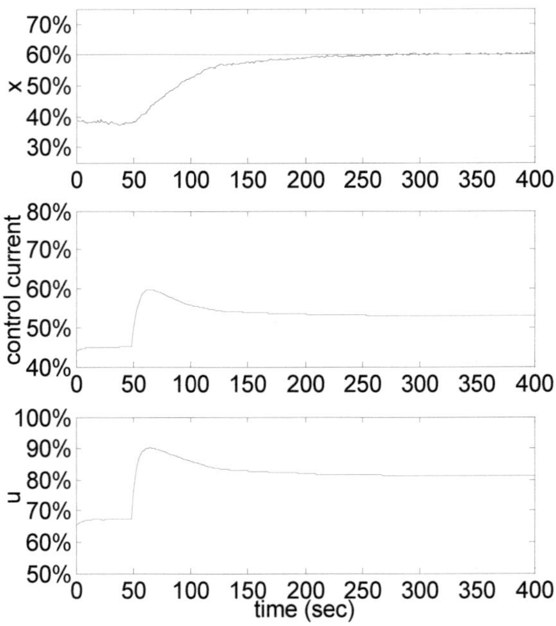

Fig. 15. Experiment of one-step NMPC with setpoint $x_{sp} = 60\%$ (p=10 and and No overshooting)

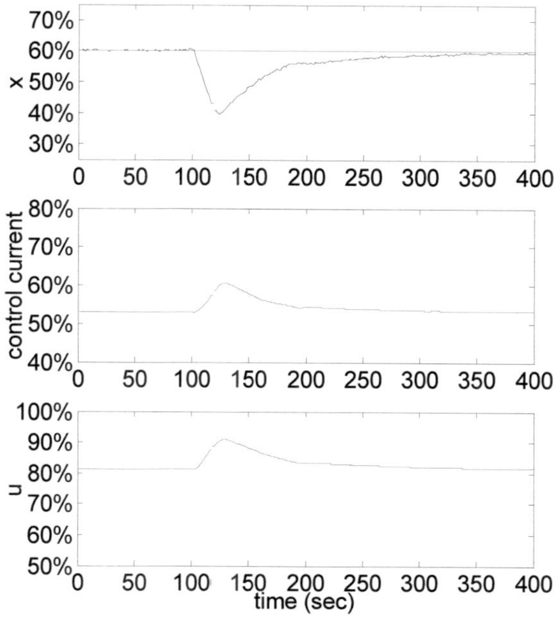

Fig. 16. Experiment of one-step NMPC under disturbance

4. Conclusion and acknowledgement

Using a series of approximate one-step predictions instead of the traditional multi-step prediction, the proposed multi-step NMPC leaded to an analytic result for nonlinear control of affine system. The use of stair-like control strategy caused a very little computational load and the feedback compensation brought robustness of model mismatch to it.

The simulations and experiments verify the practicability and efficiency of this multi-step NMPC for affine system, while the theoretical stability and other analysis will be the future work with considerable value.

This work is supported by National Natural Science Foundation of China (Youth Foundation, No. 61004082) and Special Foundation for Ph. D. of Hefei University of Technology (No. 2010HGBZ0616, from the Fundamental Research Funds for the Central Universities).

5. References

Cannon M. (2004). Efficient nonlinear model predictive control algorithms. *Annual Reviews in Control*, Vol.28, No.2, January, 2004, pp. 229–237, ISSN 1367-5788.

Chen W.; Wu G. (2006). Modeling of nonlinear two-tank system and model predictive control. *Journal of System Simulation*, Vol.18, No.8, August, 2006, pp2078-2081, ISSN 1004-731X

Ferreau H. J.; Lorini G.; Diehl M. (2006). Fast nonlinear model predictive control of gasoline engines. *Proceedings of the 2006 IEEE International Conference on Control Applications*,

ISBN0-7803-9795-9, pp.2754-2759, Munich, Germany, October, 2006, IEEE Publishing, Piscataway, USA.

Qin S. J.; Badgwell T. A. (2003). A survey of industrial model predictive control technology. *Control Engineering Practice*, Vol.11, No.7, July, 2003, pp. 733–764, ISSN 0967-0661.

Yuzgec U. ; Becerikli Y, ; Turker M. (2006). Nonlinear predictive control of a drying process using genetic algorithms. *ISA Transactions*, Vol. 45, No. 4, October, 2006, pp.589-602, ISSN 0019-0578.

Zheng A. (1997). A Computationally Efficient Nonlinear MPC Algorithm. *Proceedings of the American Control Conference*, ISBN 0-7803-3832-4, pp. 1623-1627, Albuquerque, USA, June, 1997, IEEE Publishing, Piscataway, USA.

Zheng T.; Wu G.; He D. F. (2007). Nonlinear Model Predictive Control Based on Lexicographic Multi-objective Genetic Algorithm. *Proceedings of International Conference on Intelligent & Advanced Systems 2007*, ISBN 1-4244-1355-9, pp. 61-65, Kuala Lumpur, Malaysia, November, 2007, IEEE Publishing, Piscataway, USA.

6

Implementation of Multi-dimensional Model Predictive Control for Critical Process with Stochastic Behavior

Jozef Hrbček and Vojtech Šimák
University of Žilina , Faculty of Electrical Engineering, Department of Control and Information Systems, Žilina Slovak Republic

1. Introduction

Model predictive control (MPC) is a control method (or group of control methods) which make explicit use of a model of the process to obtain the control signal by minimizing an objective function. Control law is easy to implement and requires little computation, its derivation is more complex than that of the classical PID controllers. The main benefit of MPC is its constraint handling capacity: unlike most other control strategies, constraints on inputs and outputs can be incorporated into the MPC optimization (Camacho, E., 2004). Another benefit of MPC is its ability to anticipate to future events as soon as they enter the prediction horizon. The implementation supposes good knowledge of system for the purpose of model creation using the system identification. Modeling and identification as a methodology dates back to Galileo (1564-1642), who also is important as the founder of dynamics (Johanson, R., 1993). Identification has many aspects and phases. In our work we use the parametric identification of real system using the measured data from control centre. For the purpose of identification it is interesting to describe the sought process using input-output relations. The general procedure for estimation of the process model consists of several steps: determination of the model structure, estimation of parameters and verification of the model. Finally we can convert the created models to any other usable form. This chapter gives an introduction to model predictive control, and recent development in design and implementation. The controlled object is an urban tunnel tube. The task is to design a control system of ventilation based on traffic parameters, i.e. to find relationship between traffic intensity, speed of traffic, atmospheric and concentration of pollutants inside the tunnel. Nowadays the control system is designed as tetra - positional PID controller using programmable logic controllers (PLC). More information about safety requirements for critical processes control is mentioned in the paper (Ždánsky, J., Rástočný, K. and Záhradník, J., 2008). The ventilation system should be optimized for chosen criteria. Using of MPC may lead to optimize the control way for chosen criteria. Even more we can predict the pollution in the tunnel tube according to appropriate model and measured values. This information is used in the MPC controller as measured disturbances. By introducing predictive control it will be made possible to greatly reduce electric power consumption while keeping the degree of pollution within the allowable limit.

2. Tunnel ventilation system

Emissions from cars are determined not only by the way they are built but also by the way they are driven in various traffic situations. Various gases are emitted by combustion engine. They consist largely of oxides of nitrogen (NO_x), carbon monoxide (CO), steam (H_2O) and particles (opacity). Because of these dangerous gases, it is necessary to provide fresh air in longer tunnels. The fresh air which is used to lower the concentration of CO also serves to improve visibility. The purpose of ventilation is to reduce the noxious fumes in a tunnel to a bearable amount by introducing fresh air. Every tunnel has some degree of natural ventilation. But a mechanical ventilation system should have to be installed. In order to create air stream, jet fans are installed on the ceiling or side walls of the tunnel. The fans take in tunnel air and blow it out at higher speed along the axis of the tunnel. In the tunnel is mounted several fans with about hundred kW each.
The design and industrial implementation of automatic control systems requires powerful and economic techniques together with efficient tools. In order to solve a control problem it is necessary to first describe somehow the dynamic behaviors of the system to be controlled. Traditionally this is done in terms of a mathematical model. Mathematical models are mathematical expressions of essential characteristics of an existing system that describe knowledge about the system in a usable form.
Ventilation control system will base on model and sensors information about NO_x, opacity, velocity, number of vehicles and CO measurements. According to the amounts of pollutants in exhaust gas, air flow driven by the vehicles and degree of pollution inside the tunnel, optimized operation commands will given to the jet fans. "Optimum" means that pollutant concentration is kept within the allowable limit (for CO, 75 ppm or less), and at the same time electric power consumption is minimized. Another criterion is also possible, for example: number of switching the jet fans.

2.1 Tunnel description

The ventilation system in urban tunnel Mrázovka in Prague represents one function unit designed as longitudinal ventilation with a central efferent shaft and protection system avoiding spread of harmful pollutants into the tunnel surround area. Ventilation is longitudinal facing in direction of traffic with air suction at the south opening of the eastern tube (ETT) and at the branch B, with air being transferred at the north opening to the western tunnel tube (WTT) (Pavelka, M. and Přibyl, P., 2006). The task is to design a control system of ventilation based on traffic parameters, i.e. to find relationship between traffic intensity, speed of traffic, atmospheric and concentration of pollutants inside the tunnel. To do that the eastern tunnel tube (ETT) has been chosen as a model example due to principle of mixing polluted air from ETT to WTT (measured concentrations of pollutants in the WTT are also influenced by traffic intensity in the ETT).
To get the required description, the following data has been taken from the tunnel control centre: traffic intensity of trucks and cars, their speed, concentrations of CO (carbon monoxide), NO_x (oxides of nitrogen), OP (opacity-visibility) from the ETT, atmospheric pressure etc. These values are measured by sensors installed inside the tunnel (at five different places of the ETT). Traffic parameters are measured at three places, air flow at three places and concentrations of NO_x and atmospheric pressure in the north portal.
Traffic intensity is sensed by a camera system and the cars are then counted and sorted by categories in database system. More information about monitoring of the traffic can be found in (Pirník, R., Čapka, M. and Halgaš, J., 2010). About the security by transferring the data is discussed in (Holečko, P., Krbilová, I., 2006).

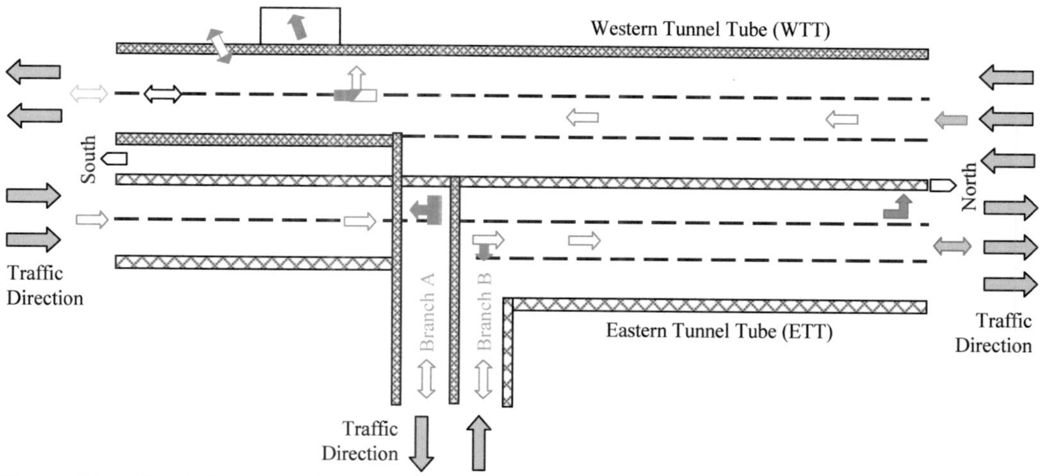

Fig. 1. Simplified diagram of road tunnel (length of 1230m)

3. Mathematical models for MPC

Mathematical models are mathematical expressions of essential characteristics of an existing or designed system that describe knowledge about the system in a usable form. Turbulence inside the tunnel, variety of traffic and atmospherics make the system behavior stochastic. To make the models we use the parametric identification. The MATLABs tools allowed a conversion between several types of models. The main tasks of system identification were the choice of model type and model order. For Single-input Multiple-output discrete time linear systems we can write the metrics equation for jet fan characteristics:

$$\begin{bmatrix} y_1(k) \\ y_2(k) \\ y_2(k) \end{bmatrix} = \begin{bmatrix} G_{11}(k) \\ G_{21}(k) \\ G_{31}(k) \end{bmatrix} \cdot [u_1(k)] \tag{1}$$

The "Jet Fan Model" is a model in the MPC format that characterizes effect of ventilator on CO concentration, NO_x concentration and visibility (opacity). It is a system with 1 input (u) and 3 outputs (dilution of CO-Out1, NO_x concentration-Out2 and opacity OP-Out3). One of the main advantages of predictive control is incorporation of limiting conditions directly to the control algorithm. The "Jet Fan Model" characteristics are shown in Fig. 2.
The "Disturbance Model" is a model of the tunnel tube in state space representation:

$$\begin{aligned} \bar{\mathbf{x}}'(t) &= \bar{\mathbf{A}}(t)\cdot\bar{\mathbf{x}}(t)+\bar{\mathbf{B}}(t)\cdot\bar{\mathbf{u}}(t) \\ \bar{\mathbf{y}}(t) &= \bar{\mathbf{C}}(t)\cdot\bar{\mathbf{x}}(t)+\bar{\mathbf{D}}(t)\cdot\bar{\mathbf{u}}(t) \end{aligned} \tag{2}$$

This model is used to predict the pollutions inside the tunnel tube. These data enter to the measured disturbances input (MD) of MPC controller for the purpose of switching the jet fans before the limit will be exceeded.
Created model was validated by several methods (Hrbček, J., 2009). The purpose of model validation is to verify that identified model fulfills the modeling requirements according to

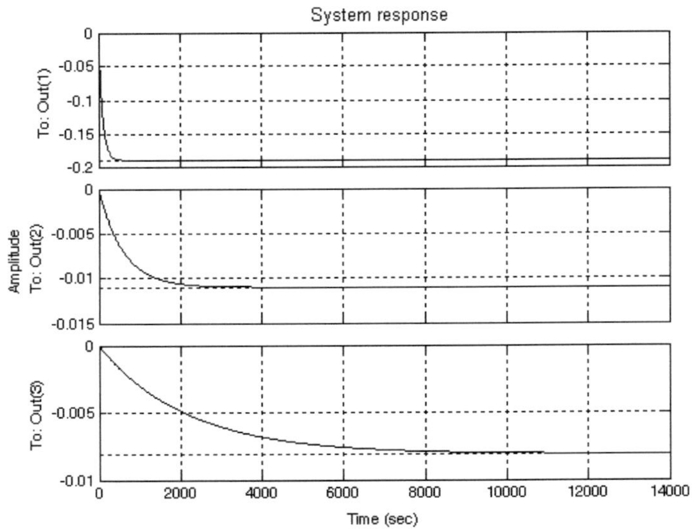

Fig. 2. Jet fan characteristics

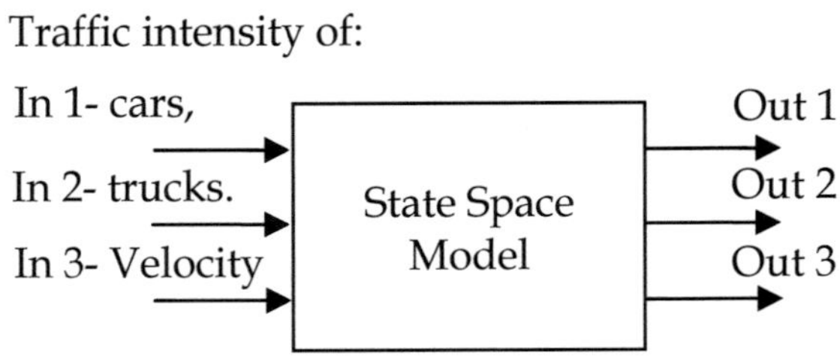

Fig. 2. Model of the tunnel tube. Out 1 - Concentration of CO, Out 2 -OP-opacity inside the tunnel, Out 3 - Concentration of NOx

subjective and objective criteria of good model approximation. In method "Model and parameter accuracy" we compare the model performance and behavior with real data. A deterministic simulation can be used, where real data are compared with the model response to the recorded input signal used in the identification. This test should ascertain whether the model response is comparable to real data in magnitude and response delay.

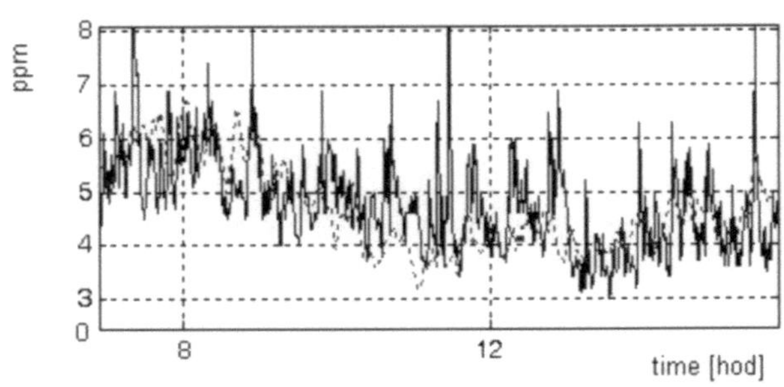

Fig. 3. Model and parameter accuracy test for CO concentrations. Measured data is shown by black line and simulated data are shown by gray dashed line.

This method showed graphically accuracy between simulated values and measured values. Although the simulated and measured data not fit precisely, this result is sufficient for most stochastic system like pollution inside the road tunnel.
An Akaike Final Prediction Error (FPE) for estimated model was also determinate. The average prediction error is expected to decrease as the number of estimated parameters increase. One reason for this behavior is that the prediction errors are computed for the data set that was used for parameter estimation. It is now relevant to ask what prediction performance can be expected when the estimated parameters are applied to another data set. This test shows the flexibility of the model structure. We are looking for minimum value of FPE coefficient.

4. Model predictive control of ventilation system

Under the term "Model Predictive Control" we understand a class of control methods that have certain characteristic features. MPC refers to a class of computer control algorithms that utilize an explicit process model to predict the future response of a plant. From this model the future behaviour of the system is predicted over a finite time interval, usually called prediction horizon, starting at the current time t.

4.1 The basic idea of predictive control

The receding horizon strategy is shown in Fig. 4.

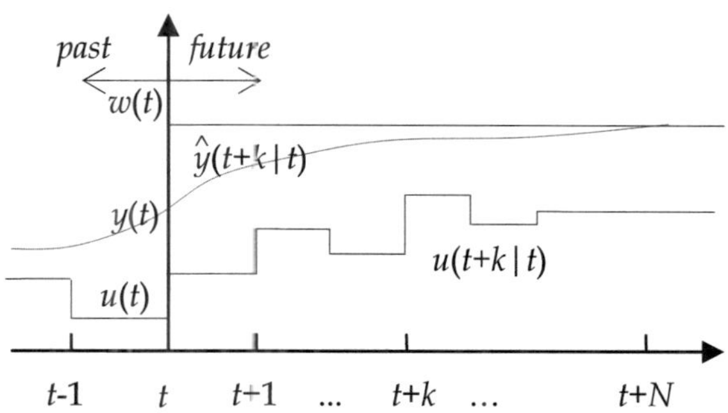

Fig. 4. The receding horizon strategy, the basic idea of predictive control.

The future outputs for a determined horizon N, called the prediction horizon, are predicted at each instant t using the process model. These predicted outputs $y(t + k \mid t)$ 1 for $k = 1 \ldots N$ depend on the known values up to instant t (past inputs and outputs) and on the future control signals $u(t+k \mid t)$, k = 0... N-1, which are those to be sent to the system and calculated.
The set of future control signals is calculated by optimizing a determined criterion to keep the process as close as possible to the reference trajectory $w(t + k)$ (which can be the setpoint itself or a close approximation of it). This criterion usually takes the form of a quadratic function of the errors between the predicted output signal and the predicted reference trajectory. The control effort is included in the objective function in most cases. An explicit solution can be obtained if the criterion is quadratic, the model is linear, and there are no constraints; otherwise an iterative optimization method has to be used. Some assumptions about the structure of the future control law are also made in some cases, such as that it will be constant from a given instant.

The control signal $u(t\,|\,t)$ is sent to the process whilst the next control signals calculated are rejected, because at the next sampling instant $y(t+1)$ is already known and step 1 is repeated with this new value and all the sequences are brought up to date. Thus the $u(t+1\,|\,t+1)$ is calculated (which in principle will be different from the $u(t+1\,|\,t)$ because of the new information available) using the receding horizon concept (Camacho, E. F., Bordons, C., 2004). The notation indicates the value of the variable at the instant $t + k$ calculated at instant t.

4.2 Application to real system

Tunnel ventilation is expected to fulfil the following requirements at least (Godan, J. at all. 2001):

- Concentration of emissions in the tunnel kept within the acceptable limits for the monitored harmful pollutants, in consideration of time spent by persons inside the tunnel;
- Good visibility for through passage of vehicles under polluted air inside the tunnel;
- Reduction of effects of smoke and heat on persons in the case of vehicle fire;
- Regulation of dispersion of pollutants in the air caused by petrol fumes from vehicles into the surround environment of the tunnel.

Model Predictive Control (MPC) comes from the late seventies when it became significantly developed (Camacho, E. F., Bordons, C., 2004) and several methods were defined. In this work we have applied the Dynamic Matrix Control (DMC) method which is one of the most spread approaches and creates the base of many commercially available MPC products. It is based on the model obtained from the real system:

$$y(k) = \sum_{i=1}^{N} h_i u(k-i), \tag{3}$$

where h_i are FIR (Finite Impulse Response) coefficients of the model of the controlled system. Predicted values may be expressed:

$$\begin{aligned}\hat{y}(n+k\,|\,n) &= \sum h_i \Delta u(n+k-i) + \hat{d}(n+k\,|\,n) = \\ &= \sum h_i \Delta u(n+k-i) + \sum h_i \Delta u(n+k-i) + \hat{d}(n+k\,|\,n)\end{aligned} \tag{4}$$

We assume that the additive failure is constant during the prediction horizon:

$$\hat{d}(n+k\,|\,n) = \hat{d}(n\,|\,n) = y_m(n) - \hat{y}(n\,|\,n) \tag{5}$$

Response can be decomposed to the component depending on future values of control and to the component determined by the system state in time n:

$$\hat{y}(n+k\,|\,n) = \sum h_i \Delta u(n+k-i) + f(n+k), \tag{6}$$

where $f(n+k)$ is that component which does not depend on future values of action quantity:

$$f(n+k) = y_n(n) - \sum (h_{k+i} - h_i)\Delta u(n-i). \tag{7}$$

Predicted values within the prediction horizon p (usually $p>>N$) can be arranged to the relation (8):

$$\begin{aligned} \hat{y}(n+1|n) &= h_1\Delta u(n) + f(n+1) \\ \hat{y}(n+2|n) &= h_2\Delta u(n) + h_1\Delta u(n+1) + f(n+2) \\ &\dots \\ \hat{y}(n+p|n) &= \sum_{i=p-m+}^{p} h_i\Delta u(n+p-i) + f(n+p) \ , \end{aligned} \tag{8}$$

where the prediction horizon is k=1...p, with respect to m control actions. Regulation circuit is stable if the prediction horizon is long enough. The values may be arranged to the dynamic matrix **G**:

$$G = \begin{vmatrix} h_1 & 0 & \dots & 0 \\ h_2 & h_1 & \dots & 0 \\ \vdots & \vdots & & \vdots \\ h_p & h_{p-1} & \dots & h_{p-m+1} \end{vmatrix} , \tag{9}$$

and expression used for prediction can be written in the matrix form:

$$\hat{y} = Gu + f \ , \tag{10}$$

where $\hat{y}$ is a vector of contributions of action quantity and f are free responses.
The MATLAB's Model Predictive Control Toolbox uses linear dynamic modeling tools. We can use transfer functions, State-space matrices, or its combination. We can also include delays, which are in the real system. The model of the plant is a linear time-invariant system described by the equations:

$$\begin{aligned} x(k+1) &= Ax(k) + B_u u(k) B_v v(k) + B_d d(k) \\ y_m(k) &= C_m x(k) + D_{vm} v(k) + D_{dm} d(k) \\ y_u(k) &= C_u x(k) + D_{vu} v(k) + D_{du} d(k), \end{aligned} \tag{11}$$

where $x(k)$ is the n_x -dimensional state vector of the plant, $u(k)$ is the n_u-dimensional vector of manipulated variables (MV), i.e., the command inputs, $v(k)$ is the n_v-dimensional vector of measured disturbances (MD), $d(k)$ is the n_d-dimensional vector of unmeasured disturbances (UD) entering the plant, y_m (k) is the vector of measured outputs (MO), and $y_u(k)$ is the vector of unmeasured outputs (UO). The overall n_y-dimensional output vector $y(k)$ collects $y_m(k)$ and $y_u(k)$. In the above equations $d(k)$ collects both state disturbances ($B_d \neq 0$) and output disturbances ($D_d \neq 0$).
The unmeasured disturbance $d(k)$ is modeled as the output of the linear time invariant system:

$$\begin{aligned} x_d(k+1) &= \bar{A}x_d(k) + \bar{B}n_d(k) \\ d(k) &= \bar{C}x_d(k) + \bar{D}n_d(k). \end{aligned} \tag{12}$$

The system described by the above equations is driven by the random Gaussian noise $n_d(k)$, having zero mean and unit covariance matrix. For instance, a step-like unmeasured disturbance is modeled as the output of an integrator. In many practical applications, the matrices **A**, **B**, **C**, **D** of the model representing the process to control are obtained by linearizing a nonlinear dynamical system, such as

$$\begin{aligned} x' &= f(x,u,v,d) \\ y &= h(x,u,v,d). \end{aligned} \tag{13}$$

at some nominal value $x=x_0$, $u=u_0$, $v=v_0$, $d=d_0$. In these equations x' denotes either the time derivative (continuous time model) or the successor $x(k+1)$ (discrete time model).
The MPC control action at time k is obtained by solving the optimization problem:

$$\begin{aligned} &\min\Big\{ \Delta u(k\mid k),...,\Delta u(m-1+k\mid k),\ \varepsilon\Big[\sum_{i=0}^{p-1}\Big(\sum_{j=1}^{n_z}\Big|w^{u}_{i+1,j}(y_j(k+i+1\mid k)-r_j(k+i+1))\Big|^2 \\ &+\sum_{i=1}^{n_u}\Big|w^{\Delta u}_{i,j}\Delta u_j(k+i\mid k)\Big|^2+\sum_{i=1}^{n_u}\Big|w_{i,j}(u_j(k+i\mid k)-u_{j\text{target}}(k+i))\Big|^2\Big)+\rho_\varepsilon\varepsilon^2\Big]\Big\}, \end{aligned} \tag{14}$$

where the subscript "( )*j*" denotes the *j*-th component of a vector, "$(k+i\mid k)$" denotes the value predicted for time $k+i$ based on the information available at time k; $r(k)$ is the current sample of the output reference, subject to

$$\begin{aligned} &u_{j\min}(i)-\varepsilon V^{u}_{j\min}(i)\le u_j(k+i\mid k)\le u_{j\max}(i)+\varepsilon V^{u}_{j\max}(i); \\ &\Delta u_{j\min}(i)-\varepsilon V^{\Delta u}_{j\min}(i)\le \Delta u_j(k+i\mid k)\le \Delta u_{j\max}(i)+\varepsilon V^{\Delta u}_{j\max}(i); \\ &y_{j\min}(i)-\varepsilon V^{y}_{j\min}(i)\le y_j(k+i+1\mid k)\le y_{j\max}(i)+\varepsilon V^{y}_{j\max}(i), \\ &\Delta u(k+h\mid k)=0, \\ &\textit{where} \\ &i=0,...,p-1, \\ &h=m,...,p-1, \\ &\varepsilon\ge 0, \end{aligned} \tag{15}$$

with respect to the sequence of input increments $\{\Delta u(k|k),...,\Delta u(m-1+k|k)\}$ and to the slack variable ε, and by setting $u(k)=u(k-1)+\Delta u(k\mid k)$, where $\Delta u(k\mid k)$ is the first element of the optimal sequence. Note that although only the measured output vector $y_m(k)$ is fed back to the MPC controller, $r(k)$ is a reference for all the outputs. When the reference r is not known in advance, the current reference $r(k)$ is used over the whole prediction horizon, namely $r(k+i+1)=r(k)$ in Equation 14.
In Model Predictive Control the exploitation of future references is referred to as anticipative action (or look-ahead or preview). A similar anticipative action can be performed with respect to measured disturbances $v(k)$, namely $v(k+i)=v(k)$ if the measured disturbance is not known in advance (e.g. is coming from a Simulink block) or $v(k+i)$ is obtained from the workspace. In the prediction, $d(k+i)$ is instead obtained by setting $n_d(k+i)=0$. The $w^{\Delta u}ij$, $w^{u}ij$, $w^{y}ij$, are nonnegative weights for the corresponding variable. The smaller w, the less important is the behavior of the corresponding variable to the overall performance index. And $u_{j,min}$, $u_{j,max}$, $\Delta u_{j,min}$, $\Delta u_{j,max}$, $y_{j,min}$, $y_{j,max}$ are lower/upper bounds on the corresponding variables. The constraints on u, Δu, and y are relaxed by introducing the slack variable $\varepsilon\ge 0$. The weight ρ_ε on the slack variable ε

penalizes the violation of the constraints. The larger $\rho\varepsilon$ with respect to input and output weights, the more the constraint violation is penalized. The Equal Concern for the Relaxation vectors $V^{u}_{min}, V^{u}_{max}, V^{\Delta u}_{min}, V^{Du}_{max}, V^{y}_{min}, V^{y}_{max}$ have nonnegative entries which represent the concern for relaxing the corresponding constraint; the larger V, the softer the constraint. V=0 means that the constraint is a hard one that cannot be violated (Bemporad A., Morari M., N. Lawrence Ricker., 2010).

4.3 Constraints

In many control applications the desired performance cannot be expressed solely as a trajectory following problem. Many practical requirements are more naturally expressed as constraints on process variables. There are three types of process constraints: Manipulated Variable Constraints: these are hard limits on inputs $u(k)$ to take care of, for example, valve saturation constraints; Manipulated Variable Rate Constraints: these are hard limits on the size of the manipulated variable moves $\Delta u(k)$ to directly influence the rate of change of the manipulated variables; Output Variable Constraints: hard or soft limits on the outputs of the system are imposed to, for example, avoid overshoots and undershoots (Maciejovski, J.M., 2002). We use the Output constraints and Manipulated Variable Constraints.

5. Simulation in MATLAB

Models of the tunnel and ventilator have been obtained through identification of real equipments. Higher traffic intensity causes increase of pollutant concentrations in the tunnel. This intensity is expressed as a vector containing really measured data. The MATLAB environment is used to simulate behavior of the system according to the Fig. 5.

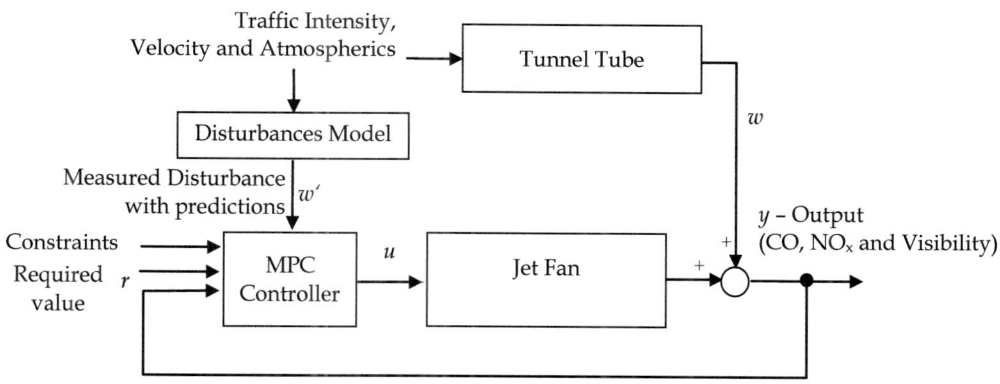

Fig. 5. Improved MPC of multi-dimensional ventilation system

It is a closed-loop control (regulation) of the system with limitations imposed to control quantity and outputs. It uses the internal model and solves optimization problem with the use of quadratic programming. We can choose the prediction horizon P and the control horizon M. The output constraints were set to 6, because this is the maximum input for three pairs of jet fans corresponding with real system. Weight matrix is selected as a diagonal matrix, with each element weighting the corresponding control signal. For instance, if the influence of particular control is to be reduced, then the corresponding diagonal element will be increased to reflect this intention. Weight tuning is the essential task to set the controller. In Fig. 7 we can see the results in comparison to Fig. 6.

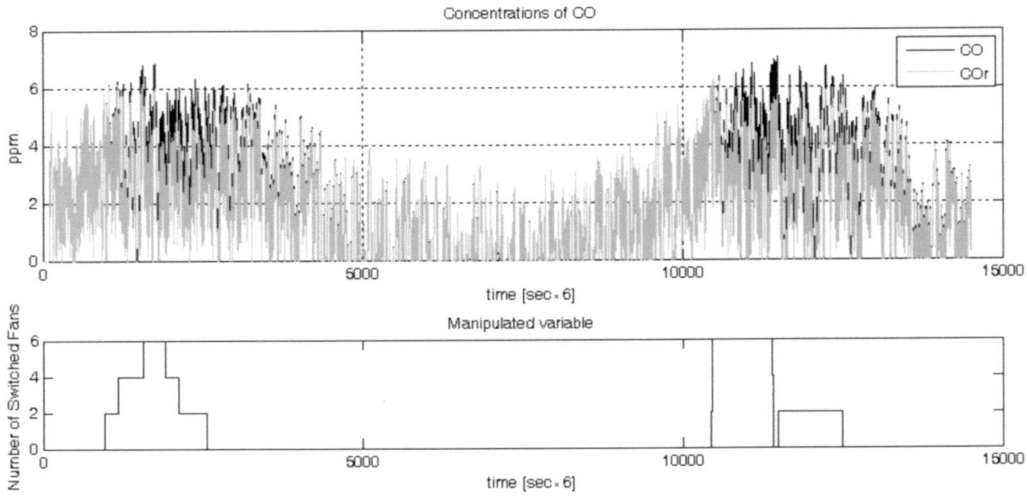

Fig. 6. Simulated values of CO pollution inside the tunnel with (grey line) and without (black line) MPC controller and fan number (number of acting jet fans)

5.1 Simulation results

The presented simulation results are obtained for the following concentration limits: 6 ppm for CO concentrations, 0.02 ppm for NOx concentrations and 0.05 ppm for visibility concentrations. These values are below really defined maximum limits. According to the curve of the output quantity (Fig. 6) it is apparent, that no emission value has extended the defined limit. However, the value of under-set maximum limit may be extended since one ventilator need not be able to dilute CO concentration sufficiently.

The abbreviation ppm is a way of expressing very dilute concentrations of substances. Just as per cent means out of a hundred, so parts per million or ppm means out of a million. It describes the concentration of something in air.

For this simulation we have six acting jet fans in this part of road tunnel. In the next simulations we have used a possibility to set weighing matrices (*uwt*) for tuning the controller. The control quantity u is adapted to the input of Jet Fan control unit. The black lines represent the concentrations of pollution without using the controller. They are named CO. The grey lines represent the concentrations of pollution with using the controller. They are named COr. Opacity and concentration of NO_x is below the dangerous limits. The jet fans were switched on two times per day for chosen limits. We can see how affect the ventilation system to reduce the pollution. In this paper we pointed out only to concentration of CO, because this type of pollution is most dangerous for human organism.

As it was mentioned in the previous section the weight tuning is also important part of controller creation.

Well tuned controller leads to optimal control. After changing the weights, the jet fans were switched on only once per day, furthermore the next day all the fans were not switched on in the same conditions.

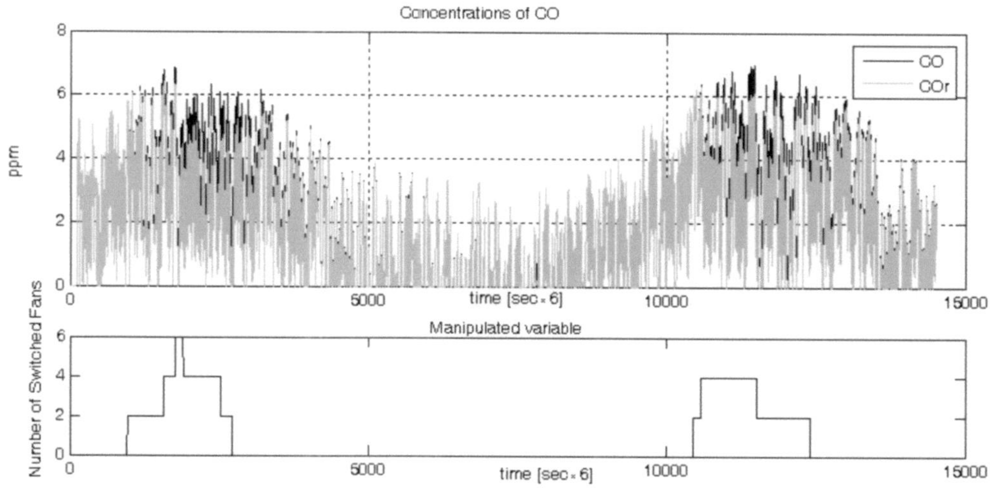

Fig. 7. Weight tuning. Simulated values of CO pollution inside the tunnel with (grey line) and without (black line) MPC controller and fan number (number of acting jet fans)

Opacity and concentration of NO_x is below the dangerous limits. The jet fans were switched on once per day for chosen limits. The concentrations of NO_x and opacity (OP) are shown in Fig. 8.

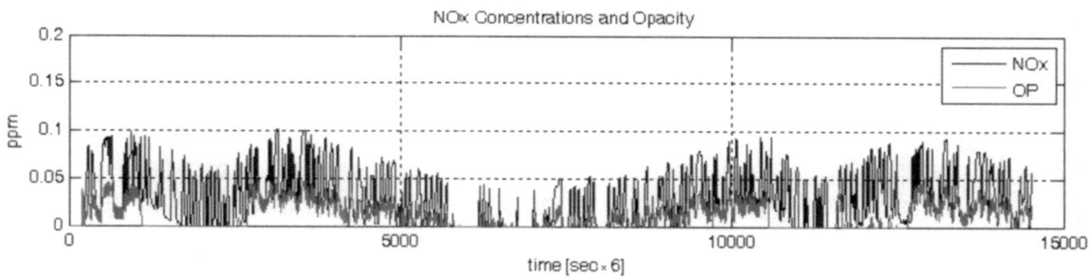

Fig. 8. Simulated values of NO_x concentrations inside the tunnel tube (black line) and opacity (grey line)

For this simulation the NO_x concentrations and opacity was below defined maximum limits. When the jet fans are switched on these pollutions are also decreased.

5.2 Implementation

The biggest advantage of Automatic Code Generation affects those developers who already use MATLAB and Simulink for simulation and solutions design and to developers who used to tediously rework implemented structures in a language supported by Automation Studio in the past. In the procedures listed below the Automatic Code Generation tool provided by B&R represents an innovation with endless possibilities that help to productively reform the development of control systems. The basic principle is simple: The module created in Simulink is automatically translated using Real-Time Workshop and Real-Time Workshop Embedded Coder into the optimal language for the B&R target system guaranteeing maximum performance of the generated source code. Seamless integration into an Automation Studio project makes the development process perfect (B&R Automation Studio Target for Simulink. 2011). Since the tunnel ventilation system use programmable logic controllers (PLC) it is suitable for real implementation. In our department we have appropriate equipment for this solution.

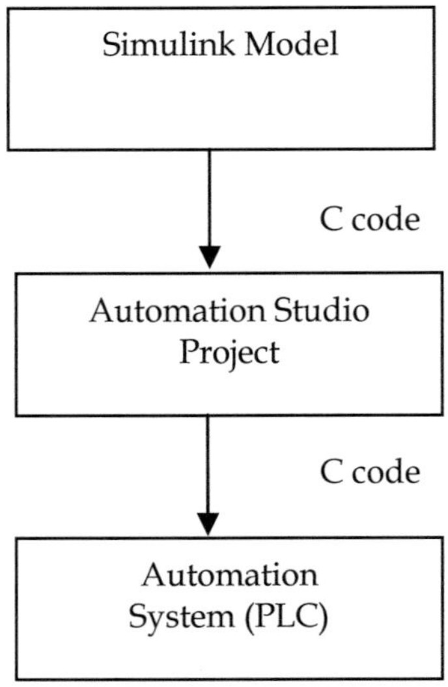

Fig. 9. Workflow of the Automatic Code Generation

The elimination of extensive reengineering in Automation Studio allows simple transfer of complex and sophisticated Simulink models to the PLC (Hardware-in-the-Loop). Closed-loop controllers can also be easily tested and optimized on the target system without requiring the user to adjust large amounts of code and run the risk of creating coding errors (Rapid Prototyping). Rapid prototyping: Automatic Code Generation makes it possible to quickly and easily transform sophisticated Simulink based control systems into source code and integrate them into an Automation Studio project. Many potentially successful ideas have been immediately rejected due to the large amount of time required for conversion into executable machine code and the risk of developing a dead end solution.

The "Rapid Prototyping" concept brings an end to this. Using Simulink and the Automatic Code Generation tool provided by B&R, any system, no matter how complex, can be intuitively built, compiled and tested in a short amount of time. This practically eliminates implementation errors as the Automatic Code Generation tool has been well-proven over several years in critical fields like aviation or automotive industry (B&R Automation Studio Target for Simulink. 2011). Nowadays the control algorithm is implemented and awaiting for connection to the real system. Fig. 11. shows the model in Simulink.

We created the model in Simulink according to model for simulations. We replaced the simulated inputs by "B&R IN" blocks and simulated output by the "B&R OUT" block. The Real-time Workshop provides utilities to convert the SIMULINK embedded models in C code and then, with the compiler, compile the code into a real-time executable file. Although the underlying code is compiled into a real-time executable file via the C compiler, this conversion is performed automatically without much input from the user. The concept in

Fig. 10. shows that a simulation model can be used in the simulation testing of the predictive control system, and after completing the test, then with simple modification to the original Simulink programs, the same real-time predictive control system can be connected to the actual plant for controlling the plant.

Fig. 10. The control system

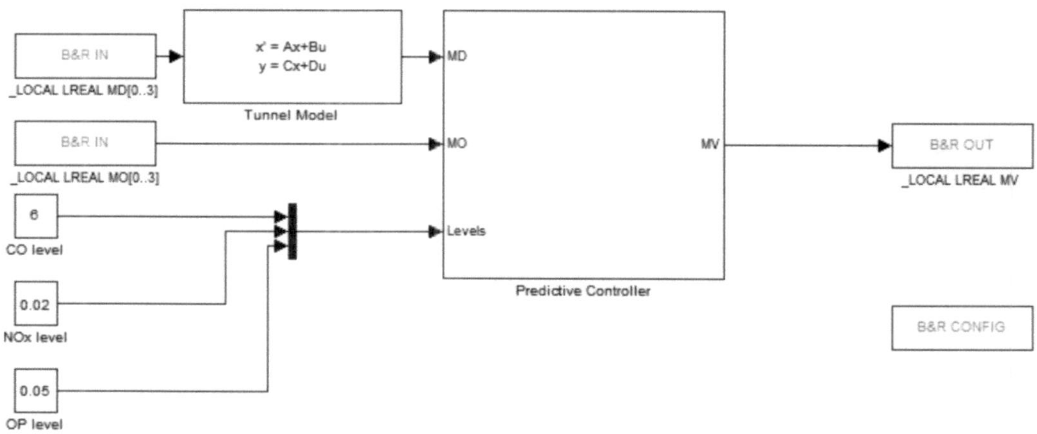

Fig. 11. The control system in Simulink

6. Conclusion

The paper presents a methodology that has been used for design parametric models of the road tunnel system. We needs identification of system based on data obtained from the real ventilation system. Model from one week data has been created and verified in MATLAB environment. This part is the ground for best design of ventilation control system. Presented results point out that created model by identification method should be validate by several method. Model of a three-dimensional system has been created and simulated in MATLAB environment using the predictive controller. Presented results confirm higher effectiveness of predictive control approach. The weight tuning is important part of controller creation as the simulation results had proved. The predictive controller was successfully implemented to programmable logic controller.

7. Acknowledgment

This paper was supported by the operation program "Research and development" of ASFEU-Agency within the frame of the project "Nové metódy merania fyzikálnych dynamických parametrov a interakcií motorových vozidiel, dopravného prúdu a vozovky". ITMS 26220220089 OPVaV-2009/2.2/03-SORO co financed by European fond of regional development. "We support research activities in Slovakia / The project is co financed by EU".

8. References

Camacho, E. F., Bordons, C. (2004). *Model Predictive Control*. 2nd ed., Springer-Verlag LondonLimited, 405 p. ISBN 1-85233-694-3

Maciejovski, J.M. (2002). *Predictive Control with constrains*, Prentice Hall, 331p, ISBN: 0-201-39823-0

Ždánsky, J., Hrbček, J., Zelenka, J. (2008) Trends in Control Area of PLC Reliability and Safety Parameters, *ADVANCES in Electrical and Electronic Engineering*, ISSN 1336-1376, Vol.7/2008, p. 239-242

Johanson, R. (1993). *System modeling and identification*, Prentice-Hall, 512p., ISBN: 0-13-482308-7

Hrbček, J. and Janota, A. (2008). Improvement of Road Tunnel Ventilation through Predictive Control, *Communications*, 2/2008, p.15-19, ISSN: 1335-4205

Pavelka, M.- Přibyl, P. (2006). *Simulation of Air Motion and pollutions inside the Road Tunnel– Mathematical Model. OPTUN 228/06-EEG*

EN 61508. (2002). Functional safety of electrical/electronic/programmable electronic safety-related systems.

Bemporad A., Morari M. N. Lawrence Ricker. (2010). Model Predictive Control Toolbox. Available from: http://www.mathworks.com/help/toolbox/mpc

Vojtech Veselý and Danica Rosinová. (2010). Robust Model Predictive Control Design, In: *Model Predictive Control,* 87-108, Scyio, ISBN 978-953-307-102-2, Available from http://www.intechopen.com/books/show/title/model-predictive-control

Godan, J. at all. (2001). *Tunnels, Road Tunnels and Railway Tunnels*, p. 202. 135882/p-UK/Z

Liuping Wang. (2009) *Model Predictive Control System Design and Implementation Using MATLAB*, Springer, p. 371, ISBN 978-1-84882-330-3

Hrbček, J. (2009). Parametric Models of Real System and the Validation Tools. *Proceeding of 8th European Conference of Young Research and Science Workers in Transport and Communications technology*. p. 93-96. Žilina: June, 22. - 24. 2009. ISBN 978-80-554-0027-3

Pirník, R., Čapka, M. and Halgaš, J. (2010). Non-invasive monitoring of calm traffic. *Proceedings of international symposium on advanced engineering & applied management – 40th anniversary in higher education*, CD ver. S. II-107-112, ISBN 978-973-0-09340-7

B&R Automation Studio Target for Simulink. (2011). Available from: http://www.mathworks.com/products/connections/product_detail/product_35953.html

Bubnicki, Z. (2005). *Modern Control Theory*. Springer, 2005, 422 p., ISBN 3-540-23951-0

Tammi, K. (2007). *Active control of radial rotor vibrations*, Picaset Oy, Helsinki 2007, VTT, ISBN 987-951-38-7007-2

Hrbček, J., Spalek J. and Šimák, V. (2010). Process Model and Implementation the Multivariable Model Predictive Control to Ventilation System. *Proceeding of 8th International Symposium on Applied Machine Intelligence and Informatics*, CD, p. 211-214, Herľany, Slovakia, January 28-30, 2010, ISBN 978-1-4244-6423-4

Rossiter J. A. (2003). *Model-Based Predictive Control: A Practical Approach*. Crc press, 318 p., ISBN 0-8493-1291-4

Holečko, P., Krbilová, I. (2006). IT Security Aspects of Industrial Control Systems. *Advances in Electrical and Electronic Engineering*, No. 1-2 Vol. 5/2006, pp. 136-139, ISSN 1336-1376

Ždánsky, J., Rástočný, K. and Záhradník,J. (2008). Problems Related to the PLC Application in the Safety Systems. *Trudy rostovskogo gosudarstvennogo universiteta putej soobščenija*, No. 2(6) 2008, pp. 109–116, ISSN 1818–5509

Lewis, P. - Yang, Ch. (1997). Basic Control Systems Engineering, 1997, Prentice-Hall, ISBN 0-13-597436-4

Přibyl, P., Janota, A. and Spalek, J. (2008). *Analýza a řízení rizik v dopravě. Pozemní komunikace a železnice. (Analysis and risk control in transport. Highway and railway)*. BEN Praha, ISBN 80-7300-214-0

Noskievič, P. (1999). *Modelování a identifikace systémů. (Systems modeling and identification).*, Ostrava: MONTANEX, a.s., 1999, 276 p., ISBN 80-7225-030-2

Balátě, J. (2004). *Automatické řízení. (Automation control)*. BEN, Praha, 2004, ISBN 80-7300-148-9

Zelenka, J. - Matejka, T. (2010). The application of discrete event systems theory for the real manufacturing system analysis and modeling. *Proceedings of the conference of cybernetics and informatics 2010*, Vyšná Boca 2010, ISBN 978-80- 227-3241-3

Yinghua He, Hong Wang, and Bo Zhang. (2004). Color-Based Road Detection in Urban Traffic Scenes, IEEE Transactions on intelligent transportation systems, vol. 5, no. 4, december 2004. Available from:
http://ieeexplore.ieee.org/stamp/stamp.jsp?tp=&arnumber=1364007

Harsányi L., Murgaš J., Rosinová D., Kozáková A. (1998). *Teória automatického riadenia.(Automation system theory)*. STU Bratislava, Bratislava 1998, ISBN 80-227-1098-9

Hrbček, J. (2007). *Active Control of Rotor Vibration by Model Predictive Control - a simulation study*. Report 153, Picaset Oy, Helsinki 2007, ISSN: 0356-0872, ISBN: 978-951-22-8824-3

Ma Y., Soatto S., Košecká J., Sastry S. S. (2004). *An Invitation to 3-D Vision - From Images to Geometric Models*. Springer - Verlag New York, Inc., New York 2004, ISBN 978-0387-00893-6

Šimák, V., Hrbček, J., Pirník, R. (2010). Traffic flow videodetection, *International conference KYBERNETIKA A INFORMATIKA ´10, Vyšná Boca,* Slovakia, February 10-13, 2010, ISBN 978-80-227-3241-3

7

Fuzzy–neural Model Predictive Control of Multivariable Processes

Michail Petrov, Sevil Ahmed, Alexander Ichtev and Albena Taneva
Technical University Sofia, Branch Plovdiv/Control Systems Department
Bulgaria

1. Introduction

Predictive control is a model-based strategy used to calculate the optimal control action, by solving an optimization problem at each sampling interval, in order to maintain the output of the controlled plant close to the desired reference. Model predictive control (MPC) based on linear models is an advanced control technique with many applications in the process industry (Rossiter, 2003). The next natural step is to extend the MPC concept to work with nonlinear models. The use of controllers that take into account the nonlinearities of the plant implies an improvement in the performance of the plant by reducing the impact of the disturbances and improving the tracking capabilities of the control system.

In this chapter, Nonlinear Model Predictive Control (NMPC) is studied as a more applicable approach for optimal control of multivariable processes. In general, a wide range of industrial processes are inherently nonlinear. For such nonlinear systems it is necessary to apply NMPC. Recently, several researchers have developed NMPC algorithms (Martinsen et al., 2004) that work with different types of nonlinear models. Some of these models use empirical data, such as artificial neural networks and fuzzy logic models. The model accuracy is very important in order to provide an efficient and adequate control action. Accurate nonlinear models based on soft computing (fuzzy and neural) techniques, are increasingly being used in model-based control (Mollov et al., 2004).

On the other hand, the mathematical model type, which the modelling algorithm relies on, should be selected. State-space models are usually preferred to transfer functions, because the number of coefficients is substantially reduced, which simplifies the computation; systems instability can be handled; there is no truncation error. Multi-input multi-output (MIMO) systems are modelled easily (Camacho et al., 2004) and numerical conditioning is less important.

A state-space representation of a Takagi-Sugeno type fuzzy-neural model (Ahmed et al., 2010; Petrov et al., 2008) is proposed in the Section 2. This type of models ensures easier description and direct computation of the gradient control vector during the optimization procedure. Identification procedure of the proposed model relies on a training algorithm, which is well-known in the field of artificial neural networks.

Obtaining an accurate model is the first stage of the of the NMPC predictive control strategy. The second stage involves the computation of a future control actions sequence. In order to obtain the control actions, a previously defined optimization problem has to be solved. Different types of objective and optimization algorithms (Fletcher, 2000) can be used

in the optimization procedure. Two different approaches for NMPC are proposed in Section 3. They consider the unconstrained and constrained model predictive control problem. Both of the approaches use the proposed Takagi-Sugeno fuzzy-neural predictive model.
The proposed techniques of fuzzy-neural MPC are studied in Section 4, by experimental simulations in Matlab® environment in order to control the levels in a multi tank system (Inteco, 2009). The case study is capable to show how the proposed NMPC algorithms handle multivariable processes control problem.

2. Multivariable fuzzy-neural predictive model

The Takagi-Sugeno fuzzy-neural models are powerful modelling tools for a wide class of nonlinear systems. Fuzzy reasoning is capable of handling uncertain and imprecise information while neural networks can learn from samples. Fuzzy-neural networks combine the advantages of both artificial intelligent techniques and incorporate them in adaptive features. Those futures, based on a real time learning algorithm are the main advantage of the fuzzy-neural models.
The importance of the used in MPC strategy models and their adaptive characteristics is obvious. The accuracy of the model determines the accuracy of the control action. The proposed fuzzy-neural model is implemented in a classical NMPC scheme (Fig. 1) as a predictor (Camacho et al., 2004).

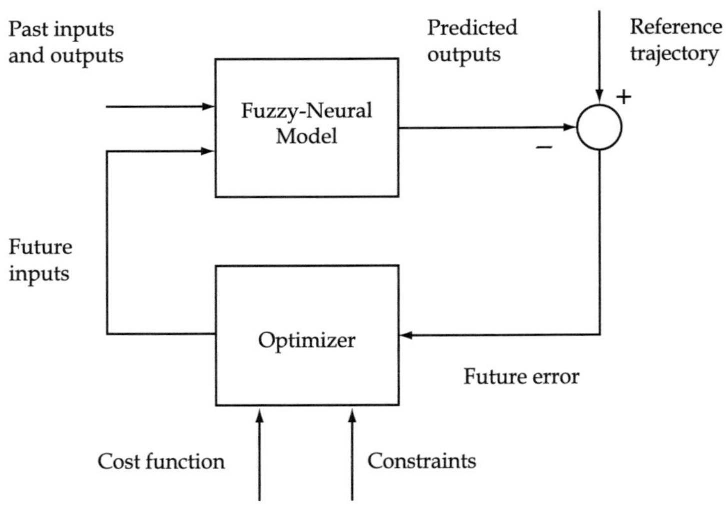

Fig. 1. Basic structure of the proposed Fuzzy-Neural NMPC

In this chapter a nonlinear discrete time state-space implementation is considered to represent the system dynamic:

$$\begin{aligned} x(k+1) &= f_x(x(k), u(k)) \\ y(k) &= f_y(x(k), u(k)) \end{aligned} \tag{1}$$

where $x(k) \in \Re^n$, $u(k) \in \Re^m$ and $y(k) \in \Re^q$ are state, control and output variables of the system, respectively. The unknown nonlinear functions f_x and f_y can be approximated by Takagi-Sugeno type fuzzy rules in the next form:

$$R_l : if \quad z_1(k)\, is\, M_{l1} \quad and \quad \ldots \quad and \quad z_i(k)\, is\, M_{li} \quad and \quad \ldots \quad z_p(k)\, is\, M_{lp}$$
$$then \quad \begin{cases} x_l(k+1) = A_l x(k) + B_l u(k) \\ y_l(k) = C_l x(k) + D_l u(k) \end{cases} \tag{2}$$

where R_l is the l-th rule of the rule base. Each rule is represented by an *if-then* conception. The antecedent part of the rules has the following form "$z_i(k)$ *is* M_{li}" where $z_i(k)$ is an i-th linguistic variable (i-th model input) and M_{li} is a membership function defined by a fuzzy set of the universe of discourse of the input z_i. Note that the input regression vector $\mathbf{z}(k) \in \Re^p$ in this chapter contains the system states and inputs $\mathbf{z}(k)=[x(k)\ u(k)]^T$. The consequent part of the rules is a mathematical function of the model inputs and states. A state-space implementation is used in the consequent part of R_l, where $A_l \in \Re^{n\times n}$, $B_l \in \Re^{n\times m}$, $C_l \in \Re^{q\times n}$ and $D_l \in \Re^{q\times m}$ are the state-space matrices of the model (Ahmed et al., 2009).
The states in the next sampling time $\hat{x}(k+1)$ and the system output $\hat{y}(k)$ can be obtained by taking the weighted sum of the activated fuzzy rules, using

$$\begin{aligned} \hat{x}(k+1) &= \sum_{l=1}^{L} \bar{\mu}_{yl}(k)(A_l x(k) + B_l u(k)) \\ \hat{y}(k) &= \sum_{l=1}^{L} \bar{\mu}_{yl}(k)(C_l \hat{x}(k) + D_l u(k)) \end{aligned} \tag{3}$$

On the other hand the state-space matrices A, B, C, and D for the global state-space plant model could be calculated as a weighted sum of the local matrices A_l, B_l, C_l, and D_l from the activated fuzzy rules (2):

$$\begin{aligned} A(k) &= \sum_{l=1}^{L} A_l \bar{\mu}_{yl}(k) \qquad \mathrm{B}(k) = \sum_{l=1}^{L} B_l \bar{\mu}_{yl}(k) \\ \mathrm{C}(k) &= \sum_{l=1}^{L} C_l \bar{\mu}_{yl}(k) \qquad \mathrm{D}(k) = \sum_{l=1}^{L} D_l \bar{\mu}_{yl}(k) \end{aligned} \tag{4}$$

where $\bar{\mu}_{yl} = \mu_{yl} \Big/ \sum_{l=1}^{L} \mu_{yl}$ is the normalized value of the membership function degree μ_{yl} upon the l-th activated fuzzy rule and L is the number of the activated rules at the moment k.

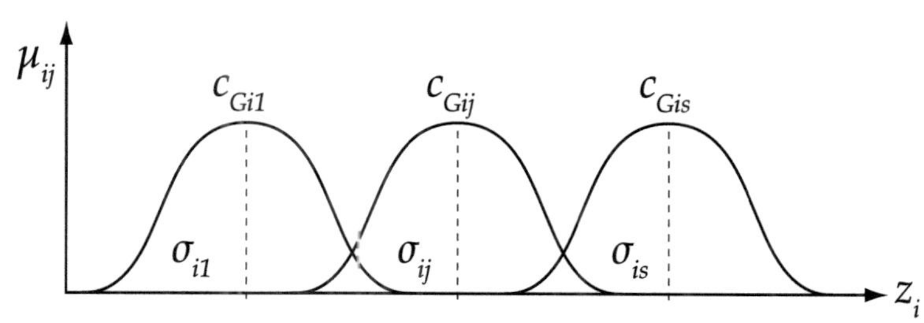

Fig. 2. Gaussian membership functions of the i-th input

Fuzzy implication in the l-th rule (2) can be realized by means of a product composition

$$\mu_{yl} = \prod_{i=1}^{p} \mu_{ij} \tag{5}$$

where μ_{ij} specifies the membership degree (Fig. 2) upon the activated j-th fuzzy set of the corresponded i-th input signal and it is calculated according to the chosen here Gaussian membership function (6) for the l-th activated rule:

$$\mu_{ij}(z_i) = \exp - \frac{(z_i - c_{Gij})^2}{2\sigma_{ij}^{\;2}} \tag{6}$$

where z_i is the current input value of the i-th model input, c_{Gij} is the centre (position) and σ_{ij} is the standard deviation (wide) of the j-th membership function (j=1, 2, .., s) (Fig.2).

2.1 Identification procedure for the fuzzy–neural model

The proposed identification procedure determines the unknown parameters in the Takagi-Sugeno fuzzy model, i.e. the parameters of membership functions, according to their shape and the parameters of the functions f_x and f_y in the consequent part of the rules (2). It is realised by a five-layer fuzzy-neural network (Fig. 3). Each of the layers performs typical fuzzy logic strategy operations:

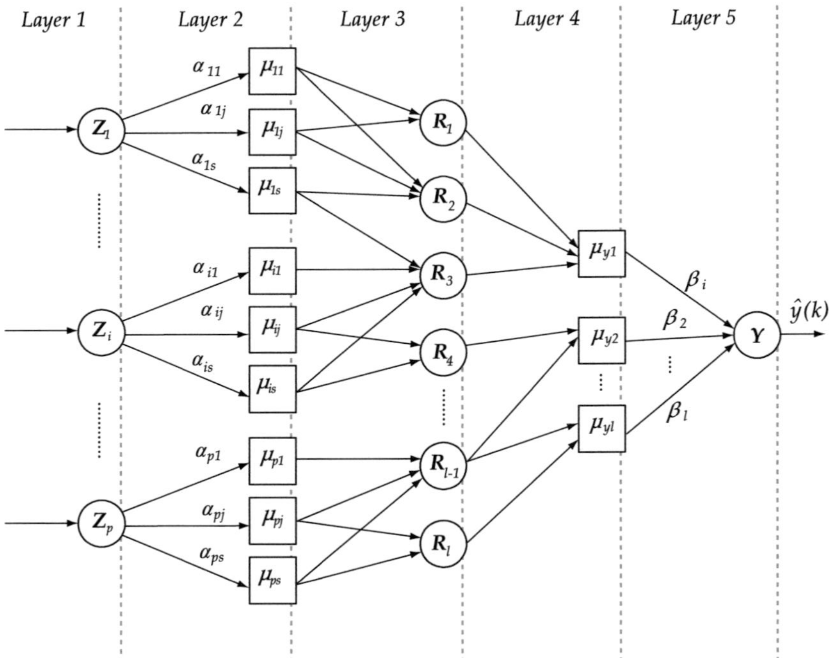

Fig. 3. The structure of the proposed fuzzy - neural model

Layer 1. The first layer represents the model inputs through its own input nodes Z_1, Z_2, …, Z_p. The network synaptic weights are set to one, so the model inputs are directly passed through the nodes to the next layer. Neurons here are represented by the elements of the regression vector $\mathbf{z}(k)$.

Layer 2. The fuzzification procedure of the input variables is performed in the second layer. The weights in this layer are the parameters of the chosen membership functions. Their number depends on the type and the number of the applied functions. All these parameters α_{ij} are adjustable and take part in the premise term of the Takagi-Sugeno type fuzzy rule base (2). In that section the membership functions for each model input variable are represented by Gaussian functions (Fig. 2). Hence, the adjustable parameters α_{ij} are the centres c_{Gij} and standard deviations σ_{ij} of the Gaussian functions (6). The nodes in the second layer of the fuzzy-neural architecture represent the membership degrees $\mu_{ij}(z_i)$ of the activated membership functions for each model input $z_i(k)$ according to (6). The number of the neurons depends on the number of the model inputs p and the number of the membership functions s in corresponding fuzzy sets. It is calculated as $p \times s$.

Layer 3. The third layer of the network interprets the fuzzy rule base (2). Each neuron in the third layer has as many inputs as the input regression vector size p. They are the corresponding membership degrees for the activated membership functions calculated in the previous layer. Therefore, each node in the third layer represents a fuzzy rule R_l, defined by Takagi-Sugeno fuzzy model. The outputs of the neurons are the results of the applied fuzzy rule base.

Layer 4. The fourth layer implements the fuzzy implication (5). Weights in this layer are set to one, in case the rule R_l from the third layer is activated, otherwise weights are zeros.

Layer 5. The last layer (one node layer) represents the defuzzyfication procedure and forms the output of the fuzzy-neural network (3). This layer also contains a set of adjustable parameters – β_l. These are the parameters in the consequent part of Takagi-Sugeno fuzzy model (2). The single node in this layer computes the overall model output signal as the summation of all signals coming from the previous layer.

$$I^5 = \sum_{l=1}^{L} f_{yl}\mu_{yl} \text{ or } I^5 = \sum_{l=1}^{L} f_{xl}\mu_{yl} \quad O^5 = \frac{\sum_{l=1}^{L} f_{yl}\mu_{yl}}{\sum_{l=1}^{L} \mu_{yl}} \text{ or } O^5 = \frac{\sum_{l=1}^{L} f_{xl}\mu_{yl}}{\sum_{l=1}^{L} \mu_{yl}} \tag{7}$$

where $f_{xl} = A_l x(k) + B_l u(k)$ and $f_{yl} = C_l x(k) + D_l u(k)$.

2.2 Learning algorithm of the fuzzy–neural model

Two-step gradient learning procedure is used as a learning algorithm of the internal fuzzy-neural model. It is based on minimization of an instant error function E_{FNN}. At time k the function is obtained from the following equation

$$E_{FNN}(k) = \varepsilon^2(k)/2 \tag{8}$$

where the error $\varepsilon(k)$ is calculated as a difference between the controlled process output $y(k)$ and the fuzzy-neural model output $\hat{y}(k)$:

$$\varepsilon(k) = y(k) - \hat{y}(k) \tag{9}$$

During step one of the procedure, the consequent parameters of Takagi-Sugeno fuzzy rules are calculated according to summary expression (10) (Petrov et al., 2002).

$$\beta_l(k+1) = \beta_l(k) + \eta\left(-\frac{\partial E_{FNN}}{\partial \beta_l}\right) \tag{10}$$

where η is a learning rate and β_l represents an adjustable coefficient a_{ij}, b_{ij}, c_{ij}, d_{ij} (11) for the activated fuzzy rule R_l (2). The coefficients take part in the state matrix A_l, control matrix B_l and output matrices C_l and D_l of the l-th activated rule (Ahmed et al., 2009). The matrices approximate the unknown nonlinear functions f_x and f_y according to defined fuzzy rule model (2). The matrix dimensions are specified by the system parameters - numbers of inputs m, outputs q and states n of the system.

$$A_l = \begin{bmatrix} a_{11} & \cdots & a_{1n} \\ \vdots & \ddots & \vdots \\ a_{n1} & \cdots & a_{nn} \end{bmatrix} \quad B_l = \begin{bmatrix} b_{11} & \cdots & b_{1m} \\ \vdots & & \vdots \\ b_{n1} & \cdots & b_{nm} \end{bmatrix} \quad C_l = \begin{bmatrix} c_{11} & \cdots & c_{1n} \\ \vdots & & \vdots \\ c_{q1} & \cdots & c_{qn} \end{bmatrix} \quad D_l = \begin{bmatrix} d_{11} & \cdots & d_{1m} \\ \vdots & & \vdots \\ d_{q1} & \cdots & d_{qm} \end{bmatrix} \tag{11}$$

In order to find a weight correction for the parameters in the last layer of the proposed fuzzy-neural network the derivative $\frac{\partial E_{FNN}}{\partial \beta_l}$ of the instant error should be determined. Following the chain rule, the derivative is calculated considering the expressions (7) and (8)

$$\frac{\partial E_{FNN}}{\partial \beta_l} = \frac{\partial E_{FNN}}{\partial \hat{y}} \cdot \frac{\partial \hat{y}}{\partial I^5} \cdot \frac{\partial I^5}{\partial \beta_l} \tag{12}$$

After the calculation of the partial derivatives, the matrix elements for each matrix of the state-space equations corresponding to the l-th activated rule (2) are obtained according to the summary expression (12) (Petrov et al., 2002; Ahmed et al., 2010):

$$\begin{aligned} a_{ij}(k+1) &= a_{ij}(k) + \eta\varepsilon(k)\bar{\mu}_{yl}(k)x_i(k) & i &= j = 1 \div n \\ b_{ij}(k+1) &= b_{ij}(k) + \eta\varepsilon(k)\bar{\mu}_{yl}(k)u_j(k) & i &= 1 \div n, j = 1 \div m \\ c_{ij}(k+1) &= c_{ij}(k) + \eta\varepsilon(k)\bar{\mu}_{yl}(k)x_j(k) & i &= 1 \div q, j = 1 \div n \\ d_{ij}(k+1) &= d_{ij}(k) + \eta\varepsilon(k)\bar{\mu}_{yl}(k)u_j(k) & i &= 1 \div q, j = 1 \div m \end{aligned} \tag{13}$$

The proposed fuzzy-neural architecture allows the use of the previously calculated output error (8) in the next step of the parameters update procedure. The output error E_{FNN} is propagated back directly to the second layer, where the second group of adjustable parameters are situated (Fig. 3). Depending on network architecture, the membership degrees calculated in the fourth and the second network layer are related as $\mu_{yl} \rightarrow \mu_{ij}$. Therefore, the learning rule for the second group adjustable parameters can be done in similar expression as (10):

$$\alpha_{ij}(k+1) = \alpha_{ij}(k) + \eta\left(-\frac{\partial E_{FNN}}{\partial \alpha_{ij}}\right) \tag{14}$$

where the derivative of the output error E_{FNN} is calculated by the separate partial derivatives:

$$\frac{\partial E_{FNN}}{\partial \alpha_{ij}} = \frac{\partial E_{FNN}}{\partial \hat{y}} \cdot \frac{\partial \hat{y}}{\partial \mu_{ij}} \cdot \frac{\partial \mu_{ij}}{\partial \alpha_{ij}} \tag{15}$$

The adjustable premise parameters of the fuzzy-neural model are the centre c_{Gij} and the deviation σ_{ij} of the Gaussian membership function (6). They are combined in the representative parameter α_{ij}, which corresponds to the i-th model input and its j-th activated fuzzy set. Following the expressions (14) and (15) the parameters are calculated as follows (Petrov et al., 2002; Ahmed et al., 2010)

$$c_{Gij}(k+1) = c_{Gij}(k) + \eta\varepsilon(k)\bar{\mu}_{yl}(k)[f_{yl} - \hat{y}(k)]\frac{[z_i(k) - c_{Gij}(k)]}{\sigma_{ij}^2(k)} \tag{16}$$

$$\sigma_{ij}(k+1) = \sigma_{ij}(k) + \eta\varepsilon(k)\bar{\mu}_{yl}(k)[f_{yl} - \hat{y}(k)]\frac{[z_i(k) - c_{Gij}(k)]^2}{\sigma_{ij}^3(k)} \tag{17}$$

The proposed identification procedure for the fuzzy-neural model could be summarized in the following steps (Table 1).

Step 1. Initialize the membership functions - number, shape, parameters;
Step 2. Assign initial values for the network inputs;
Step 3. Start the algorithm at the current moment k;
Step 4. Fuzzify the network inputs and calculate the membership degrees upon the activated fuzzy set of the membership functions according to (6);
Step 5. Perform fuzzy implication according to (5);
Step 6. Calculate the fuzzy-neural network output, which is represented by state-space description of the modelled system - (3) and (4);
Step 7. Calculate the instant error according to (8) and (9);
Step 8. Start training procedure for fuzzy-neural network;
Step 9. Adjust the consequent parameters according to (13);
Step 10. Adjust the premise parameters according to (16) and (17).
Repeat the algorithm from Step 3 for each sampling time.

Table 1. Fuzzy-neural model identification procedure

3. Optimization algorithm of multivariable model predictive control strategy

The model provided by the Takagi-Sugeno type fuzzy-neural network is used to formulate the objective function for the optimization algorithm and to calculate the future control actions. The second stage of the predictive control strategy includes an optimization procedure. It utilizes the obtained results during the first (modelling) stage predictive model of the system. Using the Takagi-Sugeno fuzzy-neural model (3), the optimization algorithm computes the future control actions at each sampling period, by minimizing the typical for MPC strategy (Generalized Predictive Control - GPC) cost function (Akesson, 2006):

$$J(k) = \sum_{i=H_w}^{H_p+H_w-1} \|\hat{y}(k+i) - r(k+i)\|_{\bar{Q}}^2 + \sum_{i=0}^{H_u-1} \|\Delta u(k+i)\|_{\bar{R}}^2 \tag{18}$$

where $\hat{y}(k)$, $r(k)$ and $\Delta u(k)$ are the predicted outputs, the reference trajectories, and the predicted control increments at time k, respectively. The length of the prediction horizon is

H_p, and the first sample to be included in the horizon is H_w. The control horizon is given by H_u. $\tilde{Q} \geq 0$ and $\tilde{R} > 0$ are weighting matrices representing the relative importance of each controlled and manipulated variable and they are assumed to be constant over the H_p.
The cost function (18) may be rewritten in a matrix form as follows

$$J(k) = \|Y(k) - T(k)\|_Q^2 + \|\Delta U(k)\|_R^2 \tag{19}$$

where $Y(k)$, $T(k)$, $\Delta U(k)$, Q and R are predicted output, system reference, control variable increment and weighting matrices, respectively,

$$Y(k) = \begin{bmatrix} \hat{y}(k|k) \\ \vdots \\ \hat{y}(k+H_p-1|k) \end{bmatrix}, \; T(k) = \begin{bmatrix} r(k|k) \\ \vdots \\ r(k+H_p-1|k) \end{bmatrix}, \; \Delta U(k) = \begin{bmatrix} \Delta u(k|k) \\ \vdots \\ \Delta u(k+H_u-1|k) \end{bmatrix}$$

$$Q = \begin{bmatrix} \tilde{Q}(1) & \cdots & 0 \\ \vdots & \ddots & \vdots \\ 0 & \cdots & \tilde{Q}(H_p) \end{bmatrix} \qquad R = \begin{bmatrix} \tilde{R}(1) & \cdots & 0 \\ \vdots & \ddots & \vdots \\ 0 & \cdots & \tilde{R}(H_u) \end{bmatrix}$$

The linear state-space model used for Takagi-Sugeno fuzzy rules (2) could be represented in the following form:

$$\begin{aligned} \hat{x}(k+1) &= Ax(k) + Bu(k-1) + B\Delta u(k) \\ \hat{y}(k) &= C\hat{x}(k) + Du(k-1) + D\Delta u(k) \end{aligned} \tag{20}$$

Based on the state-space matrices A, B, C and D (4), the future state variables are calculated sequentially using the set of future control parameters:

$$\begin{aligned}
&\hat{x}(k+1) = Ax(k) + Bu(k-1) + B\Delta u(k) \\
&\hat{x}(k+2) = A^2x(k) + (AB+B)u(k-1) + (AB+B)\Delta u(k) + B\Delta u(k+1) \\
&\hat{x}(k+3) = A^3x(k) + (A^2B+AB+B)u(k-1) + (A^2B+AB+B)\Delta u(k) + (AB+B)\Delta u(k+1) + B\Delta u(k+2) \\
&\cdots\cdots \\
&\hat{x}(k+j) = A^jx(k) + \sum_{i=0}^{j-1} A^iBu(k-1) + \sum_{i=0}^{j-1} A^iB \sum_{m=0}^{j-1-i} \Delta u(k+m) \\
&\cdots\cdots \\
&\hat{x}(k+H_p) = A^{H_p}x(k) + \sum_{i=0}^{H_p-1} A^iBu(k-1) + \sum_{i=0}^{H_p-1} A^iB\Delta u(k) + \sum_{i=0}^{H_p-2} A^iB\Delta u(k+1) + \cdots + A^{H_p-H_u}B\Delta u(k+H_u-1)
\end{aligned}$$

The predictions of the output $\hat{y}$ for j steps ahead could be calculated as follows

$$\begin{aligned}
&\hat{y}(k+1) = Cx(k+1) + Du(k+1) = CAx(k) + (CB+D)u(k-1) + (CB+D)\Delta u(k) + D\Delta u(k+1) \\
&\hat{y}(k+2) = CA^2x(k) + (CAB+CB+D)u(k-1) + (CAB+CB+D)\Delta u(k) + (CB+D)\Delta u(k+1) + D\Delta u(k+2) \\
&\cdots\cdots
\end{aligned}$$

$$\hat{y}(k+j)=CA^{j}x(k)+\left(C\sum_{i=0}^{j-1}A^{i}B+D\right)u(k-1)+\left(C\sum_{i=0}^{j-1}A^{i}B+D\right)\Delta u(k)+(CB+D)\Delta u(k+j-1)+D\Delta u(k+j)$$

....

$$\hat{y}(k+H_p-1)=CA^{H_p-1}x(k)+\left(C\sum_{i=0}^{H_p-2}A^{i}B+D\right)u(k-1)+\left(C\sum_{i=0}^{H_p-2}A^{i}B+D\right)\Delta u(k)+$$

$$+\left(C\sum_{i=0}^{H_p-3}A^{i}B+D\right)\Delta u(k+1)+\cdots+\left(C\sum_{i=0}^{H_p-H_u-1}A^{i}B+D\right)\Delta u(k+H_u-1)$$

The recurrent equation for the output predictions $\hat{y}(k+j_p)$, where $j_p = 1, 2, \ldots, H_p-1$, is in the next form:

$$\hat{y}(k+j_p)=CA^{j_p}x(k)+\left(C\sum_{i=0}^{j_p-1}A^{i}B+D\right)u(k-1)+\begin{cases}\sum_{i=0}^{j_p}\left(C\sum_{j=0}^{j_p-1}A^{j}B+D\right)\Delta u(k+i),\ j_p<H_u\\ \sum_{i=0}^{H_u-1}\left(C\sum_{j=0}^{j_p-i-1}A^{j}B+D\right)\Delta u(k+i),\ j_p>H_u\end{cases} \quad (21)$$

The prediction model defined in (21) can be generalized by the following matrix equality

$$Y(k) = \Psi x(k) + \Gamma u(k-1) + \Theta\Delta U(k) \quad (22)$$

where

$$\Psi=\begin{bmatrix}C\\CA\\CA^2\\\vdots\\CA^{Hp-1}\end{bmatrix}\quad \Gamma=\begin{bmatrix}D\\CB+D\\CAB+CB+D\\\vdots\\C\sum_{i=0}^{H_p-2}A^{i}B+D\end{bmatrix}\quad \Theta=\begin{bmatrix}D & 0 & \cdots & 0\\ CB+D & D & \cdots & \vdots\\ CAB+CB+D & CB+D & \ddots & \vdots\\ \vdots & & \ddots & 0\\ C\sum_{i=0}^{H_u-2}A^{i}B+D & & \cdots & D\\ \vdots & & \ddots & \vdots\\ C\sum_{i=0}^{H_p-2}A^{i}B+D & & \cdots & C\sum_{i=0}^{H_p-H_u-1}A^{i}B+D\end{bmatrix}$$

All matrices, which take part in the equations above, are derived by the Takagi-Sugeno fuzzy-neural predictive model (4).
It is also possible to define the vector

$$E(k) = T(k) - \Gamma u(k-1) - \Theta\Delta U(k) \quad (23)$$

This vector can be thought as a *tracking error*, in the sense that it is the difference between the future target trajectory and the *free response* of the system, namely the response that would occur over the prediction horizon if no input changes were made, i.e. $\Delta U(k)=0$. Hence, the quantity of the so called *free response F(k)* is defined as follows

$$F(k) = \Psi x(k) + \Gamma u(k-1) \quad (24)$$

3.1 Unconstrained model predictive control

In this section, the study is focused on the optimization problem of the unconstrained nonlinear predictive control with the quadratic cost function (18). The section presents an approximate solution of the problem where the information given by the obtained fuzzy-neural model is used to solve the problem.
The unconstrained optimization problem can be formulated in a matrix form. First, the predictor can be constructed as follows

$$Y(k) = \Theta \Delta U(k) + F(k) \tag{25}$$

Second, the cost function (19) can be rewritten as

$$J(\Delta U) = (T - Y)^T Q (T - Y) + \Delta U^T R \Delta U \tag{26}$$

Hence, substituting the predictive model (25) into expression (26), the cost function of the model predictive optimization problem can be specified as follows:

$$J(\Delta U) = \Delta U^T(\Theta^T Q\Theta + R)\Delta U + 2(F - T)^T Q\Theta\Delta U + (T - F)^T Q(T - F) \tag{27}$$

The minimum of the function $J(\Delta U)$ can be obtained by calculating the input sequence ΔU so that $\partial J/\partial \Delta U = 0$:

$$\frac{\partial}{\partial \Delta U} J(\Delta U) = \frac{\partial}{\partial \Delta U}\left[\Delta U^T(\Theta^T Q\Theta + R)\Delta U + 2(F - T)^T Q\Theta\Delta U + (F - T)^T Q(F - T)\right] = 0 \tag{28}$$

Then the optimal sequence ΔU^* is

$$\Delta U^* = (\Theta^T Q\Theta + R)^{-1}\Theta^T Q(T - F) \tag{29}$$

The input applied to the controlled plant at time k is computed according to the receding horizon principle, i.e. the first element from the control sequence $\Delta u^*(k)$ of the vector ΔU^* is taken. Then, control signal is calculated from:

$$u(k) = u(k - 1) + \Delta u^*(k) \tag{30}$$

It is evident that the expression given by the matrix equation (29) is the same as expression obtained for the generalized predictive control. However, in the GPC formulation the components involved in the calculation of the formula (29) are obtained from a linear model. In the present case the components introduced in this expression are generated by the designed nonlinear fuzzy-neural model. A more rigorous formulation of (29) will be representation of the components as time-variant matrices, as they are shown in the expression (22). In this case the matrix $\Theta(k)$ and the vectors $\Psi(k)$, $T(k)$ are being reconstructed at each sampling time. The vector $\Psi(k)$ is obtained by simulating the fuzzy model with the current input $u(k)$; the matrix $\Theta(k)$ is also rebuilt using a method described below.

$$\Delta U^*(k) = \left[\Theta^T(k) Q\Theta(k) + R\right]^{-1}\Theta^T(k) Q\left[T(k) - F(k)\right] \tag{31}$$

The proposed method solves the problem of unconstrained MPC. A system of equations is solved at each sampling time k. The proposed approach decreases computational burden avoiding the necessity to inverse the gain matrix in (31) at each sampling time k.

Applying this method, minimization of the GPC criterion (18) is based on a calculation of the gradient vector of the criterion cost function J at the moment k subject to the predicted control actions:

$$\nabla J(k) = \left[\frac{\partial J(k)}{\partial \Delta u(k)}, \frac{\partial J(k)}{\partial \Delta u(k+1)}, \ldots, \frac{\partial J(k)}{\partial \Delta u(k+H_u-1)} \right]^T \tag{32}$$

Each element of this gradient vector (32) can be calculated using the following derivative matrix equation:

$$\frac{\partial J(k)}{\partial \Delta U(k)} = \left[2[T(k) - Y(k)]^T Q \frac{\partial Y(k)}{\partial \Delta U(k)} + 2\Delta U(k)^T R \frac{\partial \Delta U(k)}{\partial \Delta U(k)} \right] \tag{33}$$

From the above expression (33) it can be seen that it is necessary to obtain two groups of partial derivatives. The first one is $\left[\frac{\partial Y(k)}{\partial \Delta U(k)} \right]$, and the second one is $\left[\frac{\partial \Delta U(k)}{\partial \Delta U(k)} \right]$. The first partial derivatives in (33) have the following matrix form:

$$\frac{\partial Y(k)}{\partial \Delta U(k)} = \begin{bmatrix} \frac{\partial \hat{y}(k+H_w)}{\partial \Delta u(k)} & \cdots & \frac{\partial \hat{y}(k+H_w)}{\partial \Delta u(k+H_u-1)} \\ \vdots & & \vdots \\ \frac{\partial \hat{y}(k+H_p+H_w-1)}{\partial \Delta u(k)} & \cdots & \frac{\partial \hat{y}(k+H_p+H_w-1)}{\partial \Delta u(k+H_u-1)} \end{bmatrix} \tag{34}$$

For computational simplicity assume that H_w=0 (18). Then each element of the matrix (34) is calculated by the expressed equations according to the Takagi-Sugeno rules consequents (2). For example the derivatives from first column of the matrix (34) have the following form:

$$\frac{\partial \hat{y}(k)}{\partial \Delta u(k)} = \sum_{l=1}^{L} D_l \bar{\mu}_{yl}(k) \tag{35}$$

$$\frac{\partial \hat{y}(k+1)}{\partial \Delta u(k)} = \sum_{l=1}^{L} (C_l B_l + D_l) \bar{\mu}_{yl}(k+1) \tag{36}$$

$$\frac{\partial \hat{y}(k+2)}{\partial \Delta u(k)} = \sum_{l=1}^{L} (C_l A_l B_l + C_l B_l + D_l) \bar{\mu}_{yl}(k+2) \tag{37}$$

……………………………………………….

$$\frac{\partial \hat{y}(k+H_p-1)}{\partial \Delta u(k)} = \sum_{l=1}^{L} \left(C_l \sum_{j=0}^{H_p-2} A_l^j B_l + D_l \right) \bar{\mu}_{yl}(k+H_p-1) \tag{38}$$

The second group partial derivatives in (33) has the following matrix form:

$$\frac{\partial \Delta U(k)}{\partial \Delta U(k)} = \begin{bmatrix} \dfrac{\partial \Delta u(k)}{\partial \Delta u(k)} & \cdots & \dfrac{\partial \Delta u(k)}{\partial \Delta u(k+H_u-1)} \\ \vdots & & \vdots \\ \dfrac{\partial \Delta u(k+H_u-1)}{\partial \Delta u(k)} & \cdots & \dfrac{\partial \Delta u(k+H_u-1)}{\partial \Delta u(k+H_u-1)} \end{bmatrix} \quad (39)$$

Since $\Delta u(k)=u(k)-u(k-1)$, the matrix (39) has the following form:

$$\frac{\partial \Delta U(k)}{\partial \Delta U(k)} = \begin{bmatrix} 1 & 0 & \cdots & 0 \\ -1 & 1 & & \\ \vdots & -1 & \ddots & \vdots \\ 1 & \vdots & \ddots & 1 \\ -1 & 1 & \cdots & -1 \end{bmatrix} \quad (40)$$

Following this procedure it is possible to calculate the rest column elements of the matrix (34) which belongs to the next gradient vector elements (32). Finally, each element of the gradient-vector (32) could be obtained by the following system of equations:

$$\begin{aligned} \frac{\partial J(k)}{\partial \Delta u(k+1)} &= 2\hat{e}(k+1)\tilde{Q}(1)\frac{\partial \hat{y}(k+1)}{\partial \Delta u(k)} + \ldots + 2\hat{e}(k+H_p)\tilde{Q}(H_p)\frac{\partial \hat{y}(k+H_p)}{\partial \Delta u(k)} + \\ &+2\tilde{R}(1)\Delta u(k) - 2\tilde{R}(2)\Delta u(k+1) + \ldots - 2\tilde{R}(H_u)\Delta u(k+H_u-1) = 0 \end{aligned} \quad (41)$$

$$\begin{aligned} \frac{\partial J(k)}{\partial \Delta u(k+2)} &= 2\hat{e}(k+1)\tilde{Q}(1)\frac{\partial \hat{y}(k+1)}{\partial \Delta u(k+1)} + \ldots + 2\hat{e}(k+H_p)\tilde{Q}(H_p)\frac{\partial \hat{y}(k+H_p)}{\partial \Delta u(k+1)} + \\ &+2\tilde{R}(2)\Delta u(k+1) - 2\tilde{R}(3)\Delta u(k+2) + \ldots - 2\tilde{R}(H_u)\Delta u(k+H_u-1) = 0 \end{aligned} \quad (42)$$

$$\begin{aligned} \frac{\partial J(k)}{\partial \Delta u(k+H_u-2)} &= 2\hat{e}(k+1)\tilde{Q}(1)\frac{\partial \hat{y}(k+1)}{\partial \Delta u(k+H_u-2)} + \ldots + 2\hat{e}(k+H_p)\tilde{Q}(H_p)\frac{\partial \hat{y}(k+H_p)}{\partial \Delta u(k+H_u-2)} + \\ &+2\tilde{R}(H_u-1)\Delta u(k+H_u-2) - 2\tilde{R}(H_u)\Delta u(k+H_u-1) = 0 \end{aligned} \quad (43)$$

$$\begin{aligned} \frac{\partial J(k)}{\partial \Delta u(k+H_u-1)} &= 2\hat{e}(k+1)\tilde{Q}(1)\frac{\partial \hat{y}(k+1)}{\partial \Delta u(k+H_u-1)} + \ldots + 2\hat{e}(k+H_p)\tilde{Q}(H_p)\frac{\partial \hat{y}(k+H_p)}{\partial \Delta u(k+H_u-1)} + \\ &+2\tilde{R}(H_u)\Delta u(k+H_u-1) = 0 \end{aligned} \quad (44)$$

where $\hat{e}(k+j) = r(k+j) - \hat{y}(k+j), \quad j = 1,2,\ldots,H_p$ is the predicted system error.

The obtained system of equations (41)-(44) can be solved very easily, starting from the last equation (44) and calculating the last control action $u(k+H_u-1)$ first. Then, the procedure can continue with finding the previous control action $u(k+H_u-2)$ from (43). The calculations continue until the whole number of the control actions over the horizon H_u is obtained. The calculation order of the control actions is very important, since the calculations should contain only known quantities. After that, only the first control action *u(k)* (30) will be used at the moment *k* as an input to the controlled process. The software implementation of the proposed algorithm is realized easily according to the following equations:

$$\Delta u(k+H_u-1)=\tilde{R}(H_u)^{-1}\left[\hat{e}(k+1)\tilde{Q}(1)\frac{\partial\hat{y}(k+1)}{\partial\Delta u(k+H_u-1)}+\cdots+\hat{e}(k+H_p)\tilde{Q}(H_p)\frac{\partial\hat{y}(k+H_p)}{\partial\Delta u(k+H_u-1)}\right] \quad (45)$$

$$\Delta u(k+H_u-2)=\Delta u(k+H_u-1)+ \\ +\tilde{R}(H_u-1)^{-1}\left[\hat{e}(k+1)\tilde{Q}(1)\frac{\partial\hat{y}(k+1)}{\partial\Delta u(k+H_u-2)}+\cdots+\hat{e}(k+H_p)\tilde{Q}(H_p)\frac{\partial\hat{y}(k+H_p)}{\partial\Delta u(k+H_u-2)}\right] \quad (46)$$

$$\Delta u(k+1)=\Delta u(k+2)+\tilde{R}(2)^{-1}\left[\hat{e}(k+1)\tilde{Q}(1)\frac{\partial\hat{y}(k+1)}{\partial\Delta u(k+1)}+\cdots+\hat{e}(k+H_p)\tilde{Q}(H_p)\frac{\partial\hat{y}(k+H_p)}{\partial\Delta u(k+1)}\right] \quad (47)$$

$$\Delta u(k)=\Delta u(k+1)+\tilde{R}(1)^{-1}\left[\hat{e}(k+1)\tilde{Q}(1)\frac{\partial\hat{y}(k+1)}{\partial\Delta u(k)}+\cdots+\hat{e}(k+H_p)\tilde{Q}(H_p)\frac{\partial\hat{y}(k+H_p)}{\partial\Delta u(k)}\right] \quad (48)$$

The proposed unconstrained predictive control algorithm could be summarized in the following steps (Table 2).

Step 1.	Initial identification of the Takagi-Sugeno fuzzy-neural predictive model;
Step 2.	Start the algorithm at the sample *k* with the initial parameters;
Step 3.	Calculate the predicted model output $\hat{y}(k+j)$ using the tuned fuzzy-neural model (2);
Step 4.	Calculate the derivatives for the matrix (34) according to the equations (35)-(38);
Step 5.	Calculate predicted control actions according to (45)-(48) and update the sequence;
Step 6.	Apply the first optimal control action *u(k)*;
Step 7.	Modify the model parameters into the rule (3) and update them for the next step 3 for the next sample *k*

Table 2. Basic fuzzy-neural model unconstrained predictive control algorithm

3.2 Constrained model predictive control

The constrained nonlinear predictive control problem can be described as a problem of finding the "optimal" input sequence to move a dynamic system to a desired state, taking into account the constraints on the inputs and the outputs of the control systems. This section reveals the formulation of the constrained control problem for MPC uses. Essentially, the problem becomes a quadratic programming problem with linear inequality constraints (LICQP). It follows by the nature of the operational constraints, which are usually described by linear inequalities of the control and plant variables.

The problem of nonlinear constrained predictive control is formulated as a nonlinear quadratic optimization problem. By means of local linearization (20) the problem can be solved using QP. That way the solution to the linear constrained predictive control problem is obtained. At each sampling time the LICQP is solved with new parameters, which are obtained by the Takagi-Sugeno fuzzy-neural model. An active set method is used for solving the constructed quadratic programming problem.

3.2.1 Constraint types in model predictive control

The operational constraints may be classified in three major types according to the type of the system variables, which they are imposed on. The first two types of constraints deal with the control variable incremental variation $\Delta u(k)$ and control variable $u(k)$. The third type is concerned with output $y(k)$ or state variable $x(k)$ constraints.

Related to the origin model predictive control problem, the constraints are expressed in a set of linear equations. All types of constraints are taken into consideration for each moving horizon window.

$$\begin{aligned} U_{min}(k) &\leq U(k) \leq U_{max}(k) \\ \Delta U_{min}(k) &\leq \Delta U(k) \leq \Delta U_{max}(k) \\ Y_{min}(k) &\leq Y(k) \leq Y_{max}(k) \end{aligned} \tag{49}$$

Where

$$U_{\max}(k) = \begin{bmatrix} u_{\max}(k) \\ u_{\max}(k+1) \\ \vdots \\ u_{\max}(k+N_u-1) \end{bmatrix} \quad \Delta U_{\max}(k) = \begin{bmatrix} \Delta u_{\max}(k) \\ \Delta u_{\max}(k+1) \\ \vdots \\ \Delta u_{\max}(k+N_u-1) \end{bmatrix}$$

$$U_{\min}(k) = \begin{bmatrix} u_{\min}(k) \\ u_{\min}(k+1) \\ \vdots \\ u_{\min}(k+N_u-1) \end{bmatrix} \quad \Delta U_{\min}(k) = \begin{bmatrix} \Delta u_{\min}(k) \\ \Delta u_{\min}(k+1) \\ \vdots \\ \Delta u_{\min}(k+N_u-1) \end{bmatrix}$$

$$Y_{\max}(k) = \begin{bmatrix} y_{\max}(k) \\ y_{\max}(k+1) \\ \vdots \\ y_{\max}(k+N_p-1) \end{bmatrix}$$

$$Y_{\min}(k) = \begin{bmatrix} y_{\min}(k) \\ y_{\min}(k+1) \\ \vdots \\ y_{\min}(k+N_p-1) \end{bmatrix}$$

Therefore, for multi-input case the number of the constraints for the change of the control variable $\Delta u(k)$ is $m \times N_u$. Similarly, the number of the constraints for the control variable amplitude is also $m \times N_u$ and for the output constraints it is $q \times N_p$.

3.2.2 Quadratic programming in use of constrained MPC

Since the cost function $J(k)$ (19) is quadratic and the constraints are linear inequalities, the problem of finding an optimal predictive control becomes one of finding an optimal solution to a standard quadratic programming problem with linear inequality constraints

$$\begin{aligned} & min J(x) = \frac{1}{2} x^T H x + f^T x \\ & subject\ to\ Ax \le b \end{aligned} \tag{50}$$

where H and f are the Hessian and the gradient of the Lagrange function, x is the decision variable. Constraints on the QP problem (50) are specified by $Ax \le b$ according to (49).
The Lagrange function is defined as follows

$$L(x,\lambda) = J(x) + \sum_{i=1}^{N} \lambda_i a_i, \ i = 1,2,\ldots N, \tag{51}$$

where λ_i are the Lagrange multipliers, a_i are the constraints on the decision variable x, N is the number of the constraints considered in the optimization problem.
Several algorithms for constrained optimization are described in (Fletcher, 2000). In this chapter a primal active set method is used. The idea of active set method is to define a set S of constraints at each step of algorithm. The constraints in this active set are regarded as equalities whilst the rest are temporarily disregarded and the method adjusts the set in order to identify the correct active constraints on the solution to (52)

$$\begin{aligned} & \min J(x) = \frac{1}{2} x^T H x + f^T x \\ & subject\ to\ a_i x = b_i \\ & \qquad a_i x \le b_i \end{aligned} \tag{52}$$

At iteration k a feasible point $x(k)$ is known which satisfies the active constraints as equalities. Each iteration attempts to locate a solution to an equality problem (EP) in which only the active constraints occur. This is most conveniently performed by shifting the origin to $x(k)$ and looking for a correction $\delta(k)$ which solves

$$\begin{aligned} & \min_{\delta} \left\{ \frac{1}{2} \delta^T H x + f^T \delta \right\} \\ & subject\ to\ a_i \delta = 0 \quad a_i \in S \end{aligned} \tag{53}$$

where $f(k)$ is defined by $f(k) = f + Hx(k)$ and is $\nabla J(x(k))$ for the function defined by (52). If $\delta(k)$ is feasible with regard to the constraints not included in S, then the feasible point in next iteration is taken as $x(k+1) = x(k) + \delta(k)$. If not, a line search is made in the direction of $\delta(k)$ to find the best feasible point. A constraint is active if the Lagrange multipliers $\lambda_i \ge 0$, i.e. it is at the boundary of the feasible region defined by the constraints. On the other hand, if there exist $\lambda_i < 0$, the constraint is not active. In this case the constraint is relaxed from the active constraints set S and the algorithm continues as before by solving the resulting equality constraint problem (53). If there is more than one constraint with corresponding $\lambda_i < 0$, then the $\min_{i \in S} \lambda_i(k)$ is selected (Fletcher, 2000).
The QP, described in that way, is used to provide numerical solutions in constrained MPC problem.

3.2.3 Design the constrained model predictive problem

The fuzzy-neural identification procedure from the Section 2 provides the state-space matrices, which are needed to construct the constrained model predictive control optimization problem. Similarly to the unconstrained model predictive control approach, the cost function (18) can be specified by the prediction expressions (22) and (23).

$$\begin{aligned} J(k) &= [\Psi x(k) + \Gamma u(k-1) + \Theta\Delta U(k) - T(k)]^T Q[\Psi x(k) + \Gamma u(k-1) + \Theta\Delta U(k) - T(k)] + \Delta U^T(k) R\Delta U(k) = \\ &= [\Theta\Delta U(k) - E(k)]^T Q[\Theta\Delta U(k) - E(k)] + \Delta U^T(k) R\Delta U(k) = \\ &= \Delta U^T(k)\left[\Theta^T Q\Theta + R\right]\Delta U(k) + E^T(k)QE(k) - 2\Delta U^T(k)\Theta^T QE(k) \end{aligned}$$

Assuming that

$$H = \Theta^T Q\Theta + R \text{ and } \Phi = 2\Theta^T QE(k), \tag{54}$$

the cost function for the model predictive optimization problem can be specified as follow

$$J(k) = \Delta U^T(k) H \,\Delta U(k) - \Delta U^T(k)\Phi + E^T(k)QE(k) \tag{55}$$

The problem of minimizing the cost function (55) is a quadratic programming problem. If the Hessian matrix H is positive definite, the problem is convex (Fletcher, 2000). Then the solution is given by the closed form

$$\Delta U = \frac{1}{2}H^{-1}\Phi \tag{56}$$

The constraints (49) on the cost function may be rewritten in terms of $\Delta U(k)$.

$$\begin{gathered} U_{\min}(k) \le I_u u(k-1) + I_{\Delta u}\Delta U(k) \le U_{\max}(k) \\ \Delta U_{\min}(k) \le \Delta U(k) \le \Delta U_{\max}(k) \\ Y_{\min}(k) \le \Psi x(k) + \Gamma u(k-1) + \Theta\Delta U(k) \le Y_{\max}(k) \end{gathered} \tag{57}$$

where $I_m \in \Re^{m\times m}$ is an identity matrix, $I_u = \begin{bmatrix} I_m \\ I_m \\ \vdots \\ I_m \end{bmatrix} \in \Re^{mN_u \times m}$, $I_{\Delta u} = \begin{bmatrix} I_m & 0 & \cdots & 0 \\ I_m & I_m & \cdots & 0 \\ \vdots & \vdots & \ddots & \vdots \\ I_m & I_m & \cdots & I_m \end{bmatrix} \in \Re^{mN_u \times mN_u}$.

All types of constraints are combined in one expression as follows

$$\begin{bmatrix} -I_{\Delta u} \\ I_{\Delta u} \\ -I \\ I \\ -\Theta \\ \Theta \end{bmatrix} \Delta U \le \begin{bmatrix} -U_{\min} + I_u u(k-1) \\ U_{\max} - I_u u(k-1) \\ \Delta U_{\min} \\ \Delta U_{\max} \\ -Y_{\min} + (\Psi x(k) + \Gamma u(k-1)) \\ Y_{\max} - (\Psi x(k) + \Gamma u(k-1)) \end{bmatrix} \tag{58}$$

where $I \in \Re^{mN_u \times mN_u}$ is an identity matrix.
Finally, following the definition of the LIQP (50), the model predictive control in presence of constraints is proposed as finding the parameter vector ΔU that minimizes (55) subject to the inequality constraints (58).

$$\begin{aligned} &\min\ J(k) = \Delta U^T H\ \Delta U - \Delta U^T \Phi + E^T Q E \\ &\textit{subject to } \Omega \Delta U \leq \omega \end{aligned} \tag{59}$$

In (59) the constraints expression (58) has been denoted by $\Omega\Delta U \leq \omega$, where Ω is a matrix with number of rows equal to the dimension of ω and number of columns equal to the dimension of ΔU. In case that the constraints are fully imposed, the dimension of ω is equal to $4 \times m \times N_u + 2 \times q \times N_p$, where m is the number of system inputs and q is the number of outputs. In general, the total number of constraints is greater than the dimension of the ΔU. The dimension of ω represents the number of constraints.
The proposed model predictive control algorithm can be summarized in the following steps (Table 3).

At each sampling time:
Step 1. Read the current states, inputs and outputs of the system;
Step 2. Start identification of the fuzzy-neural predictive model following Algorithm 1;
Step 3. With $A(k)$, $B(k)$, $C(k)$, $D(k)$ from Step 2 calculate the predicted output $Y(k)$ according to (17);
Step 4. Obtain the prediction error $E(k)$ according to (23);
Step 5. Construct the cost function (55) and the constraints (58) of the QP problem;
Step 6. Solve the QP problem according to (59);
Step 7. Apply only the first control action $u(k)$.

Table 3. State-space implementation of fuzzy-neural model predictive control strategy

At each sampling time, LIQP (59) is solved with new parameters. The Hessian and the Lagrangian are constructed by the state-space matrices $A(k)$, $B(k)$, $C(k)$ and $D(k)$ (4) obtained during the identification procedure (Table 1). The problem of nonlinear constrained predictive control is formulated as a nonlinear quadratic optimization problem. By means of local linearization a relaxation can be obtained and the problem can be solved using quadratic programming. This is the solution of the linear constrained predictive control problem (Espinosa et al., 2005).

4. Fuzzy-neural model predictive control of a multi tank system. Case study

The case study is implemented in MATLAB/Simulink® environment with Inteco® Multi tank system. The Inteco® Multi tank System (Fig. 4) comprises from three separate tanks fitted with drain valves (Inteco, 2009). The additional tank mounted in the base of the set-up acts as a water reservoir for the system. The top (first) tank has a constant cross section, while others are conical or spherical, so they are with variable cross sections. This causes the main nonlinearities in the system. A variable speed pump is used to fill the upper tank. The liquid outflows the tanks by the gravity. The tank valves act as flow resistors C_1, C_2, C_3. The area ratio of the valves is controlled and can be used to vary the outflow characteristic. Each tank is equipped with a level sensor PS_1, PS_2, PS_3 based on hydraulic pressure measurement.

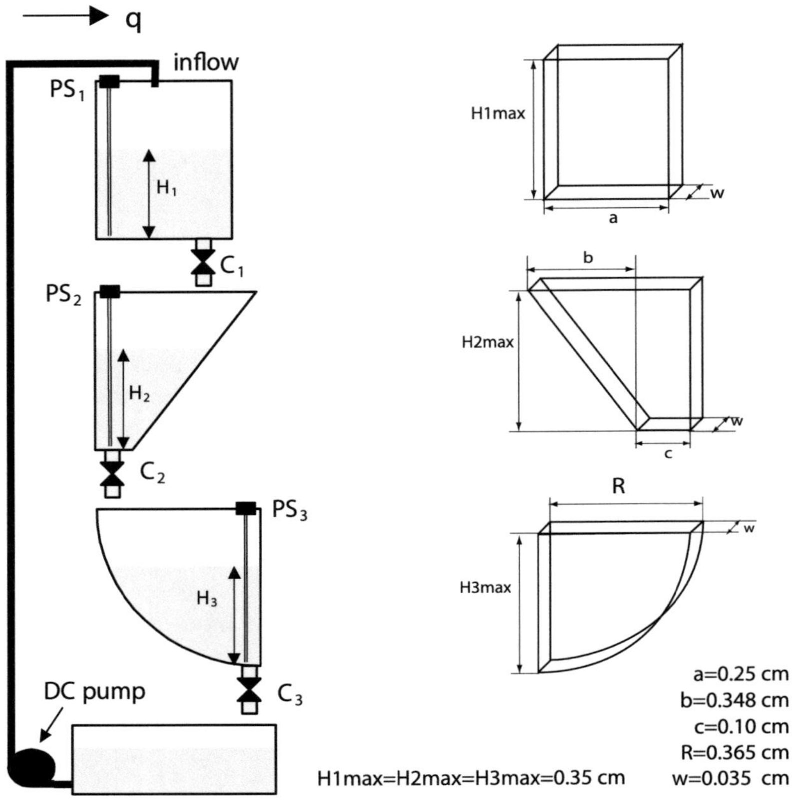

Fig. 4. Controlled laboratory multi tank system

The linearized dynamical model of the triple tank system could be described by the linear state-space equations (2) where the matrices *A*, *B*, *C* and *D* are as follow (Petrov et al., 2009):

$$A=\begin{bmatrix} \dfrac{-\alpha_1}{awH_1^{1-\alpha_1}} & 0 & 0 \\ \dfrac{\alpha_1}{w\left(c+bH_1/H_{1\max}\right)H_1^{1-\alpha_1}} & \dfrac{-\alpha_2}{w\left(c+bH_2/H_{2\max}\right)H_2^{1-\alpha_2}} & 0 \\ 0 & \dfrac{\alpha_2}{w\sqrt{R^2-(H_{3\max}-H_3)^2}\,H_2^{1-\alpha_2}} & \dfrac{-\alpha_2}{w\sqrt{R^2-(H_{3\max}-H_3)^2}\,H_3^{1-\alpha_3}} \end{bmatrix}$$

$$B=\begin{bmatrix} \dfrac{1}{aw} & \dfrac{-1}{awH_1^{1-\alpha_1}} & 0 & 0 \\ 0 & 0 & \dfrac{-1}{w\left(c+bH_2/H_{2\max}\right)H_2^{1-\alpha_2}} & 0 \\ 0 & 0 & 0 & \dfrac{-1}{w\sqrt{R^2-(H_{3\max}-H_3)^2}\,H_3^{1-\alpha_3}} \end{bmatrix}$$

$$C = \begin{bmatrix} 1 & 0 & 0 \\ 0 & 1 & 0 \\ 0 & 0 & 1 \end{bmatrix}$$

$$D = \begin{bmatrix} 0 & 0 & 0 & 0 \\ 0 & 0 & 0 & 0 \\ 0 & 0 & 0 & 0 \end{bmatrix} \tag{60}$$

The parameters α_1, α_2 and α_3 are flow coefficients for each tank of the model. The described linearized state-space model is used as an initial model for the training process of the fuzzy-neural model during the experiments.

4.1 Description of the multi tank system as a multivariable controlled process

Liquid levels H_1, H_2, H_3 in the tanks are the state variables of the system (Fig. 4). The Inteco Multi Tank system has four controlled inputs: liquid inflow q and valves settings C_1, C_2, C_3. Therefore, several models of the tanks system can be analyzed (Fig. 5), classified as pump-controlled system, valve-controlled system and pump/valve controlled system (Inteco, 2009).

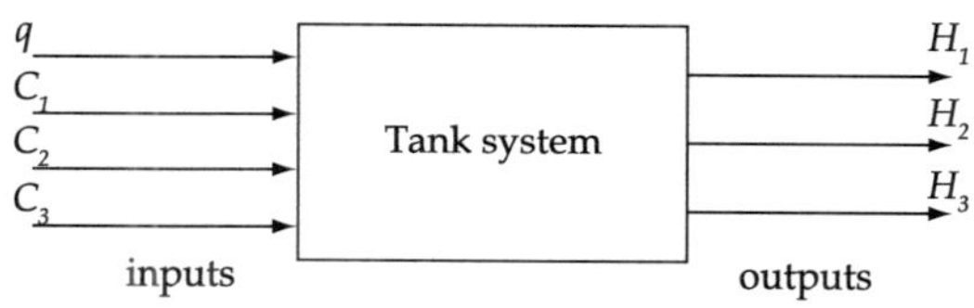

Fig. 5. Model of the Multi Tank system as a pump and valve-controlled system

In this case study a multi-input multi-output (MIMO) configuration of the Inteco Multi Tank system is used (Fig. 5). This corresponds to the linearized state-space model (60). Several issues have been recognized as causes of additional nonlinearities in plant dynamics:

- nonlinearities (smooth and nonsmooth) caused by shapes of tanks;
- saturation-type nonlinearities, introduced by maximum or minimum level allowed in tanks;
- nonlinearities introduced by valve geometry and flow dynamics;
- nonlinearities introduced by pump and valves input/output characteristic curve.

The simulation results have been obtained with random generated set points and following initial conditions (Table 4):

Model predictive controller parameters	Prediction horizon H_p=*10* First included sample of the prediction horizon H_w=*1* Control horizon H_u=*3*
Inteco Multi tank system parameters	Flow coefficients for each tank α1=0.29; α2=0.2256; α3=0.2487
Operational constraints on the system	Constraints on valve cross section ratio $0 \le C_i \le 2e\text{-}04$, $i=1,2,3$ Constraint on liquid inflow $0 \le q \le 1e\text{-}04\ m^3/s$ Constraints on liquid level in each tank $0 \le H_i \le 0.35\ m$, $i=1,2,3$
Simulation parameters	Time of simulation *600 s* Sample time T_s=*1 s*

Table 4. Simulation parameters for unconstrained and constrained fuzzy-neural MPC

Figures below show typical results for level control problem. The reference value for each tank is changed consequently in different time. The proposed fuzzy-neural identification procedure ensures the matrices for the optimization problem of model predictive control at each sampling time T_s. The plant modelling process during the unconstrained and constrained MPC experiments are shown in Fig. 6 and Fig. 9, respectively.

4.2 Experimental results with unconstrained model predictive control

The proposed unconstrained model predictive control algorithm (Table 2) with the Takagi-Sugeno fuzzy-neural model as a predictor has been applied to the level control problem. The experiments have been implemented with the parameters in Table 4. The weighting matrices are specified as follow: $\tilde{Q} = 0.01 * diag(1,\ 1,\ 1)$ and $\tilde{R} = 10e4 * diag(1,\ 1,\ 1,\ 1)$. Note that the weighting matrix $\tilde{R}$ is constant over all prediction horizon, which allows to avoid matrix inversion at each sampling time with one calculation of $\tilde{R}^{-1}$ at time *k=0*.

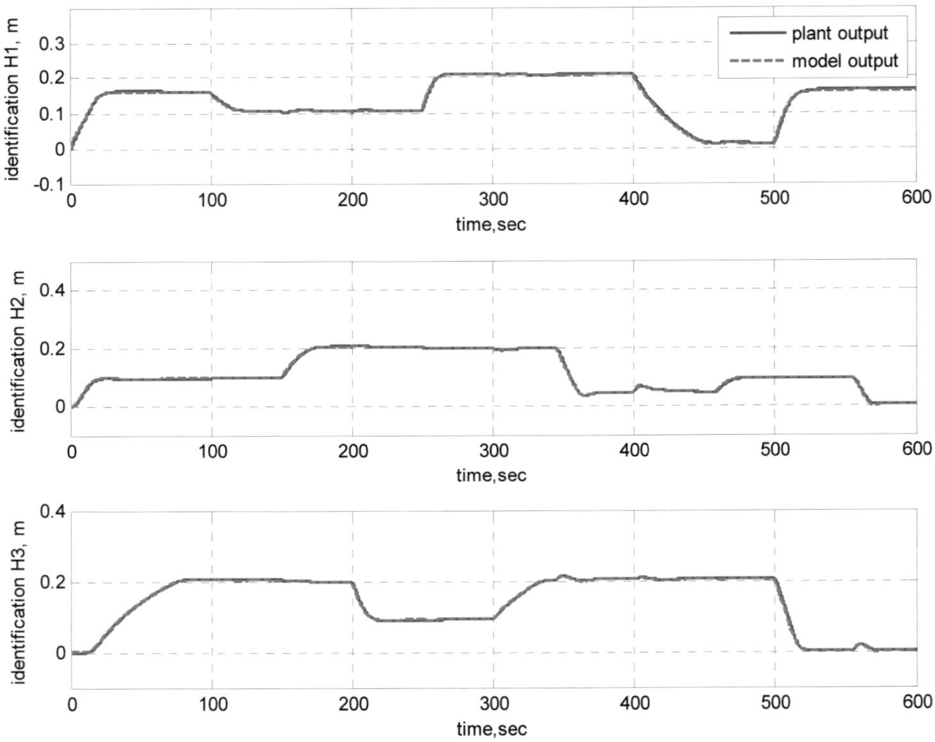

Fig. 6. Fuzzy-neural model identification procedure of the multi tank system – unconstrained NMPC

The next two figures - Fig. 7 and Fig. 8, show typical results regarding level control, where the references for H_1, H_2 and H_3 are changed consequently in different time. The change of every level reference behaves as a system disturbance for the other system outputs (levels). It is evident that the applied model predictive controller is capable to compensate these disturbances.

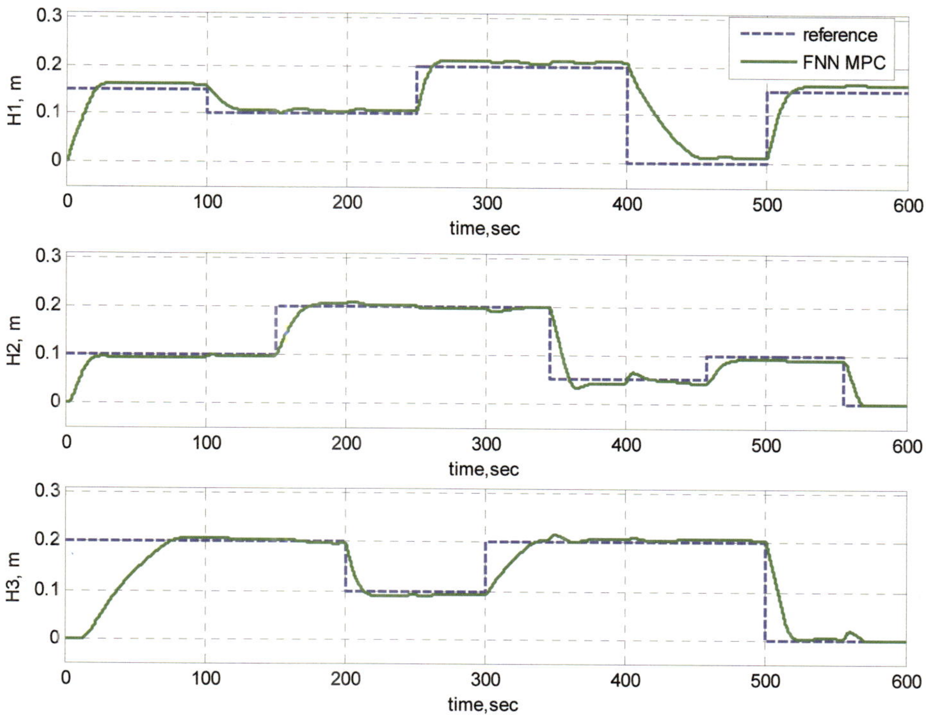

Fig. 7. Transient responses of multi tank system outputs - unconstrained NMPC

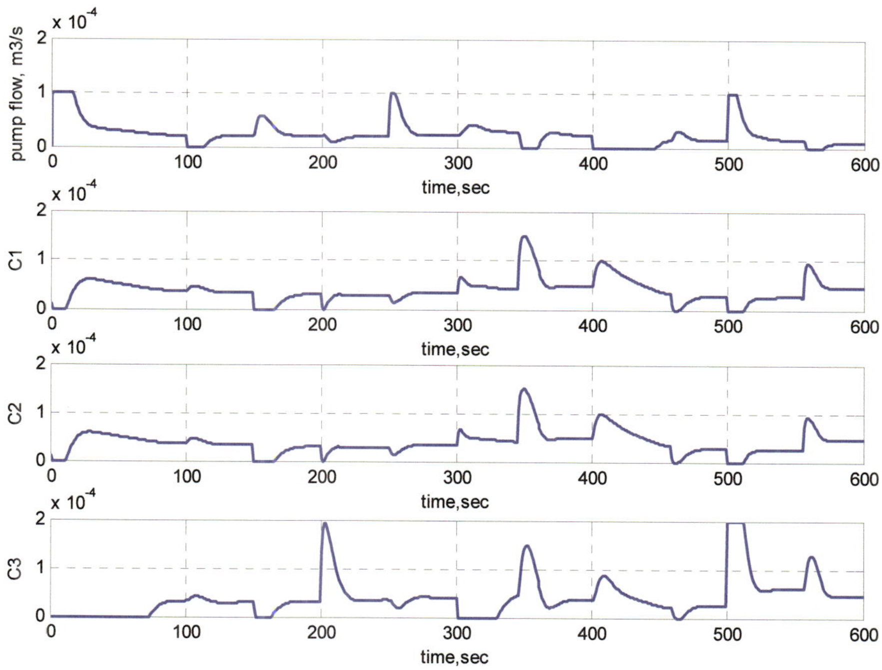

Fig. 8. Transient responses of multi tank system inputs - unconstrained NMPC

4.3 Experimental results with fuzzy-neural constrained predictive control

The experiments with the proposed constrained model predictive control algorithm (Table 3) have been made with level references close to the system outputs constraints. The weighting matrices in GPC cost function (19) are specified as $\tilde{Q} = diag(1,\ 1,\ 1)$ and $\tilde{R} = 15e4 * diag(1,\ 1,\ 1,\ 1)$. System identification during the experiment is shown on Fig. 9. The proposed identification procedure uses the linearized model (60) of the Multi tank system as an initial condition.

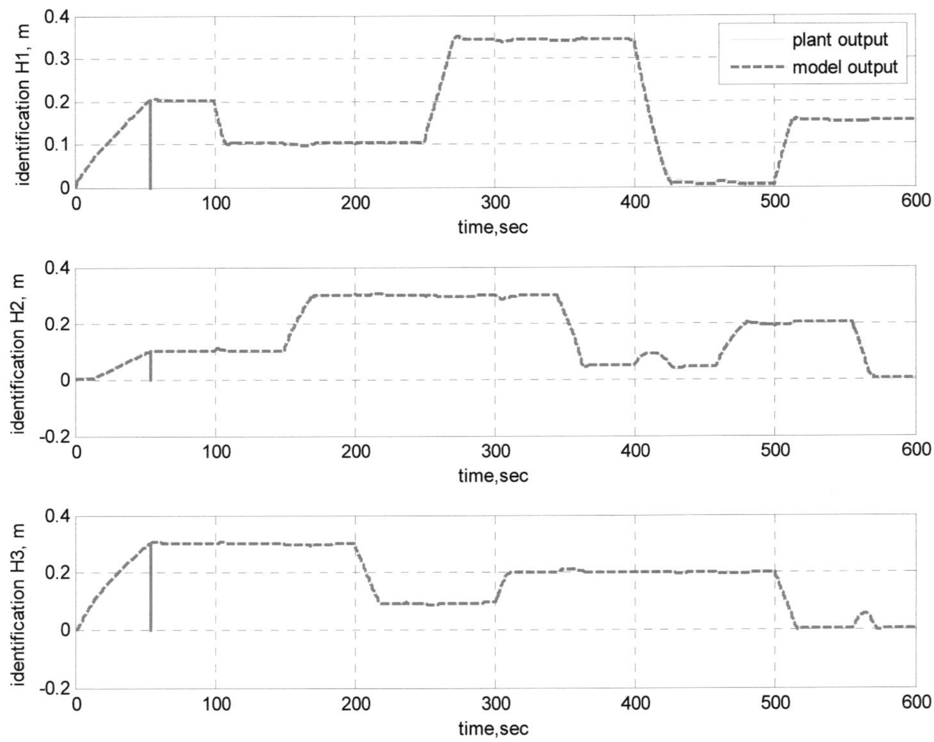

Fig. 9. Fuzzy-neural model identification procedure of the multi tank system – constrained NMPC

The proposed constrained fuzzy-neural model predictive control algorithm provides an adequate system response as it can be seen on Fig. 10 and Fig. 11. The references are achieved

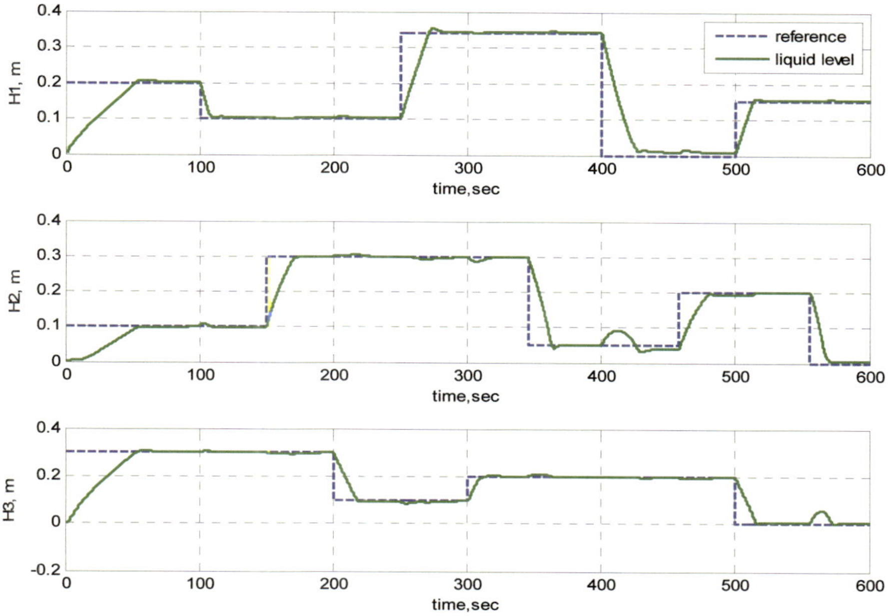

Fig. 10. Transient responses of the multi tank system outputs - constrained NMPC

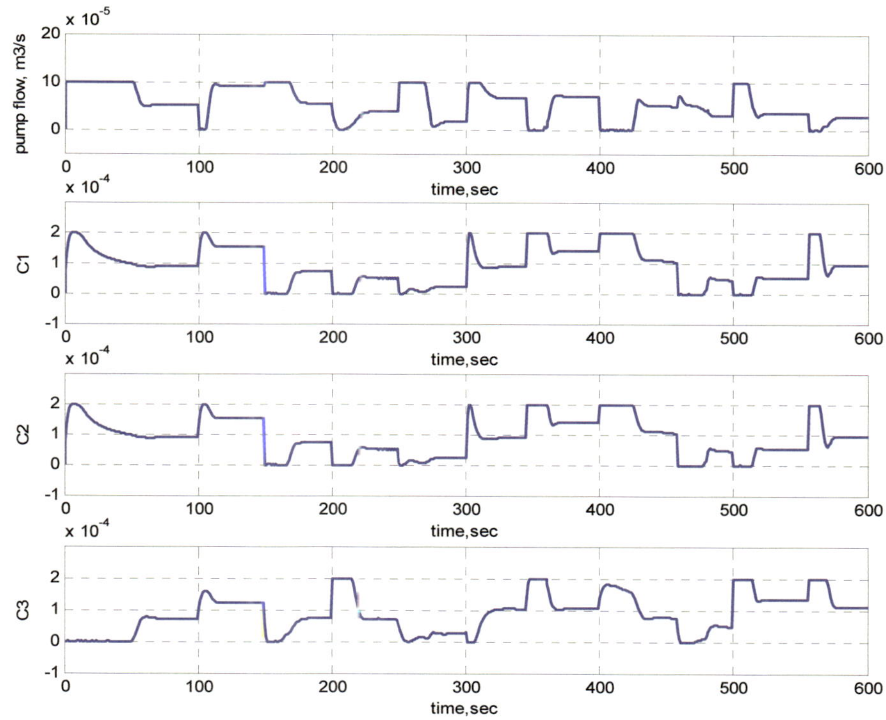

Fig. 11. Transient responses of the multi tank system inputs - constrained NMPC

without violating the operational constraints specified in Table 4. Similarly to the unconstrained case, the Takagi-Sugeno type fuzzy-neural model provides the state-space matrices *A*, *B* and *C* (the system is strictly proper, i.e. *D*=0) for the optimization procedure of the model predictive control approach. Therefore, the LIQP problem is constructed with "fresh" parameters at each sampling time and improves the adaptive features of the applied model predictive controller. It can be seen on the next figures that the disturbances, which are consequences of a sudden change of the level references, are compensated in short time without violating the proper system work.

5. Conclusions

This chapter has presented an effective approach to fuzzy model-based control. The effective modelling and identification techniques, based on fuzzy structures, combined with model predictive control strategy result in effective control for nonlinear MIMO plants. The goal was to design a new control strategy - simple in realization for designer and simple in implementation for the end user of the control systems.

The idea of using fuzzy-neural models for nonlinear system identification is not new, although more applications are necessary to demonstrate its capabilities in nonlinear identification and prediction. By implementing this idea to state-space representation of control systems, it is possible to achieve a powerful model of nonlinear plants or processes. Such models can be embedded into a predictive control scheme. State-space model of the system allows constructing the optimization problem, as a quadratic programming problem. It is important to note that the model predictive control approach has one major advantage – the ability to solve the control problem taking into consideration the operational constraints on the system.

This chapter includes two simple control algorithms with their respective derivations. They represent control strategies, based on the estimated fuzzy-neural predictive model. The two-stage learning gradient procedure is the main advantage of the proposed identification procedure. It is capable to model nonlinearities in real-time and provides an accurate model for MPC optimization procedure at each sampling time.

The proposed consequent solution for unconstrained MPC problem is the main contribution for the predictive optimization task. On the other hand, extraction of a "local" linear model, obtained from the inference process of a Takagi–Sugeno fuzzy model allows treating the nonlinear optimization problem in presence of constraints as an LIQP.

The model predictive control scheme is employed to reduce structural response of the laboratory system - multi tank system. The inherent instability of the system makes it difficult for modelling and control. Model predictive control is successfully applied to the studied multi tank system, which represents a multivariable controlled process. Adaptation of the applied fuzzy-neural internal model is the most common way of dealing with plant's nonlinearities. The results show that the controlled levels have a good performance, following closely the references and compensating the disturbances.

The contribution of the proposed approach using Takagi–Sugeno fuzzy model is the capacity to exploit the information given directly by the Takagi–Sugeno fuzzy model. This approach is very attractive for systems from high order, as no simulation is needed to obtain the parameters for solving the optimization task. The model's state-space matrices can be generated directly from the inference of the fuzzy system. The use of this approach is very attractive to the industry for practical reasons related with the capacity of this model structure to combine local models identified in experiments around the different operating points.

6. Acknowledgment

The authors would like to acknowledge the Ministry of Education and Science of Bulgaria, Research Fund project BY-TH-108/2005.

7. References

Ahmed S., M. Petrov, A. Ichtev (July 2010). Fuzzy Model-Based Predictive Control Applied to Multivariable Level Control of Multi Tank System. *Proceedings of 2010 IEEE International Conference on Intelligent Systems (IS 2010),* London, UK. pp. 456 - 461.

Ahmed S., M. Petrov, A. Ichtev, "Model predictive control of a laboratory model - coupled water tanks," in *Proceedings of International Conference Automatics and Informatics'09,* October 1-4, 2009, Sofia, Bulgaria. pp. VI-33 - VI-35.

Åkesson Johan. *MPCtools 1.0 – Reference Manual.* Technical report ISRN LUTFD2/TFRT--7613--SE, Department of Automatic Control, Lund Institute of Technology, Sweden, January 2006.

Camacho E. F., C. Bordons (2004). Model Predictive Control (Advanced Textbooks in Control and Signal Processing). Springer-Verlag London, 2004.

Espinosa J., J. Vandewalle and V. Wertz. *Fuzzy Logic, Identification and Predictive Control. (Advances in industrial control).* © Springer-Verlag London Limited, 2005.

Fletcher R. (2000). *Practical Methods of Optimization.* 2nd.ed., Wiley, 2000.

Inteco Ltd. (2009). *Multitank System - User's Manual.* Inteco Ltd., http://www.inteco.com.pl.

Lee, J.H.; Morari, M. & Garcia, C.E. (1994). State-space interpretation of model predictive control, *Automatica,* 30(4), pp. 707-717.

Maciejowski J. M. (2002). *Predictive Control with Constraints.* Prentice Hall Inc., NY, USA, 2002.

Martinsen F., Lorenz T. Biegler, Bjarne A. Foss (2004). A new optimization algorithm with application to nonlinear MPC, *Journal of Process Control,* vol.14, pp 853–865, 2004.

Mendonça L.F., J.M. Sousa J.M.G. Sá da Costa (2004). Optimization Problems in Multivariable Fuzzy Predictive Control, *International Journal of Approximate Reasoning,* vol. 36, pp. 199–221, 2004 .

Mollov S, R. Babuska, J. Abonyi, and H. Verbruggen (October 2004). Effective Optimization for Fuzzy Model Predictive Control. *IEEE Transactions on Fuzzy Systems,* Vol. 12, No. 5, pp. 661 - 675.

Petrov M., A. Taneva, T. Puleva, S. Ahmed (September, 2008). Parallel Distributed Neuro-Fuzzy Model Predictive Controller Applied to a Hydro Turbine Generator. *Proceedings of the Forth International IEEE Conference on "Intelligent Systems",*

Golden Sands resort, Varna, Bulgaria. ISBN 978-1-4244-1740-7, Vol. I, pp. 9-20 - 9-25.

Petrov M., I. Ganchev, A. Taneva (November 2002). Fuzzy model predictive control of nonlinear processes. *Preprints of the International Conference on "Automation and Informatics 2002"*, Sofia, Bulgaria, 2002. ISBN 954-9641-30-9, pp. 77-80.

Rossiter J.A. (2003). *Model based predictive control – A practical Approach.* CRC Press, 2003.

8

Using Subsets Sequence to Approach the Maximal Terminal Region for MPC

Yafeng Wang[1,2], Fuchun Sun[2], Youan Zhang[1], Huaping Liu[2] and Haibo Min[2]
[1]*Department of Control Engineering, Naval Aeronautical Engineering University, Yantai,*
[2]*Department of Computer Science and Technology, Tsinghua University, Beijing,*
China

1. Introduction

Due to the ability to handle control and state constraints, MPC has become quite popular recently. In order to guarantee the stability of MPC, a terminal constraint and a terminal cost are added to the on-line optimization problem such that the terminal region is a positively invariant set for the system and the terminal cost is an associated Lyapunov function [1, 9].
As we know, the domain of attraction of MPC can be enlarged by increasing the prediction horizon, but it is at the expense of a greater computational burden. In [2], a prediction horizon larger than the control horizon was considered and the domain of attraction was enlarged. On the other hand, the domain of attraction can be enlarged by enlarging the terminal region. In [3], an ellipsoidal set included in the stabilizable region of using linear feedback controller served as the terminal region. In [4], a polytopic set was adopted. In [5], a saturated local control law was used to enlarge the terminal region. In [6], SVM was employed to estimate the stabilizable region of using linear feedback controller and the estimated stabilizable region was used as the terminal region. The method in [6] enlarged the terminal region dramatically. In [7], it was proved that, for the MPC without terminal constraint, the terminal region can be enlarged by weighting the terminal cost. In [8], the enlargement of the domain of attraction was obtained by employing a contractive terminal constraint. In [9], the domain of attraction was enlarged by the inclusion of an appropriate set of slacked terminal constraints into the control problem.
In this paper, the domain of attraction is enlarged by enlarging the terminal region. A novel method is proposed to achive a large terminal region. First, the sufficient conditions to guarantee the stability of MPC are presented and the maximal terminal region satisfying these conditions is defined. Then, given the terminal cost and an initial subset of the maximal terminal region, a subsets sequence is obtained by using one-step set expansion iteratively. It is proved that, when the iteration time goes to infinity, this subsets sequence will converge to the maximal terminal region. Finally, the subsets in this sequence are separated from the state space one by one by exploiting SVM classifier (see [10,11] for details of SVM).

2. Model predictive control

Consider the discrete-time system as follows

$$x_{k+1} = f(x_k, u_k) \tag{1}$$

where $x_k \in R^n$, $u_k \in R^m$ are the state and the input of the system at the sampling time k respectively. $x_{k+1} \in R^n$ is the successor state and the mapping $f: R^{n+m} \mapsto R^n$ satisfying $f(\mathbf{0},\mathbf{0}) = \mathbf{0}$ is known. The system is subject to constraints on both state and control action. They are given by $x_k \in X$, $u_k \in U$, where X is a closed and bounded set, U is a compact set. Both of them contain the origin.

The on-line optimization problem of MPC at the sample time k, denoted by $P_N(x_k)$, is stated as

$$\begin{aligned} &\min_{u(i,x_k)\in U} J_N(\boldsymbol{u}, x_k) = \sum_{i=0}^{N-1} q(x(i,x_k), u(i,x_k)) + F(x(N,x_k)) \\ &\quad s.t. \;\; x(i+1,x_k) = f(x(i,x_k), u(i,x_k)) \\ &\qquad\quad x(i+1,x_k) \in X, u(i,x_k) \in U, x(N,x_k) \in X_f \end{aligned} \tag{2}$$

where $x(0,x_k) = x_k$ is the state at the sample time k, $q(x,u)$ denotes the stage cost and it is positive definite, N is the prediction horizon, X_f denotes the terminal region and it is closed and satisfies $\mathbf{0} \in X_f \subseteq X$, $F(\cdot)$ satisfying $F(\mathbf{0}) = 0$ is the terminal cost and it is continuous and positive definite.

Consider an assumption as follows.

Assumption 1. For the terminal region and the terminal cost, the following two conditions are satisfied [1]:

(C1) $F(\cdot)$ is a Lyapunov function. For any $x \in X_f$, there exists

$$F(x) \geq \min_{u\in U}\{q(x,u) + F(f(x,u))\}.$$

(C2) X_f is a positively invariant set. For any $x \in X_f$, by using the optimal control resulting from the minimization problem showed in (C1), denoted by u_{opt} , we have $f(x, u_{opt}) \in X_f$.

Let $J_N^*(x_k)$ be the minimum of $P_N(x_k)$ and $\boldsymbol{u}_N^*(x_k) = \{u_N^*(0,x_k), \cdots, u_N^*(N-1,x_k)\}$ be the optimal control trajectory. The control strategy of MPC is that, at the sample time k, $u_N^*(0,x_k)$ is inputted into the real system and at the sample time $k+1$, the control inputted into the system is not $u_N^*(1,x_k)$ but the first element of the optimal control trajectory resulting from the similar on-line optimization problem. At the sample time $k+1$, the state is $x_{k+1} = f(x_k, u_N^*(0,x_k))$ and the on-line optimization problem, denoted by $P_N(x_{k+1})$, is same as (2) except that x_k is replaced by x_{k+1}. Similarly, let $J_N^*(x_{k+1})$ be the minimum of $P_N(x_{k+1})$ and $\boldsymbol{u}_N^*(x_{k+1}) = \{u_N^*(0,x_{k+1}), \cdots, u_N^*(N-1,x_{k+1})\}$ be the optimal control trajectory. The control inputted into the system at the sample time $k+1$ is $u_N^*(0,x_{k+1})$. So, the control law of MPC can be stated as $u_{RH}(x_k) = u_N^*(0,x_k), k = 0,1,2,\cdots,\infty$.

The closed-loop stability of the controlled system is showed in lemma 1.

Lemma 1. For any $x_0 \in X$, if x_0 satisfies $x_N^*(N,x_0) \in X_f$ and assumption 1 is satisfied, it is guaranteed that, x_0 will be steered to $\mathbf{0}$ by using the control law of MPC.

The proof can be found in [1].

Proof. The proof of lemma 1 is composed of two parts: the existence of feasible solution; the monotonicity of $J_N^*(\cdot)$.

Part 1. At the sample time 1, $x_1 = x^*(1,x_0) = f(x_0, u^*(0,x_0))$ is obtained by inputting $u^*(0,x_0)$ into the system, where $u^*(0,x_0)$ denotes the first element of the optimal solution of $P_N(x_0)$. It is obvious that, $\boldsymbol{u}(x_1) = \{u^*(1,x_0),\cdots,u^*(N-1,x_0),u_{opt}(x^*(N,x_0))\}$ is a feasible solution of $P_N(x_1)$ since $x^*(N,x_0) \in X_f$ and $f(x^*(N,x_0),u_{opt}(x^*(N,x_0))) \in X_f$ as assumption 1 shows.

Part 2. When $\boldsymbol{u}(x_1)$ is used, we have

$$\begin{aligned}
&J_N(\boldsymbol{u}(x_1),x_1) - J_N^*(x_0) \\
&\quad = q(x^*(N,x_0),u_{opt}(x^*(N,x_0))) + F(f(x^*(N,x_0),u_{opt}(x^*(N,x_0)))) \\
&\qquad - q(x^*(0,x_0),u^*(0,x_0)) - F(x^*(N,x_0)) \\
&\quad \le -q(x^*(0,x_0),u^*(0,x_0)) \\
&\quad \le 0
\end{aligned}$$

Since $J_N^*(x_1) \le J_N(\boldsymbol{u}(x_1),x_1)$, it follows that, $J_N^*(x_1) - J_N^*(x_0) \le 0$.

Endproof.

3. Using subsets sequence to approach the maximal terminal region

Using SVM classifier to estimate the terminal region is not a new technology. In [6], a large terminal region was achieved by using SVM classifier. However, the method in [6] is somewhat conservative. The reason is that, the obtained terminal region actually is the stabilizable region of using a predetermined linear feedback controller.

In this section, a novel method of computing a terminal region is proposed. Given the terminal cost and a subset of the maximal terminal region, a subsets sequence is constructed by using one-step set expansion iteratively and SVM is employed to estimate each subset in this sequence. When some conditions are satisfied, the iteration ends and the last subset is adopted to serve as the terminal region.

3.1 The construction of subsets sequence

Consider an assumption as follows.

Assumption 2. A terminal cost is known.

If the stage cost is a quadratic function as $q(x,u) = x^T Q x + u^T R u$ in which Q, R are positive definite, a method of computing a terminal cost for continuous-time system can be found in [3]. In this paper, the method in [3] is extended to discrete-time system. Consider the linearization of the system (1) at the origin

$$x_{k+1} = A_d x_k + B_d u_k$$

with $A_d = (\partial f / \partial x)(0,0)$ and $B_d = (\partial f / \partial u)(0,0)$.

A terminal cost can be obtained through the following procedure:

Step 1. Solving the Riccati equation to get G_0,

$$G_0 = A_d^T G_0 A_d - \left(A_d^T G_0 B_d\right)\left(B_d^T G_0 B_d + R\right)^{-1}\left(B_d^T G_0 A_d\right) + Q$$

Step 2. Getting a locally stabilizing linear state feedback gain K,

$$K = -\left(B_d^T G_0 B_d + R\right)^{-1}\left(B_d^T G_0 A_d\right)$$

Step 3. Computing G_K by solving the following Riccati equation,

$$(\alpha A_K)^T G_K (\alpha A_K) - G_K = -Q_K$$

where $A_K = A_d + B_d K$, $Q_K = Q + K^T R K$, and $\alpha \in [1, +\infty)$ is an adjustable parameter satisfying $\alpha\left|\lambda_{\max}(A_K)\right| < 1$. Then, $F(x) = x^T G_K x$ can serve as a terminal cost.
Given $F(\cdot)$ and from conditions (C1,C2), the terminal region X_f can be defined as

$$X_f := \left\{x \in X \mid F(x) \geq F_{X_f}^*(x)\right\} \tag{3}$$

where $F_{X_f}^*(x)$ is the minimum of the following optimization problem

$$\begin{aligned} &\min_{u \in U} F_{X_f}(x) = q(x,u) + F\left(f(x,u)\right) \\ &\text{s.t.} \quad f(x,u) \in X_f \end{aligned} \tag{4}$$

Remark 1. The construction of X_f has two meanings: (I) the optimization problem (4) has feasible solution, that is to say, $\exists u \in U$, s.t. $f(x,u) \in X_f$; (II) the minimum of the optimization problem satisfies that $F_{X_f}^*(x) \leq F(x)$.
Remark 2. From the definition of X_f, it is obvious that, the terminal region is essentially a positively invariant set of using the optimal control resulting from the optimization problem (4) when $F(\cdot)$ is given.
Remark 3. In [3,4,6], the linear feedback control is attached to the construction of X_f and X_f is the stabilizable region of using the linear feedback controller. In [5], a saturated local control law was used. But, in this paper, there is no explicit control attached to the definition of X_f. So, the requirement on X_f is lower than that in [3-6] while guaranting the stability of the controlled system.
From the definition of X_f, it can not be determined whether a state point belongs to X_f. The difficulty lies in that, the X_f itself acts as the constraint in the optimization problem (4). To avoid this problem, the method of using one-step set expansion iteratively is adopted. Define $X_{f,\max}$ as the largest terminal region and consider an assumption.

Assumption 2. A subset of $X_{f,\max}$, denoted by X_f^0 and containing the origin, is known.

Assumption 3. X_f^0 is a positively invariant set, that is to say, for any $x \in X_f^0$, $\exists u \in U$, s.t. $F(x) \geq q(x,u) + F\left(f(x,u)\right)$ and $f(x,u) \in X_f^0$.
Given X_f^0, another subset of $X_{f,\max}$, denoted by X_f^1, can be constructed as

$$X_f^1 := \left\{ x \in X \mid F(x) \geq F_{X_f^0}^*(x) \right\} \tag{5}$$

where $F_{X_f^0}^*(x)$ is the minimum of

$$\begin{aligned} &\min_{u \in U} F_{X_f^0}(x) = q(x,u) + F(f(x,u)) \\ &\quad s.t. \quad f(x,u) \in X_f^0 \end{aligned} \tag{6}$$

As mentioned in remark 1, the construction of X_f^1 contains two meanings: (I) for any $x \in X_f^1$, $\exists u \in U$, s.t. $f(x,u) \in X_f^0$; (II) the minimum of (6) satisfies $F_{X_f^0}^*(x) \leq F(x)$. The constructions of X_f^j in sequel have the similar meanings.

Lemma 2. If assumption 3 is satisfied, there is $X_f^0 \subseteq X_f^1$.

Proof. If assumption 3 is satisfied, it is obvious that, for any $x \in X_f^0$, $\exists u \in U$, s.t. $F(x) \geq q(x,u) + F(f(x,u))$ and $f(x,u) \in X_f^0$. It follows that, $F(x) \geq F_{X_f^0}^*(x)$. From the construction of X_f^1, we can know $x \in X_f^1$, namely, $X_f^0 \subseteq X_f^1$.
Endproof.

Remark 4. From the construction of X_f^1, it is obvious that, if assumption 3 is satisfied, X_f^1 is a positively invariant set. We know that, for any $x \in X_f^1$, $\exists u \in U$, s.t. $F(x) \geq q(x,u) + F(f(x,u))$ and $f(x,u) \in X_f^0$. Because of $X_f^0 \subseteq X_f^1$ as showed in lemma 2, we have $f(x,u) \in X_f^1$.

Similarly, by replacing X_f^0 with X_f^1 in the constraint of (6), another subset, denoted by X_f^2, can be obtained as follows

$$X_f^2 := \left\{ x \in X \mid F(x) \geq F_{X_f^1}^*(x) \right\} \tag{7}$$

where $F_{X_f^1}^*(x)$ is the minimum of

$$\begin{aligned} &\min_{u \in U} F_{X_f^1}(x) = q(x,u) + F(f(x,u)) \\ &\quad s.t. \quad f(x,u) \in X_f^1 \end{aligned} \tag{8}$$

Repeatedly, X_f^j, $j = 3,4,\cdots,\infty$ can be constructed as

$$X_f^j := \left\{ x \in X \mid F(x) \geq F_{X_f^{j-1}}^*(x) \right\} \tag{9}$$

where $F_{X_f^{j-1}}^*(x)$ is the minimum of

$$\begin{aligned} &\min_{u \in U} F_{X_f^{j-1}}(x) = q(x,u) + F(f(x,u)) \\ &\quad s.t. \quad f(x,u) \in X_f^{j-1} \end{aligned} \tag{10}$$

This method of constructing X_f^j given X_f^{j-1} is defined as one-step set expansion in this paper. By employing it iteratively, a subsets sequence of largest terminal region, denoted by $\{X_f^j\}$, $j=1,2,\cdots,\infty$, can be achived.

Remark 5. Similar with lemma 2 and remark 4, any subset in this sequence is positively invariant and any two neighbouring subsets satisfy $X_f^{j-1} \subseteq X_f^j$.

As j increases, $\{X_f^j\}$ will converge to a set, denoted by $X_f^{+\infty}$. Theorem 1 will show that, $X_f^{+\infty}$ is equal to the largest terminal region.

Theorem 1. *If assumption 2 and assumption 3 are satisfied, for X_f^j constructed in (9) and (10), when j goes to infinity, $\{X_f^j\}$ will converge to $X_{f,\max}$.*

Proof. This theorem is proved by contradiction.

(A) Assume that, there exists a set which is denoted by X_{spo} satisfying $X_{spo} \subset X_{f,\max}$ and $X_f^j \to X_{spo}$ when $j \to +\infty$. From remark 5, we can know $X_f^0 \subseteq X_{spo}$. It is obvious that $\mathbf{0} \in X_{spo}$ because of $\mathbf{0} \in X_f^0$ as showed in assumption 2. It follows that, $\mathbf{0} \notin X_{f,\max} \setminus X_{spo}$ and for any $x \in X_{f,\max} \setminus X_{spo}$, we have $F(x) > 0$ since $F(\cdot)$ is positive definite. Define ξ as the infimum of $\{F(x) \mid x \in X_{f,\max} \setminus X_{spo}\}$, it is satisfied that, $\xi > 0$.

From the construction of X_f^j, we know that, for any $x_0 \in X_{f,\max} \setminus X_{spo}$, there exists no such a $u \in U$ satisfying $F(x_0) \ge q(x_0,u) + F(f(x_0,u))$ and $f(x_0,u) \in X_{spo}$ because of $X_{spo} \subset X_{f,\max}$. However, from (C1) and (C2), we know that, $\exists u(x_0) \in U$, s.t. $F(x_0) \ge q(x_0,u(x_0)) + F(x_1)$ and $x_1 \in X_{f,\max}$, where $x_1 = f(x_0,u(x_0))$. It is obvious that, $x_1 \notin X_{spo}$. So we have, $x_1 \in X_{f,\max} \setminus X_{spo}$. Similarly, we can know, $\exists u(x_1) \in U$, s.t. $F(x_1) \ge q(x_1,u(x_1)) + F(x_2)$ and $x_2 \in X_{f,\max} \setminus X_{spo}$, where $x_2 = f(x_1,u(x_1))$, since $x_1 \in X_{f,\max} \setminus X_{spo}$.

Repeatly, for $x_i \in X_{f,\max} \setminus X_{spo}, \exists u(x_i) \in U$, s.t. $F(x_i) \ge q(x_i,u(x_i)) + F(x_{i+1})$ and $x_{i+1} \in X_{f,\max} \setminus X_{spo}$, where $x_{i+1} = f(x_i,u(x_i))$, $i=2,\cdots,\infty$. It is clear that, $F(x_i) \to 0$ when $i \to \infty$. We know that, for the infimum of $\{F(x) \mid x \in X_{f,\max} \setminus X_{spo}\}$, defined as ξ, there is a positive real number δ satisfing $0 < \delta < \xi$. Since $F(x_i) \to 0$ when $i \to \infty$, $\exists N_\delta > 0$, s.t. for any $i \ge N_\delta$, we have $F(x_i) < \delta$. Obviously, this is contradicted with that ξ is the infimum of $\{F(x) \mid x \in X_{f,\max} \setminus X_{spo}\}$.

(B) Similarly, assume that, there exists a X_{spo} satisfying $X_{spo} \supset X_{f,\max}$ and $X_f^j \to X_{spo}$ when $j \to +\infty$. For any $x \in X_{spo}$, we have that $F(x) \ge \min_{u \in U}\{q(x,u) + F(f(x,u))\}$ and $f(x,u) \in X_{spo}$. Obviously, this is contradicted with that $X_{f,\max}$ is the largest one satisfying (C1) and (C2).

Endproof.

Remark 6. In this paper, the largest terminal region means the positively invariant set satisfying conditions (C1) and (C2). But, (C1) and (C2) are sufficient conditions to guarantee the stability of the controlled system, not the necessary conditions. There may be a set larger than $X_{f,\max}$ and the stability of the controlled system can be guaranteed by using this set as the terminal region.

Remark 7. In the calculation of $X_{f,\max}$, it is impossible to keep iteration computation until $j \to +\infty$. When the iteration time goes to $j = E$ (E is a positive integer), if X_f^E is equal to X_f^{E-1} in principle, it can be deemed that $\{X_f^j\}$ converges to X_f^E in rough. Hence, X_f^E can be taken as the terminal region and it is a good approximation to $X_{f,\max}$.

Remark 8. If the iteration time does not go to infinity, the obtained set may be just a large positively invariant subset of $X_{f,\max}$. This has no effect on the stability of the controlled

system. The only negative influence is that its corresponding domain of attraction is smaller than that corresponding to $X_{f,\max}$.
Untill now, it seems that we can choose any X_f^j in the subsets sequence as the terminal region. This is infeasible. Since X_f^j is not described in explicit expression, it can not serve as the terminal constraint in the optimization problem (2) directly. Then, an estimated one described in explicit expression is needed. Due to the strong optimizing ability of SVM, SVM is exploited to separate each X_f^j from the state space.

3.2 Support vector machine

SVM is the youngest part in the statistical learning theory. It is an effective approach for pattern recognition. In SVM approach, the main aim is to obtain a function, which determines the decision boundary or hyperplane. This hyperplane optimally separates two classes of input data points.
Take the example of separating X into A and $X \setminus A$. For each $x_i \in A$, an additional variable $y_i = +1$ is introduced. Similarly, for each $x_i \in X \setminus A$, $y_i = -1$ is introduced. Define $I^+ := \{i : y_i = +1\}$ and $I^- := \{i : y_i = -1\}$, SVM will find a separating hyperplane, denoted by $O(x) := w \cdot \phi(x_i) + b = 0$, between A and $X \setminus A$. Therefore, A can be estimated as $\hat{A} = \{x \in X \mid O(x) \geq 0\}$, where $O(x)$ is determined by solving the following problem:

$$\begin{aligned} \min_{\alpha} \quad & \frac{1}{2}\sum_i \sum_j \alpha_i \alpha_j y_i y_j ker(x_i, x_j) - \sum_i \alpha_i \\ \text{s.t.} \quad & \sum_i \alpha_i y_i = 0 \\ & 0 \leq \alpha_i \leq C, \ \forall i \in I^+; \ \alpha_i \geq 0, \forall i \in I^- \end{aligned} \tag{11}$$

where $ker(\cdot,\cdot)$ denotes the kernel function and the Gaussian kernel as follows is adopted in this paper:

$$ker(x, x_i) = \exp\left(-\frac{\|x - x_i\|^2}{2\sigma^2}\right) \tag{12}$$

with σ being the positive Gaussian kernel width.
When $\{\alpha_i\}$ are computed out, some support vectors are chosen from $\{x_i\}$ and the optimal hyperplane can be determined with these support vectors and their relevant weights. Denote P_s as the number of support vectors and X_s as the support vectors set, the optimal hyperplane is described as:

$$O(x) = \sum_{i=1}^{P_s} w_i \cdot ker(x_i, x) + b \tag{13}$$

where $x_i \in X_s$ is the support vector and $w_i = \alpha_i y_i$ satifying $\sum_{i=1}^{P_s} w_i = 0$ is the relevant weight. There are many software packages of SVM available on internet. They can be downloaded and used directly. To save space, it is not introduced in detail in this paper. For more details, please refer to [10] and [11].

3.3 Estimating the subset by employing SVM

From subsection 3.2, we know that, SVM find a separating hyperplane between $\{x_i \mid i\in I^+\}$ and $\{x_i \mid i\in I^-\}$.This hyperplane is used to separate X into A and $X\setminus A$. All of $\{x_i\}$ and their relevant $\{y_i\}$ compose a set, named the training points set. This subsection will show how to achieve the training points set when estimating X_f^j and how to determine X_f^j when the separating hyperplane is known.

Firstly, choose arbitrary points $x_i\in X$, $i=1,2,...,P$ (P is the number of training points); then, assign y_i to each x_i by implementing the following procedure:

IF (I) the following optimization problem has feasible solution

$$\begin{aligned} &\min_{u\in U} F_{X_f^j}(x_i)=q(x_i,u)+F(f(x_i,u)) \\ &s.t.\quad f(x_i,u)\in \hat{X}_f^{j-1} \end{aligned}.$$

(When $j=1$, $\hat{X}_f^0=X_f^0$.)

(II) its minimum satisfies

$$F(x_i)\ge F_{X_f^j}^*(x_i).$$

THEN $y_i=+1$

ELSE $y_i=-1$

ENDIF.

By implementing this procedure for every x_i, each y_i is known. Input $\{x_i\}$ and $\{y_i\}$ into SVM classifier, an optimal hyperplane $O^j(x)=0$ will be obtained. Therefore, the estimated set of X_f^j can be achieved as $\hat{X}_f^j=\{x\in X \mid O^j(x)\ge 0\}$.

When $\hat{X}_f^j$ is known, the training points for separating X_f^{j+1} from X can be computed by the similar procedure. By inputting them into SVM classifier, a hyperplane $O^{j+1}(x)=0$ and an estimated set of X_f^{j+1}, denoted by $\hat{X}_f^{j+1}=\{x\in X \mid O^{j+1}(x)\ge 0\}$ will be obtained. Repeatedly, $\{O^j(x)\},j=1,2,\cdots,\infty$ and $\{\hat{X}_f^j\}$ can be be achieved by the similar technology.

4. Estimating the terminal region

Section 3 showed how to achieve the subsets sequence by employing SVM. Theoretically, the larger the iteration time j, the higher the precision of $\hat{X}_f^j$ approaching to $X_{f,\max}$. But, it is impossible to keep computation until $j\to+\infty$. To avoid this problem, the iteration should be ended when some conditions are satisfied.

When $j=E$, if it is satisfied that, for $x_i\in X_{s,E-1}$, $i=1,2,\cdots,P_{s,E-1}$, there exists

$$\sum_{i=1}^{P_{s,E-1}} \left\|O^E(x_i)-O^{E-1}(x_i)\right\| \le \varepsilon P_{s,E-1}, \tag{14}$$

it can be deemed that $\hat{X}_f^E$ is equal to $\hat{X}_f^{E-1}$ in principle and $\hat{X}_f^j$ converges to $\hat{X}_f^E$. In (14), $X_{s,E-1}$ is the support vectors set at $j=E-1$, $P_{s,E-1}$ is the number of support vectors and ε is a tunable threshold. The smaller ε is, the higher the precision of $\hat{X}_f^E$ approximating to $X_{f,\max}$ is. Finally, $\hat{X}_f^E$ is used to serve as the terminal region.

Remark 9. Here, we used the information that, in SVM classifier, the hyperplanes are only determined on the support vectors.

Now, the concrete algorithm of estimating the largest terminal region is displayed as follows.

Step 4. Step 1 Set the number of training points P used in SVM and the tunable threshold ε.

Step 5. Step 2 For $j=1,2,\cdots,\infty$, use SVM to achieve the optimal hyperplane $O^j(x)=0$ and the estimated set of X_f^j, denoted by $\hat{X}_f^j$.

Substep 1. Choose arbitrary points $x_i \in X$, $i=1,2,...,P$.

Substep 2. Assign y_i to each x_i by implementing the procedure in subsection 3.3.

Substep 3. Input $\{x_i, y_i\}$ into the SVM. An optimal hyperplane $O^j(x)=0$ will be obtained and X_f^j can be approximated by $\hat{X}_f^j = \{x \in X \mid O^j(x) \geq 0\}$, where

$$O^j(x) = \sum_{i=1}^{P_{s,j}} w_i \cdot ker(x_i, x) + b_j$$

with $P_{s,j}$ denoting the number of support vectors, x_i being the support vector, w_i denoting its relevant weight and b_j denoting the classifier threshold.

Step 6. Step 3 Check the iteration status. When $j=E$, if inequality (14) is satisfied, end iteration and take $\hat{X}_f^E$ as the largest terminal region.

Remark 10. It is obvious that, $\hat{X}_f^j$ is achieved one by one. Namely, $\hat{X}_f^j$ can only be achieved when $\hat{X}_f^{j-1}$ is known.

5. Simulation experiment

The model is a discrete-time realization of the continuous-time system used in [3, 6]:

$$\begin{bmatrix} x_1(k+1) \\ x_2(k+1) \end{bmatrix} = \begin{bmatrix} 1 & T \\ T & 1 \end{bmatrix} \begin{bmatrix} x_1(k) \\ x_2(k) \end{bmatrix} + \begin{bmatrix} T\mu \\ T\mu \end{bmatrix} u(k) + \begin{bmatrix} T(1-\mu) & 0 \\ 0 & -4T(1-\mu) \end{bmatrix} \begin{bmatrix} x_1(k) \\ x_2(k) \end{bmatrix} u(k)$$

where $\mu=0.5$, $T=0.1s$, and the state constraint and control constraint are $X=\{x \mid \|x\|_1 \leq 4\}$, $U=\{u \mid |u| \leq 2\}$, respectively.

The stage cost is $q(x,u)=x^T Q x + u^T R u$ where $Q=0.5I$ and $R=1$. The terminal cost is chosen as $F(x)=x^T G x$ where $G=[1107.356 \quad 857.231; 857.231 \quad 1107.356]$ and X_f^0 is given as the terminal region in [3] which is

$$X_f^0 = \left\{ x \in X \mid x^T \begin{bmatrix} 16.5926 & 11.5926 \\ 11.5926 & 16.5926 \end{bmatrix} x \le 0.7 \right\}.$$

To estimate each X_f^j, 4000 training points are generated. Set $\varepsilon = 2.5$, when $j = 22$, there exists

$$\sum_{i=1}^{P_{s,21}} \left\| O^{22}(x_i) - O^{21}(x_i) \right\| \le \varepsilon P_{s,21},$$

where $x_i \in X_{s,21}$, $X_{s,21}$ is the support vectors set and $P_{s,21}$ is the number of support vectors at $j = 21$. Then, it is deemed that, $\hat{X}_f^{22}$ is equal to $\hat{X}_f^{21}$ in principle and $\hat{X}_f^{22}$ can be taken as the final estimation of $X_{f,\max}$. Figure 1 shows the approximation process of $X_{f,\max}$.

In figure 1, the blue ellipsoid is the terminal region in [3], which serves as X_f^0 in the estimation of $X_{f,\max}$ in this paper. The regions surrounded by black solid lines are $\{\hat{X}_f^j\}, j = 1,2,\cdots 22$ in which the smallest one is $\hat{X}_f^1$, the largest one is $\hat{X}_f^{22}$ and the regions between them are $\{\hat{X}_f^j\}, j = 2,3,\cdots 21$ satisfying $\hat{X}_f^{j-1} \subseteq \hat{X}_f^j$. The time cost of employing SVM to estimate each $\hat{X}_f^j$ is about 44 minutes and the total time cost of computing the final estimation of $X_{f,\max}$, namely, $\hat{X}_f^{22}$ is about 16 hours.

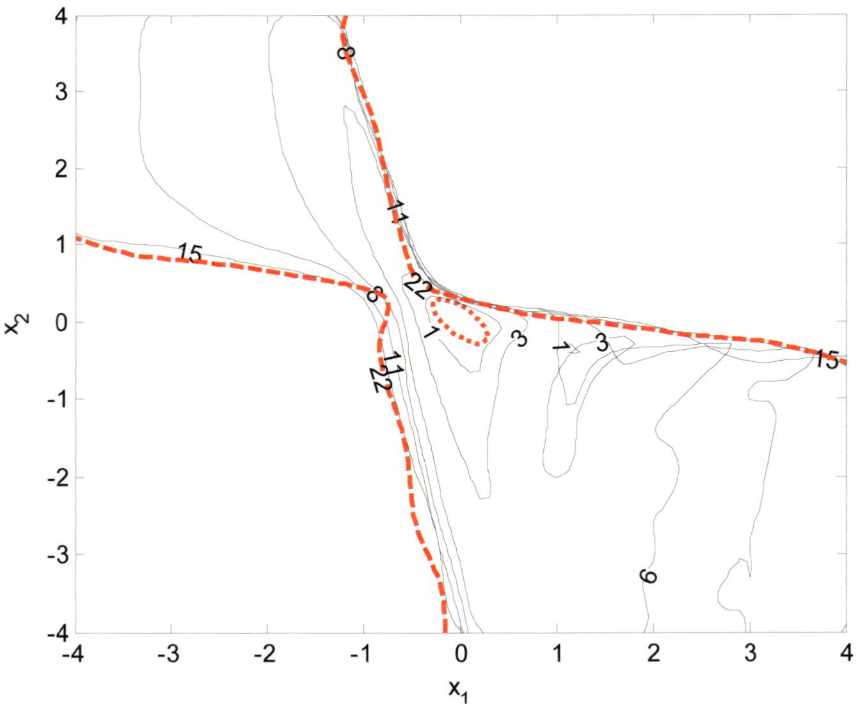

Fig. 1. The approximation process

Set the prediction horizon as $N = 3$, some points in the region of attraction (this example is very exceptional, the region of attraction is coincident with the terminal region in rough. Therefore, these points are selected from the terminal region) are selected and their closed-loop trajectories are showed in Figure 2.

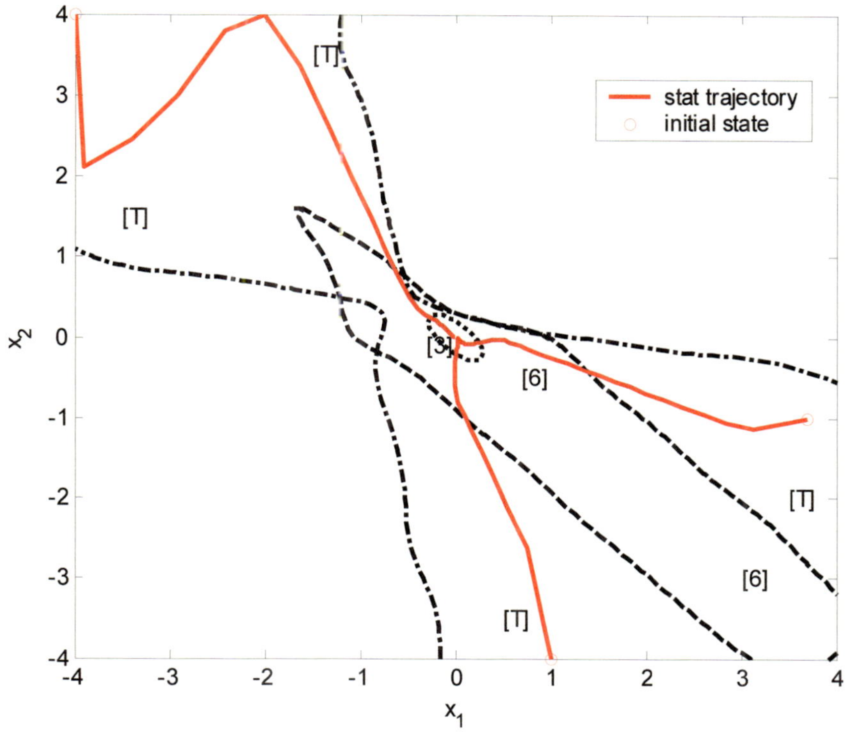

Fig. 2. The closed-loop trajectories of states

In figure 2, the blue ellipsoid is the terminal region in [3] and the region encompassed by black dash lines is the result in [6]. The region encompassed by black solid lines is the terminal region in this paper. We can see, the terminal region in this paper contain the result in [3], but not contain the result in [6] although it is much larger than that in [6]. The reason is that, the terminal region in this paper is the largest one satisfying conditions (C1) and (C2). However, (C1) and (C2) are just the sufficient conditions to guarantee the stability of the controlled system, not the necessary conditions as showed in remark 6. The red solid lines denote the closed-loop trajectories of the selected points. Note that, with the same sampling interval and prediction horizon as those in this paper, these points are not in the regions of attraction of MPC in [3] and [6]. But, they can be leaded to the orgin by using the control law of MPC in this paper.

6. Conclusion

Given the terminal cost, a sequence of subsets of the maximal terminal region are extracted from state space one by one by employing SVM classifier. When one of them is equal to its succesive one in principle, it is used to serve as the terminal region and it is a good approximation to the maximal terminal region.

7. References

[1] D. Q. Mayne, J. B. Rawlings, C. V. Rao, P. O. M. Scokaert. Constrained model predictive control: stability and optimality. *Automatica*, 2000, 36(6): 789-814.

[2] L. Magni, G. De Nicolao, L. Magnani and R. Scattolini. A stabilizing model based predictive control algorithm for nonlinear systems. *Automatica*, 2001, 37: 1351-1362.

[3] Chen H. and Allgower F.. A quasi-infinite horizon nonlinear model predictive control scheme with guaranteed stability. *Automatica*, 1998, 34: 1205-1217.

[4] M. Cannon, V. Deshmukh and B. Kouvaritakis. Nonlinear model predictive control with polytopic invariant sets. *Automatica*, 2003, 39:1487-1494.

[5] J.A. De Doná, M.M. Seron, D.Q. Mayne, G.C. Goodwin. Enlarged terminal sets guaranteeing stability of receding horizon control. *Systems & Control Letters*, 2002, 47: 57-63.

[6] C.J. Ong，D. Sui and E.G. Gilbert. Enlarging the terminal region of nonlinear model predictive control using the support vector machine method. *Automatica*, 2006, 42: 1011-1016.

[7] D. Limon, T. Alamo, F. Salas and E. F. Camacho. On the stability of constrained MPC without terminal constraint. *IEEE transactions on automatic control*, 2006, 51(6): 832-836.

[8] D. Limon, T. Alamo, E.F. Camacho. Enlarging the domain of attraction of MPC controllers, *Automatica*, 2005, 41: 629-635.

[9] Alejandro H. González, Darci Odloak, Enlarging the domain of attraction of stable MPC controllers, maintaining the output performance, *Automatica*, 2009, 45: 1080-1085.

[10] Burges C. C.. A tutorial on support vector machines for pattern recognition. *Data Mining and Knowledge Discovery*, 1998, 2(2): 121–167.

[11] Vapnik V. The nature of statistical learning theory. New York, Springer, 1995.

9

Model Predictive Control for Block-oriented Nonlinear Systems with Input Constraints

Hai-Tao Zhang*
Department of Control Science and Engineering
State Key Laboratory of Digital Manufacturing Equipment and Technology
Huazhong University of Science and Technology, Wuhan
P.R.China

1. Introduction

In process industry, there exist many systems which can be approximated by block-oriented nonlinear models, including Hammerstein and Wiener models. Hammerstein model consists of the cascade connection of a static (memoryless) nonlinear block followed by a dynamic linear block while Wiener model the reverse. Moreover, these systems are usually subjected to input constraints, which makes the control of block-oriented nonlinearities challenging.

In this chapter, a Multi-Channel Identification Algorithm (*MCIA*) for Hammerstein systems is first proposed, in which the coefficient parameters are identified by least squares estimation (*LSE*) together with singular value decomposition (*SVD*) technique. Compared with traditional single-channel identification algorithms, the present method can enhance the approximation accuracy remarkably, and provide consistent estimates even in the presence of colored output noises under relatively weak assumptions on the persistent excitation (PE) condition of the inputs.

Then, to facilitate the following controller design, the aforementioned *MCIA* is converted into a Two Stage Single-Channel Identification Algorithm (*TS-SCIA*), which preserves most of the advantages of *MCIA*. With this *TS-SCIA* as the inner model, a dual-mode Nonlinear Model Predictive Control (NMPC) algorithm is developed. In detail, over a finite horizon, an optimal input profile found by solving a open-loop optimal control problem drives the nonlinear system state into the terminal invariant set, afterwards a linear output-feedback controller steer the state to the origin asymptotically. In contrast to the traditional algorithms, the present method has a maximal stable region, a better steady-state performance and a lower computational complexity. Finally, a case study on a heat exchanger is presented to show the efficiency of both the identification and the control algorithms.

On the other hand, for Wiener systems with input constraints, since most of the existing control algorithms cannot guarantee to have sufficiently large regions of asymptotic stability, we adopted a subspace method to separate the nonlinear and linear blocks in a constrained multi-input/multi-output (MIMO) Wiener system and then developed a novel dual-mode

*H. T. Zhang acknowledges the support of the National Natural Science Foundation of China (NNSFC) under Grant Nos. 91023034 and 51035002, and Program for New Century Excellent Talents in University of China under Grant No. 2009343

nonlinear model predictive control algorithm to maximize the region of the asymptotic stability. Simulation results are presented to demonstrate the superiority of this new control algorithm.

In sum, this chapter developed some new NMPC methods for block-oriented nonlinearities with input constraints. Meanwhile, these approaches can effectively enlarge the closed-loop stable area so as to extend the feasible working region and improve the reliability of the control systems in real process industrial applications.

2. Model Predictive Control for Hammerstein systems with input constraints

2.1 Introduction

In industrial processes (1), most dynamical systems can be better represented by nonlinear models, which are able to describe the systems over large operation ranges, rather than by linear ones that are only able to approximate the systems around given operation points (23; 48). One of the most frequently studied classes of nonlinear models is the Hammerstein model (17; 48), which consists of the cascade connection of a static (memoryless) nonlinear block followed by a dynamic linear block. Under certain considerations such as fading memory assumption (10) the Hammerstein approximation could be a good representation. Thus, this model structure has been successfully applied to chemical processes (heat exchanger (17), distillation (5; 17; 35)), biological processes (20; 30) signal processing (3; 55), and communications (3; 25)). In recent years, identification and control of Hammerstein systems has become one of the most needed and yet very difficult tasks in the field of the process industry.

In MPC (Model Predictive Control) framework (20; 32), the input is calculated by on-line minimization of a performance index based on model predictions. It is well known that the control quality relies on the accuracy of the model. In recent years, extensive efforts were devoted to modelling of Hammerstein nonlinearities (2; 17; 23; 26; 27; 31). For example, Bai (2) studied SISO (Single Input/ Single Output) systems subject to external white noise. Gómez and Baeyens (23) designed a non-iterative identification with guaranteed consistent estimation even in the present of coloured output noise. Both of their works use only one channel to identify the system, therefore, owing to the SVD (singular value decomposition) nature of their methods, the identification errors usually can not be minimized. A basic reason is that the error is determined by the second largest singular value (for SISO system) or the st largest singular value (for MIMO system with inputs) of the estimated coefficients matrix. For a SISO system, if the sampling set is not big enough or the PE (persistent excitation) conditions are not fulfilled, the second largest singular value can not be neglected, making the identification accuracy unsatisfactory or even unacceptable. On the other hand, the research on the control of Hammerstein systems is still on the midway so far. Most of the existent control algorithms have some of the following disadvantages

- Reliance on prior knowledge;
- Insufficiently large closed-loop stable regions;
- Limited capacity of handling input constraints.

In detail, Haddad and Chellaboina (28) suggested a design that can guarantee global asymptotic closed-loop stability for nonlinear passive systems by embedding a nonlinear dynamic compensator with a suitable input nonlinearity, which requires the memoryless nonlinear block to be partially known or measurable without considering input constraints. Patwardhan *et al.* (51) used a PLS (Partial Least Square) framework to decompose the modelling problem into a series of univariate problems in the latent subspace while preserving optimality of the input constraints. In this way, they can extend the SISO formulation into a constrained MIMO scenario. In this approach, however, the computational complexity is prohibitive, and the reliance on prior knowledge can not be eliminated. Knohl *et al.* (40) slightly alleviated this reliance by an ANN (Artificial Neural Network) inverse compensation, which makes the control scheme more flexible, but its stable region is still small. Fruzzetti *et al.* (18) and *Zhu et al.* (71) developed GPC (Generalized Predictive Control) and MPC algorithms respectively by taking input constraints into account. These schemes still can not ensure a large stable region in general, and require prior knowledge of the real plant such as order, structure, partial coefficients, etc. Bolemen *et al.* (9) extended their own work (8) which preserves the convex property of the optimization problem, but does not consider input constraints. In order to enlarge the asymptotically stable region for constrained nonlinear systems, Chen and Allgöwer (14) developed a quasi-infinite horizon Nonlinear Model Predictive Control (NMPC) algorithms based on a dual-mode (or two-step) technique, which has opened a new avenue in this fascinating field. Among the various following works of Chen and Allgöwer's work (14), there are three important investigations made by Kouvartakis *et al.* (41) Lin *et al.* (44) and Ding *et al.* (16). More precisely, Kouvartakis *et al.* (41) proposed a new approach that deployed a fixed state-feedback law with the assistance of extra degrees of freedom through the use of perturbations, which led to a significant reduction in computational cost. More generally, for linear systems with actuator rate constraint, Lin *et al.* (44) designed both state-feedback and output-feedback control laws that achieve semi-global asymptotic stabilization based on the assumption of detectability of the system. For input saturated Hammerstein systems, Ding *et al.* (16) designed a two-step MPC by solving nonlinear algebraic equation group and deconstraint. The stable region is enlarged and its domain of attraction is designed applying semi-global stabilization techniques. Unfortunately, this nice work is still based on the measurability of the state of the linear block.

Based on the above analysis, two important tasks are formulated as follows:

- *Task one*: Develop a better identification algorithm to separate the nonlinear/linear blocks of the Hammerstein system more effectively so that some mature linear control theories can be used to facilitate the nonlinear control algorithm design.
- *Task two*: Develop a more efficient control algorithm for constrained Hammerstein systems.

Bearing these tasks in mind, we propose a NMPC algorithm based on a Two Stage Single-Channel Identification Algorithm (TS-SCIA) (68). More precisely:

- A Multi-Channel Identification Algorithm (MCIA) is developed for Hammerstein systems which eliminates requirement of prior knowledge about the plant and minimizes the identification errors. The MCIA is then converted to a TS-SCIA thereby facilitating the controller design. A sufficient condition for the convergence and approximation capability is given for the new algorithm.
- A dual-mode NMPC algorithm is developed by taking the above mentioned Two Stage Single-Channel Identification Model (TS-SCIM) as the internal model. The closed-loop stable region is maximized by using ellipsoidal invariant set theory together with linear matrix inequality (LMI) techniques.

2.2 Model identification

The key problem on this issue is how to efficiently separate the coefficients of the linear and nonlinear blocks, namely nonlinear/linear separation. A number of approaches are previously proposed: these include the singular value decomposition (SVD) combined with least square estimation (LSE) (23), iterative finite response (FIR) method (45), separable LSE (63), Hunter-Korenberg iteration (35) , correlation analysis (4) and so on. Among them, SVD-LSE approach is one of the most extensively studied and most widely applied methods. In this approach, the system output $y(t)$ is expanded as

$$\begin{aligned} y(t) &= G(z^{-1})\mathcal{N}(u(t)) + \xi(t) \\ &= \textstyle\sum_{k=1}^{N} c_k x_k(z^{-1}) \sum_{i=1}^{r} a_i g_i(u(t)) + \xi(t), \end{aligned} \tag{1}$$

where $u(t) \in \mathbb{D} \subset \mathbb{R}^n$, $v(t) = \mathcal{N}(u(t)) \in \mathbb{R}^n$, $y(t) \in \mathbb{R}^m$ and $\xi(t) \in \mathbb{R}^m$ are the input, intermediate variable, output and external noise vector at time t, respectively. External noise $\xi(t)$ can be white or colored noise sequence induced by measurement or external disturbances, and input signal $u(t)$ can be random or stationary. $G(z^{-1})$ and $\mathcal{N}(\cdot)$ denote the linear and nonlinear blocks expanded by suitable orthonormal and nonlinear bases $x_k(z^{-1})$ and $g_i(\cdot)$, respectively. The sequences $\{c_k \in \mathbb{R}^{m\times n}\}_{k=1}^{N}$ and $\{a_i \in \mathbb{R}^{n\times n}\}_{i=1}^{r}$ are the coefficients of the linear and nonlinear blocks, respectively, and z^{-1} is the one-step backward shifting operator, i.e. $z^{-1}u(t) = u(t-1)$. The state $x_k(z^{-1})$ could be Jacobi series [13], spline functional series (**?**), orthonormal functional series (OFS, including Laguerre series (19; 60; 66; 67; 69), Kautz series (19; 33) and so on) or some others.

Actually, in recent years, extensive efforts (23) were devoted to this kind of SVD-LSE approaches. For example, Bai (2) studied SISO (Single Input/ Single Output) systems subject to external white noise. Gómez and Baeyens (23) designed a non-iterative identification with guaranteed consistent estimation even in the present of colored output noise. Both of their works use merely one channel to identify the system, therefore, owing to the SVD nature of their methods, the identification errors usually can not be minimized. A basic reason is that, for the Hammerstein system (1), the error is determined by the $(n+1)$th largest singular value the estimated coefficients matrix. If the sampling set is not big enough or the PE (persistent excitation) conditions are not fulfilled, the $(n+1)$th largest singular value can not be neglected, making the identification accuracy unsatisfactory or unacceptable, especially for small numbers of truncation lengths of the nonlinear/linear basis series, i.e. r and N (see Eq. (1)). In brief, it is an urgent task to develop a better identification algorithm to separate the nonlinear/linear blocks of the Hammerstein system more effectively.

In this section, we argue that the single-channel separation is a bottleneck to better modeling accuracy, and adding more identification channels can effectively enhance the performance, for they have the capability to compensate the residuals of the single-channel nonlinear/linear separation.

Fig. 1(a) shows the implementary details on a single identification channel of the present modeling method. First, system input $u(t)$ is fed into parallel weighted nonlinear bases to produce the intermediate variable $v(t)$. According to *Weierstrass Theorem* (42) (every continuous function defined on a finite region can be uniformly approximated as closely as desired by a polynomial function), the bases are generally chosen as polynomial bases. Then $v(t)$ is injected into the linear OFS filter, through which the filter output sequence can be yielded.

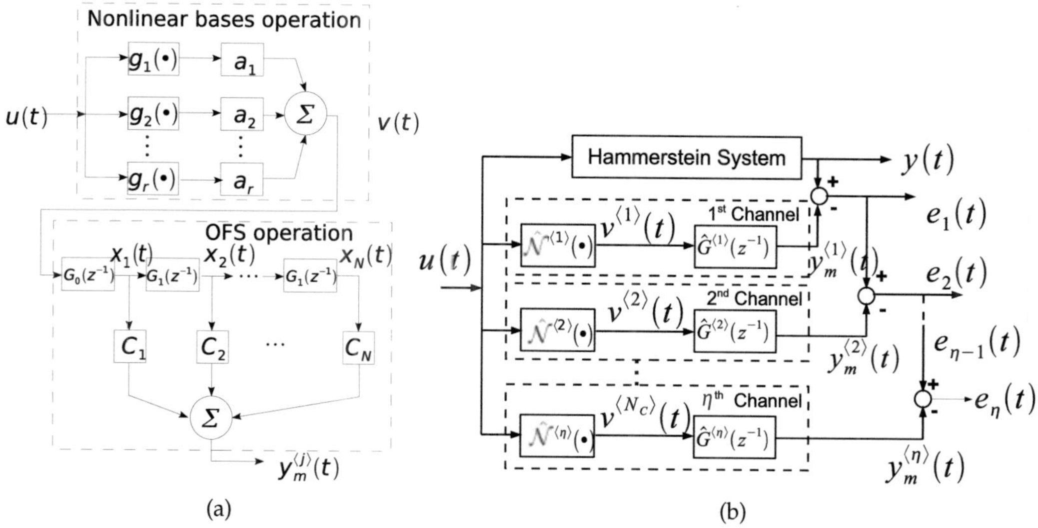

Fig. 1. (Color online) (a): Implementary details of each modeling channel; (b): Multi-channel identification

Since $\{x_k(t)\}_{k=1}^{\infty}$ forms a complete orthonormal set in functional space $\mathcal{L}_2(\mathbb{R}^+)$ (33; 66–68), each stable linear system can be approximately represented as $h(t) = \sum_{k=1}^{N} c_k \varphi_k(t)$, where $\{c_k\}_{k=1}^{N}$ are the coefficients of the linear block, the Laguerre function $\varphi_k(t)$ is given in (33; 66), and the kth order filter output is calculated by $x_k(t) = \int_0^{\infty} \varphi_k(\tau)v(t-\tau)d\tau$. To get the OFS filter output sequence, one should pre-calculate all the OFS according to the state equation $x(t+1) = Ax(t) + Bv(t)$, where A and B are pre-optimized matrices (33; 66). As shown in Fig. 1, the first-order filters is the Laguerre series, in which

$$\begin{aligned} G_0(z^{-1}) &= z^{-1}\sqrt{1-p^2}/(1-z^{-1}p), \\ G_1(z^{-1}) &= (z^{-1}-p)/(1-z^{-1}p), \end{aligned} \tag{2}$$

where p is the filter pole. The second-order OFS is the Kautz Series, in which $G_0(z^{-1})$ and $G_1(z^{-1})$ are the second order OFS transfer functions. Analogically, Heuberger *et al.* (33) introduced the higher-order OFS models. As the order increases, OFS model can handle more complex dynamics. Finally, the model is obtained by synthesizing the OFS filter output sequence and their corresponding coefficients according to $y_m(t) = \sum_{k=1}^{N} c_k x_k(t)$ which leads to Eq. (2). Consequently, considering the S-point data set suffered by external noise sequence $\{\xi(t)\}_{t=1,\cdots,S}$, Eq. (1) can be rewritten in a linear regressor form as

$$Y_S = \Phi_S^T \theta + \Upsilon_S, \tag{3}$$

with

$$\begin{aligned} Y_S &\triangleq [y(1),\cdots,y(S)]^T, \\ \Upsilon_S &\triangleq [\xi(1),\cdots,\xi(S)]^T, \\ \theta &\triangleq [c_1a_1,\cdots,c_1a_r,\cdots,c_Na_1,\cdots,c_Na_r]^T, \\ \Phi_S &\triangleq [\phi(1),\cdots,\phi(S)]^T, \end{aligned}$$

and

$$\phi(t) \triangleq \left[x_1(z^{-1})g_1^T(u(t)), \cdots, x_1(z^{-1})g_r^T(u(t)), \cdots, \right. \\ \left. x_N(z^{-1})g_1^T(u(t)), \cdots, x_N(z^{-1})g_r^T(u(t))\right]^T, \tag{4}$$

where the superscript 'T' means transpose. Note that in real applications, the orthonormal basis $x_k(z^{-1})g_i^T(u(t))$ ($k = 1, \cdots, N$; $i = 1, \cdots, r$) and the system output $y(t)$ in Eq. (1) are calculated according to the state-space equations in (62; 66; 67). As shown in Fig. 1(a), kernel matrix Φ_S is obtained by carrying out nonlinear bases and OFS operations on the input sequence, and θ is the coefficient vector of Φ_S. Then, provided the indicated inverse exists, it is well known that the LSE $\hat{\theta}$ of minimizing the prediction errors $\varepsilon_s = Y_S - \Phi_S^T\theta$ is calculated by (46)

$$\hat{\theta} \triangleq (\Phi_S\Phi_S^T)^{-1}\Phi_S Y_S. \tag{5}$$

Define $\Theta_{ac} \triangleq \left\{a_i^T c_j^T\right\}_{1\leq i\leq r; 1\leq j\leq N} = ac^T$ with $a \triangleq [a_1, \cdots, a_r]^T$ and $c \triangleq [c_1^T, \cdots, c_N^T]^T$, it can be seen that θ is the block column matrix obtained by stacking the block columns of Θ_{ac} on the top one by one. Now, the problem is how to separate the nonlinear/linear coefficient parameter matrices a and c from the LSE $\hat{\Theta}_{ac}$ of Θ_{ac} . It is clear that feasible estimates $\hat{a}$ and $\hat{c}$ are the solutions of the optimization problem

$$(\hat{a}, \hat{c}) = \arg\min_{a,c} \left\|\hat{\Theta}_{ac} - ac^T\right\|_2^2. \tag{6}$$

This problem can be solved by the standard SVD (22) with the prerequisite $\|a_i\|_2 = 1$ ($i = 1, \cdots, r$). However, bearing the spectral nature of SVD in mind, one can easily find that the closest estimates of $\{a, c\}$ are not a single pair $\{\hat{a}, \hat{c}\}$ but a series of pairs $\left\{\hat{a}^{\langle j\rangle}, \hat{c}^{\langle j\rangle}\right\}_{j=1}^{\eta}$, which solves the optimization problem

$$(\hat{a}^{\langle j\rangle}, \hat{c}^{\langle j\rangle})_{j=1}^{\eta} = \arg\min_{a^{\langle j\rangle}, c^{\langle j\rangle}} \left\|\hat{\Theta}_{ac} - \Sigma_{j=1}^{\eta} a^{\langle j\rangle}(c^{\langle j\rangle})^T\right\|_2^2. \tag{7}$$

From now on, the pair $\left\{\hat{a}^{\langle j\rangle}, \hat{c}^{\langle j\rangle}\right\}$ is defined as the jth **identification channel**, with j and η denoting the sequence index and number of the identification channels, respectively. Therefore, in order to separate the nonlinear/linear blocks more effectively, more channels should be used to compensate the separation residuals of the single-channel method (23). To explain it more clearly, we will give a lemma and a theorem as follows. Note that, for the Hammerstein system (1), the multi-channel estimates $\hat{a}^{\langle j\rangle} \in \mathbb{R}^{rn\times n}$ and $\hat{c}^{\langle j\rangle} \in \mathbb{R}^{Nm\times n}$. In special, for SISO case, i.e. $m = 1$ and $n = 1$, the estimates $\hat{a}^{\langle j\rangle}$ and $\hat{c}^{\langle j\rangle}$ are all column vectors.

Lemma 1. *Let rank*$(\Theta_{ac}) = \gamma$*, here* $\hat{\Theta}_{ac}$ *is the estimate of* Θ_{ac}*, then the SVD of* Θ_{ac} *is*

$$\hat{\Theta}_{ac} = U_\gamma \Sigma_\gamma V_\gamma^T = \Sigma_{j=1}^{\gamma} \sigma_j \mu_j v_j^T \tag{8}$$

such that the singular value matrix $\Sigma_\gamma = diag\left\{\sigma_j\right\}$ $(j = 1, \cdots, \min(r, N))$ *satisfies* $\sigma_1 \geq \cdots \geq \sigma_\gamma > 0$ *and* $\sigma_l = 0$ $(l > \gamma)$*, where* μ_j *and* v_j $(j = 1, \cdots, \gamma)$ *are pairwise orthogonal vectors. If* $\|a^{\langle j\rangle}\|_2 = 1$*, then* $\forall\, \eta$ $(1 \leq \eta \leq \gamma)$*, each identification channel can be calculated as below according to the optimization problem (7)*

$$(\hat{a}^{\langle j \rangle}, \hat{c}^{\langle j \rangle}) = (\mu_j, \sigma_j v_j) \ (j = 1, \cdots, \eta) \tag{9}$$

with approximation error e_η given by

$$e_\eta = \left\| \hat{\Theta}_{ac} - \sum_{j=1}^{\eta} a^{\langle j \rangle} (c^{\langle j \rangle})^T \right\|_2^2 = \sum_{j=\eta+1}^{\gamma} \sigma_j^2. \tag{10}$$

It can be seen from Lemma 1 that after the SVD operation, $\hat{\Theta}_{ac}$ is decomposed into a series of pairs (or channels) $(\hat{a}^{\langle j \rangle}, \hat{c}^{\langle j \rangle})$. More precisely, as shown in Fig. 1(b), first the 1st channel model is estimated using the basic identification algorithm (30) from input-output data $\{u(t), y(t)\}_{t=1}^{S}$. Afterwards, the 1st channel model error $e_1(t) = y(t) - y_m^{\langle 1 \rangle}$ is used to identify the 2nd channel model. Analogously, $e_2(t), \cdots, e_{\eta-1}(t)$ determine the 3rd, $\cdots$, ηth channel models, respectively. The approximation accuracy enhancement will be proven by the following theorem.

Theorem 1. *For the Hammerstein system (1), with the identification matrix calculated by Eq. (22), if rank$(\hat{\Theta}_{ac}) = \gamma$, then, with the identification pairs $(\hat{a}^{\langle j \rangle}, \hat{c}^{\langle j \rangle})$ obtained by Eqs. (8) and (9) and the identification error index defined by Eq. (10), one has*

$$e_1 > e_2 > \cdots > e_\gamma = 0.$$

In other words, the the identification error decreases along with the increasing η.

Proof: This can be easily drawn from Lemma 1. ■

In principle, one can select a suitable η according to the approximation error tolerance $\bar{e}$ and Eq. (10). Even for the extreme case that $\bar{e} = 0$, one can still set $\eta = \gamma$ to eliminate the approximation error, thus such suitable η is always feasible. For simplicity, if $\gamma \geq 3$, the general parameter setting $\eta = 2$ or 3 works well enough.

According to the conclusions of Lemma 1 and Theorem 1, multi-channel model $y_m(t) = \sum_{j=1}^{\eta} \hat{G}^{\langle j \rangle}(z^{-1}) \hat{\mathcal{N}}^{\langle j \rangle}(u(t))$ outperforms single-channel model $y_m(t) = \hat{G}(z^{-1}) \hat{\mathcal{N}}(u(t))$ in modeling accuracy. We hereby design a Multi-Channel Identification Algorithm (MCIA) based on Theorem 1 as follows. As shown in Fig. 1(b), the Multi-Channel Identification Model (MCIM) is composed of η parallel channels, each of which consists of a static nonlinear block described by a series of nonlinear basis $\{g_1(\cdot), \cdots, g_r(\cdot)\}$, followed by a dynamic linear block represented by the discrete Laguerre model (33; 60; 67; 69) in the state-space form (62; 66; 67). Without loss of generality, the nonlinears bases are chosen as polynomial function bases. Thus, each channel of the MCIM, as shown in Fig. 1, is described by

$$x^{\langle j \rangle}(t+1) = Ax^{\langle j \rangle}(t) + B \sum_{i=1}^{r} \hat{a}_i^{\langle j \rangle} g_i(u(t)) \tag{11}$$

$$y_m^{\langle j \rangle} = (\hat{c}^{\langle j \rangle})^T x^{\langle j \rangle}(t) \ (j = 1, \cdots, \eta), \tag{12}$$

where $y_m^{\langle j \rangle}(t)$ and $x^{\langle j \rangle}(t)$ denote the output and state vector of the jth channel, respectively. Finally, the output of the MCIM can be synthesized by

$$y_m(t) = \sum_{j=1}^{\eta} y_m^{\langle j \rangle}(t) \tag{13}$$

Next, we will give a convergence theorem to support the MCIA.

Theorem 2. *For a Hammerstein system (1) with* $\|a_i\|_2 = 1$ $(i = 1, \cdots, r)$*, nominal output* $\bar{y}(t) = \sum_{k=1}^{N} c_k x_k(z^{-1}) \sum_{i=1}^{r} a_i g_i(u(t))$ *and allowable input signal set* $\mathbb{D} \subset \mathbb{R}^n$ *. If the regressor* $\phi(t)$ *given by Eq. (4) is PE in the sense that for an arbitrary positive integer* t_0 *there exist some integer* N_1 *and positive constants* α_1 *and* α_2 *such that*

$$0 < \alpha_1 I \leq \sum_{t=t_0}^{t_0+N_1} \phi^T(t)\phi(t) \leq \alpha_2, \tag{14}$$

then

$$\sum_{j=1}^{\eta} \hat{a}^{\langle j \rangle} (\hat{c}^{\langle j \rangle})^T \xrightarrow{\text{a.s.}} \Theta_{ac} \tag{15}$$

$$y_m \xrightarrow{\text{a.s.}} \bar{y}(t) \tag{16}$$

where the symbol ' $\xrightarrow{\text{a.s.}}$ *' denotes 'converge with probability one as the number of the data points* S *tends to infinity', and the model output* $y_m(t)$ *is determined by Eqs. (11), (12) and (13).*

Proof: Since the linear block is stable, and $g_i(u(t))$ $(i = 1, \cdots, r)$ is bounded (because $u(t) \in \mathbb{D}$ is bounded and $g(\cdot)$ are nonlinear basis functions), the model output $y_m(t)$ is also bounded. Taking Eqs. (3) and (11) into consideration, one has that $\|\phi(t)\|_2$ is bounded, i.e. $\exists \delta_L > 0$, such that $\|\phi(t)\|_2 \leq \delta_L$. On the other hand, $\forall\ \varepsilon > 0$, $\exists \varepsilon_1, \varepsilon_2 > 0$ such that $\varepsilon = \varepsilon_1 + \varepsilon_2$. Let $\varepsilon_3 = \varepsilon_1 / (\delta_L \max(r, N))$ and $\varepsilon_4 = \varepsilon_2 / \delta_L$. Since the regressor $\phi(t)$ is PE in the sense of Eq. (14), one has that the estimate θ is strongly consistent in the sense that $\theta \to \hat{\theta}$ with probability one as $S \to \infty$ (denoted $\hat{\theta} \xrightarrow{\text{a.s.}} \theta$) (46), in other words, $\forall \varepsilon_4 > 0$, $\exists N_0 > 1$ such that $\|\hat{\theta} - \theta\|_2^2 \leq \varepsilon_4$ with probability one for $S > N_0$. Moreover, the consistency of the estimate $\hat{\theta}$ holds even in the presence of colored noise ξ (23). The convergence of the estimate $\hat{\theta}$ implies that

$$\hat{\Theta}_{ac} \xrightarrow{\text{a.s.}} \Theta_{ac} \tag{17}$$

Note that the consistency of the estimation $\hat{\theta}$ holds even in the presence of colored output noise (23).
Using Lemma 1 and assuming $\text{rank}(\hat{\Theta}_{ac}) = \gamma$, one gets from Theorem 1 that $\forall \varepsilon_3 > 0$, $\exists \eta \leq \gamma$ such that $\sum_{\eta+1}^{\gamma} \sigma_j \mu_j \varphi \nu_j^T \leq \varepsilon_3$, in other words, $\left\| \sum_{j=1}^{\eta} \hat{a}^{\langle j \rangle} (\hat{c}^{\langle j \rangle})^T - \hat{\Theta}_{ac} \right\|_2^2 \leq \varepsilon_3$ or

$$\sum_{j=1}^{\eta} \hat{a}^{\langle j \rangle} (\hat{c}^{\langle j \rangle})^T \to \hat{\Theta}_{ac}. \tag{18}$$

Thereby, substituting Eq. (18) into Eq. (17) yields Eq. (15). Furthermore, define

$$\hat{\theta}^{\langle j \rangle} \triangleq [\hat{c}_1^{\langle j \rangle} \hat{a}_1^{\langle j \rangle}, \cdots, \hat{c}_1^{\langle j \rangle} \hat{a}_r^{\langle j \rangle}, \hat{c}_2^{\langle j \rangle} \hat{a}_1^{\langle j \rangle},$$
$$\cdots, \hat{c}_2^{\langle j \rangle} \hat{a}_r^{\langle j \rangle}, \cdots, \hat{c}_N^{\langle j \rangle} \hat{a}_1^{\langle j \rangle}, \cdots, \hat{c}_N^{\langle j \rangle} \hat{a}_r^{\langle j \rangle}]^T,$$

then $\forall S > N_0$, the following inequality holds with probability one

$$\begin{aligned}
[y_m(t) - \bar{y}(t)]^2 &= \left[\varphi^T(t)(\Sigma_{j=1}^{\eta}\hat{\theta}^{\langle j\rangle} - \hat{\theta} + \hat{\theta} - \theta)\right]^2 \\
&\leq \delta_L \left\|\Sigma_{j=1}^{\eta}\hat{\theta}^{\langle j\rangle} - \hat{\theta}\right\|_2^2 + \delta_L \left\|\bar{\theta} - \theta\right\|_2^2 \\
&\leq \delta_L \max(r, N) \left\|\Sigma_{j=1}^{\eta}\sigma_j \mu_j \varphi v_j^T - \hat{\Theta}_{ac}\right\|_2^2 + \delta_L \left\|\bar{\theta} - \theta\right\|_2^2 \\
&= \delta_L \max(r, N)\varepsilon_3 + \delta_L \varepsilon_4 = \varepsilon,
\end{aligned}$$

where the definition of matrix 2-norm is given in (23). Thus, the conclusion (16) holds. This completes the proof.

■

Thus, it is drawn from Theorems 1 and 2 that the increase of the identification channel number will help decrease the identification errors, which is the main theoretical contribution of this section.

2.3 Controller design

A Hammerstein system consists of the cascade connection of a static (memoryless) nonlinear block $\mathcal{N}(\cdot)$ followed by a dynamic linear block with state-space expression (A, B, C) as below

$$\begin{aligned} \dot{x} = Ax + Bv, y = Cx, \\ v = \mathcal{N}(u), \end{aligned} \tag{19}$$

with $u(t) \in [-\bar{u}, \bar{u}]$. Naturally, a standard output feedback control law can be derived by (13)

$$\begin{aligned} v = K\hat{x} \\ u = \mathcal{N}^{-1}(v), \end{aligned} \tag{20}$$

where $\hat{x}$ is the estimation of x by some state observer L, $\mathcal{N}^{-1}(\cdot)$ is the inverse of $\mathcal{N}(\cdot)$, and the closed-loop state matrix $A + BK$ and cbserver matrix $A_L C$ are designed Hurwitz. Now, the *problem* addressed in this section becomes optimize such an output-feedback controller for the Hammerstein system (19) such that the closed-loop stability region is maximized and hence the settling time is substantially abbreviated.
The nonlinear block $\mathcal{N}(\cdot)$ can be described as (68):

$$\mathcal{N}(z(t)) = \sum_{r=1}^{N} a_i g_i(z(t)), \tag{21}$$

where $g_i(\cdot) : \mathbb{R} \to \mathbb{R}$ are known nonlinear basis functions, and a_i are unknown matrix coefficient parameters. Here, $g_i(\cdot)$ can be chosen as polynomials, radial basis functions (RBF), wavelets, etc. At the modeling stage, the sequence $v(t_j)$ $(j = 1, \cdots, N)$ is obtainable with a given input sequence $u(t_j)$ $(j = 1, \cdots, N)$ and an arbitrary initial state $x(0)$. Thereby, according to Lease Square Estimation (LSE), the coefficient vector $a := [a_1, \cdots, a_N]^T$ can be identified by

$$\hat{a} = (G^T G)^{-1} G^T v \tag{22}$$

with

$$G = \begin{bmatrix} g_1(t_1) & \cdots & g_N(t_1) \\ & \vdots & \\ g_1(t_s) & \cdots & g_N(t_s) \end{bmatrix},$$

$v = [v(t_1), \cdots, v(t_N)]^T$ and $s \geq N$. Note that $\hat{a}$ is the estimation of a, which is an consistent one even in the presence of colored external noise.
Now the intermediate variable control law $v(t)$ in Eq. (19) can be designed based on the linear block dynamics. Afterwards, one can calculate the control law $u(t)$ according to the inverse of $v(t)$. Hence, for the Hammerstein system (19), suppose the following two assumptions hold:
A1 The nonlinear coefficient vector a can be accurate identified by the LSE (22), i.e., $\hat{a} = a$;
A2 For $|u(t)| \leq \bar{u}$, the inverse of $\mathcal{N}(\cdot)$ exists such that

$$\mathcal{N}(\mathcal{N}_Z^{-1}(v(t))) := \tilde{v}(t) = (1 + \delta(v(t)))v(t),$$

where $\delta(v(t))) < \sigma$ ($\sigma \in \mathbb{R}^+$), and $\mathcal{N}_Z^{-1}$ denotes the inverse of $\mathcal{N}(\cdot)$ calculated by some suitable nonlinear inverse algorithm, such as Zorin method (21).
For conciseness, we denote $\delta(v(k))$ by $\delta(\cdot)$, and hence, after discretization, the controlled plant is described as follows:

$$\begin{aligned} x(k+1) = Ax(k) + B\tilde{v}(k) = Ax(k) + B(1 + \delta(\cdot))v(k), \\ y(k) = Cx(k). \end{aligned} \tag{23}$$

Afterwards, a state observer $L \in R^N$ is used to estimate $x(k)$ as follows:

$$\hat{x}(k+1) = A\hat{x}(k) + B(1 + \delta(\cdot))v(k) + LCe(k), \tag{24}$$

$$e(k+1) = \Phi e(k), \tag{25}$$

where $\hat{x}$ is the estimation of x, $e(k) := x(k) - \hat{x}(k)$ is the state estimation error, and the matrix $\Phi = A - LC$ is designed as Hurwitz. Then, an NMPC law is designed with an additional term $D(k+i|k)$ as follows:

$$\begin{aligned} v(k+i|k) = K\hat{x}(k+i|k) + ED(k+i|k), \\ u(k|k) = \mathcal{N}_Z^{-1}((v(k|k))), \end{aligned} \tag{26}$$

where $E := [1, 0, \cdots, 0]_{1\times M}$, $\hat{x}(k|k) := \hat{x}(k)$, $v(k|k) := v(k)$, and $D(k|k) := D(k) = [d(k), \cdots, d(k+M-1)]^T$ is defined as a perturbation signal vector representing extra degree of freedom. Hence the role of $D(k)$ is merely to ensure the feasibility of the control law (26), and $D(k+i|k)$ is designed such that

$$D(k+1|k) = TD(k+i-1|k) \; (i = 1, \cdots, M),$$

where

$$T = \begin{bmatrix} \mathbf{0} & I_{(M-1)\times(M-1)} \\ 0 & \mathbf{0}^T \end{bmatrix}_{M\times M},$$

$M \geq 2$ is the prediction horizon and $\mathbf{0}$ is compatible zero column vector. Then, substituting Eq. (26) into Eq. (24) yields

$$\begin{aligned}\hat{z}(k+i|k)) &= \Pi\hat{z}(x+k-i|k)\\ &+[(\delta(\cdot)B\bar{K})^T,0]^T\hat{z}(k+i-1/k)\\ &+[(LC)^T,0]^T e(k+i-1|k),\ (i=1,\cdots,M)\end{aligned} \tag{27}$$

with $\hat{z}(k+i|k) = [\hat{x}^T(k+i|k), D(k+i|k)^T]^T$, $\Pi = \begin{bmatrix} \Psi & BE \\ 0 & T \end{bmatrix}$, $\bar{K} = [K, E]$, where $\Psi = A + BK$ is designed as Hurwitz. In order to stabilize the closed-loop system (27), we define two ellipsoidal invariant sets (39) of the extended state estimations $\hat{z}(k)$ and error $e(k)$, respectively, by

$$S_x := \{\hat{z}|\hat{z}^T(k)P_z\hat{z}(k) \leq 1\}, \tag{28}$$

and

$$S_e := \{e(k)|e^T(k)P_e e(k) \leq \bar{e}\},\ (0 < \bar{e} \leq 1), \tag{29}$$

where P_z and P_e are both positive-definite symmetric matrices and the perturbation signal vector $D(k)$ (see Eq. (26)) is calculated by solving the following optimization problem

$$\begin{aligned}&\min_{D(k)} J(k) = D^T(k)D(k),\\ &s.t.\ \hat{z}^T(k)P_z\hat{z}(k) \leq 1.\end{aligned} \tag{30}$$

2.4 Stability analysis

To guarantee the feasibility and stability of the control law (26), it is required to find the suitable matrices P_z and P_e assuring the invariance of S_z and S_e (see Eqs. (28) and (29)) by the following lemma.

Lemma 2. *Consider a closed-loop Hammerstein system (23) whose dynamics is determined by the output feedback control law (26) and (30) and subject to the input constraints $|u| \leq \bar{u}$, the ellipsoidal sets S_z and S_e are invariant in the sense of (28) and (29), respectively, and the control law (26) and (30) is feasible provided that Assumptions A1, A2 and the following three Assumptions A3–A5 are all fulfilled.*

A3 *The matrices Φ and Ψ are both Hurwitz;*

A4 *There exist $\tau_{1,2} > 1$, $0 < \bar{e} < 1$ such that*

$$\Phi^T P_z \Phi \leq P_e, \tag{31}$$

$$\eta_1 C^T L^T E_x^T P_z E_x LC \leq P_e, \tag{32}$$

$$\begin{aligned}&\tau_1\tau_2\Pi^T P_z \Pi + \tau_1\eta_2\sigma^2\bar{K}^T B^T E_x^T P_z E_x B\bar{K},\\ &\leq (1-\bar{e}^2)P_x,\end{aligned} \tag{33}$$

where $\eta_1 = 1 + (\tau_1 - 1)^{-1}$, $\eta_2 = 1 + (\tau_2 - 1)^{-1}$ and

$$E_x^T = \begin{bmatrix} 1 & 0 & \cdots & \cdots & 0 \\ \vdots & \ddots & 0 & \cdots & \vdots \\ 0 & 0 & 1 & \cdots & 0 \end{bmatrix}_{N\times(N+M)}$$

is the projection matrix such that $E_x^T\hat{z}(k) = \hat{x}(k)$;
A5 There exist $\mu > 0$ and $\lambda \in (0, \bar{u})$ such that

$$|u(k)| \le \mu|v(k)| + \lambda \tag{34}$$

(local Lipschitz condition) and

$$\begin{bmatrix} -(\bar{u}-\lambda)^2/\mu^2 & \bar{K} \\ \bar{K}^T & -P_z \end{bmatrix} \le 0. \tag{35}$$

Proof: We start the proof by a fact that (68), for $\forall\tau > 1$ and $\eta = 1 + (\tau - 1)^{-1}$,

$$\begin{aligned} &(A_1 + A_2)^T P(A_1 + A_2) \\ &\le \tau A_1^T P A_1 + (1 + (\tau - 1)^{-1}) A_2^T A_2. \end{aligned} \tag{36}$$

Thereby, one has $\forall\tau_{1,2} > 0$ and $\eta_{1,2} = 1 + (\tau_{1,2} - 1)^{-1}$, such that

$$\begin{aligned} \hat{z}^T(k+i|k)P_z\hat{z}(k+i|k) \le \tau_1(&\Pi\hat{x}(k+i-1|k) \\ &+[\delta(\cdot)(B\bar{K})^T, 0]^T x(k+\hat{i}-1|k))^T P_z \\ \cdot(\Pi\hat{x}(k+i-1|k) &+ [\delta(\cdot)(B\bar{K})^T, 0]^T x(k+\hat{i}-1|k)) \\ &+\eta_1([(LC)^T, 0]^T e(k+i-1|k))^T P_z \\ &\cdot([(LC)^T, 0]^T e(k+i-1|k)) \\ \le \tau_1\tau_2 \hat{z}^T&(k+i-1|k)\Pi^T P_z \Pi\hat{x}(k+i-1|k)) \\ +\tau_1\eta_2\hat{x}^T(k+i-1|k)&\sigma^2\bar{K}^T B^T E_x^T P_z E_x \bar{K}\hat{x}(k+i-1|k) \\ +\eta_1 e^T(k+i-1|k)&C^T L^T E_x^T P_z L C e(k+i-1|k). \end{aligned}$$

Thereby, if Eqs. (32) and (33) hold and $\hat{z}^T(k+i-1|k)P_z\hat{z}(k+i-1|k) \le 1$, then $\hat{z}^T(k+i|k)P_z\hat{z}(k+i|k) \le 1$, i.e., S_z is an invariant set (39).
Analogously, if Eq. (31) hold and $e^T(k+i-1|k)P_e e(k+i-1|k) \le \bar{e}$, then $e^T(k+i|k)P_e e(k+i|k) \le \bar{e}$, i.e., S_e is an invariant set.
On the other hand, $|v(k)| = |\bar{K}\hat{z}(k)| = |\bar{K}P_z^{-1/2}P_z^{1/2}\bar{z}(k)| \le \|\bar{K}P_z^{-1/2}\| \cdot \|P_z^{1/2}\bar{z}(k)\| \le \|\bar{K}P_z^{-1/2}\|$. Taking Eq. (35) into consideration, one has

$$|v(k)| \le (\bar{u} - \lambda_1)/\mu_1, \tag{37}$$

and substituting Eq. (37) into Eq. (34) yields $|u(k)| \le \bar{u}$, or $u(k)$ is feasible. This completes the proof. □
Let us explain the *dual-mode* NMPC algorithm determined by Lemma 2 as below. First, let us give the standard output feedback control law as

$$\begin{aligned} v(k) &= K\hat{x}(k) \\ u(k) &= \mathcal{N}_z^{-1}(v(k)), \end{aligned} \tag{38}$$

and then the invariant set shrinks to

$$S_x := S_z(M = 0) = \{\hat{x}(k)|\hat{x}^T(k)P_x\hat{x}(k) \le 1\}. \tag{39}$$

If the current $\hat{x}(k)$ moves outside of S_x, then the controller enters the *first mode*, in which the dimension of $\hat{x}(k)$ is extended from N to $N+M$ by D(k) (see Eq. (27)). Then, $\hat{x}(k)$ will be driven into S_x in no more than M steps, i.e., $\hat{x}(k+M) \in S_x$, which will also be proven later. Once $\hat{x}(k)$ enters S_x, the controller is automatically switched to the *second mode*, in which the initial control law (38) is feasible and can stabilize the system.

It has been verified by extensive experiments that assumptions A4 and A5 are not difficult to fulfil, and most of the time-consuming calculations are done off-line. First, the stable state-feedback gain K (see Eq. (26)) and observer gain L (see Eq. (24)) are pre-calculated by MATLAB. Then, compute P_e based on Eq. (29). Afterwards, pick $\mu \in (0,1)$ and $\lambda \in (0,\bar{u})$ satisfying the local Lipschitz condition (34). Finally, pick $\tau_{1,2}$ (generally in the range (1, 1.5)), and calculate P_x off-line by MATLAB according to assumptions A4 and A5.

The aforementioned controller design is for regulator problem, or making the system state to settle down to zero. But it can be naturally extended to address the tracking problem with reference signal $r(t) = a \neq 0$. More precisely, the controller (26) is converted to $v(k) = \bar{K}\hat{z}(k) + a\rho$ with $1/\rho = \lim_{z\to 1}(\tilde{C}(zI-\Pi)^{-1}\tilde{B})$, $\tilde{C} := [C, \mathbf{0}]_{1\times(N+M)}$ and $\tilde{B} := [B^T, \mathbf{0}]^T_{(N+M)\times 1}$. Moreover, if $I - Pi$ is nonsingular, a coordinate transformation $\hat{z}(k) - z_c \to \hat{z}(k)$ with $z_c = (I-\Pi)^{-1}\tilde{B}a\rho$ can be made to address the problem. Even if $I - \Pi$ is singular, one can still make some suitable coordinate transformation to obtain Eq. (27).

Next we will show that the dual-mode method can enlarge the closed-loop stable region. First, rewrite P_z by

$$P_z = \begin{bmatrix} (P_x)_{N\times N} & P_{xD} \\ PxD^T & (P_D)_{M\times M} \end{bmatrix},$$

and hence the maximum ellipsoid invariant set of $x(k)$ is given as

$$S_{xM} := \{\hat{x} | \hat{x}^T (P_x - P_{xD}P_D^{-1}P_{xD}^T)\hat{x}(k) \leq 1\}. \tag{40}$$

Bearing in mind that $P_x - P_{xD}P_D^{-1}P_{xD}^T = (E_x^T P_z^{-1} E_x)^{-1}1$, it can be obtained that

$$\mathrm{vol}(S_{xM}) \propto \det(E_x^T P_z^{-1} E_x), \tag{41}$$

where vol($\cdot$) and det($\cdot$) denote the volume and matrix determinant. It will be verified later that the present dual-mode controller (26) can substantially enlarge the $\det(E_x^T P_z^{-1} E_x)$ with the assistance of the perturbation signal $D(k)$ and hence the closed-loop stable region S_{xM} is enlarged. Based on the above mentioned analysis of the size of the invariant set S_{xM}, we give the closed-loop stability theorem as follows.

Theorem 3. *Consider a closed-loop Hammerstein system (23) whose dynamics is determined by the output-feedback control law (26) and (30) and subject to the input constraints $|u| \leq \bar{u}$, the system is closed-loop asymptotically stable provided that assumptions A1–A5 are fulfilled.*

Proof: Based on assumptions A1–A5, one has that there exists $D(k+1)$ such that $z(k+1) \in S_z$ for arbitrary $x(k) \in S_x$; then by invariant property, at next sampling time $D(k+1|k) = TD(k)$ provides a feasible choice for $D(k+1)$ (only if $D(k) = 0$, $J(k+1) = J(k)$, otherwise $J(k+1) < J(k)$). Thus, the present NMPC law (26) and (30) generates a sequence of $D(k+i|k) = TD(k+i-1|k)$ $(i = 1, \cdots, M)$ which converges to zero in M steps and ensures the input magnitudes constraints satisfaction. Certainly, it is obvious that $TD(k)$ need not have the optimal value of $D(k+1)$ at the current time, hence the cost $J(k+1)$ can be reduced further still. Actually, the optimal $D^*(k+1)$ is obtained by solving Eq. (30), thus $J^*(k+1)| \leq J(k+1|k) < J(k)$ $(D(k) \neq 0)$. Therefore, as the sampling time k increases, the optimization index function $J(k)$ will decrease monotonously and $D(k)$ will converge to zero in no more than M steps. Given constraints satisfaction, the system state $\hat{x}(k)$ will enter the invariant set S_x in no more than M steps. Afterwards, the initial control law will make the closed-loop system asymptotically stable. This completes the proof. □

2.5 Case study

2.5.1 Modeling

Consider a widely-used heat exchange process in chemical engineering as shown in Fig. 2 (17), the stream condenses in the two-pass shell and tube heat exchanger, thereby raising the temperature of process water. The relationship between the flow rate and the exit-temperature of the process water displays a Hammerstein nonlinear behavior under a fixed rate of steam flow. The condensed stream is drained through a stream trap which lets out only liquid. When the flow rate of the process water is high, the exit-temperature of stream drops below the condensation temperature at atmospheric pressure. Therefore, the steam becomes subcooled liquid, which floods the exchanger, causing the heat transfer area to decrease. Therefore, the heat transfer per unit mass of process water decreases. This is the main cause of the nonlinear dynamics.

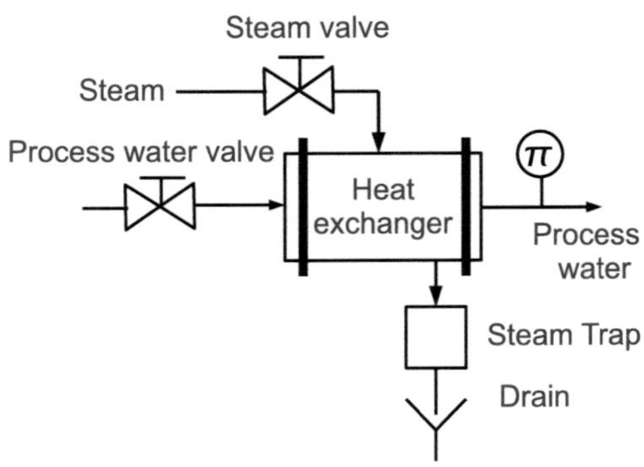

Fig. 2. Heat exchange process

The mathematical Hammerstein model describing the evolution of the exit-temperature of the process water VS the process water flow consists of the following equations (17):

$$\begin{aligned} v(t) = &-31.549u(t) + 41.732u^2(t) \\ &-24.201u^3(t) + 68.634u^4(t), \end{aligned} \tag{42}$$

$$y(t) = \frac{0.207z^{-1} - 0.1764q^{-2}}{1 - 1.608z^{-1} + 0.6385q^{-2}} v(t) + \xi(t), \tag{43}$$

where $\xi(t)$ is a white external noise sequence with standard deviation 0.5. To simulate the fluctuations of the water flow containing variance frequencies, the input is set as periodical signal $u(t) = 0.07\cos(0.015t) + 0.455\sin(0.005t) + 0.14\sin(0.01t)$. In the numerical calculation, without loss of generality, the OFS is chosen as Laguerre series with truncation length $N = 8$, while the nonlinear bases of the nonlinear block $\mathbb{N}(\cdot)$ are selected as polynomials with $r = 9$. The sampling number $S = 2000$, and sampling period is 12s.

Note that we use odd-numbered data of the S-point to identify the coefficients $\left\{a_i^{\langle j\rangle}\right\}_{i=1}^{r}$ and $\left\{c_k^{\langle j\rangle}\right\}_{k=1}^{N}$ $(j = 1, \cdots, \eta)$, and use the even-numbered data to examine the modeling accuracy.

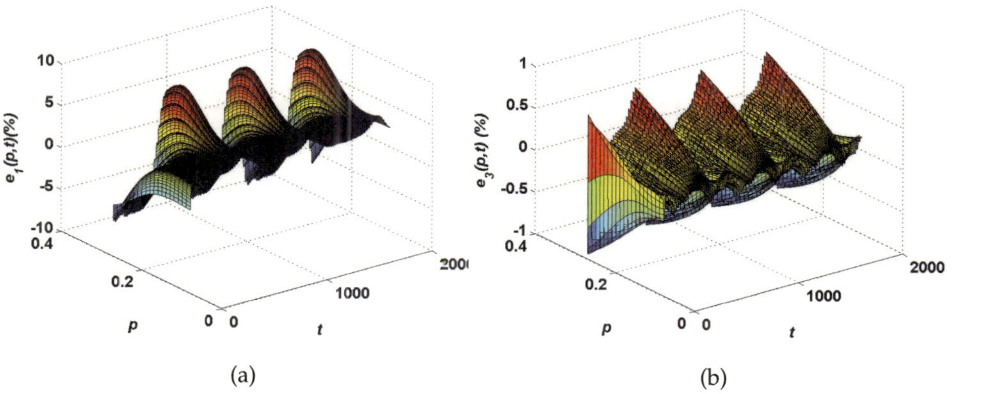

Fig. 3. (Color online) (a): Modeling error of the traditional single-channel method; (b): Modeling error of the present multi-channel method (triple channels)

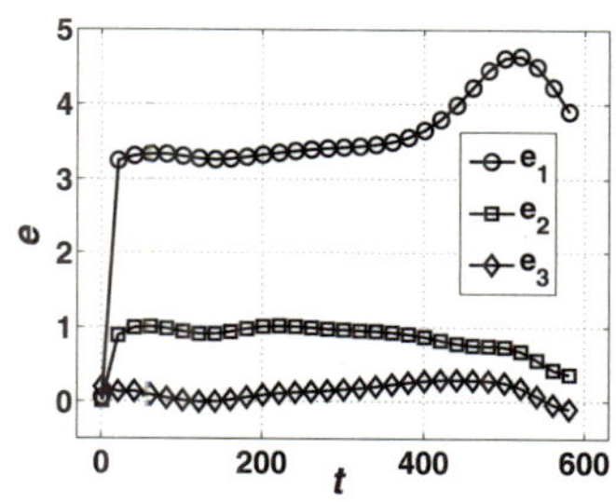

Fig. 4. Modeling error for $p = 0.1$

Denoted by $e(t)$ is the modeling error. Since the filter pole p (see Eq. (2)) plays an important role in the modeling accuracy, in Fig. 3(a) and (b), we exhibit the average modeling errors of the traditional single-channel and the present multi-channel methods along with the increase of Laguerre filter pole p. For each p, the error is obtained by averaging over 1000 independent runs. Clearly, the method proposed here has remarkably smaller modeling error than that of the traditional one. To provide more vivid contrast of these two methods, as shown in Fig. 4, we fix the Laguerre filter pole $p = 0.1$ and then calculate the average modeling errors of the single-channel ($\eta = 1$), double-channel ($\eta = 2$), and triple-channel models ($\eta = 3$) averaged over 1000 independent runs for each case. This is a standard error index to evaluate the modeling performances. The modeling error of the present method ($\eta = 3$) is reduced by more than 10 times compaired with those of the traditional one ($\eta = 1$), which vividly demonstrates the advantage of the present method.

Note that, in comparison with the traditional method, the modeling accuracy of the present approach increased by $10 - 17$ times with less than 20% increase of the computational time. So a trade off between the modeling accuracy and the computational complexity must be made. That is why here we set the optimal channel number as $\eta = 3$. The underlying reason for the obvious slow-down of the modeling accuracy enhancement rate after $\eta = 4$ is that the 4th largest singular value σ_4 is too small compared with the largest one σ_1 (see Eq. (8)). This fact also supports the validity of the present method.

2.5.2 Control

The present dual-mode NMPC is performed in the Heat Exchanger System model (55?57) with the results shown in Figures 7,8 (Regulator Problem, $N = 2$), Figures 9?11 (Regulator Problem, $N = 3$) and Figure 12 (Tracking Problem, $N = 3$), respectively. The correspondence parameter settings are presented in Table 1.

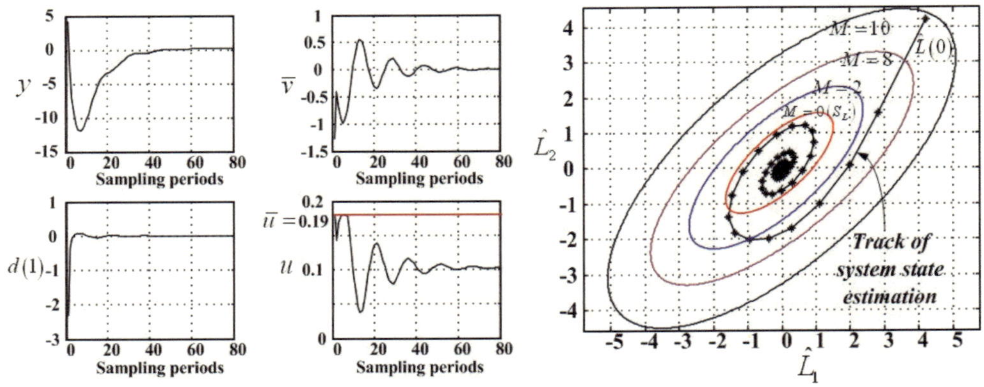

Fig. 5. (Color online) Left panel: Control performance of regulator problem ; Right panel: state trajectory L and its invariant set. Here, $N = 2$.

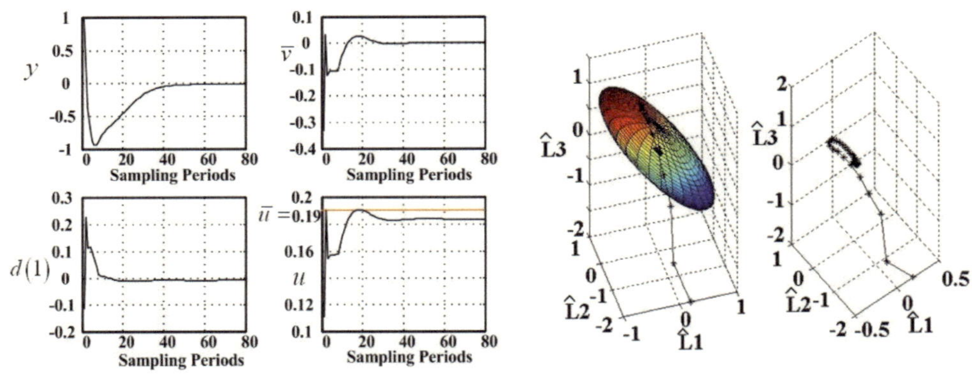

Fig. 6. (Color online) Left panel: Control performance of regulator problem ; Right panel: state trajectory L and its invariant set. Here, $N = 3$.

In these numerical examples, the initial state-feedback gain K and state observer gain Γ are optimized offline via *DLQR* and *KALMAN* functions of MATLAB6.5, respectively. The curves of $y(k)$, $u(k)$, $\bar{v}(k)$ and the first element of $D(k)$, i.e. , $d(1)$, are shown in Figure 7 (N=2) and Figure 8 (N=3), respectively. To illustrate the superiority of the proposed dual-mode NMPC, we present the curve of $\hat{L}(k)$, the invariant sets of and in Figure 8 ($N = 2$, M=$\{2,8,10\}$) and Figure 10 and 11 ($N = 3$, $M = \{0, 5, 10\}$). One can find that $\hat{L}(0)$, is outside the feasible initial invariant set S_L(referred to (48), see the red ellipse in Figure 10 and the left subfigure of Figure

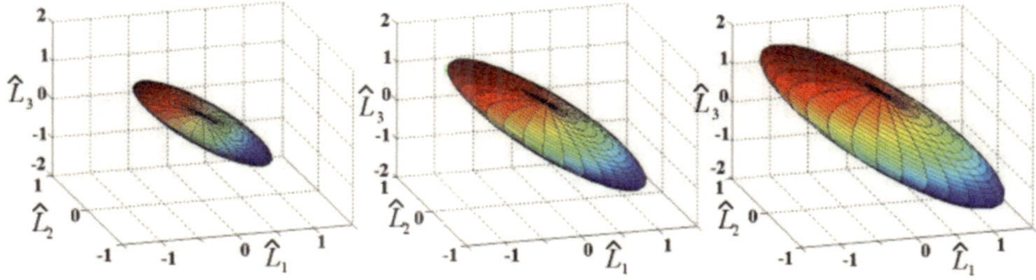

Fig. 7. (Color online) Invariant sets S_L (left) and S_{LM}, $M = 5$ (middle), $M = 3$ (right). Here, $N = 3$.

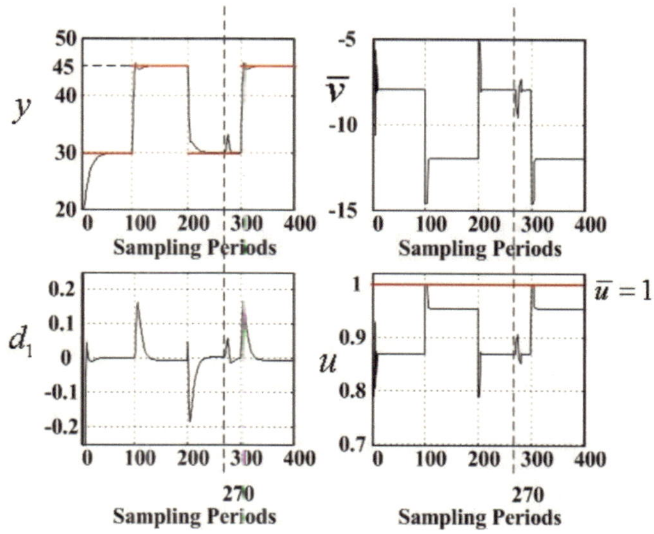

Fig. 8. (Color online) Control performances of Tracking problem.

11). Then the state extension with $M = 10$ is used to enlarge S_L to S_{LM} (referred to (52), see the black ellipse in Figure 8 and the right subfigure of Figure 11) containing $\hat{L}(0)$. After eight (Figure 8) or six steps (Figure 10), $\hat{L}(k)$ enters S_L. Afterwards, the initial control law (47) can stabilize the system and leads the state approach the origin asymptotically. Lemma 2 and Theorem 3 are thus verified. Moreover, the numerical results of Figures 8 and 11 also have verified the conclusion of the ellipsoid volume relation (54), i.e. the size of S_{LM} increases along with the enhancement of the prediction horizon M.

As to the tracking problem (see Figure 12), one should focus on the system state response to the change of the set-point. In this case, $\hat{L}(k)$ moves outside S_L, thus $D(k)$ is activated to enlarge S_L to S_{LM} and then to drive $\hat{L}(k)$ from S_L to S_{LM} in no more than steps. After 60 sampling periods, the overshooting, modulating time and steady-state error are 2.2%, 15 and 0.3% respectively. Moreover, robustness to the time-delay variations is examined at the 270-th sampling period, while the linear block of this plant is changed from (58) to

$$y(k+1) = \frac{0.207z^{-1} - 0.1764z^{-2}}{1 - 1.608z^{-1} + 0.6385z^{-2}} v(k) \tag{44}$$

dual-mode NMPC can still yield satisfactory performances, thanks to the capability of the Laguerre series in the inner model. The feasibility and superiority of the proposed control algorithm are thus demonstrated by simulations on both regulator and tracking problems.
Still worth mentioning is that some other simulations also show that the size of increases as decreases. In other words, more accurate identification and inverse solving algorithms would help further enlarge the closed-loop stable region. Fortunately, the proposed TS-SCIA can do this job quite well.
To further investigate the proposed dual-mode NMPC, a number of experiments were carried out to yield statistical results. More precisely, $\{\lambda, \mu, \tau\}$ are fixed to $\{0.70, 0.35, 1.12\}$, and N, M and σ are selected from the sets $\{2,3,4\}$, and $\{8,9,\cdots,18\}$, $\{0.001, 0.002, \cdots, 0.005\}$, respectively. The set-point is the same as Figure 12. In this set-up, 165 experiments were performed. The statistical results, such as expectations and optimal values for the settling time, overshooting, steady-state error and computational time of 400 steps are shown in Table 2. In addition, the corresponding optimal parameters are given. The statistical results further illustrate the advantages of the proposed algorithm regarding transient performance, steady-state performance and robustness to system uncertainties.

Remark 1. *The increase of the Laguerre truncation length can help enhancing the modelling and control accuracy at the cost of an increasing computational complexity. Therefore, a tradeoff must be made between accuracy and computational complexity. Note that the general parameter setting procedure is given in Remark 3.*

Parameter	Regular problem($N = 2$)	Regular problem ($N = 3$)	Tracking problem($N = 3$)
$\hat{L}(0)$	$[4.2, 4.2]^T$	$[0.02, -2, -2]^T$	$[0.02, -2, -2]^T$
K	$[0.327, -0.570]$	$[-0.328, -0.200, 0.200]$	$[-0.328, 0.200, -0.120]$
Γ	$[0.187, 0.416]^T$	$[0.217, 0.132, 0.276]^T$	$[0.217, 0.132, 0.276]^T$
$\bar{u}$	0.19	0.19	1.0
σ	0.001	0.001	10.001
μ	0.05	0.02	0.35
τ_1	1.05	1.12	1.12
τ_2	1.30	1.25	1.37
$\bar{e}$	0.5	0.5	0.5

Table 1. Parameter settings

Control indexes	STVSP (steps)	STVTD (steps)	overshooting (%)	steady-state error (%)	computational time of 400 step(s)
Optimal $\{N, M, \sigma\}$	$\{4,7,0.003\}$	$\{3,12,0.002\}$	$\{4,8,0.002\}$	$\{4,10,0.001\}$	$\{2,8,0.005\}$
Expectation value	16.7	12.4	± 2.78	± 0.41	21.062

Table 2. Statistical control performance of tracking problems, (Computation platform: 2.8G-CPU and 256M-RAML; STVSP, STVTD denote the settling times for the variations of set-point and time delay, respectively.)

2.6 Section conclusion

In this section, a novel multi-channel identification algorithm has been proposed to solve the modelling problem for constrained Hammerstein systems. Under some weak assumptions on the persistent excitation of the input, the algorithm provides consistent estimates even in the presence of colored output noise, and can eliminate any needs for prior knowledge about the system. Moreover, it can effectively reduce the identification errors as compared to the traditional algorithms. To facilitate the controller design, the *MCIA* is converted to a two-stage identification algorithm called *TS-SCIA*, which preserve almost all the advantages of the former. In addition, to support these two algorithms, systematical analyses about their convergence and approximation capability has been provided. Based on the *TS-SCIA*, a novel dual-mode *NMPC* is developed for process control. This approach is capable of enlarging the closed-loop stable region by providing extra degrees of design freedom. Finally, modelling and control simulations have been performed on a benchmark Hammerstein system, i.e., a heat exchanger model. The statistical results have demonstrated the feasibility and superiority of the proposed identification and control algorithms for a large class of nonlinear dynamic systems often encountered in industrial processes.

3. Model Predictive Control for Wiener systems with input constraints

3.1 Introduction

The Wiener model consists of a dynamic linear filter followed by a static nonlinear subsystem. This model can approximate, with arbitrary accuracy, any nonlinear time-invariant systems (10; 23) with fading memory, thus it appears in a wide range of applications. For example, in wireless communications, the Wiener model has been shown to be appropriate for describing nonlinear power amplifiers with memory effects (15; 47). In chemistry, regulation of the pH value and identification of the distillation process have been dealt with by using the Wiener model (7; 38; 58). In biology, the Wiener model has been extensively used to describe a number of systems involving neural coding like the neural chain (47), semicircular canal primary neurons (53) and neural spike train impulses (37). Moreover, applications of the Wiener model in other complex systems such as chaotic systems have been explored (12). In fact, the control of Wiener systems has become one of the most urgently needed and yet quite difficult tasks in many relevant areas recently.

To address various control problems of Wiener systems, extensive efforts have been devoted to developing suitable MPC (model predictive control) methods. Under the MPC framework, the input is calculated by on-line minimization of the performance index based on model predictions. MPC has been practiced in industry for more than three decades and has become an industrial standard mainly due to its strong capability to deal with various constraints (23). However, to design an effective MPC, an accurate data-driven model of Wiener systems is required. A large volume of literature has been devoted to studying this issue; see (6; 24; 29; 64) for comprehensive reviews. More recently, some research interests have been focused on extending the linear subspace identification method for this typical class of nonlinear systems (23) (52) (59). Among them, Gómez's approach (23) is one of the most efficient methods since it has good prediction capabilities, and guarantees stability over a sufficiently wide range of models with different orders. In addition, this subspace method delivers a Wiener model in a format that can be directly used in a standard MPC strategy, which makes it very suitable to act as the internal model of our proposed NMPC (Nonlinear MPC) method to be further discussed below.

Nevertheless, due to its specific structure, the achievements on the control of the Wiener model are still fairly limited so far. Most of the existent control algorithms have some, if not all, of the following disadvantages:

- small asymptotically stable regions;

- limited capacity in handling input constraints;
- reliance on the detectability of the intermediate output.

For instance, Nesic (49) designs an output feedback stabilization control law for Wiener systems, but this work does not address input constraints; moreover, some rigorous conditions such as 0-state detectable are required to guarantee the global stability. Norquay *et al.* (50) and Bolemen *et al.* (7) develop NMPC strategies with ARX/polynomial and polytopic internal models, respectively, but neither considers stable region enlargement. Gómez *et al.* (23) use a subspace internal model to develop an NMPC strategy mainly accounting for unmeasurable disturbances; however, it merely inherits the stability properties of a standard linear MPC with linear constraints and quadratic cost function. Motivated by all the above-mentioned backgrounds and existing problems, the main task of this section is to develop a new efficient control algorithm for constrained Wiener systems, which can maximize the region of asymptotic stability and eliminate the reliance on the measurability of the intermediate output.

To accomplish this task, Gómez's modelling approach (23) is first used to separate the nonlinear and linear blocks of the underlying system, and then a dual-mode mechanism (14) is combined with our proposed NMPC approach to enlarge the stable region. More specifically, over a finite horizon, an optimal input profile found by solving an open-loop optimal control problem drives the nonlinear system state into the terminal invariant set (39); to that end, a linear output-feedback controller steers the state to the origin asymptotically. The main contribution of this section is the development of an algorithm that can effectively maximize the asymptotic stability region of a constrained Wiener system, by using the *dual-mode NMPC* technique, which can also eliminate the reliance on the detectability of the intermediate output (70). As a byproduct, since the nonlinear/linear blocks are separated at first and the online calculation is mainly done on the linear block, the computational complexity is remarkably reduced compared with some traditional nonlinear empirical model-based NMPCs (7; 50). Moreover, since the subspace identification method can directly yield the estimate of the nonlinear block inverse, the complex inverse-solving method is avoided in the new NMPC algorithm. Furthermore, some rigorous sufficient conditions are proposed here to guarantee the feasibility and stability of the control system.

3.2 Problem description

Consider a discrete MIMO Wiener system with a linear time-invariant (LTI) block described by

$$x(k+1) = Ax(k) + Bu(k), \tag{45}$$

$$\eta(k) = Cx(k), \tag{46}$$

and a nonlinear block by

$$y(k) = f(\eta(k)), \tag{47}$$

where $f(\cdot)$ is an invertible memoryless nonlinear function, $u(k) \in \mathbb{R}^p$, $y(k) \in \mathbb{R}^m$ are the input and output, respectively, $x(k) \in \mathbb{R}^n$ is the state vector, and $\eta(k) \in \mathbb{R}^m$ is the unmeasurable intermediate output. This Wiener system is subject to an input constraint:

$$|u_i| \leq \bar{u}_i,\ i = 1, \cdots, p. \tag{48}$$

Typically, there are two kinds of problems to consider:

- *Regulator problem*: Design an output-feedback control law such that the response of the initial conditions will die out at a desired rate;
- *Tracking problem*: Design an output-feedback control law to drive $y(t)$ to a set-point $r(k) = a$ asymptotically.

In general, for unconstrained systems with measurable $\eta(k)$, to address these two problems, one can respectively design a stable state observer,

$$\hat{x}(k+1) = A\hat{x}(k) + Bu(k) + L(\eta(k) - C\hat{x}(k)),$$

in combination with a stable state-feedback control law $u(k) = K\hat{x}(k)$ or with a stable state-feedback control law having offset (13) $u(k) = K\hat{x}(k) + a\theta$, where $1/\theta = \lim_{z\to 1}(C(zI - \Psi)^{-1}B)$ and $\Psi = A + BK$. However, for constrained Wiener systems with unmeasurable intermediate output $\eta(t)$, these basic control methods will be infeasible and the problems will become much more complex. This section develops a novel algorithm that can handle such challenging situations.

3.3 Control algorithm design

For the constrained Wiener system (45)–(48), in order to focus on the main idea of this section, i.e. dual-mode predictive mechanism, it is assumed that the system state matrices (A, B) can be estimated accurately while the identification error only appears in the output matrix estimate $\hat{C}$:

Assumption A1) the LTI matrices (A, B) can be precisely identified.

This identification can be implemented with the efficient subspace methods (23; 52; 59). In general, subspace methods give estimates of the system matrices (A, B, C). The robustness issue with estimate errors of (A, B) is beyond the scope of the current chapter, hence will not be discussed.

First, use a stable observer L to estimate $x(k)$ as follows:

$$\hat{x}(k+1) = A\hat{x}(k) + Bu(k) + L(\tilde{\eta}(k) - \hat{C}\hat{x}(k)), \tag{49}$$

where $\hat{x}(k)$ is the state estimate, and $\tilde{\eta}(k) \triangleq \hat{f}^{-1}(y(k))$ with $\hat{f}^{-1}(\cdot)$ denoting the inverse of f calculated by Gómez's subspace method (23). The state estimate error is defined as

$$e(k) \triangleq x(k) - \hat{x}(k). \tag{50}$$

Since the identified inverse $\tilde{\eta}(k)$ rather than $\eta(k)$ is used to estimate the state $x(k)$, the state estimate error $e(k)$ is caused by both the identification error of $f^{-1}(\cdot)$ and the initial condition mismatch. Therefore, the intermediate output estimate error $\hat{C}e(k)$ can be separated into two parts as follows:

$$\begin{aligned}\hat{C}e(k) &= \hat{C}x(k) - \hat{C}\hat{x}(k)\\ &= \underbrace{(\hat{C}x(k) - \tilde{\eta}(k))}_{part1} + \underbrace{(\tilde{\eta}(k) - \hat{C}\hat{x}(k))}_{part2}.\end{aligned} \tag{51}$$

Clearly, *part 1* equals $\Delta Cx(k) + \eta(k) - \tilde{\eta}(k)$ with $\Delta C = \hat{C} - C$. For a fixed nonlinear block f, *part 1* is yielded by the subspace method (23) based on state calculation, and hence the proportion of *part 1* to the whole estimate error $\hat{C}e(k)$ is solely determined by the current state

$x(k)$. It can be assumed that *part 1* of Eq. (51) satisfies the following equality:

$$\hat{C}x(k) - \tilde{\eta}(k) = \delta(x(k))\hat{C}e(k). \tag{52}$$

with $\delta(x(k)) = \text{diag}\{\delta_1(x(k)), \cdots, \delta_m(x(k))\}_{m\times m}$. For conciseness, from now on $\delta(x(k))$ and $\delta_i(x(k))$ are denoted by $\delta(\cdot)$ and $\delta_i(\cdot)$, respectively.
Next, recall the input constraint (48), and extend $\hat{x}(k)$ to $\hat{z}(k) \triangleq [\hat{x}^T(k), D^T(k)]^T$ with $D(k) \triangleq [d_1(k), \cdots, d_{H_p}(k)]^T$. Here, H_p is defined as the *prediction horizon*, and $D(k)$ represents the *auxiliary state* to be computed. Then, an extended state-feedback control law is set as

$$u(k) = K\hat{x}(k) + ED(k), \tag{53}$$

or

$$u_i(k) = K_i\hat{x}(k) + FD(k),\ i = 1, \cdots, p, \tag{54}$$

with $K = [K_1^T, \cdots, K_p^T]^T$, $E = [F^T, \cdots, F^T]^T$ and $F = [1, 0, \cdots, 0]_{1\times H_p}$.
Note that the novelty here lies in $D(k)$, and the reason will be demonstrated later. When the current state $x(k)$ is not in the asymptotic stability region of the constrained Wiener system (45)–(48) governed by (49) and (53), the auxiliary state $D(k)$ will be activated to drive $x(k)$ back into the asymptotic stability region in less than H_p steps, and $D(k)$ will vanish thereafter.
Now, substituting (52) and (53) into (49) yields

$$\hat{z}(k+1) = \Omega\hat{z}(k) + \begin{bmatrix} L(I_m - \delta(\cdot))\hat{C} \\ \mathbf{0} \end{bmatrix} e(k), \tag{55}$$

where

$$\Omega = \begin{bmatrix} \Psi & BE \\ \mathbf{0} & M \end{bmatrix},\ \Psi = A + BK,\ M = \begin{bmatrix} \boldsymbol{0} & I_{H_p-1} \\ 0 & \boldsymbol{0}^T \end{bmatrix}_{H_p\times H_p},$$

I_n is the n-dimensional identity matrix, $\mathbf{0}$ and $\boldsymbol{0}$ denote compatible zero matrix and zero column vector, respectively.
In order to stabilize the closed-loop system (55), define two ellipsoidal initial state invariant sets of $\hat{z}(k)$ and $e(k)$ as follows:

$$S \triangleq \{\hat{z}|\hat{z}^T P\hat{z} \le 1\}, \tag{56}$$

$$S_e \triangleq \{e|e^T P_e e \le \bar{e}^2, 0 \le \bar{e} \le 1\}, \tag{57}$$

where P and P_e are both positive definite and symmetric matrices to be computed, and $\bar{e}$ is a pre-defined constant. The auxiliary state $D(k)$ (see Eq. (53)) can be calculated by solving the following optimization problem:

$$\min_{D(k)} J(k) \triangleq D^T(k)D(k) \tag{58}$$

subject to Eq. (56). The following lemma and theorem guarantee the existence of S and S_e, and the feasibility of the control law (53).

Lemma 3. *For any constant matrices A_1, A_2 with compatible dimensions, and for any $\mu > 1$, one has*

$$(A_1 + A_2)^T P(A_1 + A_2) \le \mu A_1^T P A_1 + \tau A_2^T P A_2,$$

where $\tau = 1 + (\mu - 1)^{-1}$ and P is a positive definite matrix.

Proof: For any $\mu > 1$, one has

$$
\begin{aligned}
&(A_1 + A_2)^T P(A_1 + A_2) = \mu A_1^T P A_1 \\
&+(1 + (\mu - 1)^{-1}) A_2^T P A_2 - (\mu - 1)(A_1 - (\mu - 1)^{-1} A_2)^T \\
&\cdot P(A_1 - (\mu - 1)^{-1} A_2) \\
&\leq \mu A_1^T P A_1 + (1 + (\mu - 1)^{-1}) A_2^T P A_2. \qquad \blacksquare
\end{aligned}
$$

Theorem 4. *For the constrained Wiener system (45)–(48) governed by the control law (49), (53) and (58), if Assumption* **A1** *and the following assumption* **A2** *hold, then S and S_e defined respectively by (56) and (57) are invariant sets and the control law (53) satisfies (48).*
Assumption A2) *there exist $\sigma > 0$, $\mu_1 > 1, \mu_2 > 1$ and positive definite and symmetric matrices P, P_e, such that*

$$\delta(\cdot) L^T P_e L \delta(\cdot) \leq \sigma^2 L^T P_e L \tag{59}$$

$$(I_m - \delta(\cdot)) L^T E_x^T P E_x L (I_m - \delta(\cdot)) \leq (1 + \sigma)^2 L^T E_x^T P E_x L, \tag{60}$$

$$\mu_2 \Omega^T P \Omega \leq (1 - \bar{e}^2) P, \tag{61}$$

$$\begin{bmatrix} -\bar{u}_i^2 & K_i \\ K_i^T & -P \end{bmatrix} \leq 0, \ i = 1, \cdots, p, \tag{62}$$

$$\tau_2 (1 + \sigma)^2 \hat{C}^T L^T E_x^T P E_x L \hat{C} \leq P_e, \tag{63}$$

$$\mu_1 \Phi^T P_e \Phi + \tau_1 \sigma^2 \hat{C}^T L^T P_e L \hat{C} \leq P_e, \tag{64}$$

where $\tau_1 = 1 + (\mu_1 - 1)^{-1}$, $\tau_2 = 1 + (\mu_2 - 1)^{-1}$, $\Phi = A - L\hat{C}, \bar{K}_i \triangleq [K_i, F]$, (Here, E_x is a matrix satisfying $\hat{x} = E_x^T \hat{z}$ and K_i and F are defined in Eq. (54)).

Proof: For any $\hat{z}(k) \in S$, it can be verified from (56) and (61) that

$$\mu_2 \hat{z}(k) \Omega^T P \Omega \hat{z}(k) \leq (1 - \bar{e}^2) \hat{z}^T(k) P \hat{z}(k) \leq 1 - \bar{e}^2. \tag{65}$$

Moreover, for any $e(k) \in S_e$, from (57), (60) and (63), one has

$$
\begin{aligned}
&\tau_2 e^T(k) \hat{C}^T (I_m - \delta(\cdot)) L^T E_x^T P E_x L (I_m - \delta(\cdot)) \hat{C} e(k) \\
&\leq \tau_2 (1 + \sigma)^2 e^T(k) \hat{C}^T L^T E_x^T P E_x L \hat{C} e(k) \\
&\leq e^T(k) P_e e(k) \leq \bar{e}^2.
\end{aligned}
\tag{66}
$$

It follows from *Lemma 1* and (55) that

$$
\begin{aligned}
&\hat{z}^T(k+1) P \hat{z}(k+1) \leq \mu_2 \hat{z}^T(k) \Omega^T P \Omega \hat{z}(k) \\
&+\tau_2 e(k)^T \hat{C}^T (1 - \delta(\cdot)) L^T E_x^T P E_x L (1 - \delta(\cdot)) \hat{C} e(k).
\end{aligned}
\tag{67}
$$

Substituting (65) and (66) into (67) yields $\hat{z}^T(k+1) P \hat{z}^T(k+1) \leq 1$, which implies that S is an invariant set.

On the other hand, for any $e(k) \in S_e$, (49) and (50) yield that $e(k+1) = (\Phi + L\mathbf{ffi}(\cdot)\hat{C})e(k)$, thus by *Lemma 1*, for any $\mu_1 > 1$, the following inequality holds:

$$\begin{aligned} e^T(k+1)P_e e(k+1) &= ((\Phi + L\mathbf{ffi}(\cdot)\hat{C})e(k))^T P_e \\ &\cdot((\Phi + L\mathbf{ffi}(\cdot)\hat{C})e(k)) \le \mu_1 e^T(k)\Phi^T P_e \\ &\cdot\Phi e(k) + \tau_1 e^T(k)\hat{C}^T \mathbf{ffi}(\cdot)^T L^T P_e L\delta(\cdot)\hat{C}e^T(k) \\ &\le e^T(k)(\mu_1 \Phi^T P_e \Phi + \tau_1 \hat{C}^T \mathbf{ffi}(\cdot)^T L^T P_e L\mathbf{ffi}(\cdot)\hat{C})e(k). \end{aligned}$$

By taking (59) and (64) into account, one has $e^T(k+1)P_e e(k+1) \le \bar{e}^2$, i.e. S_e is an invariant set.

To satisfy the input constraint (48), rewrite the control law and get

$$\begin{aligned} |u_i(k)| &= |\bar{K}_i \hat{z}(k)| = |\bar{K}_i P^{-1/2} \cdot P^{1/2}\hat{z}(k)| \\ &\le \|\bar{K}_i P^{-1/2}\| \cdot \|P^{1/2}\hat{z}(k)\|. \end{aligned}$$

Since $\|P^{1/2}\hat{z}(k)\| \le 1$ from (56) and $\|\bar{K}_i P^{-1/2}\| \le \bar{u}_i$ from (62), it immediately follows that $|u_i(k)| \le \bar{u}_i$, $i = 1, \cdots, p$. ■

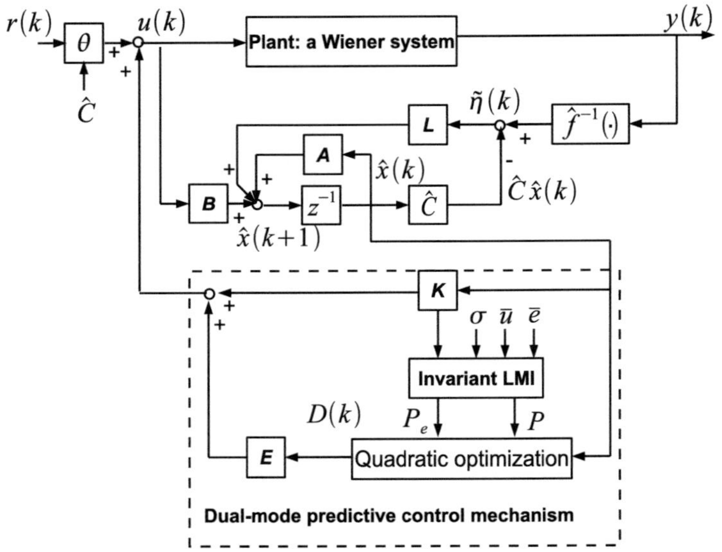

Fig. 9. Dual-mode predictive controller structure.

Remark 2. *Both the internal matrices $L^T P_e L$ and $L^T E_x^T P E_x L$ in (59) and (60) of Assumption **A2** are positive definite and symmetric matrices. Hence these two inequalities provide a measure for the upper bound of the identification error of the nonlinear block inverse $f^{-1}(\cdot)$. Specifically, take the single-input/single-output (SISO) case as an instance for discussion. Inequalities (59) and (60) of Assumption **A2** can be simplified to*

$$|\delta(\cdot)| \le \sigma, \tag{68}$$

*which is independent of P, P_e and L. Since (A, B) is estimated precisely (see Assumption **A1**), the subspace method yields sufficiently accurate local approximations of $f^{-1}(\cdot)$ and C. Consequently, as shown in (51), compared with the second part, the first part is much smaller, and it is thus a logical* *assumption that $\delta(\cdot)$ is locally bounded as given in (68) for SISO systems, or (59) and (60) for MIMO systems. As a consequence, the estimate error $\hat{C}e(k)$ is mainly caused by part2, i.e. the initial condition mismatch, and (59) and (60) can be easily satisfied in general. They will be further illustrated by a case study later. Moreover, σ is a parameter determining the stability performance of the new our approach. Note that even if the subspace can not yield sufficiently accurate $f^{-1}(\cdot)$ and C, i.e. σ is not sufficiently small, the present algorithm still works, but with shrinking invariant sets S_{xm} and S_e.*

Remark 3. *It is shown by a number of simulations that Assumption* **A2** *is not difficult to fulfill. A detailed procedure to obtain P and P_e is given here: First, the stable observer gain L (see (49)) and state-feedback gain K (see (53)) are pre-calculated by MATLAB, which is quite fast. Then, select $\mu_2 > 1$ (generally, it can be selected in the range $(1, 1.5)$), and compute P by the LMI toolbox of MATLAB according to (61), (62) and the pre-determined constant $\bar{e}$. Note that even for some unfavorable μ_2, one can still adjust the state-feedback gain K to guarantee the feasibility of P. Finally, select a suitable $\mu_1 > 1$ and calculate P_e by the LMI toolbox according to (63), (64), the pre-determined constant σ, and the pre-calculated matrix P. Furthermore, the obtained parameters L, P and P_e are substituted into (59) and (60) to verify their feasibility according to $\delta(\cdot)$, which is now available. If they can not be fulfilled, one should increase σ until it can be satisfied. Of course, for SISO systems, since (59) and (60) has been simplified to (68), this condition can be directly verified, which is independent of L, P and P_e. As a consequence, the computation time is mainly taken by the above-mentioned two LMIs composed of (61), (62) and (63), (64), respectively, whose computational complexities are both $O(N^2)$. Moreover, since L and K can be designed separately for the linear block, even for some unfavorable values of μ_2, one can still assure the feasibility of P_e by adjusting the feasible state-feedback gain L or moderately increasing σ. Certainly, suitable selections of μ_1 and μ_2 ($\mu_1 \in [1, 1.5]$ and $\mu_1 \in [1, 2]$ for example according to our numerical simulations) will accelerate the searching of P and P_e.*

Based on Theorem 4 and Eq. (58), a block diagram depicting the control structure is illustrated in Fig. 9.

Now, the main advantages of the present algorithm can be demonstrated. If $H_p = 0$, i.e. the system state dimension is not extended, then the control law (53) reduces to the standard output-feedback control law

$$u(k) = K\hat{x}(k), \tag{69}$$

and the invariant set S reduces to

$$S_x \triangleq \left\{\hat{x} | \hat{x}^T P_x \hat{x} \leq 1\right\}. \tag{70}$$

Here, (69) is called the initial control law, which drives $\hat{x}(k)$ to the origin asymptotically provided that the initial state estimate $\hat{x}(0)$ is inside S_x. However, S_x is the minimal case of S with $H_p = 0$, so it is very likely that $\hat{x}(0) \notin S_x$. Fortunately, it will shown later that the extension of the system state by (53) can enlarge S_x effectively so as to include $\hat{x}(0)$ in general. More precisely, if the current $\hat{x}(k)$ moves to outside S_x, then the controller enters the *First Mode*, in which the optimal input profile (49), (53), found by solving the open-loop optimal control problem (58), drives the nonlinear system state $\hat{x}$ into the terminal invariant set S_x over a finite horizon H_p, i.e. $\hat{x}(k + H_p) \in S_x$. To that end, the controller is automatically switched to the *Second Mode*, in which the local linear output-feedback control law (69) steers the state $\hat{x}(k)$ to the origin asymptotically. This approach is called the *dual-mode NMPC* method (14), and H_p is hereby called the prediction horizon.

Remark 4. *The present method focuses on the regulator problem. To address the tracking problem, as shown in the top-left part of Fig. 9, Equation (53) should be converted into a state-feedback control law with an offset, in the form of*

$$u(k) = \bar{K}\hat{z}(k) + a\theta,$$

where $1/\theta = \lim_{z\to 1}(\bar{C}(zI - \Omega)^{-1}\bar{B})$, $\bar{C} = [\hat{C}, \mathbf{0}]_{m\times(n+H_p)}$, $\bar{B} = [B^T, \mathbf{0}]^T_{(n+H_p)\times p}$, *and the set-point* $r(k) = a$. *Furthermore, if* $(I - \Omega)$ *is nonsingular, then one can make a coordinate transformation* $\hat{z}(k) - \alpha \longrightarrow \tilde{z}(k)$, *with* $\alpha = (I - \Omega)^{-1}Ba\theta$, *so as to convert the tracking problem to the regulator problem. Even if* $(I - \Omega)$ *is singular, one can still use some suitable coordinate transformation to convert it to the regulator problem. In turn, the terminal invariant set* S_x *has hereby moves to a new place in the old state space spanned by* $\hat{x}$, *and the center of* S_x *is thus shifted from the origin to* α. *In this sense, these two control problems are equivalent, and hence the terminal set* S_x *should be recalculated once a new set-point variation occurs.*

Remark 5. *In many industrial applications, the constraints on the changing rate of the input, i.e.* $|\Delta u_i(k)| \le \bar{v}_i$ *with* $\Delta u_i(k) \triangleq u_i(k+1) - u_i(k)$, $i = 1, \cdots, p$. *are very common. This kind of constraints can also be handled by the present method. More precisely, taking into consideration of (55), and letting* $\Upsilon \triangleq \begin{bmatrix} L(1-\delta(\cdot))\hat{C} \\ \mathbf{0} \end{bmatrix}$, *one has*

$$\begin{aligned} |\Delta u_i(k)| &= \bar{K}_i(z(k+1) - z(k)) = |\bar{K}_i(\Omega - I)\hat{z}(k)| \\ &+ |\bar{K}_i\Upsilon e(k)| = |\bar{K}_i(\Omega - I)P^{-1/2}P^{1/2}\hat{z}(k)| \\ &+ |\bar{K}_i\Upsilon P_e^{-1/2}P_e^{1/2}e(k)|. \end{aligned}$$

Since $\|P^{1/2}\hat{z}(k)\| \le 1$ *from (56) and* $\|P_e^{1/2}e(k)\| \le \bar{e}$ *from (57), one has*

$$|\Delta u_i(k)| \le \|\bar{K}_i(\Omega - I)P^{-1/2}\| + \bar{e}\|\bar{K}_i\Upsilon P_e^{-1/2}\|. \tag{71}$$

Thus, one can first compute K, L *and* P *according to Remark 3, and then substitute them into (71) to obtain a new matrix inequality of* P_e *as follows:*

$$\bar{K}_i\Upsilon P_e^{-1}\Upsilon^T\bar{K}_i^T \le 1/\bar{e}^2 \cdot (\bar{v}_i - \|\bar{K}_i(\Omega - I)P^{-1/2}\|)^2, \tag{72}$$

or

$$\begin{bmatrix} -1/\bar{e}^2 \cdot (\bar{v}_i - \|\bar{K}(\Omega - I)P^{-1/2}\|)^2 & \bar{K}\Upsilon \\ \Upsilon^T\bar{K}^T & -P_e \end{bmatrix} \le 0. \tag{73}$$

In this way, the constraints on the changing rate of the input, i.e. $|\Delta u_i(k)| \le \bar{v}_i$, can be handled by introducing (73) into *Assumption* ***A2***.

3.4 Stability analysis

Theorem 4 , provides the initial state invariant set guaranteeing the existence of $D(k)$ (see (53)), and this section focuses on maximizing the invariant set, i.e. the asymptotic stability region. Rewrite the matrix P as

$$\begin{bmatrix} (P_x)_{n\times n} & P_{xD} \\ P_{xD}^T & (P_D)_{H_p\times H_p} \end{bmatrix}$$

and define an extended invariant set of $\hat{x}$ as

$$S_{xm} \triangleq \left\{\hat{x}|\hat{x}^T(P_x - P_{xD}P_D^{-1}P_{xD}^T)\hat{x} \leq 1\right\}. \tag{74}$$

Moreover, denote the volume of the ellipsoidal invariant set S_{xm} by vol(S_{xm}) and let det$(\cdot)$ be the determinant and "$\propto$" denotes "proportional to". Accordingly, a new theorem is established to show the quantitative correlation between the size of the invariant set of $\hat{x}$ and the matrix P.

Theorem 5. *The following two conclusions hold:*
1) If $\hat{z}(k) \in S$ (see (56)), then $\hat{x}(k) \in S_{xm}$.
2) $vol(S_{xm}) \propto \det(E_x^T P^{-1} E_x)$, here $vol(\cdot)$ denotes the size of a set.

Proof: Since $\hat{z} \in S$, it follows from (56) that

$$\begin{aligned} &\hat{x}^T(k)P_x\hat{x}(k) \\ &\quad \leq 1 - 2\hat{x}^T(k)P_{xD}D(k) - D^T(k)P_DD(k). \end{aligned} \tag{75}$$

In addition, it can be easily verified that

$$\begin{aligned} -P_D^{-1}P_{xD}^T\hat{x}(k) = \arg\max\nolimits_{D(k)}\{&1 - 2\hat{x}^T(k) \\ &\times P_{xD}D(k) - D^T(k)P_DD(k)\}. \end{aligned} \tag{76}$$

Thus, substituting (76) into (75) yields $\hat{x}^T(k)(P_x - P_{xD}P_D^{-1}P_{xD}^T)\hat{x}(k) \leq 1$, which proves Conclusion 1). Furthermore, using $P_x - P_{xD}P_D^{-1}P_{xD}^T = (E_x^T P^{-1} E_x)^{-1}$ and (74), one can easily verify Conclusion 2). ■

It can be seen from *Theorem 2* that the initial invariant set of $\hat{x}$ is effectively enlarged from S_x to S_{xm}. Moreover, in order to maximize S_{xm}, one may maximize $\det(E_x^T P^{-1} E_x)$ by the method detailed in (10), with calculation carried out using the *MATLAB LMI* toolbox. This maximized S_{xm} guarantees the existence of $D(k)$, or the feasibility of the present control law (53), with the largest possible probability. Consequently, based on Theorems 4 and 5, the closed-loop stability is guaranteed by the following theorem.

Theorem 6. *For the constrained Wiener system (45)–(48), if the control law (49), (53) and (58) is implemented along with stable state-feedback gain K and a stable state-observer L, and Assumptions* ***A1–A2*** *hold, then the closed-loop system is asymptotically stable.*

Proof: If the current state estimate $\hat{z}(k) \in S$, then it follows from Theorem 4 that there exists $D(k+1)$ such that $\hat{z}(k+1) \in S$. Additionally, from (55) it is obvious that $\bar{D}(k+1) \triangleq MD(k)$ is always a candidate for $D(k+1)$, since it satisfies $\bar{J}(k+1) = \bar{D}^T(k+1)\bar{D}(k+1) \leq J(k)$, and one has $\bar{J}(k+1) = J(k)$ only if $D(k) = 0$. Indeed, it can be easily seen that the feasible sequence $\bar{D}(k+i)$ $(i = 1, \cdots, H_p)$ decreases to zero in H_p steps. Moreover, it can be seen that $\bar{D}(k+1)$ is not always the optimal value $D(k+1)$ (or $D^*(k+1)$) calculated by (58). Thus, one has $J^*(k+1) \leq \bar{J}(k+1) < J(k)$ for $D(k) \neq 0$, and the control law with the optimal auxiliary state $D^*(k+1)$ will converge to the initial control law in no more than H_p steps. Thereafter, controller (69) will make the system asymptotically stable. ■

3.5 Case study

Consider a Wiener system described by (45)–(48) with

$$A = \begin{bmatrix} 2.3 & -1.2 \\ 1.0 & 0 \end{bmatrix}, \quad B = \begin{bmatrix} 1 \\ 0 \end{bmatrix}, \quad C = [1, 0],$$
$$f(\eta) = \eta^4 \sin(\eta) - \eta^5, \; \bar{u} = 1.5.$$

It can be easily verified that $f(\cdot)$ is a monotonically increasing function and is thus invertible. First, Gómez's subspace method (23) is used to identify the parameters of the linear block (A, B, C) and the inverse of the nonlinear block $f^{-1}(\eta)$. In order to demonstrate the merits of our proposed *dual-mode NMPC* algorithm, as shown in Fig. 10, we show a comparison to a traditional NMPC based on the NAARX (Nonlinear Additive AutoRegressive models with eXogeneous inputs) model (34), which is a special class of the well-known NARMAX models. This model is defined as

$$y(t) = \sum_{i=1}^{s} h_i(y(t-i)) + \sum_{j=0}^{q} g_j(u(t-j)) + \xi(t) \tag{77}$$

where $\{h_i(\cdot)\}$ and $\{g_i(\cdot)\}$ are scalar nonlinear functions, generally polynomials, and $\xi(t)$ is external white noise.

In this numerical example, $p = m = 1$, and the parameters of the present *dual-mode NMPC* are selected as follows: the initial state-feedback gain $K = [-2.3536, 1.1523]$ and the state observer gain $L = [1.0765, 0.3678]^T$ are optimized via *DLQR* and *KALMAN* functions of *MATLAB 6.5*, respectively. Prediction horizon $H_p = 6$, $\bar{e} = 0.4$, $\bar{x}(0) = [2.2, 2.2]^T$, $x(0) = [4.3, 4.8]^T$, $\mu_1 = 1.1$, $\mu_2 = 1.4$ and $\sigma = 0.1$. The estimate $\hat{C} = [1.01, 0]$. The parameters of the above traditional NMPC are: prediction horizon $H_p = 7$, control horizon $H_u = 7$, $s = 5$ and $q = 3$ (see Eq. (77)).

In Fig. 10, the most interesting part is the system state response to the change of the set-point. The trajectories of $\{y, \eta\}$, u, and d_1 (see (53)) are shown in the upper, middle and lower parts of Fig. 10, respectively. In this case, after 200 sampling periods, the overshoot, settling time and steady-state error of the present *Dual-mode NMPC* are 12.2%, 15 steps, and 0.3%, respectively. The first two transient performance indexes are much smaller than the counterparts of the traditional NMPC as shown by the read curves in Fig. 10. These merits root in the dual-mode mechanism of our proposed NMPC, which can effectively enlarge the closed-loop stability region thereby improving the transient performances.

To illustrate the superiority of the proposed method more vividly, we present the curves of $x(k)$ (star-line), $\hat{x}(k)$ (circle-line) and the invariant sets S_x (see Eq. (70)), S_{xm} (see Eq. (74)) in Fig. 11. It should be noted that x and $\hat{x}$ in these two figures were implemented with coordinate transforms according to *Remark* 3. One can see that $\hat{x}(0)$ is outside the feasible initial invariant set S_x (see the solid ellipse in Fig. 11), thus $D(k)$ is activated to enlarge S_x to S_{xm} (see the dashed ellipse in Fig. 11 containing $\hat{x}(0)$), and then to drive $\hat{x}(k)$ back to S_x in no more than H_p steps. Thereafter, the initial control law (69) stabilizes the system and leads the state to approach the origin asymptotically. Remarkably, for favorable parameters like $H_p = 6$, $\bar{e} = 0.4$ and $\sigma = 0.1$, as shown in Fig. 11, the attraction region S_{xm} is much larger than the counterpart of the standard NMPC. Moreover, in Fig. 12 one can observe the dynamics of the state estimate error $e(k)$ (see (50) and the star-curve). One can observe that, when the trajectory of $e(k)$ starts from outside of S_e (see (57) and the ellipse), it will move back into S_e after no more than H_p steps and then converge to the origin asymptotically. Theorem 4 is thus verified.

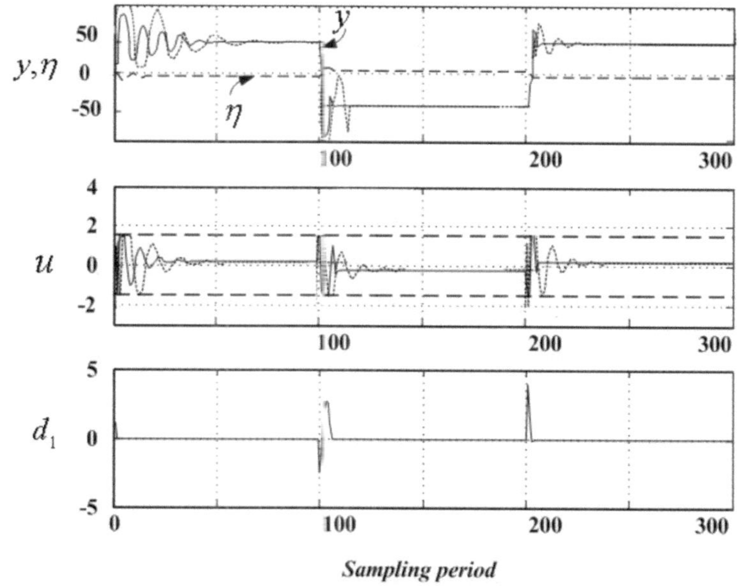

Fig. 10. Control performance comparison of the tracking problem. Solid curve: dual-mode NMPC; dotted curve: traditional NMPC; dashed curve: intermediate output (upper sub-figure) and input constraints (middle sub-figure); set-point: $\{-40, 40\}$.

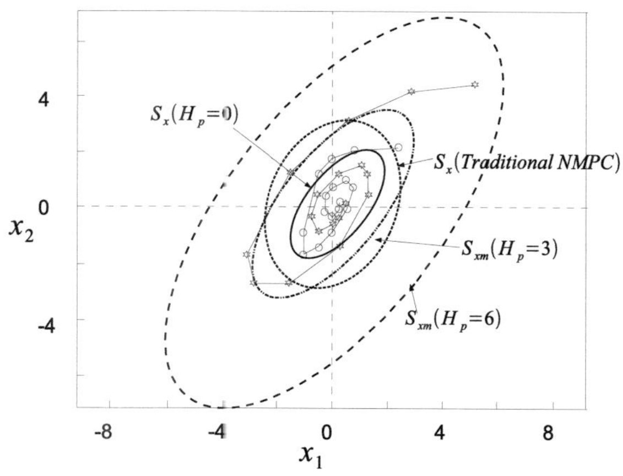

Fig. 11. Trajectory and invariant set of system states.

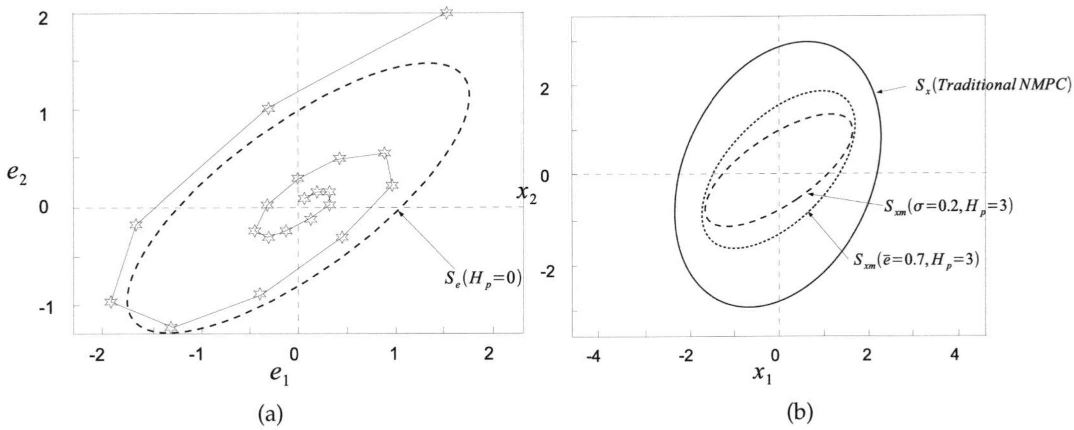

Fig. 12. (Color online) (a): Trajectory and invariant set of state estimate error; (b): Comparison of the invariant sets with $\sigma = 0.2$ or $\bar{e} = 0.7$.

In addition, the size comparison of S_{xm} with different $H_p \in \{0, 3, 6\}$ is presented in Fig. 11. The numerical results have verified *Theorem* 5, i.e. the size of S_{xm} increases along with the enhancement of $\det(E_x^T P^{-1} E_x)$, and the increase of H_p helps enlarge the stable region. However, this enhancement will also increase the computational complexity. Therefore, a tradeoff must be made between stability enhancement and computational load. Moreover, in order to compare the stability of the present algorithm and the traditional NMPC, their attraction regions are shown together in Fig. 11. It can be observed that the attraction region S_{xm} of the present algorithm is much larger than that of the traditional NMPC with favorable parameters.

Still worth mentioning is that some other simulations also show that the size of S_{xm} increases as σ decreases. In other words, more accurate identification algorithms would help further enlarge the asymptotic stability region. Fortunately, Gómez's method (23) is helpful in this regard. Furthermore, to verify the feasibility of (68) in *Assumption* ***A2***, the values of $\hat{C}e(k)$ and $\hat{C}x(k) - \tilde{\eta}(k)$ are compared throughout the whole simulation process, and it is found that $| \hat{C}x(k) - \tilde{\eta}(k)| \leq \sigma|\hat{C}e(k)|$ always holds with $\sigma = 0.1$. The feasibility of Theorem 4 is thus verified.

Finally, it is remarked that the performance of the present algorithm highly depends on the effectiveness of the subspace method and the state observer L, and hence the present algorithm is not always better than NMPC. In other words, if σ or $\bar{e}$ can not be guaranteed small enough, the performance of the proposed dual-mode algorithm becomes worse than the traditional NMPC although the former is simpler and has a lower computational burden thanks to its block-oriented internal model. For instance, if $\bar{e} = 0.7$ or $\sigma = 0.2$, as shown in Fig. 12(b) (all the other parameters are the same as those in Fig. 11), the invariant set S_{xm} shrinks and becomes even smaller than that of the traditional NMPC.

3.6 Section conclusion

This section has developed an effective control method for MIMO Wiener systems with input constraints. First, the nonlinear and linear blocks of the system are separated by a subspace method. Then, a novel *dual-mode NMPC* algorithm is developed and used for the remaining process control. This approach is capable of maximizing the asymptotic stability region by a dual-mode control mechanism, and eliminating the reliance on the measurability about the intermediate output. Finally, control simulations have demonstrated the feasibility and superiority of the proposed control algorithm for a large class of nonlinear dynamic systems. It is believed that this novel approach has promising potential in handling many complex systems often encountered in industrial control processes.

4. References

[1] Adel M. E., Makoudi M., and Radouane L., Decentralized adaptive control of linear interconnected systems based on Lauguerre series representation, *Automatica*, vol. 35, pp. 1873–1881, 1999.

[2] Bai E., An optimal two-stage identification algorithm for Hammerstein–Wiener nonlinear systems, *Automatica* vol. 34, pp. 333–338, 1998.

[3] Bai E., and Fu M., A blind approach to Hammerstein model identification, *IEEE Transactions on Acoustics, Speech, and Signal Processing* vol. 50, pp. 1610–1619, 2002.

[4] Billings S., Fakhouri Y. S., Identification of non-linear systems using correlation analysis and pseudorandom inputs, *Int. J. Systems Science* vol. 11, no. 3, pp. 261–279, 1980.

[5] Bhandari N., and Rollins D., Continuous?time Hammerstein nonlinear modelling applied to distillation, *A.I.Ch.E. Journal* vol. 50, pp. 530–533, 2004.

[6] Billings S, Fakhouri S, Identification of systems containing linear dynamic and static nonlinear elements, *Automatica* vol. 18, no. 1, pp. 15–26, 1981.

[7] Bloemen HHJ, Chou CT, van den Boom HHJ, Verdult V, Verhaegen M, Backx TC, Wiener model identification and predictive control for dual composition control of a distillation column, *Journal of Process Control* vol. 11, no. 6, pp. 601–620, 2001.

[8] Bolemen H. H. J. and Van Den Boom T. T. J., Model-based predictive control for Hammerstein systems, *Proc. 39th IEEE Conference on Decision and Control* pp. 4963–4968, 2000

[9] Bolemen H. H. J., Van Den Boom T. T. J., and Verbruggen H.B., Model-based predictive control for Hammerstein-Wiener systems, *International Journal of Control* vol. 74, pp.482–495.

[10] Boyd E., and Chua L, Fading memory and the problem of approximating nonlinear operators with Volterra series, *IEEE Transactions on Circuits and Systems* vol. 32, pp. 1150–1161, 1985.

[11] Cao M. L., Wu Z. M., Ding B. C., and Wang C. X., On the stability of two-step predictive controller based-on state observer, *Journal of Systems Engineering and Electronics*, vol. 17, pp. 132–137, 2006.

[12] Chen G, Chen Y, Ogmen H, Identifying chaotic systems via a Wiener-type cascade model, *IEEE Control System Magazine* vol. 17, no. 10, pp. 29–36, 1997.

[13] Chen C. T., *Linear System Theory and Design*, 3rd edition, Oxford University Press, 1998.

[14] Chen H., and Allgöwer F., A quasi-infinite horizon nonlinear model predictive control scheme with guaranteed stability, *Automatica* vol. 34, pp. 1205–1217, 1998.

[15] Clark C. J. , Chrisikos G., Muha M. S., Moulthrop A. A. and Silva C. P., Time-domain envelope measurement technique with application to wideband power amplifier modeling, *IEEE Transactions on Microwave Theory Tech.* vol. 46, no. 12, pp. 2531–2540, 1988.

[16] Ding B. C. and Xi Y. G., A two-step predictive control design for input saturated Hammerstein systems., *International Journal of Robust and Nonlinear Control* vol. 16, pp. 353–367, 2006.

[17] Eskinat E., and Johnson S. H., Use of Hammerstein Models in identification of Nonlinear systems, *AI.Ch.E. Journal* vol. 37, pp. 255–268, 1991.

[18] Fruzzetti K. P., Palazoglu A., and McDonald K. A., Nonlinear model predictive control using Hammerstein models, *Journal of Process Control*, vol. 7, pp. 31–41, 1997.

[19] Fu Y., and Dumont G. A., An optimum time scale for discrete Laguerre network, *IEEE Transactions on Automatic Control* vol. 38, pp. 934–938, 1993.

[20] Garcia C. E., Prett D. M., and Morari M., Model predictive control: theory and practice Đa survey, *Automatica* vol. 25, pp. 335–348, 1989.

[21] George E. F., Malcolm M. A., and Moler C. B., *Computer methods fro mathematical computations*, Prentice Hall, Inc., Englewood Cliffs, New Jersey, 1977.

[22] Golub G., and Van L. C., *Matrix Computations*, 2nd Edition, The Johns Hopkins University Press, Baltimore and London, 1989.

[23] Gómez J. C., and Baeyens E., Identification of block-oriented nonlinear systems using orthonormal basis, *Journal of Process Control* vol. 14, 685–697, 2004.

[24] Greblicki W, Nonparametric identification of Wiener systems by orthogonal series, *IEEE Transactions on Autom. Control* , vol. 39, no. 10, pp. 2077–2086, 1994.

[25] Greblicki W., Nonlinearity estimation in Hammerstein systems based on ordered observations, *IEEE Transactions on Signal Processing* vol. 44, pp. 1224–1233, 1996.

[26] Greblicki W., Stochstic approximation in nonparametric identification of Hammerstein Systems, *IEEE Transactions on Automatic Control* vol. 47, pp. 1800–1810, 2002.

[27] Greblicki W., and Pawlak M., Nonparametric identification of Hammerstein systems, *IEEE Transactions on Information Theory* vol. 35, 409–418, 1989.

[28] Haddad W. M. and Chellaboina V. S., Nonlinear control of Hammerstein systems with passive nonlinear dynamics, *IEEE Transactions on Automatic Control* vol. 46, pp. 1630–1634, 2001.

[29] Hagenblad A., *Aspects of the identification of Wiener models*, Dept. Elect. Eng., Linköping Univ., Linköping, Sweden, Tech. Rep. Licentiate, Thesis 793, Nov. 1999.

[30] Hasiewicz Z., Hammerstein system identification by the Haar multiresolution approximation, *International Journal of Adaptive Control and Signal Processing* vol. 13, pp. 691–717, 1999.

[31] Hasiewicz Z., and Mzyk G, Combined parametric-nonparametric identification of Hammerstein systems, *IEEE Transactions on Automatic Control* vol. 49, 1370–1375, 2004.

[32] Henson M. A., Nonlinear model predictive control: current status and future directions, *Computers and Chemical Engineering* vol. 23, pp. 187–202, 1998.

[33] Heuberger P. S. C., Van den Hof P. M. J., and Bosgra O. H., A generalized orthonormal basis for linear dynamical system, *IEEE Transactions on Automatic Control* vol. 40, 451–465, 1995.

[34] Henson H. A., *Nonlinear Process Control*, Prentice Hall, New Jersey, 1997.

[35] Huner I. W., and Korenberg M. J., The identification of nonlinear biological systems Wiener and Hammerstein cascade models, *Biological Cybernetics* vol. 55, pp. 135–144, 1986.

[36] Juditsky A., Hjalmarsson H., Benveniste A., Delyon B., Ljung J., Sjöberg J., and Zhang Q., Nonlinear black-box modelling in system identification: mathematical foundations, *Automatica* vol. 31, pp.1725–1750, 1995.

[37] Joeken S. and Schwegler H., Predicting spike train responses of neuron models, In *Proc. 3rd Eur. Symp. Artificial Neural Networks*, pp. 93–98, 1995.

[38] Kalafatis A., Arifin N., Wang L. and Cluett W. R., A new approach to the identification of pH processes based on the Wiener model, *Chem. Eng. Sci.* vol. 50, no. 23, pp. 3693–3701, 1995.

[39] Khalil H. K., *Nonlinear Systems* (Prentice Hall), 2002.

[40] Knohl T., Xu W. M. and H. Hubelhauen, Indirect adaptive dual control for Hammerstein systems using ANN, *Control Engineering Practice* vol. 11, pp. 277–285, 2003.

[41] Kouvaritakis B., Rossiter J. A., and Schuurmans J., Efficient robust predictive control, *IEEE Transactions on Automatic Control* vol. 45, pp. 1545–1549, 2000.

[42] Krantz S. G., *A Handbook of Real Variables: With Applications to Differential Equations and Fourier Analysis*, Birkhauser, 2004.

[43] Lakshminarayanan S. , Shah S. L. and Nandakumar K. Identification of Hammerstein models using multivariate statistical tools, *Chemical Engineering Science* vol. 50, no. 22, pp. 3599–3613, 1995.

[44] Lin Z. L. , Semi-global stabilization of discrete-time linear systems with position and rate-limited actuators. *Systems and Control Letters*, vol. 34, pp. 313–322, 1998.

[45] Liu Y, Bai E.W., Iterative identification of Hammerstein systems, *Automatica* vol. 43, no. 2, pp. 346–354, 2007.

[46] Ljung L., 1999, *System Identification: Theory for the User*, 2nd edition, Prentice-Hall, Inc. Englewood Cliffs, NJ, 1999

[47] Marmarelis VZ and Naka KI, White-noise analysis of a neuron chain: an application of the Wiener theory, *Science* vol. 175, no. 4027 pp. 1276–1278, 1972

[48] Minesh A. S., and Matthew A. F., Frequency-based controller design for a class of nonlinear systems, *International Journal of Robust and Nonlinear Control* vol. 9, pp. 825–840, 1999.

[49] Nešić D, Output feedback stabilization of a class of Wiener systems, *IEEE Transactions on Automatic Control* vol. 45, no. 9, pp. 1727–1731, 2000

[50] Norquay S. J., Palazoglu A. and Romagnoli J. A., Application of Wiener model predictive control to a pH neutralization experment, *IEEE Transactions on Control System Technology* vol. 7, no. 4, pp. 437–445, 1999.

[51] Patwardhan R., Lakshminarayanan S., and Shah S. L., 1998, Constrained nonlinear MPC using Hammerstein, *AICh.E. Journal* vol. 44, pp. 1611–1622, 1998.

[52] Raich R., Zhou G. T. and Viberg M., Subspace based approaches for Wiener system identification, *IEEE Transactions on Autom. Control* vol. 50, no.10, pp.1629–1634, 2005

[53] Segal BN, Outerbridge JS, Vestibular (semicircular canal) promary neurons in bullfrog: nonnlinearity of individual and population response to rotation, *Journal of Neurophysics* vol.47, no.4, pp. 545–562, 1982.

[54] Sjoberg J., Zhang Q., Ljung L., Benveniste A., Delyon B., Glorennec P., Hjalmarsson H., and Juditsky A., 1995, Nonlinear black-box modelling in system identification: a unified approach, *Automatica*, vol. 31, pp. 1691–1724, 1995.

[55] Stapleton J., and Bass S., Adaptive noise cancellation for a class of nonlinear dynamic reference channels, *IEEE Transactions on Circuits and Systems* vol. CAS-32, pp. 143–150, 1985

[56] Tanguy N., Morvan R., VilbŐ P and Calvez L. P., Online optimization of the time scale in adaptive Laguerre-based filters. *IEEE Transactions on Signal Processing,* vol. 48, pp. 1184–1187, 2000.

[57] Tanguy N., Vilbé P., and Calvez L. C., Optimum choice of free parameter in orthonormal approximations, *IEEE Transactions on Automatic Control* vol. 40, pp. 1811–1813, 1995.

[58] Tan W. W., Lu F. and Loh A. P., An application of genetic algorithm for designing a Wiener-model controller to regulate the pH value in a pilot plant, in *Proc. 2001 Congr. Evolutionary Computation* vol. 2, pp. 1055–1061, 2001.

[59] Verhaegen M., Identification of the deterministic part of MIMO state space models given in innovations form from input-out data, *Automatica* vol. 30, no. 1, pp. 61–74, 1994.

[60] Wahlberg B. and Mäkilä P. M., On approximation of stable linear dynamic system using Laguerre and Kautz functions. *Automatica* vol. 32, pp. 693–708, 1996.

[61] Wang L., Cluett W.R., Optimal choice of time-scaling factor for linear system approximation using Laguerre models. *IEEE Transactions on Automatic Control* vol. 39, no. 7, pp. 1463–1467, 1995.

[62] Wang L. P., 2004, Discrete model predictive controller design using Laguerre functions, *Journal of Process Control* vol. 14, pp. 131–142, 2004.

[63] Westwick D. T. and Kearney R. E., Separable Least Squares Identification of Nonlinear Hammerstein Models: Application to Stretch Reflex Dynamics, *Ann. Biomed. Eng.* vol. 29, no. 8, pp. 707–718, 2001.

[64] Wigren T., Recursive prediction error identification using the nonlinear Wiener model, *Automatica* vol. 29, no. 4, pp. 1011–1025, 1993.

[65] Yuan H., Westwick D. T., Ingenito E. P., Lutchen K. R. and Suki B., Parametric and Nonparametric Nonlinear System Identification of Lung Tissue Strip Mechanics, *Ann. Biomed.Eng.* vol. 27, no. 4, pp. 548–562, 1999.

[66] Zhang H. T., Chen Z. G., Wang Y. J., Qin T., and Li M., Adaptive predictive control algorithm based on Laguerre functional model, *International Journal of Adaptive Control and Signal Processing* vol. 20, pp. 53–76, 2006.

[67] Zhang H. T., and Li H. X., A general control horizon extension method for nonlinear model predictive control, *Industrial & Engineering Chemistry Research,* vol. 46, pp. 9179–9189, 2007.

[68] Zhang H. T. , Li H. X. and Chen G. , A novel predictive control algorithm for constrained Hammerstein systems, *International Journal of Control,* vol. 81, pp. 1609–1625, 2008.

[69] Zhang H. T., Chen M. Z. Q. and Chen Z., Nonlinear Laguerre-Volterra observer-controller and its application to process control, *International Journal of Robust and Nonlinear Control,* vol.20, pp. 412–423, 2010.

[70] Zhang H. T., Chen G. and Chen M. Z. Q., A novel dual-mode predictive control strategy for constrainedWiener systems, *International Journal of Robust and Nonlinear Con- trol,*vol.20, pp. 975–986, 2010.

[71] Zhu Q. M., Warwick K., and Douce J. L., Adaptive general predictive controller for nonlinear systems, *IEE Proceedings-D* vol. 138, pp. 33–41, 1991.

10

A General Lattice Representation for Explicit Model Predictive Control

William Sun [1] and Thomas Winn [2]
[1] *School of Business Administration, Duquesne University*
[2] *Emerson Process Management, Power & Water Solutions*
USA

1. Introduction

Model predictive control (MPC) is one of the most successful techniques to control multi-variable constraint systems. The MPC uses a mathematical model to predict the future effects of control inputs to system behaviors. The optimal control policy is obtained by solving an optimization problem that minimizes/maximizes a performance objective subject to inputs and outputs constraints over a future time horizon (Morari & Lee, 1999, Mayne et al., 2000, Dua et al, 2008). The MPC control laws are open-loop optimal with respect to the corresponding objective function. However, the conventional MPC may require intensive online computation due to the repetitive solutions of an optimization problem. This limits its applications to large and slowly varying systems. In addition, the MPC control strategy is hard to validate, especially for safety critical system (Grancharova & Johansen, 2005, Pistikopoulos, et al. 2001, Alessio & Bemporad 2009).

In 2002, Bemporad, Morari, Dua & Pistikopolous introduced the concept of explicit MPC (eMPC). The eMPC reformulates the online optimization in a MPC into a mutli-parametric linear/quadratic program (mpLP/mpQP). The optimal control action is calculated off-line as a continuous piecewise-affine (PWA) function of the state and reference vectors (Saffer et al, 2004, Bemporad et al. 2002a, 2002b). The eMPC has several advantages: i) The online computational time can be reduced to the microsecond-millisecond range. It makes the eMPC attractive for fast systems; ii) The MPC functionality is achieved in an easily verifiable way. The eMPC control policies can be validated before real online operations; iii) The eMPC solutions can be implemented with low-cost hardware (Johansen et al., 2007, Wen & Ma, 2008a, 2008b). The eMPC is then promising to be used in embedded systems, small/medium process systems, where the control systems should not be more expensive than the process systems. The eMPC found successful applications in many areas, e.g. AC-DC converters (Beccuti et al. 2009), autonomous vehicle steering, air separation unit (Grancharova et al. 2004), active valve train, hybrid separation, air condition (Pistikopoulos, et al. 2001), biomedical systems and drug delivery systems (Dua et al, 2004), scheduling (Ryu, & Pistikopoulos, 2007), spacecraft attitude control (Hegrenas et al, 2000)and crude distillation unit (Pannocchia et al, 2007).

The eMPC is essentially a strategy of trading time for space. A continuous PWA control map is calculated offline, and stored in memory for online usage (Pannocchia et al, 2007). The efficiency of eMPC method depends critically on finding an efficient representation model for eMPC controllers. However, it is not easy to develop a general model set to represent continuous PWA functions. The PWA representations usually have two conflicting criteria: description and evaluation complexities. The description complexity deals with the number of parameters in the representation model, while the evaluation complexity specifies the time for online calculation of function values.

When an eMPC controller is executed, one needs to solve a point-location problem with two steps: i) identify which polyhedral region the measured state lies in; ii) compute the control action using the corresponding affine control law. The simplest point-location solver is the sequential search (SS) algorithm, which is implemented by substituting a given state variable into the constraint inequalities of different regions. This method requires to store all the polyhedral regions and affine functions individually. Due to the combinatorial nature, the number of polyhedral regions in an eMPC control law can grow exponentially with the size of the optimal control problem (Wen et al., 2005a). Hence, the online function evaluation is computationally expensive when an eMPC control consists of a large number of regions or is defined over a complicated domain partition (Wen et al., 2005b).

Many researchers developed alternative ways to represent the eMPC solutions with appropriate data structures. In 2001, Borrelli, Baotić, Bemporad & Morari propose a search algorithm based on the convexity of the piecewise affine value function. This convex value function (CVF) algorithm reduces the storage space significantly. In some cases, this method might be time consuming because it requires a kind of sequential search. A binary search tree (BST) algorithm is proposed by Tondel, Johansen & Bemporad (2003) on the basis of the geometric structure of the polyhedral partition. This method can deal with the fully general PWA functions, including the discontinuous ones defined on overlapping regions or holes. In this scheme, the auxiliary hyper-planes are introduced, which may subdivide existing regions. This might lead to a worst case combinatorial number of subdivided regions. Then the following search procedure has to consider an additional increase in the number of regions, which may imply a prohibitive pre-processing time or online memory requirements. In 2007, Christophersen, Kvasnica, Jones & Morari developed an efficient search tree algorithm by utilizing the concept of bounding boxes and interval trees. This bounding box tree (BBT) algorithm can deal with the PWA functions defined over a large number of polyhedral regions. But the storage demand and online search time are still linear in the number of polyhedral regions in the original PWA functions. In Geyer et al, 2008, an efficient approach was developed to reduce the number of partitions in eMPC controllers by optimally merging the polyhedral regions where the affine gain is the same. All these algorithms are successful in dealing with the PWA functions defined over a large number of polyhedral regions. However, their storage demands and/or online evaluation computation are dependent on the number of polyhedral regions in the original PWA functions. These algorithms do not utilize the global structure information in eMPC controllers. Then their efficiency may be reduced substantially for some large-scale and complicated eMPC solutions.

The PWA approximation technique presents another efficient way to deal with the computation and description complexities of eMPC solutions. In Johansen, Petersen & Slupphaug (2002), an approach is proposed, which calculates the sub-optimal solutions by predetermining a small number of sampling data when the active set or input is allowed to change on the horizon. An alternative sub-optimal approach was developed in Bemporad & Filippi (2003) where small slacks are introduced on the optimality conditions and the mp-QP is used for the relaxed problem. In 2003, an algorithm is suggested that can determine a suboptimal explicit MPC control on a hypercubic partition (Johansen & Grancharova, 2003). In this partition, the domain is divided into a set of hypercubes separated by orthogonal hyperplanes. In 2006, Jones, Grieder & Rakovic interpret the PWA value function as weighted power diagrams (extended Voronoi diagrams). By using the standard Voronoi search methods, the online evaluation time is solved in logarithmic time (Jones, Grieder, & Rakovic, 2006; Spjotvold, Rakovic, Tondel, & Johansen, 2006). Dynamic programming can also be used to calculate the approximate explicit MPC laws (Bertsekas & Tsitsiklis, 1998; Lincoln & Rantzer, 2002, 2006). The main idea of these approaches is to find the sub-optimal solutions with known error bounds. The prescribed bounds can achieve a good trade-off between the computation complexity and accuracy. These approximation algorithms are very efficient regarding the storage and online calculation time. However, the approximate PWA functions usually have different domain partitions from the original explicit MPC laws. This deviation may hinder the controller performance and closed-loop stability.

The established representation and approximation algorithms have found many successful applications in a variety of fields. However, they can only evaluate the control actions for discrete measured states. None of them can provide the exact analytical expression of the PWA control laws. An analytical expression will ease the process of closed-loop performance analysis, online controller tuning and hardware implementations. The analytic expression also provides the flexibility of tailoring the PWA controllers to some specific applications, e.g. to develop different sub-optimal controllers in different zones (a union of polyhedral regions), and to smooth the PWA controllers at region boundaries or vertices (Wen et al. 2009a). In addition, the canonical PWA (CPWA) theory shows that the continuous PWA functions often consist of many redundant parameters. A global and compact analytical expression can significantly increase the computation and description complexity of eMPC solutions (Wen, et al., 2005a). An ideal representation algorithm should describe and evaluate the simplified MPC solutions after removing the redundant parameters.

In 1977 Chua & Kang proposed the first canonical representation for continuous PWA functions. A canonical PWA (CPWA) function is the sum of an affine function and one or more absolute values of affine functions. All continuous PWA functions of one variable can be expressed in the canonical form. However, if the number of variables is greater than one, only a subset of PWA functions have the CPWA representations (Chua & Deng, 1988). In 1994, Lin, Xu & Unbehauen proposed a generalized canonical representation obtained by nesting several CPWA functions. Such a representation is available for any continuous PWA function provided that the nesting level is sufficiently high. The investigations (Lin & Unbehauen, 1995; Li, et al. 2001, Julian et al., 1999) showed that for a continuous PWA function, the nesting level does not exceed the number of its variables. However, the nested absolute value functions often have implicit functional forms and are defined over complicated boundary configurations. In 2005, Wen, Wang & Li proposed a basis function CPWA (BPWA) representation theorem. It is shown that any continuous PWA function of n variables can be expressed by a BPWA function, which is formulated as the sum of a suitable number of the maximum/minimum of n+1 affine functions.

The class of lattice PWA functions is a different way to represent a continuous PWA function (Tarela & Martínez, 1999, Chikkula, et al., 1998, Ovchinnikov, 2002, Necoara et al. 2008, Boom & Schutter 2002, Wen et al, 2005c, Wen & Wang, 2005d). The lattice representation model describes a PWA function in term of its local affine functions and the order of the values of all the affine functions in each region. From theoretical point of view, the lattice PWA function has a universal representation capability for any continuous PWA function. According to the BPWA representation theorem, any BPWA function can be equivalently transformed into a lattice PWA function (Wen et al. 2005a, 2006). Then the well-developed methods to analyze and control the class of CPWA functions can be extended to that of the lattice PWA functions. From a practical point of view, it is of great significance that a lattice PWA function can be easily constructed, provided that we know the local affine functions and their polyhedral partition of the domain (Wen & Ma, 2007, Wen et al, 2009a, 2009b). Since these information on affine functions and partitions is provided in the solutions of both mp-LP and mp-QP, the lattice PWA function presents an ideal way to represent the eMPC solutions.

In this paper, we propose a general lattice representation for continuous eMPC solutions obtained by the multi-parametric program. The main advantage of a lattice expression is that it is a global and compact representation, which automatically removes the redundant parameters in an eMPC solution. The lattice representation can save a significant amount of online computation and storage when dealing with the eMPC solutions that have many polyhedral regions with equal affine control laws. Three benchmark MPC problems are illustrated to demonstrate that the proposed lattice eMPC control have a lower description complexity, comparable evaluation and preprocessing complexities, when compared to the traditional eMPC solutions without global description models.

The rest of this paper is organized as follows. Section II introduces the main features of PWA functions and eMPC problems. The lattice PWA function and representation theorem are presented in Section III. Section IV is the main part of this paper. It presents the complexity reduction theorem of lattice eMPC solutions, the lattice representation algorithm and its complexity analysis. Numerical simulation results are shown in Section V, and Section VI provides the concluding remarks.

2. PWA functions and eMPC solutions

2.1 PWA function

Definition 1. *In $\Re^n$, let $\Omega = \cup_{i=1}^{M} R_i$ be a compact set, which is partitioned into M convex polyhedrons called regions $R_i, i = 1, \ldots, M$. Then a nonlinear function $p(x) : \Omega \mapsto \Re^m$ is defined as a PWA function if*

$$p(x) = F_i x + g_i, \quad \forall x \in R_i \tag{1}$$

with $F_i \in \Re^{m\times(n+1)}, \beta_i \in \Re^m$. A PWA function is continuous if

$$F_i x + g_i = F_j x + g_j, \quad \forall x \in B_{i,j} \tag{2}$$

where $B_{i,j} = R_i \cap R_j$ is defined as boundaries and $i, j \in [1, \cdots, M]$. Specially, when $m = 1$, $p(x)$ is called as a scalar PWA function, i.e.

$$p(x) = \ell(x|\alpha_k, \beta_k) = \alpha_k^T x + \beta_k, \quad \forall x \in R_i \tag{3}$$

with $\alpha_k \in \Re^n, \beta_k \in \Re$ and $1 \le k \le M$. For convenience of statement, we simply denote $\ell(x|\alpha_k, \beta_k)$ as $\ell_k(x)$.

In Definition 1, each region R_i is a polyhedron defined by a set of inequality

$$R_i = \{x \in \Re^n | H_i x \leq K_i\} \tag{4}$$

where H_i, K_i are matrices of proper sizes with $i = 1, \cdots, M$. Geometrically, a boundary $B_{i,j}$ is a real set of an $(n-1)$-dimensional hyperplane.

2.2 Explicit MPC

Consider the linear time invariant system

$$\begin{cases} x(t+1) = Ax(t) + Bu(t) \\ y(t) = Cx(t) + Du(t) \end{cases} \tag{5}$$

which fulfills the following constraints

$$x_{min} \leq x(t) \leq x_{max},\ y_{min} \leq y(t) \leq y_{max},\ u_{min} \leq u(t) \leq u_{max}, \delta u_{min} \leq \delta u(t) \leq \delta u_{max}, \tag{6}$$

at all time instants $t \geq 0$. In (5)-(6), $x(t) \in \Re^n$ is state variable, $u(t) \in \Re^m$, $y(t) \in \Re^{n_y}$ are control input and system output, respectively. A, B, C and D are matrices of appropriate dimensions, i.e. $A \in \Re^{n \times n}, B \in \Re^{n \times m}, C \in \Re^{n_y \times n}$ and $D \in \Re^{n_y \times m}$. It is assumed that (A, B) is a controllable pair. δu_{min} and δu_{max} are rate constraints. They restrict the variation of two consecutive control inputs ($\delta u(t) = u(t) - u(t-1)$) to be within of prescribed bounds. The system is called as a single-input system when $m = 1$, and a multi-input system when $m \geq 2$. Assume that a full measurement of the state $x(t)$ is available at current time t. The MPC solves the following standard semi-infinite horizon optimal control problem:

$$J^*(x(t)) = \min_{U=\{u(t)^T,\cdots,u(t+N-1)^T\}^T} V_N(x(t+N)) + \sum_{k=0}^{N-1} V_k(x(t+k|t), u(t+k)) \tag{7}$$

subject to

$$\begin{cases} x_{min} \leq x(t+k|t) \leq x_{max}, \quad k = 1, \cdots, N_y, \\ y_{min} \leq y(t+k|t) \leq y_{max}, \quad k = 1, \cdots, N_y, \\ u_{min} \leq u(t+k|t) \leq u_{max}, \quad k = 1, \cdots, N_c, \\ \delta u_{min} \leq \delta u(t+k|t) \leq \delta u_{max}, k = 1, \cdots, N_c, \\ x(t) = x(t|t) \\ x(t+k+1|t) = Ax(t+k|t) + Bu(t+k), \quad k \geq 0 \\ y(t+k+1|t) = Cx(t+k|t) + Du(t+k), \quad k \geq 0 \\ u(t+k) = Kx(t+k|t), \quad N_u \leq k < N_y \end{cases} \tag{8}$$

at each time t, where $x(t+k|t)$ denotes the the predicted state vector at time $t+k$. It is obtained by applying the input sequence $u(t), \cdots, u(t+k-1)$ to system (5). In (7), K is the feedback gain, N_u, N_y, N_c are the input, output and constraint horizons, respectively. Normally, we have $N_u \leq N_y$ and $N_c \leq N_y - 1$.

The stage cost function is defined as

$$V_i(x(t+i|t), u(t+i)) = ||Qx(t+k|t)||_p + ||Ru(t+k)||_p \tag{9}$$
$$V_N(x(t+N_y|t)) = ||Px(t+N_y|t)||_p \tag{10}$$

where $||\cdot||$ denoted a kind of norm and $p \in \{1, 2, +\infty\}$, P, Q and R are weighting matrices of proper sizes. V_N is the terminal penalty function. In this paper, it is assumed that the parameters P, Q, R are chosen in such a way that problem (7) generates a feasible and stabilizing control law when applied in a receding horizon fashion and $J^*(x)$ is a polyhedral piecewise affine/quatratic Lyapunov function.
At each time t, the MPC control law $u(t)$ is the first item in the optimal solution $u^*(t)$, i.e.

$$u(t) = u^*(t) \tag{11}$$

where $u^*(t) = \{u^*(t), \cdots, u^*(t+N_c-1)\}$. Apply $u(t)$ as input to problem (5) and repeat the optimization (7) at time $t+1$ using the new state $x(t+1)$. This control strategy is also referred to as moving or receding horizon.
By some algebraic manipulations, the MPC problem can be formulated as a parametric Linear Program (pLP) for $p \in \{1, +\infty\}$

$$\begin{aligned} u^*(x) &= \min_u \Upsilon^T u \\ s.t.\ Gu &\leq W + Ex \end{aligned} \tag{12}$$

or a parametric quadratic Program (pQP) for $p = 2$

$$\begin{aligned} u^*(x) &= \min_u \Upsilon^T u + \frac{1}{2} u^T H u \\ s.t.\ Gu &\leq W + Ex \end{aligned} \tag{13}$$

See (Bemporad et al. 2002) for details on the computation of the matrices G, W, E, H and Υ in (12) and (13). By solving the pLP/pQP, the optimal control input $u^*(x)$ is computed for each feasible value of the state x. The features of MPC controllers and value functions are summarized in the following lemma.

Lemma 1. Kvasnica et al., 2004 *Consider the multi-parametric programming of (12) and (13). The solution $u^*(x) : \Re^n \mapsto \Re^m$ is a continuous and piecewise affine*

$$u^*(x) = F_i x + g_i, \ \forall x \in R_i \tag{14}$$

where $R_i, i = 1, \cdots, M$ is the polyhedral regions. The optimal cost $J^*(x(t))$ is continuous, convex, and piecewise quadratic ($p = 2$) or piecewise affine ($p \in \{1, \infty\}$).

3. Lattice representation of scalar eMPC solutions

3.1 Lattice PWA Function

Let $\Phi = [\phi_1, \cdots, \phi_M]^T$ be an $M \times (n+1)$ matrix and $\Psi = [\psi_{ij}]$ a $M \times M$ zero-one matrix. A lattice piecewise-affine function $P(x|\Phi, \Psi)$ may be formed as follows,

$$P(x|\Phi,\Upsilon) = \min_{1\leq i\leq M}\left\{\max_{\substack{1\leq j\leq M\\ \psi_{ij}=1}}\{\ell_j(x)\}\right\}, \ \forall x\in\Re^n. \tag{15}$$

Note that $P(x|\Phi,\Upsilon)$ is equal to one of $\ell_1(x),\cdots,\ell_M(x)$ for any $x\in\Re^n$. $P(x|\Phi,\Upsilon)$ is indeed a continuous PWA function whose local affine functions are just $\ell_j(x), 1\leq j\leq M$. The parameter vectors of these affine functions are exactly the row vectors of Φ. Hence the matrix Φ is called a parameter matrix. The matrix Υ is defined as a structure matrix, if its elements are calculated as

$$\psi_{ij} = \begin{cases} 1 & if \quad \ell_i(x)\geq \ell_j(x) \\ 0 & else \end{cases} \tag{16}$$

with $x\in R_i$ and $1\leq i,j\leq M$. Similarly, a dual structure matrix $\hat{\Upsilon} = [\hat{\psi}_{ij}]^{M\times M}$ is defined by

$$\hat{\psi}_{ij} = \begin{cases} 1 & if \quad \ell(x|\phi_i)\leq \ell(x|\phi_j) \\ 0 & else \end{cases} \tag{17}$$

with $x\in R_i$ and $1\leq i,j\leq M$.

Lemma 2. Wen et al., 2007 *Given any n-dimensional continuous PWA function $p(x)$, there must exist a lattice PWA function $P(x|\Phi,\Upsilon)$ such that*

$$p(x) = P(x|\Phi,\Upsilon), \forall x\in\Re^n \tag{18}$$

where Φ,Υ are parameter and structure matrices, respectively.

It is shown in Lemma 2 that a continuous PWA function can be fully specified by a parameter matrix Φ and a structure matrix Υ. This provides a systematic way to represent the eMPC solutions. The lattice PWA function contains only the operators of min, max and vector multiplication. It is an ideal model structure from the online calculation point of view.
Example 1: The realization of a lattice PWA function can be made more clear using a simple example. Let $p(x)$ be a 1-dimensional PWA function with 4 affine segments,

$$p(x) = \begin{cases} \ell_1(x) = 1, & \forall x\in R_1 = [-2,-1] \\ \ell_2(x) = -x, & \forall x\in R_1 = (-1,0] \\ \ell_3(x) = x, & \forall x\in R_2 = (0,1] \\ \ell_4(x) = -x+2, & \forall x\in R_3 = (1,2] \end{cases} \tag{19}$$

where the plot of $p(x)$ is depicted in Fig. 1. It is easy to see that

$$\begin{cases} \ell_4(x) > \ell_2(x) > \ell_1(x) > \ell_3(x), & \forall x\in R_1 \\ \ell_4(x) > \ell_1(x) > \ell_2(x) > \ell_3(x), & \forall x\in R_2 \\ \ell_4(x) > \ell_1(x) > \ell_3(x) > \ell_2(x), & \forall x\in R_3 \\ \ell_3(x) > \ell_1(x) > \ell_4(x) > \ell_2(x), & \forall x\in R_4 \end{cases} \tag{20}$$

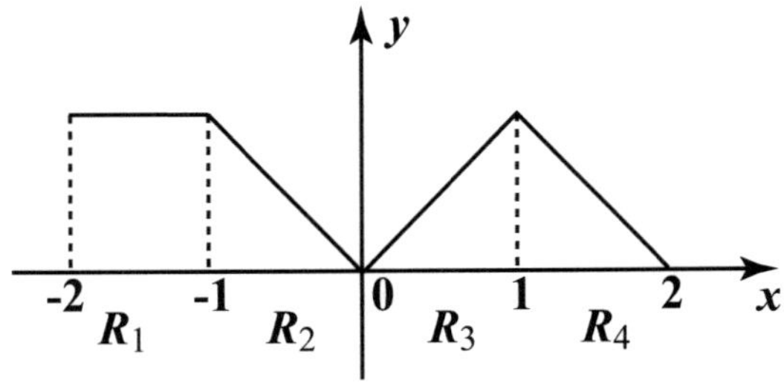

Fig. 1. Plot of 1-dimensional PWA function $p(x)$

Then the structure matrix is written as $\Psi = \begin{bmatrix} 1\,0\,1\,0 \\ 0\,1\,1\,0 \\ 0\,1\,1\,0 \\ 0\,1\,0\,1 \end{bmatrix}$. It follows from (5) that the parameter matrix is $\Phi = \begin{bmatrix} 0\;1 \\ -1\;0 \\ 1\;0 \\ -1\;2 \end{bmatrix}$. Finally, the lattice PWA function is formulated as

$$p(x) = P(x|\Phi, \Psi) = \min\{\max\{\ell_1, \ell_3\}, \max\{\ell_2, \ell_3\}, \max\{\ell_2, \ell_3\}, \max\{\ell_2, \ell_4\}\} \tag{21}$$

It is obvious that the lattice PWA function in (21) can be further simplified. The simplification algorithm will be discussed in the subsequent sections.

3.2 Lattice representation theorem of eMPC solutions

Lemma 3. *Assume that R_i, R_j are two n-dimensional convex polytopes, where $\ell_i(x), \ell_j(x)$ are their local affine functions with $i, j \in \{1, \cdots, M\}$. Then the structure matrix $\Psi = [\psi_{ij}]^{M\times M}$ can be calculated as follows:*

$$\psi_{ij} = \begin{cases} 1 & if \quad \ell_i(v_k) \geq \ell_j(v_k), 1 \leq k \leq K_i \\ 0 & if \quad \ell_i(v_k) < \ell_j(v_k), k \in \{1, \cdots, K_i\} \end{cases} \tag{22}$$

where v_k are the vertices of R_i with $1 \leq k \leq K_i$ and $K_i \in \mathbb{Z}^+$ is the number of vertices of R_i.

Proof. Since R_i is an n-dimensional polytope, it can be described by its vertices $v_1, \cdots, v_{K_i}$

$$R_i = \{x \in \Re^n | x = \sum_{k=1}^{K_i} \lambda_k v_k, 0 \leq \lambda_k \leq 1, \sum_{i=1}^{K_i} \lambda_k = 1\} \tag{23}$$

Then for any $x \in R_i$, we have

$$\ell_i(x) = \ell_i\left(\sum_{k=1}^{K_i} \lambda_k v_k\right) = \sum_{k=1}^{K_i} \lambda_k \ell_i(v_k) \tag{24}$$

$$\ell_j(x) = \ell_j\left(\sum_{k=1}^{K_i} \lambda_k v_k\right) = \sum_{k=1}^{K_i} \lambda_k \ell_j(v_k) \tag{25}$$

If $\ell_i(v_k) \geq \ell_j(v_k), \forall 1 \leq k \leq K_i$, then $\ell_i(x) \geq \ell_j(x)$ holds for all $x \in R_i$. It follows from (3) that $\psi_{ij} = 1$.
Similarly, if there exists any $k \in \{1, \cdots, K_i\}$ such that $\ell_i(v_k) < \ell_j(v_k)$, then $\ell_i(x)$ and $\ell_j(x)$ will intersect together with an $(n-1)$-dimensional hyperplane as the common boundary. This implies that $\psi_{ij} = 0$.
Using the same procedure stated above, all the elements in the structure matrix Ψ can be calculated, and this completes the proof of Lemma 3. □

Lemma 3 shows that the order of the affine function values in a convex polytope can be specified by the order of the function values at the polytope vertices. This presents a constructive way to realize the structure matrix of a given PWA function.

Theorem 1. *Any continuous eMPC solution can be represented by a lattice PWA function.*

Proof. According to Bemporad et al. 2002, an eMPC solution is presented in the form of conventional PWA representation, which lists all the parameters of the affine functions and regions in a table. Each region is a convex polytope defined by a set of inequalities. It follows from Lemma 2 that an explicit solution to MPC can be realized by a structure matrix and a parameter matrix. These two matrices specify a lattice PWA function. Then any eMPC solution can be described by a lattice PWA function. This completes the proof of Theorem 1. □

4. Simplification of scalar lattice PWA representation

4.1 Super-region

Definition 2. *Given a PWA function $p(x) : \Omega \mapsto \Re^m$ with M regions, i.e. $\Omega = \cup_{i=1}^{M} R_i$. Let*

$$\Gamma_i = \{j \in \{1, \cdots, M\} | \alpha_i x + \beta_i = \alpha_j x + \beta_j, \forall x \in \Omega\} \tag{26}$$

be a finite set with $\tilde{M}$ components. Then the set $\Pi \subseteq \Omega$ is defined as a super-region, if $\Pi = \cup_{k=1}^{\tilde{M}} R_k$ and $k \in \Gamma_i$ with $i \in \{1, \cdots, M\}$.

A super-regions is defined as a union of polyhedral regions with same affine function. It can be non-convex or even not connected. If a PWA function have many regions with the same local functions, the number of super-regions is much less than that of regions.
The concept of super-region can be clarified by an 1-dimensional PWA function shown in Fig. 2. The PWA function $p(x)$ is defined over a compact set $\Omega = AE$. The domain is partitioned into 4 regions, i.e. $\Omega = \cup_{i=1}^{4} R_i$. Each region R_i is a convex polyhedron defined by two inequalities, e.g. $R_2 = BC = \{x \in AF | x \geq x_B, x \leq x_C\}$, where x_B, x_C are the coordinates of points B, C. In Ω, there are 3 boundaries, e.g. B, C and D. Note that $p(x) = [p_1(x), p_2(x)]^T$, where

$$p_j(x) = \alpha_{i,j}^T x + \beta_{i,j} \quad \forall x \in R_i$$

with $\alpha_{i,j}, \beta_{i,j} \in \Re, i = 1, \cdots, 4, j = 1, 2$. We can get $F_i = \begin{bmatrix} \alpha_{i,1}^T \\ \alpha_{i,2}^T \end{bmatrix}$ and $g_i = \begin{bmatrix} \beta_{i,1} \\ \beta_{i,2} \end{bmatrix}$.
It follows from the plot of $p_1(x)$ that $\alpha_{1,1}^T x + \beta_{1,1} = \alpha_{4,1}^T x + \beta_{4,1}, \forall x \in AE$. Then $\Pi_1 = R_1 \cup R_2 = AB \cup DE$ is defined as a super-region. It is evident that Π_1 is not convex, because it

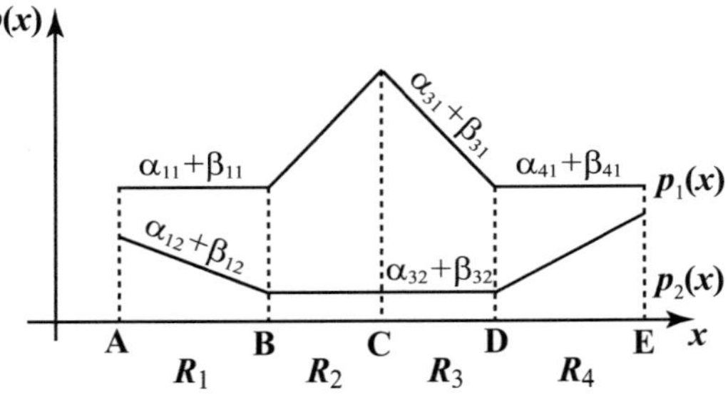

Fig. 2. Plot of a 1-dimensional vector PWA function $p(x) = [p_1(x), p_2(x)]^T$.

is composed of two disconnected line intersections. Similarly, $\Pi_2 = R_2 \cup R_3 = BD$ defines another super-region of $p_2(x)$.

4.2 Row vector simplification lemma

Lemma 4. *Assume that $P(x|\Phi, \Psi) : D \subset \Re^n \mapsto \Re$ is a PWA function with M linear segments. Let φ_i, φ_j be rows of the structure matrix. If the pointwise inequation $\varphi_i - \varphi_j \leq 0$ holds for any $i, j \in \{1, \cdots, M\}$, there exist a simplified structure matrix $\tilde{\Psi} \in \Re^{(M-1)\times M}$, such that*

$$P(x|\Phi, \Psi) = P(x|\Phi, \tilde{\Psi}) \tag{27}$$

where $\Psi \in \Re^{M\times M} = [\varphi_1, \cdots, \varphi_M]^T$ and $\tilde{\Psi} = [\varphi_1, \cdots, \varphi_{j-1}, \varphi_{j+1}, \cdots, \varphi_M]^T$.

Proof. Denote $I_i \in \Re^M$ as the index set of the local affine functions, whose values are smaller than the i-th affine function in its active region, i.e.

$$I_i = \{k | \ell_k(x) \leq \ell_i(x), \forall x \in R_i\} \tag{28}$$

with $i, k \in \{1, \cdots, M\}$. Since $\varphi_i - \varphi_j \leq 0$ holds for any pointwise inequality, we can get $I_i \subseteq I_j$. It directly follows that $\{\ell_p(x)\} \subseteq \{\ell_q(x)\}$ with $p \in I_i, q \in I_j$.
Therefore, it leads that

$$\max_{p\in I_i}\{\ell_p(x)\} \leq \max_{q\in I_j}\{\ell_q(x)\} \tag{29}$$

This implies that

$$\min\left\{\max_{p\in I_i}\{\ell_p(x)\}, \max_{q\in I_j}\{\ell_q(x)\}\right\} = \max_{p\in I_i}\{\ell_p(x)\} \tag{30}$$

Then we finally have

$$P(x|\Phi, \Psi) = \min_{1\leq i\leq M}\left\{\max_{k\in I_i}\{\ell_k(x)\}\right\} = \min_{\substack{1\leq i\leq M\\ i\neq j}}\left\{\max_{k\in I_i}\{\ell_k(x)\}\right\} = P(x|\Phi, \tilde{\Psi}) \tag{31}$$

Here we can see that the j-th row of structure matrix Ψ can be deleted without affecting the function values of $P(x|\Phi, \Psi)$. This completes the proof of Lemma 4. □

Since Lemma 4 can be used recursively, a much simplified structure matrix is obtained by deleting all the redundant rows. A single row in $\tilde{\Psi}$ corresponds to a super region, which is defined as an aggregation of several affine regions. Being a mergence of many convex polytopes, a super region can be concave or even disconnected. Then the number of super regions can be much smaller than that of regions (Wen, 2006).

4.3 Column vector simplification lemma

Lemma 5. *Assume that $P(x|\Phi,\Psi) : D \subset \Re^n \mapsto \Re$ is a PWA function with M linear segments. Denote $\Psi = [\psi_{ij}]^{M\times M}$ and $\hat{\Psi} = [\hat{\psi}_{ij}]^{M\times M}$ as the primary and dual structure matrix. Then the following results hold.*

1. *Given any $i,j,k \in \{1,\cdots,M\}$, if $k,j \in I_i$ and $\hat{\psi}_{jk} = 1$, then $\psi_{ij} = 0$, where I_i is the same as defined in (13);*
2. *If $\psi_{ij} = 0, \forall 1 \le j \le M$, then there exist a simplified structure matrix $\tilde{\Psi} \in \Re^{(M-1)\times M}$ and parameter matrix $\tilde{\Phi} \in \Re^{(M-1)\times(n+1)}$, such that*

$$P(x|\Phi,\Psi) = P(x|\tilde{\Phi},\tilde{\Psi}) \tag{32}$$

where $\Phi \in \Re^{M\times(n+1)} = [\phi_1,\cdots,\phi_M]^T$, $\tilde{\Phi} \in \Re^{(M-1)\times(n+1)} = [\phi_1,\cdots,\phi_{j-1},\phi_{j+1},\cdots,\phi_M]^T$, and $\Psi,\tilde{\Psi}$ are the same as defined in Lemma 4.

Proof. According to (17), if $\hat{\psi}_{jk} = 1$, we have

$$\ell_j(x) \le \ell_k(x), \forall x \in R_j \tag{33}$$

which implies that $\ell_j(x)$ is inactive in its own region, i.e.

$$\max\{\ell_j(x),\ell_k(x)\} = \ell_k(x), \forall x \in R_j \tag{34}$$

Note that $k,j \in I_i$ and I_i is the index set of $\ell_i(x)$. We can get

$$\max_{p\in I_i}\{\ell_p(x)\} = \max_{\substack{p\in I_i\\ p\neq j}}\{\ell_p(x)\}, \forall x \in D \tag{35}$$

This implies that $\psi_{ij} = 0$.
In addition, if $\psi_{ij} = 0$ holds for any $1 \le j \le M$, then $\ell_j(x)$ will be totally covered by other affine functions throughout the whole domain. Therefore, the j-th column of the structure matrix and j-th row of the parameter matrix can be deleted. This means that $P(x|\Phi,\Psi) = P(x|\tilde{\Phi},\tilde{\Psi})$. It should be noted that the matrix $\tilde{\Psi}$ corresponds to a simpler lattice PWA function than Ψ even without a deletion of row vectors. A lattice PWA function with less terms in max operators is produced if some elements in the structure matrix are changed from one to zero. This completes the proof of Lemma 5. □

The significance of Lemma 5 is that it can differentiate the inactive regions from the active ones in a given PWA function. The inactive regions can then be removed from the analytic expression because they do not contribute to the PWA function values. The active regions are also referred to as the lattice regions, which define the number of columns in the structure matrix $\tilde{\Psi}$.
Lemma 5 presents an efficient and constructive method to reduce the complexity of a lattice PWA function. Recalling that an eMPC controller $u(x) \in \Re$ of a single input system is a continuous scalar PWA function, in which many polyhedral regions have same feedback

gains. It implies that the number of super-regions is usually much smaller than that of polyhedral regions. The complexity reduction algorithm of Lemma 5 can produce a very compact representation of the scalar eMPC solutions.

Example 2: In order to clarify the simplification procedure, we consider the lattice PWA function of (21) derived in Example 1.
Denoting $\Psi = [\varphi_1 \ \ \varphi_2 \ \ \varphi_3 \ \ \varphi_4]^T$, we can get $\varphi_2 - \varphi_3 = [0 \ \ 0 \ \ 0 \ \ 0] \leq 0$, where " $\leq$ " is the pointwise inequality. It follows from Lemma 4 that the third row vector φ_3 can be removed. Then the structure matrix is simplified as

$$\Psi = \begin{bmatrix} 1\,0\,1\,0 \\ 0\,1\,1\,0 \\ 0\,1\,0\,1 \end{bmatrix}. \tag{36}$$

Furthermore, by using (17), we can obtain the dual structure matrix

$$\hat{\Psi} = \begin{bmatrix} 1\,1\,0\,1 \\ 1\,1\,0\,1 \\ 1\,0\,1\,1 \\ 1\,0\,1\,1 \end{bmatrix}. \tag{37}$$

According to (36), we have $I_i = \{k, j\}$ with $k = 4, j = 2$ and $i = 3$. Using (37), we further have $\hat{\psi}_{jk} = \hat{\psi}_{24} = 1$. Then it follows from Lemma 5 that the item of ψ_{32} can be put to zero. The final structure matrix is written as

$$\tilde{\Psi} = \begin{bmatrix} 1\,0\,1\,0 \\ 0\,1\,1\,0 \\ 0\,0\,0\,1 \end{bmatrix} \tag{38}$$

The corresponding lattice PWA function is

$$p(x) = \min\left\{\max\{\ell_1, \ell_3\}, \max\{\ell_2, \ell_3\}, \ell_4\right\} \tag{39}$$

4.4 Lattice PWA representation theorem

Theorem 2. *Let $P(x) : \Omega \mapsto \Re$ be a continuous scalar PWA function with $\hat{M}$ super-regions. There must exist a positive integer $\bar{M} \leq \hat{M}$, a parameter matrix $\Phi \in \Re^{\hat{M}\times(n+1)}$, a structure matrix $\Psi = [\psi_{ij}]^{\bar{M}\times\hat{M}}$ and a lattice PWA function*

$$L(x|\Phi, \Psi) = \min_{1\leq i\leq \bar{M}} \left\{ \max_{\substack{1\leq j\leq \hat{M} \\ \psi_{ij}=1}} \{\ell_j(x)\} \right\} \tag{40}$$

such that

$$P(x) = L(x|\Phi, \Psi), \forall x \in \Omega \tag{41}$$

where ψ_{ij} is a boolean variable, $\Phi = [\phi_1, \cdots, \phi_{\hat{M}}]^T$, $\phi_j = [\alpha_j^T, \beta_j]$, $\alpha_j \in \Re^n, \beta_j \in \Re$ with $1 \leq i \leq \bar{M}, 1 \leq j \leq \hat{M}$.

Theorem 2 shows that the class of lattice PWA functions provides a universal model set for continuous scalar PWA functions. The complexity of a lattice PWA function is specified by the number of super-regions instead of that of regions. Then the lattice PWA functions may present a more compact representation than the PWA models without global analytical descriptions.

The scalar lattice representation theorem can be generalized to describe a vector eMPC solution $u(x) : \Omega \mapsto \Re^m$. The main idea is to represent each component scalar eMPC feedback law individually.

Theorem 3. *Let $u(x) = [u_1(x), \cdots, u_m(x)]^T$ be a continuous vector eMPC solution with $x \in \Omega$. There must exist m lattice PWA functions $L(x|\Phi_i, \Psi_i)$ such that*

$$u_i(x) = L(x|\Phi_i, \Psi_i) \quad \forall x \in \Omega \tag{42}$$

where Φ_i, Ψ_i are parameter and structure matrices and $i = 1, \cdots, m$.

The vector lattice representation theorem is valid for continuous PWA functions. It is proved in (Spjotvold et al, 2007, Bemporad et al. 2002) that an eMPC controller is continuous from a strictly convex mpQP problem. The continuity property is further generalized to general convex mpQP problems (Spjotvold et al, 2006). An eMPC problem with a linear cost function may have discontinuous solutions because of the degeneracy of critical regions. It is proved in (Bemporad et al, 2002) that there always exists a polyhedral partition even for degenerate critical regions, such that the eMPC control is continuous. Recalling that the mpLP problems are essentially special realizations of convex mpQP problems. It is proved constructively in (Spjotvold et al, 2006) that a continuous eMPC solution can be found for LP-based MPC problems by using a minimum norm method. Therefore, the set of continuous PWA functions can cover a wide class of eMPC solutions by utilizing appropriate multi-parametric program solvers.

The continuity of eMPC controllers can be easily verified by checking the function values at the vertices of different regions. This function has been implicitly implemented in the lattice PWA representation algorithm (Wen et al., 2009a). Therefore, the lattice representation can automatically separate the continuous eMPC solutions from the discontinuous ones. In addition, the discontinuity in eMPC controls are often caused by the overlapping of critical regions (Bemporad et al, 2002). The mpt-toolbox (Kvasnica et al., 2004) has a function to detect the existence of overlapping regions. It presents another efficient way to verify the continuity of eMPC controls.

The vector lattice representation can be extended to discontinuous eMPC solutions. A discontinuous eMPC solution is usually decomposed into a set of continuous PWA functions. Recalling that each continuous PWA function has a vector lattice representation. Then the discontinuous eMPC solutions can be represented by a set of lattice PWA functions and a switch logic. The switch logic may be implemented as a binary search tree (Tondel et al. 2003) or bounding box search tree (Christophersen et al, 2007). Further research is under investigation to generalize the lattice representation method to discontinuous PWA functions.

The lattice representation has a quadratic complexity for both online evaluation and memory storage. When the eMPC solutions consist of a large number of super-regions, e.g. the eMPC problems has a large input constraint set, the BST or BBT algorithms may have a lower online computational complexity.

4.5 Representation algorithm

In Kvasnica et al. 2004, a Multi-Parametric Toolbox (MPT) for computing optimal feedback controllers of constrained linear and piecewise affine systems is developed. The toolbox offers a broad spectrum of algorithms to calculate the eMPC solutions. The proposed lattice PWA representation algorithm can be easily embedded into the MPT toolbox and provide a better performance in term of online calculation and memory space requirements.
The main steps of the representation algorithm are summarized as follows.

1. Calculate the eMPC solution using the MPT toolbox. Record the local affine functions, constrained inequalities and vertices of each region;
2. Calculate the values of each affine function at each vertex;
3. Calculate the structure matrix using Lemma 3;
4. Delete the redundant row vectors in structure matrix using Lemma 4;
5. Delete the redundant elements in structure and parameter matrices using Lemma 5;
6. Get the lattice PWA expression of an eMPC solution.

It should be noted that the multi-parametric solver may return a PWA solution that is discontinuous, even for problems where continuous PWA solution exists. Then the lattice representation algorithm is feasible for the continuous PWA solutions obtained from the multi-parametric solver.

4.6 Complexity analysis

Let $u(x) = [u_1(x), \cdots, u_m(x)]^T$ be a vector PWA function with M polydedral regions and $x \in \Re^n$. Denote $\tilde{M}_k$ as the number of lattice regions in $u_k(x)$ and $\hat{M}_k$ the number of super regions with $1 \le k \le m$.

4.6.1 Storage

The lattice representation requires the storage of a $(n+1) \times \sum_{k=1}^{m} \tilde{M}_k$ parameter matrix and a $\sum_{k=1}^{m} \hat{M}_k \times \tilde{M}_k$ structure matrix. The total memory needed is $O\big((n+1)\sum_{k=1}^{m} \tilde{M}_k\big)$ real numbers and $O\big(\sum_{k=1}^{m} \hat{M}_k \tilde{M}_k\big)$ binary numbers. The structure matrix is usually very sparse. The actual required memory space can be significantly smaller than the worst estimation through the use of appropriate sparse storage techniques.

4.6.2 Online complexity

For a given state variable, the online evaluation of a lattice PWA control law consists of 3 steps. The first step is to calculate the function value of $\tilde{M}_k$ affine functions. This requires $n\tilde{M}_k$ multiplication and $n\tilde{M}_k$ sums. In the second step, we need to calculate the maximum of $\bar{M}_k$ function values, where $\bar{M}_k$ is the number of affine functions with $\psi_{ij} = 1$ in the structure matrix. Note that $\bar{M}_k \le \tilde{M}_k$. In the worse case, this step requires $(\tilde{M}_k - 1) \times \tilde{M}_k$ by considering the $\tilde{M}_k$ maximization terms. The last step is to calculate the minimum of $\hat{M}_k$ real numbers. It needs $\hat{M}_k - 1$ comparisons. Therefore, the total online complexity is $O\big(\sum_{k=1}^{m}(2n\tilde{M} + (\tilde{M}_k - 1)\tilde{M}_k + (\hat{M}_k - 1)\big)$. It follows from Wen et al. (2009a) that $n \ll \tilde{M}_k$ and $\hat{M}_k < \tilde{M}_k$ holds for any $k \in \{1, \cdots, m\}$. The online complexity can be roughly approximated by $O\big(\sum_{k=1}^{m} \tilde{M}_k^2\big)$. It should be noted that the structure matrix is usually sparse. The estimate of online calculation is very conservative in most cases. Then the average online calculation complexity can be considerably lower than the worse case estimate.

4.6.3 Preprocessing

The preprocessing phase for the lattice representation is composed of two steps. The first step is to calculate the eMPC control law using multi-parameter program. The second one is to represent the eMPC law with a lattice PWA function. It is very difficult to present a close-form solution of the off-line complexity. However, it was observed in extensive trials that the representation step takes significantly less time than the initial computation of an eMPC controller. Therefore, the proposed lattice representation method can be applied to any system for which an explicit controller is feasible.

5. Numerical examples

Two examples are illustrated in this section. All the simulations are run in Matlab 2007a on a 2.0 GHz Core 2 CPU with 1 GB RAM.

Example 3: Consider the double integrator

$$y(t) = \frac{1}{s^2}u(t) \tag{43}$$

Its equivalent discrete-time state-space representation

$$\begin{cases} x(t+1) = \begin{bmatrix} 1 & 1 \\ 0 & 1 \end{bmatrix} x(t) + \begin{bmatrix} 0 \\ 1 \end{bmatrix} u(t) \\ y(t) = \begin{bmatrix} \frac{1}{2} & \frac{1}{2} \end{bmatrix} x(t) \end{cases}$$

is obtained by setting

$$\ddot{y}(t) \approx \frac{\dot{y}(t+T) - \dot{y}(t)}{T} \tag{44}$$

$$\dot{y}(t) \approx \frac{y(t+T) - y(t)}{T} \tag{45}$$

with $T = 1$ s. The problem of regulate the system to the origin is formulated as an optimization problem, which minimizes the following performance measure

$$\sum_{k=0}^{1} \left\| \begin{bmatrix} 1 & 1 \\ 0 & 1 \end{bmatrix} x_{k+1} \right\|_{\infty} + |0.8u_k| \tag{46}$$

subject to the input constraints $-1 \leq u_k \leq 1, k = 0, 1$. and the state constraints $-10 \leq x_k \leq 10, k = 1, 2..$ where $N_y = 2, N_u = 2, Q = \begin{bmatrix} 1 & 1 \\ 0 & 1 \end{bmatrix}, R = 0.8$. The solution of this problem is a continuous PWA function, whose surface plot is visualized in Fig. 3(a). According to Bemporad et al. 2002, the eMPC solution is given in Table 1.

Region #	Region	Controller
1	$\begin{bmatrix} 1.00 & 2.00 \\ 0.00 & 1.00 \\ -1.00 & -1.00 \\ 0.80 & -3.20 \\ 1.00 & 1.00 \\ -1.00 & -3.00 \end{bmatrix} x \leq \begin{bmatrix} 11.00 \\ 11.00 \\ 10.00 \\ -2.40 \\ 10.00 \\ -2.00 \end{bmatrix}$	-1.00
2	$\begin{bmatrix} 0.80 & 3.20 \\ -1.00 & -2.00 \\ -1.00 & -1.00 \\ 1.00 & 1.00 \\ 0.00 & -1.00 \\ 1.00 & 3.00 \end{bmatrix} x \leq \begin{bmatrix} -2.40 \\ 11.00 \\ 10.00 \\ 10.00 \\ 11.00 \\ -2.00 \end{bmatrix}$	1.00
3	$\begin{bmatrix} 0.53 & 2.13 \\ 0.67 & 0.67 \\ -1.00 & -1.00 \\ -0.33 & -1.33 \end{bmatrix} x \leq \begin{bmatrix} 0.00 \\ 0.00 \\ 10.00 \\ 1.00 \end{bmatrix}$	$[-0.33, -1.33]x$
4	$\begin{bmatrix} -0.80 & -3.20 \\ 1.00 & 3.00 \\ -1.00 & -1.00 \end{bmatrix} x \leq \begin{bmatrix} 0.00 \\ 0.00 \\ 10.00 \end{bmatrix}$	0
5	$\begin{bmatrix} -1.00 & -1.00 \\ 0.50 & 0.50 \\ -0.80 & -2.40 \\ 0.50 & 1.50 \end{bmatrix} x \leq \begin{bmatrix} 10.00 \\ 0.00 \\ 0.00 \\ 1.00 \end{bmatrix}$	$[-0.50, -1.50]x$
6	$\begin{bmatrix} 0.80 & 3.20 \\ -1.00 & -3.00 \\ 1.00 & 1.00 \end{bmatrix} x \leq \begin{bmatrix} 0.00 \\ 0.00 \\ 10.00 \end{bmatrix}$	0
7	$\begin{bmatrix} 1.00 & 1.00 \\ -0.50 & -0.50 \\ 0.80 & 2.40 \\ -0.50 & -1.50 \end{bmatrix} x \leq \begin{bmatrix} 10.00 \\ 0.00 \\ 0.00 \\ 1.00 \end{bmatrix}$	$[-0.50, -1.50]x$
8	$\begin{bmatrix} -0.53 & -2.13 \\ -0.67 & -0.67 \\ 1.00 & 1.00 \\ 0.33 & 1.33 \end{bmatrix} x \leq \begin{bmatrix} 0.00 \\ 0.00 \\ 10.00 \\ 1.00 \end{bmatrix}$	$[-0.33, -1.33]x$

Table 1. Conventional Representation of the MPC Solution

Using Lemma 3, we can get

$$\Psi = \begin{bmatrix} 1\,0\,1\,0\,1\,0\,1\,1 \\ 1\,1\,0\,1\,0\,1\,0\,0 \\ 1\,0\,1\,0\,1\,0\,1\,1 \\ 1\,0\,1\,1\,0\,1\,0\,1 \\ 1\,0\,1\,0\,1\,0\,1\,1 \\ 1\,0\,0\,1\,1\,1\,1\,0 \\ 1\,0\,0\,1\,1\,1\,1\,0 \\ 1\,0\,1\,1\,0\,1\,0\,1 \end{bmatrix} \tag{47}$$

By applying Lemma 4 and 5, we can further get a simplified structure matrix

$$\tilde{\Psi} = \begin{bmatrix} 1\,0\,1\,0\,1 \\ 0\,1\,0\,0\,0 \\ 0\,0\,1\,1\,0 \\ 0\,0\,0\,1\,1 \end{bmatrix} \tag{48}$$

The corresponding parameter matrix is

$$\tilde{\Phi} = \begin{bmatrix} 0 & 0 & -1.00 \\ 0 & 0 & 1.00 \\ -0.33 & -1.33 & 0 \\ 0 & 0 & 0 \\ -0.50 & -1.50 & 0 \end{bmatrix} \tag{49}$$

Therefore, the analytical expression of the MPC control law is written as

$$u(x) = \min\{1, \max\{-1, -0.33x_1 - 1.33x_2, -0.50x_1 - 1.50x_2\}, \tag{50}$$

$$\max\{0, -0.33x_1 - 1.33x_2\}, \max\{(0, -0.50x_1 - 1.50x_2\}\}$$

Here it is easy to see that the online MPC optimization is reduced to a lattice PWA function evaluation problem. Same as Bemporad et al. (2002), we consider the starting point $x(0) = [10, -5]^T$. This point is substituted into (50), and the corresponding control action is $u(x) = 1$, which is obtained without any optimization calculations and table searching procedures. The closed-loop response is shown in Fig. 3(b), which is exactly the same with the results from online optimization in Bemporad et al. 2002. The required memory in the analytical expression is to store a structure matrix $\tilde{\Psi} \in \Re^{4\times 5}$ and a parameter matrix $\tilde{\Phi} \in \Re^{5\times 3}$. The total memory is 35, which is much smaller than the memory space used in Table 1. The online computation requires 7 comparison operations, 8 multiplications and 5 summations. It is evident that the lattice PWA MPC control law performs better in term of online calculation and memory requirements than the conventional eMPC solution.

Example 4: This example is to demonstrate the performance of lattice representation for eMPC solutions from parametric quadratic program. The 2-norm is used in the stage cost function.

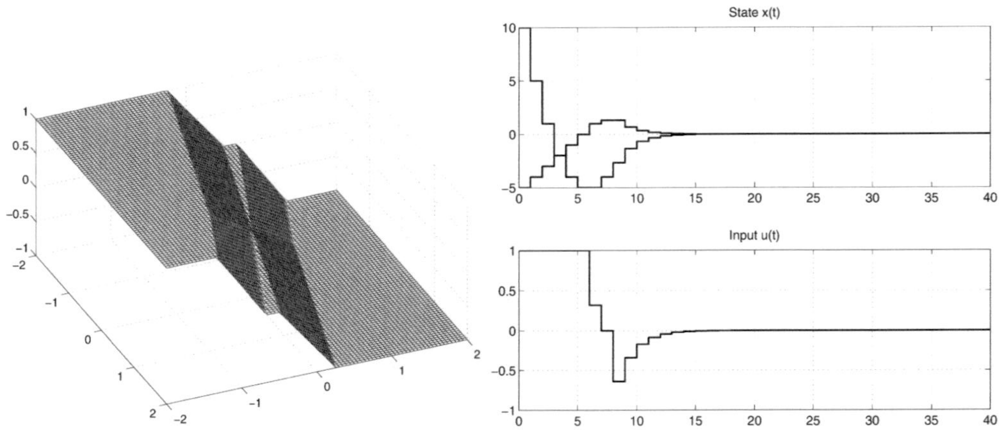

(a) Surface plot (b) Close loop response

Fig. 3. Lattice eMPC Solution

Consider the following state space representation:

$$\begin{cases} x(t+1) = \begin{bmatrix} 0.7326 & -0.0861 \\ 0.1722 & 0.9909 \end{bmatrix} x(t) + \begin{bmatrix} 0.0609 \\ 0.0064 \end{bmatrix} u(t) \\ y(t) = \begin{bmatrix} 0 & 1.4142 \end{bmatrix} x(t) \end{cases} \tag{51}$$

The constraints on input are $-2 \leq u(t) \leq 2$. The corresponding optimization problem for regulating to the origin is written as follows:

$$\begin{aligned} &\min_{u_t, u_t+1} x_{t+2|t}^T P x_{t+2|t} + \sum_{k=0}^{1} [x_{t+k|t}^T Q x_{t+k|t} + R u_{t+k}^2] \\ &s.t. -2 \leq u(t+k) \leq 2, \qquad k = 0, 1 \\ &x_{t|t} = x(t) \end{aligned} \tag{52}$$

where $P \in \Re^{2\times 2}$ is the solution of the Lyapunov equation $P = A^T P A + Q, Q = \begin{bmatrix} 1 & 0 \\ 0 & 1 \end{bmatrix}, R = 0.01, N_u = N_y = N_c = 2$. According to Bemporad et al. (2000), the eMPC solution is provided in Table 2.

Using Lemma 3, we can get the following structure matrix.

$$\Psi = \begin{bmatrix} 1&0&0&1&0&1&1 \\ 0&1&1&1&0&1&0 \\ 0&1&1&1&0&1&0 \\ 0&0&0&1&1&1&1 \\ 0&0&0&1&1&1&1 \\ 1&0&0&1&0&1&1 \\ 1&0&0&1&0&1&1 \end{bmatrix} \tag{53}$$

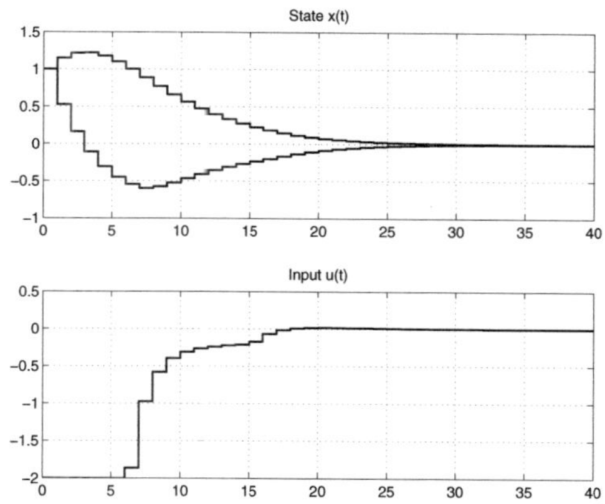

Fig. 4. Closed-loop response of the lattice PWA feedback law

By applying Lemma 4 and 5, we can further get the much simplified structure matrix by deleting all the redundant rows and columns

$$\tilde{\Psi} = \begin{bmatrix} 0\,1\,0\,0\,0 \\ 0\,0\,1\,1\,0 \\ 1\,0\,1\,0\,1 \end{bmatrix} \tag{54}$$

According to the third column of Table 2, the corresponding parameter matrix is

$$\tilde{\Phi} = \begin{bmatrix} -5.9220 & -6.8883 & 0 \\ 0 & 0 & 2 \\ 0 & 0 & -2 \\ -6.4159 & -4.6953 & 0.6423 \\ -6.4159 & -4.6953 & -0.6423 \end{bmatrix} \tag{55}$$

Finally, the analytical expression of the eMPC solution is written as

$$u(x) = \min\{2, \max\{-2, -6.4159x_1 - 4.6953x_2 + 0.6423\}, \tag{56}$$

$$\max\{-2, -5.9220x_1 - 6.8883x_2, -6.4159x_1 - 4.6953x_2 - 0.6423\}\}$$

The closed-loop response of the states is depicted in Fig. 4, which is the same as the one obtained from the online optimization in Bemporad et al. (2000). The required memory in the analytical expression is to store a structure matrix $\tilde{\Psi} \in \Re^{3\times5}$ and a parameter matrix $\tilde{\Phi} \in \Re^{5\times3}$. It is evident that the lattice representation requires much smaller memory space than Table 2. Therefore, the lattice PWA representation of an eMPC solution saves both memory space and online calculation requirements.

Now we discuss the scalability of the lattice representation algorithm. Here this system is solved using different prediction horizons. The simulation results are summarized in Table 3, where N denotes the prediction horizon, τ_{mpt} is the time in seconds to compute the original PWA control, τ_{lat} is the computation time in seconds to build the lattice representation. Note that both representation algorithms give the same lattice eMPC solutions and $\hat{\tau}_{lat} < \tau_{lat}$ hold for all prediction horizons.

No.	Region	Controller
1	$\begin{bmatrix} -5.9220 & -6.8883 \\ 5.9229 & 6.8883 \\ -1.5379 & 6.8296 \\ 1.5379 & -6.8296 \end{bmatrix} x \leq \begin{bmatrix} 2.0000 \\ 2.0000 \\ 2.0000 \\ 2.0000 \end{bmatrix}$	$-[5.9220, 6.8883]x$
2,4	$\begin{bmatrix} -3.4155 & 4.6452 \\ 0.1044 & 0.1215 \\ 0.1259 & 0.0922 \end{bmatrix} x \leq \begin{bmatrix} 2.6341 \\ -0.0353 \\ -0.0267 \end{bmatrix}$	2.0000
3	$\begin{bmatrix} 0.0679 & -0.0924 \\ 0.1259 & 0.0922 \end{bmatrix} x \leq \begin{bmatrix} -0.0524 \\ -0.0519 \end{bmatrix}$	2.0000
5	$\begin{bmatrix} -0.1259 & -0.0922 \\ -0.0679 & 0.0924 \end{bmatrix} x \leq \begin{bmatrix} -0.0519 \\ -0.0524 \end{bmatrix}$	-2.0000
6	$\begin{bmatrix} -6.4159 & -4.6953 \\ -0.0275 & 0.1220 \\ 6.4159 & 4.6953 \end{bmatrix} x \leq \begin{bmatrix} 1.3577 \\ -0.0357 \\ 2.6423 \end{bmatrix}$	$-[6.4159, 4.6953]x$ $+0.6423$
7,8	$\begin{bmatrix} 3.4155 & -4.6452 \\ -0.1044 & -0.1215 \\ -0.1259 & -0.0922 \end{bmatrix} x \leq \begin{bmatrix} 2.6341 \\ -0.0353 \\ -0.0267 \end{bmatrix}$	-2.0000
9	$\begin{bmatrix} 6.4159 & 4.6953 \\ 0.0275 & -0.1220 \\ -6.4159 & -4.6953 \end{bmatrix} x \leq \begin{bmatrix} 1.3577 \\ -0.0357 \\ 2.6423 \end{bmatrix}$	$-[6.4159, 4.6953]x$ -0.6423

Table 2. Conventional Representation of the eMPC Solution

According to Table 3, the number of affine regions in the eMPC control law increased considerably with the length of prediction horizons. However, all the eMPC laws can be represented by a single lattice PWA function with 19 lattice regions and 12 super-regions. For example, when $N = 28$, the original PWA controller consists of 894 polyhedral regions. But there are only 19 unique affine functions. Therefore, the complexity of the lattice representation is very robust to the length of prediction horizons.

Fig. 5(a) shows the domain partition of the eMPC control when $N = 8$. The corresponding surface plot is visualized in Fig. 5(b). The lattice representation requires to store a 19×3 parameter matrix and a 12×19 structure matrix. The total memory needed is 285. This implies a significant saving in the storage by considering that there are 894 affine functions in the original PWA controller. In the worst case, the online-calculation complexity is $O(234)$. This means a high computation efficiency because the complexity of the competing methods in Table 3 are all specified by the original MPC solution with 894 polyhedral regions.

In this example, the off-line computation is mainly determined by the calculation of the original PWA controllers. For the complex cases with long prediction horizons, the time for building the lattice presentation is one order magnitude less than that for calculating the initial controllers. This makes negligible the additional off-line computation cost for a lattice representation when $N \geq 20$.

Note that the more complicated eMPC laws generally have more polyhedral regions with equal affine control laws. For an instance, when $N = 28$, 875 regions can be merged into one of 19 regions. By comparison, the number of such regions is only 409 when $N = 12$. Therefore, better performance can be anticipated from the lattice representation regarding the

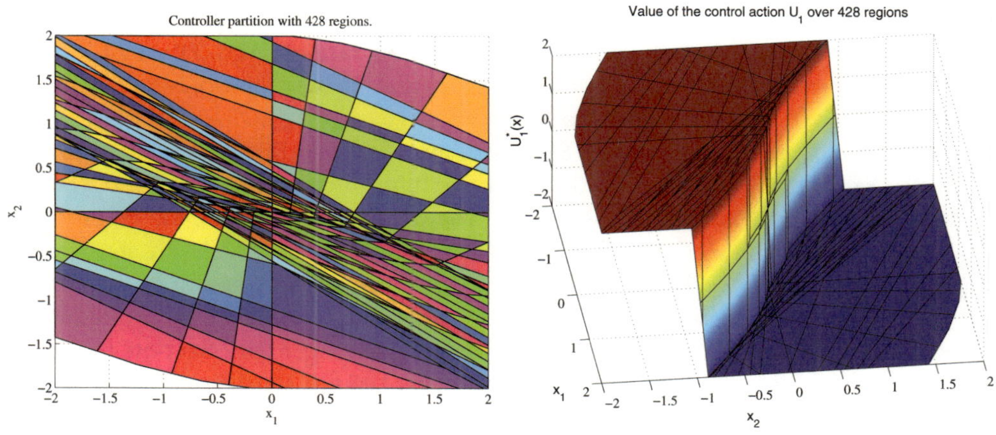

(a) Domain partition (b) Surface plot

Fig. 5. eMPC solution when $N = 8$

storage, on-line calculation and off-line preprocessing cost when dealing with the complex MPC solutions with large number of polyhedral regions.

N	M	$\tilde{M}$	$\hat{M}$	τ_{mpt}	τ_{lat}
8	254	19	12	11.73	2.06
12	428	19	12	31.19	5.46
16	697	19	12	80.59	15.77
20	863	19	12	174.66	26.23
24	892	19	12	296.05	28.25
28	894	19	12	429.55	28.39

Table 3. Performance of lattice representation for an eMPC solution with different prediction horizons

Example 5: Consider the following linear system with three states and two inputs (Christophersen et al, 2007)

$$x(t+1) = \begin{bmatrix} 0.7 & -0.1 & 0 \\ 0.2 & -0.5 & 0.1 \\ 0 & 0.1 & 0.1 \end{bmatrix} x(t) + \begin{bmatrix} 0.1 & 0 \\ 0.1 & 1 \\ 0.1 & 0 \end{bmatrix} u(t). \quad (57)$$

The system is subject to input constraints $-1 \leq u_i(t) \leq 1$ with $i = 1, 2$, and state constraints $-20 \leq x_i(t) \leq 20$ with $i = 1, 2, 3$. The constrained finite time optimal control problem is solved with $p = 1, N_y = 8$, $Q = \begin{bmatrix} 1 & 0 & 0 \\ 0 & 1 & 0 \\ 0 & 0 & 1 \end{bmatrix}$, $R = \begin{bmatrix} 0.1 & 0 \\ 0 & 0.1 \end{bmatrix}$ and $P = \begin{bmatrix} 0 & 0 & 0 \\ 0 & 0 & 0 \\ 0 & 0 & 0 \end{bmatrix}$. The MPT toolbox is used to generate the eMPC control law $u(x) = [u_1(x), u_2(x)]^T$ with 813.4 seconds. The vector eMPC control is defined over 2497 regions.

Two scalar lattice PWA functions are used to describe $u(x)$, which requires 2 structure matrices $\Psi_1 \in \Re^{18\times 21}, \Psi_2 \in \Re^{3\times 5}$ and two parameter matrices $\Phi_1 \in \Re^{21\times 4}, \Phi_1 \in \Re^{5\times 4}$. The total memory required is 497 in the worst case. In addition, the online computational complexity is calculated as 616. The average evaluation time of the lattice eMPC is 3.01×10^{-4} seconds by randomly sampling 20,000 states in the feasible region. In the preprocessing stage, it takes 11.4 seconds to build the lattice PWA function.

Table 4 lists the comparison results of SS algorithm, CVF algorithm (Christophersen et al, 2007), BST algorithm (Tondel et al. 2003), BBT algorithm (Christophersen et al, 2007) and the proposed lattice representation (LR) algorithm. The LR algorithm has the lowest description complexity. It saves more than 90% memory storage compared with the most efficient competing algorithm. The BST algorithm has the lowest online computational complexity. But it requires more than 5 millions linear programs to construct the binary search tree. The pre-processing time is 3504.2 seconds, which is much longer than the time for building the eMPC control. The SS, BBT and BST algorithms require shorter pre-processing time than the LR algorithm. However, the former algorithms have much higher evaluation and description complexities than the latter. In summary, the LR algorithm offers a significant reduction in memory storage. Therefore, the LR method has a better overall performance than other competing methods by considering description, online and off-line computational complexities.

In order to test the scalability of the LR algorithm, we solve system (57) with different horizons. The simulation results are summarized in Table 5, where $N, M, \tau_{mpt}, \hat{\tau}_{lat}$ are defined as above in Example 4. According to Table 5, the number of affine regions in the eMPC control law increased considerably with horizon N. However, the complexity of the vector lattice representation is very robust to the prediction horizons. The description and evaluation complexities of lattice representation keep constant, although the number of regions increase from 1600 ($N = 12$) to 3189 ($N = 20$) in the original eMPC controls. In all cases, the preprocessing time for building the lattice representations is negligible compared with the time for the eMPC solutions using MPT toolbox. Therefore, the LR algorithm is promising for large-scale eMPC problems with long prediction horizons.

	ϕ	ψ	τ
SS Algorithm	81661	106295	0
CVF Algorithm	9988	14982	0
BST Algorithm	21368	110	3504.2
BBT Algorithm	30104	923	10.0
LR Algorithm	497	616	11.4

Table 4. Comparison of Description, Evaluation and Preprocessing Complexities

N	M	$\bar{M}_{u1}$	$\hat{M}_{u1}$	$\bar{M}_{u2}$	$\hat{M}_{u2}$	τ_{mpt}	τ_{lat}
4	1510	14	17	3	5	86.0	6.0
8	2497	18	21	3	5	813.4	11.4
12	2600	18	21	3	5	1658.3	12.0
16	3189	18	21	3	5	3385.6	14.7
20	3189	18	21	3	5	5589.8	14.7

Table 5. Performance of LR Algorithm for eMPC Solutions with Different Horizons

6. Conclusions

This paper proposes a general lattice PWA representation theorem for continuous eMPC solutions. A constructive proof is developed to show that the global structure information of an eMPC controller can be fully represented by a set of scalar lattice PWA functions. A lattice PWA function is a global and compact representation of an eMPC solution, because the redundant parameters are automatically removed by utilizing the continuity of eMPC controllers. A lattice representation has an explicit analytical expression. This facilitates the implementation of lattice eMPC solutions using low-cost hardware.
A complexity reduction algorithm is proposed to develop a computationally efficient lattice representation of eMPC controllers obtained from multi-parametric program. The description and evaluation complexities of a lattice eMPC controller depend on the number of super-regions instead of the number of polyhedral regions. Therefore, the lattice representation of eMPC solutions can reduce computation and memory requirements significantly, when the original MPC solutions have many polyhedral regions with equal affine control laws.
The class of lattice PWA functions provides a compact and explicit model structure for continuous eMPC solutions. The global structure information in an eMPC control is utilized to reduce the complexity of its lattice representation. The lattice eMPC controllers present the first step to use a global PWA representation model to describe generic eMPC solutions. Further investigation is needed to generalize the lattice representation into the discontinuous eMPC solutions.

7. References

[1] Alessio A. & Bemporad A. (2009), "A survey on explicit model predictive control," in Proc. Int. Workshop on Assessment and Future Directions of Nonlinear Model Predictive Control (Pavia, Italy), D.M. Raimondo L. Magni, F. Allgower, Ed., *Lecture Notes in Control and Information Sciences*, 2009.

[2] Beccuti, A.G., Mariethoz, S., Cliquennois, S., Wang, S., & Morari, M. (2009) "Explicit Model Predictive Control of DC-DC Switched Mode Power Supplies with Extended Kalman Filtering", *IEEE Trans. on Industrial Electronics*, vol. 56, no. 5, To appear.

[3] Bemporad A., Borelli F. & Morari, M. (2002) "Model predictive control based on linear programming-The explicit solution." *IEEE Transactions on Automatic Control*, vol. 47 (12), 1974-1984.

[4] Bemporad, A. , Morari, M., Dua, V. & Pistikopoulos, E. N. (2002), "The explicit linear quadratic regulator for constrained systems," *Automatica*, vol. 38 (1), 3-20.

[5] Bemporad A. & Filippi, C. (2003) "Suboptimal explicit RHC via approximate quadratic programming," *Journal of Optimization Theory and Application*, vol. 117(1), 5-38.

[6] Bertsekas, D. P. & Tsitsiklis, J. N. (1998) "Neuro-dynamic prgramming." Belmont: Athena Scientific.

[7] Boom, T. J. J. & Schutter, B. D. (2002) "Model predictive control for perturbed max-plus-linear systems," *Systems & Control Letters*, vol. 45, no. 1, pp. 21-33.

[8] Borrelli, F., Baotic, M., Bemporad, A., & Morari, M. (2001) "Efficient on-Line computation of constrained optimal control." *IEEE Conference on Decision and Control*, Orlando, Florida, pp. 1187-1192.

[9] Chua, L. O. & Kang, S. M. (1977) "Section-wise piecewise-linear functions: canonical representation, properties, and applications." *IEEE Trans, Circuits Syst. I*, 24(1), 125-140.
[10] Christophersen, F. J., Kvasnica, M., Jones, C. N., & Morari, M. (2007) "Efficient evaluation of piecewise control laws defined over a large number of polyhedra", *European Control Conference*, Kos, Greece.
[11] Chikkula, Y., Lee, J. H. & Ogunnaike, B. A. (1998) "Dynamically scheduled MPC of nonlinear processes using hinging hyperplane models.", *AIChE Journal*, vol. 44, pp. 2658-2674.
[12] Dua, P., Sakizlis, V., Dua, V., Doyle III, F.J., Pistikopoulos, E. N. (2004) "Model based control for insulin delivery for Type 1 diabetics via parametric programming", *Proceedings of European Symposium on Computer Aided Process Engineering*, Lisbon, pp. 1045-1050.
[13] Dua, P., Kouramas, K., Dua, V., & Pistikopoulos, E. N. (2008) "MPC on a chip-Recent advances on the application of multi-parametric model-based control." *Computers & Chemical Engineering*, Vol. 32, No. 4-5, pp. 754-765.
[14] Geyer, T., Torrisi, F. D., Morari, M. (2008) "Optimal complexity reduction of polyhedral piecewise affine systems." *Automatica*, vol. 44, no. 7: pp. 1728-1740.
[15] Grancharova, A., Johansen, T. A. & Kocijan, J. (2004)"Explicit model predictive control of gas-liquid separation plant", *Computers & Chemical Engineering*, Vol. 28, No. 12,pp. 2481-2491.
[16] Grancharova, A. & Johansen, T. A. (2005) "Survey of explicit approaches to constrained optimal control", *Switching and Learning in Feedback Systems*, R. Shorten and R. Murray-Smith, Ed., LNCS 3355, Springer Verlag
[17] Hegrenas, O., Gravdahl, J. T. & Tondel, P. (2007) " Spacecraft attitude control using explicit model predictive control." *Automatica*, Vol. 43, No. 12, pp. 2107-2114 .
[18] Johansen, T. A., Jackson, W., Schreiber, R. & Tondel, P. (2007) "Hardware Synthesis of Explicit Model Predictive Controllers." *IEEE Transactions Control Systems Technology*, vol. 15, pp. 191-197.
[19] Johansen, T. A., Petersen I. & Slupphaug, O. (2002) "Explicit suboptimal linear quadratic regulation with input and state constraints," *Automatica*, vol. 38(7), 1099-1112.
[20] Johansen, T. A., & Grancharova, A. (2003) "Approximate explicit constrained linear model predictive control via orthogonal search tree." *IEEE Trans. Automatic Control*, vol.48, No.5, pp.810-815.
[21] Jones,C. N., Grieder, P. & Rakovic, S. (2006) "A Logarithmic-Time Solution to the Point Location Problem for Parametric Linear Programming." *Automatica*, vol. 42, no. 12, pp. 2215-2218.
[22] Julian, P., Desages, A. & Agamennoni, O. (1999). "High-level canonical piecewise linear representation using a simplicial partition." *IEEE Transactions on Circuits Systems I*, 46(4) 463-480.
[23] Kvasnica, M., Grieder P. & Baotić, M. (2004) "Multi-parametric toolbox (MPT)," http://control.ee.ethz.ch/ mpt/.
[24] Li X., Wang, S. & Yin, W. (2001) "A canonical representation of high-dimensional continuous piecewise-linear functions." *IEEE Transactions on Circuits Systems I*, 48(11), 1347-1351.
[25] Lin J. & Unbehauen, R. (1995)"Canonical Piecewise-Linear networks. *IEEE Trans. Neural Networks. I*, 6(1), 43-50.

[26] Lincoln B. & Rantzer, A. (2006) "Relaxed dynamic programming." *IEEE Transactions on Automatic Control*, vol. 51 (8), 1249-1260.

[27] Lincoln B. & Rantzer, A. (2002) "Suboptimal dynamic programming with error bounds." *Proceedings of the IEEE conference on decision and control*, Las Vegas, NV

[28] Mayne, D. Q., Rawlings, J. B., Rao, C. V., & Scokaert, P. O. M. (2000) "Constrained model predictive control: Stability and optimality." *Automatica*, Vol. 36, No. 6, pp. 789-814.

[29] Morari, M. & Lee, J (1999) "Model predictive control: past, present and future." *Computers & Chemical Engineering*, Vol. 23, No. 4-5, pp. 667-682.

[30] Necoara, I., Schutter, B. D., Boom, T. J. J. & Hellendoorn, H. (2008) "Model predictive control for uncertain max-min-plus-scaling systems," *International Journal of Control*, vol. 81, no. 5, pp. 701-713.

[31] Ovchinnikov, S. (2002) "Max-Min representation of piecewise linear functions," *Contributions to Algebra and Geometry*, vol. 43, no. 1, pp. 297Ű302.

[32] Pannocchia, G., Rawlings, J. B. & Wright, S. J. (2007) "Fast, large-scale model predictive control by partial enumeration." *Automatica*, Vol. 43, No. 5, pp. 852-860.

[33] Pistikopoulos, E. N., Dua, V., Bozinis, N. A., Bemporad A. & Morari, M. (2001) "On-line optimization via off-line parametric optimization tools." *Computers and Chemical Engineering*, vol. 26: 175-185.

[34] Ryu, J. & Pistikopoulos, E. N. (2007) " A novel approach to scheduling of zero-wait batch processes under processing time variations." *Computers & Chemical Engineering*, Vol. 31, No. 3, pp. 101-106.

[35] Saffer, D. R. II & Doyle, F. J. III, (2004) "Analysis of linear programming in model predictive control," *Computers and Chemical Engineering*, vol. 28: 2749-2763.

[36] Spjotvold, J., Rakovic, S. V., Tondel, P., Johansen, T. A. (2006) "Utilizing Reachability Analysis in Point Location Problems." *IEEE Conf. Decision and Control*, San Diego.

[37] Spjotvold, J., Kerrigan, E. C., Jones, C. N., Tondel, P. & Johansen, T. A. (2006) "On the Facet-to-Facet Property of Solutions to Convex Parametric Quadratic Programs", *Automatica*, Vol. 42, pp. 2209-2204.

[38] Spjotvold, J., Tondel, P. & Johansen, T. A. (2007) "A continuous selection and unique polyhedral representation of solutions to convex multiparametric quadratic programs", *Journal of Optimization Theory and Applications*, Vol. 134, pp. 177-189.

[39] Tarela, J. M. & Martínez, M. V. (1999) "Region configurations for realizability of lattice piecewise-linear models," *Mathematical and Computer Modelling*, Vol.30 (11-12), pp.17-27.

[40] Tondel, P., Johansen T. A. & Bemporad, A. (2003) "Evaluation of piecewise affine control via binary search tree." *Automatica*, Vol. 39, pp. 743-749.

[41] Wen, C., Wang, S., Zhang, H. & Khan, M. J. (2005a) "A novel compact piecewise-linear representation." *International Journal of Circuit Theory and Applications* Vol.33(1), 87-97.

[42] Wen, C., Wang, S., Li, F.& Khan, M. J. (2005b) "A compact $f-f$ model of high-dimensional piecewise-linear function over a degenerate intersection." *IEEE Transactions on Circuits Systems I*, Vol. 52(4), 815-821.

[43] Wen, C., Wang, S., Sun, X., Zhang, H., & Li, F. (2005c) "A global compact representation for three-dimensional continuous piecewise linear functions", *Journal of Tsinghua University*, 45 (4): 533-536.

[44] Wen, C., Wang, S. (2005d) "A complete 3-D canonical piecewise-linear representation", *International Conference on Technology and Automation- ICTA'05*, Thessaloniki, Greece.

[45] Wen, C. (2006) *A canonical piecewise-linear representation theorem and its applications in function approximation*, Ph.D Theses, Automation Department, Tsinghua University.

[46] Wen, C, Wang, S., Jin X. & Ma, X. (2007) "Identification of dynamic systems using piecewise-affine basis function models." *Automatica*, vol. 43(10): 1824-1831.

[47] Wen, C & Ma, X. (2008a) "A max-piecewise-linear neural network for function approximation." *Neurocomputing*, Vol. 71(4-6), 843-852.

[48] Wen, C.,& Ma, X. (2008b) "A Basis-Function Canonical Piecewise-Linear Approximation." *IEEE Transactions on Circuits and Systems I: Regular Papers*, Vol. 55, no. 5, pp. 1328-1334.

[49] Wen, C., Ma, X. & Ydstie, B. E. (2009a)"Analytical expression of explicit MPC solution via lattice piecewise-affine function",*Automatica*, Vol. 45, No. 4, pp. 910-917.

[50] Wen, C., & Ydstie, B. E. (2009b) "Lattice piecewise-affine representation for explicit model predictive control", *AIChE Annual Meeting*, Nashville, TN, November.

Part 2

Successful Applications of Model Predictive Control

11

Model Predictive Control Strategies for Batch Sugar Crystallization Process

Luis Alberto Paz Suárez[1], Petia Georgieva[2] and Sebastião Feyo de Azevedo[2]
[1]*Faculty of Engeneering, University of Porto,*
[2]*Institute of Electronic Engineering and Telematics of Aveiro,*
Portugal

1. Introduction

The industrial processes are governed generally by general principles of the physics and chemistry. With the aid of data acquisition systems supported in microprocessor it is possible to obtain real data of the industrial process, that it characterizes in detail his dynamics and input-output dependency. Several methods of identification allow, from these data, to obtain linear and nonlinear models of these processes (Rossiter, 2003; Morari, 1994); which are the base to predict the process behaviour within all the family of the model based predictive controllers (MPC).
Diverse algorithms MPC have demonstrated its effectiveness in those control loops characterized by strong nonlinearities, difficult dynamic, inverse answers and great delay; that they are generally those of greater influence in the final product quality and the process efficiency (Allgöwer et al., 2004; Qin & Badgwell 2003).
One of the most important steps in the implementation of a MPC is just the obtaining of the model that can predict with reliability the future behaviour of the controlled variable, like answer to a predefined optimized control action (Rawlings 2000). This work applies two kind of MPC: (i) Classical Model-Based Predictive Control and (ii) Neural Network Model Predictive Control (NNMPC).
The classical MPC strategy uses a discrete model obtained from general phenomenological model of the feed-batch crystallization process, consisting of mass, energy and population balance. The NNMPC strategy uses to obtain a neural network, the training algorithms proposed in the Neural Network Toolbox of MatLab (version 7.04) (Bemporad et al., 2005).
In this particular case it is analyzed a fed-batch sugar crystallization process, in this process there is abundant information, detailed mathematical models and real industrial data. (Chorão, 1995; Feyo de Azevedo & Gonçalves 1988; Georgieva et al., 2003). This fact motivated the use of the neural networks to model the process and to propose a neural network MPC (NNMPC) that considers the process like a gray box, of which has input-output information and the historical experience of he process behaviour.

2. Batch sugar crystallization process

2.1 General description

The operation of crystallization is applied in the sugar industry to obtain the sucrose dissolved in the extracted juice of the sugar cane or the sugar beet basically.

Typical industrial fed-batch evaporative sugar crystallization is performed in a vacuum pan crystallizer. The reactor has a cylindrical form with volume that can vary between 20-60 m3. The feed system is usually equipped with an extra water input to dilute the sugar solution if necessary. The heat transfer system is a calandria type, to permit the heat interchange between steam and suspension. The vacuum pressure in the pan is generated by the contact barometric condenser and the pan is equipped with a mechanical agitator to keep the suspension homogeneous. The operation is conducted in a fed-batch mode with an average duration of a cycle about 90 minutes.

Sugar crystallization occurs through the mechanisms of nucleation, growth and agglomeration. In the course of production, the crystallization phenomenon is driven by two mechanisms (Jancic & Grootscholten, 1984): i) mass transfer from dissolved sucrose to crystal surface and ii) heat transfer in the calandria. Shortly before the grain setting and continuing during the beginning of the crystallization phase, the available crystalline surface to deposit the molecule of sucrose is much smaller than the mass of dissolved sucrose. During this period the evaporation rate is high, the crystal area/mass of crystallized sucrose rate is very low, therefore the process is driven by the mass transfer. The supersaturation tends to increase and if not controlled, it often achieves the undesirable zone of secondary crystal nucleation. Later on, when the total crystal area and the crystallization capacity increases, the crystal area/mass of crystallized sucrose rate gets high and the process is driven by the heat transfer.

The process objectives are to maximize the speed of crystal growth, keeping high the produced sugar quality and minimizing the costs and losses. These objectives must be fulfilled without occurrence of secondary nucleation or agglomeration. The sugar quality is evaluated by the particle size distribution (PSD) at the end of the process which is quantified by two parameters - the final average (in mass) particle size (MA) and the final coefficient of particle variation (CV). The main challenge of the sugar production is the large batch to batch variation of the final PSD. This lack of process repeatability is caused mainly by improper control policy and results in product recycling and loss increase. The sugar production is heuristically operated, and while the traditionally applied PI(D) controllers are still the preferred solutions they usually lead to energy and material loss that can easily be reduced if an optimized operation policy is implemented. These problems constitute the main motivation for the operation strategy formulated in the next section.

2.2 Crystallization model

The general phenomenological model of the fed-batch crystallization process consists of mass, energy and population balances, including the relevant kinetic rates for nucleation, linear growth and agglomeration (Simoglou et al., 2005). While the mass and energy balances are common expressions in many chemical process models, the population balance is related with the crystallization phenomenon, which is still an open modelling problem. The Appendix A shows a detailed phenomenological model for crystallization process.

2.3 Problem formulation

The final values of the crystal size distribution function (CSD) parameters: mass averaged crystal size (MA) and coefficient of variation (CV) are the best indicators of the quality and efficiency of the crystallization process. The direct measurement and control of these parameters are very difficult to make actually, in fact there are no references of its industrial implementation. The most used solution in the sugar industry consists of establishing a strategy that manipulate other variables; which allows to arrive at the end of the process with acceptable values in the CSD parameters.

The batch operation imposes to the process frequent operational changes that depend of: the quality of the raw material, disturbances in the work conditions and market demand changes. The previous problem, the nonlinearities and the restrictions imposed to the process motivated the use a nonlinear MPC (NMPC).

When a NMPC algorithm is applied, the first challenge consists of obtaining the model to use, which must be viable and trustworthy. Although the sugar crystallization process has been studied in depth and efficient mathematical models exist to represent it, these must be validated and be fit before its application in a NMPC algorithm, which will cause frequent updates if the process is batch.

Like an alternative, in this work it is tried to demonstrate the efficiency that has the use of the neuronal networks in a NMPC, where the neural networks could be trained from industrial data with the input-output answer of the process.

3. Problem solution

Sugar production is characterized by strongly non-linear and non-stationary dynamics and goes naturally through a sequence of relatively independent stages: charging, concentration, seeding, setting the grain, crystallization (the main phase), tightening and discharge (Georgieva et al., 2003). Therefore the operation strategy is formulated as a cascade of individual control loops for each of the stages (Fig. 1). The feedback control policy is based on measurements of the flowrate, the temperature, the pressure, the stirrer power and the supersaturation (by a refractometer). Measurements of these variables are usually available for a conventional crystallizer.

3.1 Operation strategy

Sugar production is still a very heuristically operated process, with classical proportional integral and eventually derivative (PID) controllers being the most typical solution. The different phases of the sugar production are comparatively independent and moved by distinct driving forces, thus a single controller can hardly be effective for the complete process. Instead, individual controllers for each stage where it seems appropriate, was the adopted framework (Fig. 1). See Table 1 for more details on the formulated operation strategy.

In the present study, the control actions are performed by manipulating the valves of the liquor/syrup feed flowrates (Ff) and the steam flowrate (Fs), while the volume of massecuite (Vm), the supersaturation (S) and the current of the agitator (IA) are the controlled variables. This choice is completely inspired by the industrial practice in several refineries.

Charging (stage 1): During the first stage the crystallizer is fed with liquor until it covers approximately 40 % of the vessel height. The process starts with vacuum pressure of around 1 bar (equal to the atmospheric pressure) and reduces it up to 0.23 bar. When the vacuum pressure reaches 0.5 bar, the feed valve is completely open such that the feed flowrate is kept at its maximum value. When the liquor covers 40 % of the vessel height, the feed valve is closed and the vacuum pressure needs some time to stabilize around the value of 0.23 bar before the concentration stage starts.

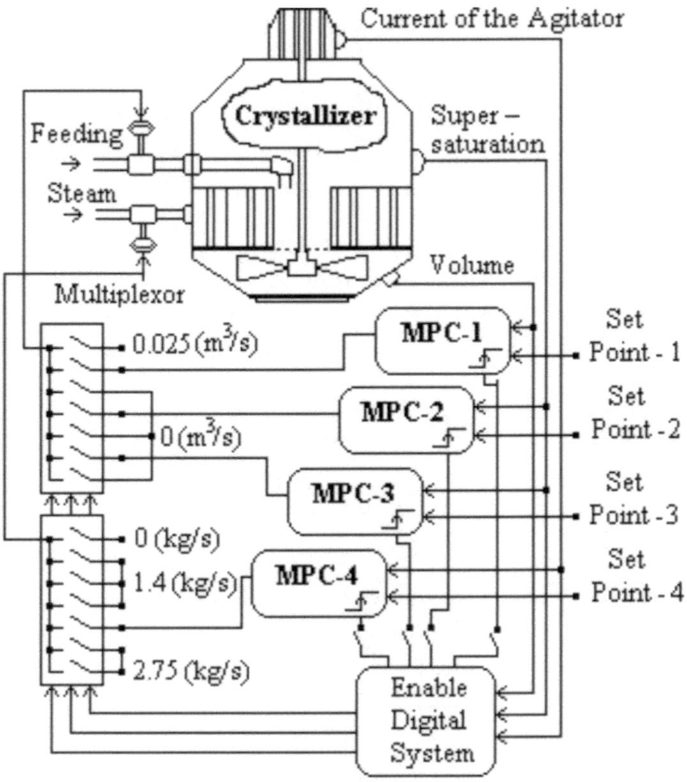

Fig. 1. Cascade MPC control - strategy

Concentration (stage 2): Once the vacuum pressure stabilizes, the stirrer is switched on and the concentration begins. In order to guarantee unperturbed operation of the barometric condenser and the steam production boiler, the steam flowrate must increase slowly (from 0 to 2 *kg/s*, in two minutes approximately). The concentration of the dissolved sucrose by evaporation under vacuum results in volume reduction. However, for technological reasons, the minimum suspension level of the pan must be above the calandria. Therefore a feed flowrate action is required to control the level (the volume) of the pan and this constitutes the *first control loop*. In this stage, the supersaturation increases rapidly (at about a rate of 0.025 per *min.*). When it reaches a value of 1.06, the feeding is stopped and the steam flowrate is reduced slowly to 1.4 *kg/s*, with the same speed as it was increased. The concentration stage is over when the supersaturation reaches the value of 1.11.

Seeding (stage 3): At this moment seed crystals are introduced into the pan to provoke crystallization. This stage is rather unstable and to prevent seed crystals from dissolution in the liquor, the feed valve must be closed and the steam flowrate kept at its minimum for a short period (about 2 *min.*). Keeping these conditions unchanged contributes to the formation of the grain and is also important for the final crystal size distribution. The supersaturation continues naturally to increase but usually no control action is required.

Crystallization with liquor (stage 4): During this stage the supersaturation is first controlled by a proper feeding to be around a set point of 1.15. This constitutes the *second control loop*. At the beginning of this stage, the mass transfer is the driving crystallization force, the crystallization rate increases and the controller usually reduces the feed flowrate

Stage	Action	Control
Charge	The steam valve is closed and the stirrer is off. The vacuum pressure changes from 1 to 0.23 *bar*. The vacuum pressure reaches 0.5 *bar*, feeding starts with max rate. Liquor covers 40 % of the vessel height.	No control The feed valve is completely open
Concentration	The vacuum pressure stabilizes around 0.23 *bar*. The stirrer is on. The volume is kept constant. The steam flowrate increases to 2 *kg/s* The supersaturation reaches 1.06, the feeding is closed, the steam flowrate is reduced to 1.4 *kg/s*	*Control loop 1* Controlled variable: Volume; Manipulated variable: liquor feed flowrate
Seeding and setting the grain	The supersaturation reaches 1.11. Seed crystals are introduced. The steam flowrate is kept at the minimum for two minutes.	No control The feed valve is closed
Crystallization with liquor (phase 1)	The steam flowrate is kept around 1.4 *kg/s*. The supersaturation is controlled at the set point 1.15.	*Control loop 2* Controlled variable: supersaturation Manipulated variable: liquor feed flowrate
Crystallization with liquor (phase 2)	The volume of crystallizer reaches ≈ 22 m^3. The feed valve is closed. The supersaturation is controlled at the set point 1.15. The stirrer power reaches 20.5 *A*.	*Control loop 3* Controlled variable: supersaturation Manipulated variable: steam flowrate
Crystallization with syrup	The steam flowrate is kept around the maximum of 2.75 *kg/s*. (*hard constraint*). The volume fraction of crystals is kept at the set point 0.45. The volume reaches its maximum value (30 m^3) The feed valve is close.	*Control loop 4* Controlled variable: volume fraction of crystals. Manipulated variable: syrup feed flowrate
Tightening	The stirrer power reaches the maximum value of 50 *A* (*hard constraint*). The steam valve is closed. The stirrer and the barometric condenser are stopped.	No control

Table 1. Summary of the sugar crystallization operation strategy.

to maintain the reference value of the supersaturation. When all liquor quantity is introduced, the feeding is stopped and the supersaturation is now kept at the same set point of 1.15 by the steam flowrate as the manipulated variable. This constitutes the *third control loop*. The heat transfer is now the driving crystallization force. A typical problem of this control loop is that at the end of this stage the steam flowrate achieves its maximum value of 2.75 *kg/s* but it is not sufficient to keep the supersaturation at the same reference value therefore a reduction of the set point is required. The stage is over when the stirrer power reaches the value 20.5 *A*.

Crystallization with syrup (stage 5): A stirrer power of 20.5*A* corresponds to a volume fraction of crystals equal to 0.4. At this moment the feed valve is reopened, but now a juice with less purity (termed syrup) is introduced into the pan until the maximum volume (30 m^3) is reached. The control objective is to maintain the volume fraction of crystals around the set point of 0.45 by a proper syrup feeding. This constitutes the *fourth control loop*.

Tightening (stage 6): Once the pan is full the feeding is closed. The tightening stage consists principally in waiting until the suspension reaches the reference consistency, which corresponds to a volume fraction of crystals equal to 0.5. The supersaturation is not a controlled variable at this stage because due to the current conditions in the crystallizer, the crystallization rate is high and it prevents the supersaturation of going out of the metastable zone. The stage is over when the stirrer power reaches the maximum value of 50 *A*. The steam valve is closed, the water pump of the barometric condenser and the stirrer are turned off. Now the suspension is ready to be unloaded and centrifuged.

4. Model based predictive control

The term model-based predictive control (MPC) does not refer to a particular control method, instead it corresponds to a general control approach (Rossiter, 2003). The MPC concept, introduced in late seventies, nowadays has evolved to a mature level and became an attractive control strategy implemented in a variety of process industries (Camacho & Bordons, 2004). The main difference between the MPC configurations is the model used to predict the future behavior of the process or the implemented optimization procedure. First the MPC based on linear models gained popularity (Morari, 1994) as an industrial alternative to the classical proportional-integral-derivative (PID) control and later on nonlinear cases as reactive distillation columns (Balasubramhanya & Doyle, 2000) and polymerization reactors (Seki et al., 2001) were reported as successfully MPC controlled processes.

4.1 Classical model based predictive control

The main difference between MPC configurations is the model used to predict the future behaviour of the process and the optimization procedure. Nonlinear model predictive control (NMPC) is an optimisation-based multivariable constrained control technique that uses a nonlinear dynamic model for the prediction of the process outputs (Qin & Badgwell, 2003). At each sampling time k the model predicts future process responses to potential control signals over the prediction horizon (H_p). The predictions are supplied to an optimization procedure, to determine the values of the control action over a specified control horizon (H_c) that minimizes the following performance index:

$$\min_{u_{\min} \leq [u_c(k), u_c(k+1), \ldots u_c(H_c)] \leq u_{\max}} J = \lambda_1 \sum_{k=1}^{H_p} \left(y_r(k) - \hat{y}(k)\right)^2 - \lambda_2 \sum_{k=1}^{H_c} \left(u_c(k-1) - u_c(k-2)\right)^2 \tag{1}$$

Subject to the following constrains

$$u_{\min} \leq u_c \leq u_{\max} \tag{2}$$

$$\Delta u_{\min} \leq \Delta u \leq \Delta u_{\max} \tag{3}$$

$$y_{\min} \leq y_p \leq y_{\max} \tag{4}$$

Where $u_{\min}$ and $u_{\max}$ are the limits of the control inputs, $\Delta u_{\min}$ and $\Delta u_{\max}$ are the minimum and the maximum values of the rate-of-change of the inputs and $y_{\min}$ and $y_{\max}$ are the minimum and maximum values of the process outputs.

H_p is the number of time steps over which the prediction errors are minimized and the control horizon H_c is the number of time steps over which the control increments are minimized, y_r is the desired response (the reference) and $\hat{y}$ is the predicted process output (Diehl et al., 2002). $u_c(k), u_c(k+1), u_c(H_c)$ are tentative future values of the control input, which are parameterized as peace wise constant. The length of the prediction horizon is crucial for achieving tracking and stability. For small values of H_p the tracking deteriorates but for high H_p values the bang-bang behavior of the process input may be a real problem. The MPC controller requires a significant amount of on-line computation, since the optimization (1) is performed at each sample time to compute the optimal control input. At each step only the first control action is implemented to the process, the prediction horizon is shifted or shrunk by usually one sampling time into the future, and the previous steps are repeated (Rossiter, 2003). λ_1 and λ_2 are the output and the input weights respectively, which determine the contribution of each of the components of the performance index (1).

4.2 Neural network model predictive control

The need for neural networks arises when dealing with non-linear systems for which the linear controllers and models do not satisfy. Two main achievements contributed to the increasing popularity of the NNs: (i) The proof of their universal approximation properties and the development of suitable algorithms for NN training as the backpropagation and (ii) The adaptation of the Levenberg-Marquard algorithm for NN optimization.

The most used NN structures are Feedforward networks (FFNN) and Recurrent (RNN) ones. The RNNs offer a better suited tool for nonlinear system modelling and is implemented in this work (Fig.2). The Levenberg-Marquard (LM) algorithm was preferred as the training method due to its advantages in terms of execution time and robustness. Since the LM algorithm requires a lot of memory, a powerful (in terms of memory) computer is the main condition for successful training. In order to solve the problem of several local minima, that is typical for all derivative based optimization algorithms (including the LM method), we have repeated several time the optimization specifying different starting points.

The individual stages of the crystallization process are approximated by different RNNs of the type shown in Fig. 2. Tangent sigmoid hyperbolic activation functions are the hidden computational nodes (Layer 1) and a linear function is located at the output (Layer 2). Each NN has two vector inputs (r and p) formed by past values of the process input and the NN output respectively. The architecture of the NN models trained to represent different process stages is summarized as follows:

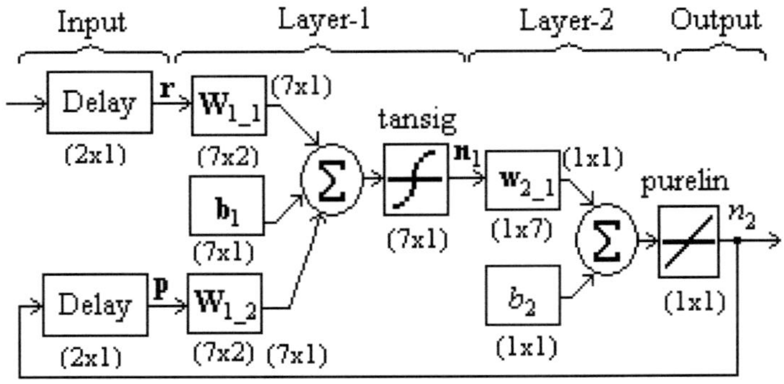

Fig. 2. Neural network architecture

$$u_{NN} = [r, p] = [u_c(k-1), u_c(k-2), y_{NN}(k-1), y_{NN}(k-2)] \tag{5}$$

$$x = W_{11}r + W_{12}p + b_1 \tag{6}$$

$$n_1 = \left(e^x - e^{-x}\right) / \left(e^x + e^{-x}\right) \tag{7}$$

$$n_2 = w_{21}n_1 + b_2 \tag{8}$$

Where $W_{11} \in R^{m\times 2}$, $W_{12} \in R^{m\times 2}$, $w_{21} \in R^{1\times m}$, $b_1 \in R^{m\times 1}$, $b_2 \in R$ are the network weights (in matrix form) to be adjusted during the NN training, m is the number of nodes in the hidden layer.

Since the objective is to study the influence of the NNs on the controller performance, a number of NN models is considered based on different training data sheets.

- **Case 1 (Generated data):** Randomly generated bounded inputs (u_i) are introduced to a simulator of a general evaporative sugar crystallization process introduced in Georgieva et al., 2003. It is a system of nonlinear differential equations for the mass and energy balances with the operation parameters computed based on empirical relations (for no stationary parameters) or keeping constant values (for stationary parameters). The simulator responses are recorded (y_i) and the respective mean values are computed ($u_{i,mean}$, $y_{i,mean}$). Then the NN is trained supplying as inputs $u_i - u_{i,mean}$ and as target outputs $y_i - y_{i,mean}$.
- **Case 2: Industrial data:** The NN is trained with real industrial data. In order to extract the underlying nonlinear process dynamics a prepossessing of the initial industrial data was performed. From the complete time series corresponding to the input signal of one stage only the portion that really excites the process output of the same stage is extracted. Hence, long periods of constant (steady-state) behavior are discarded. Since, the steady-state periods for normal operation are usually preceded by transient intervals, the data base constructed consists (in average) of 60-70% of transient period data. A number of sub cases are considered.
    - **Case 2.1:** Industrial data of two batches is used for NN training.
    - **Case 2.2:** Industrial data of four batches is used for NN training.
    - **Case 2.3:** Industrial data of six batches is used for NN training.

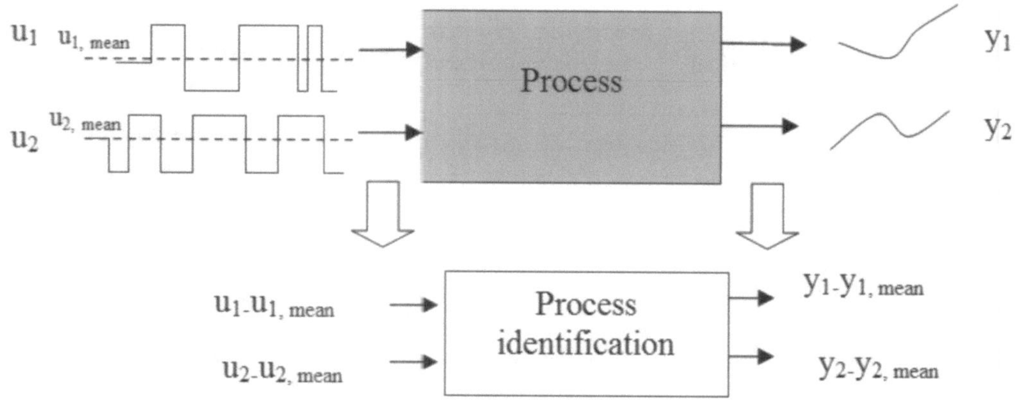

Fig. 3. Case1: NN data generation

4.3 Selection of MPC parameters: H_p, H_c, λ_2

The choice of H_p is related with the sampling period (Δt) of the digital control implementation, which in its turn is a function of the settling time t_s (the time before entering into the 5% around the set-point) of the closed loop system. As a rule of thumb, it is suggested Δt to be chosen at least 10 times smaller than t_s, (Soeterboek, 1992). Hence, the prediction horizon can be chosen as H_p = round-to-integer(t_s/ Δt). It is well known that the smaller the sampling time, the better can a reference trajectory be tracked or a disturbance rejected. However, choosing a small sampling time yields a large prediction horizon. In order to compute the optimal control input, the optimization (1) is performed at each sampling time, therefore MPC controller requires a significant amount of on-line computation. This can cause problems related with large amount of computer memory required and additional numerical problems due to the large prediction horizon. The introduction of the ET MPC as in (7) serves as a compromise between these conflicting issues and reduces significantly the computational efforts.

Parameters λ_1 and λ_2 determine the contribution (the weight) of each term of the performance index, the output error (e) and the control increments (Δu). In this work the parameter λ_1 is set to the normalized value of 1, while the choice of λ_2 is based on the following empirical expression:

$$(u_{\max} - u_{\min})^2 \cdot \lambda_2 = e_{\max} \cdot P/100 \tag{9}$$

where P defines the desired contribution of the second term in (1) (0% ≤ P ≤100%) and

$$e_{\max} = \max\left((ref - y_{\max})^2, (ref - y_{\min})^2\right) \tag{10}$$

The intuition behind (9-10) is to make the two terms of (1) compatible when they are not normalized and to overcome the problem of different numerical ranges for the two terms. Table 2 summarize the set of MPC parameters used in the four control loops define in the section 3.

Control loop (CL)	t_s (s) settling time	Δt (s) sampling period	H_p prediction horizon	H_c control horizon	λ_2 weight	Controlled variable	Set-point
CL1	40	4	10	2	1000	Volume	12.15
CL2	40	4	10	2	0.1	Supersaturation	1.15
CL3	60	4	15	2	0.01	Supersaturation	1.15
CL4	80	4	20	2	10000	Fraction of crystals	0.43

Table 2. MPC design parameters for the control loops define in Table 1

5. PID controllers

The PID parameters were tuned, where k_p, τ_i, τ_d are related with the general PID terminology as follows (Aström & Hägglund, 1995):

$$u(t+k) = K_p\left[e(t+k) + \frac{\Delta t}{\tau_i}\cdot\sum_{i=0}^{k} e(t+i) + \frac{\tau_d}{\Delta t}\cdot\left(e(t+k) - e(t+k-1)\right)\right] \tag{11}$$

Since the process is nonlinear, classical (linear) tuning procedures were substituted by a numerical optimization of the integral (or sum in the discrete version) of the absolute error (IAE):

$$IAE = \sum_{k=1}^{N}\left|ref(t+k) - y_p(t+k)\right| \tag{12}$$

Equation (12) was minimized in a closed loop framework between the discrete process model and the PID controller. For each parameter an interval of possible values was defined based on empirical knowledge and the process operator expertise. A number of gradient (Newton-like) optimization methods were employed to compute the final values of each controllers summarized in Table 1. All methods concluded that the derivative part of the controller is not necessary. Hence, PI controllers were analyzed in the next tests.

	Control loop 1	Control loop 2	Control loop 3	Control loop 4
k_p	0.05	-0.5	20	-0.01
τ_i	30	40	10	70
τ_d	0	0	0	0

Table 3. Optimized PID parameters for the control loops define in Table 1

6. Discussion of results

The operation strategy, summarized in Table 1 and implemented by a sequence of Classical-MPC, NNMPC or PI controllers is comparatively tested in Matlab environment. The output predictions are provided either by a simplified discrete model (with the main operation parameters kept constant) or by a trained ANN model (5-8). A process simulator was developed based on a detailed phenomenological model (Georgieva et al., 2003). Realistic disturbances and noise are introduced substituting the analytical expressions for the vacuum pressure, brix and temperature of the feed flow, pressure and temperature of the steam with original industrial data (without any preprocessing(Scenario-2)). The test is implemented for two different scenarios of work.

- **Scenario - 1:** The simulation uses, like process, the set of equations differentials proposed in (Georgieva et al. 2003) with empirical operation parameters.
- **Scenario - 2:** The simulation uses, like process, the set of equations differentials proposed in (Georgieva et al. 2003), but are used like operation parameter e real industrial data batch not used in neural network training.

Time trajectories of the controlled and the manipulated variables for the control loop 1, 2 and 4 of one batch (Batch 1) are depicted in Figs. 4-6. The three controllers guarantee good set point tracking. However, the quality of the produced sugar is evaluated only at the process end by the crystal size distribution (CSD) parameters, namely AM and CV. The results are summarized in Table 4 and both classical and NNPMC outperform the PI. Our general conclusion is that the main benefits of the MPC strategy are with respect to the batch end point performance.

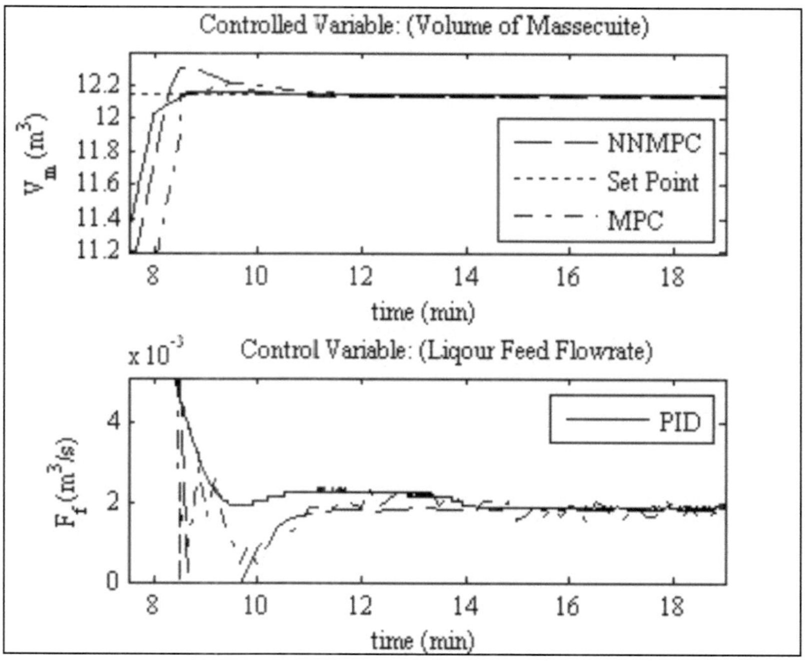

Fig. 4. Controlled (Volume of massecuite) and control variables (F_f- feed flowrate) over time for the 1st control loop.

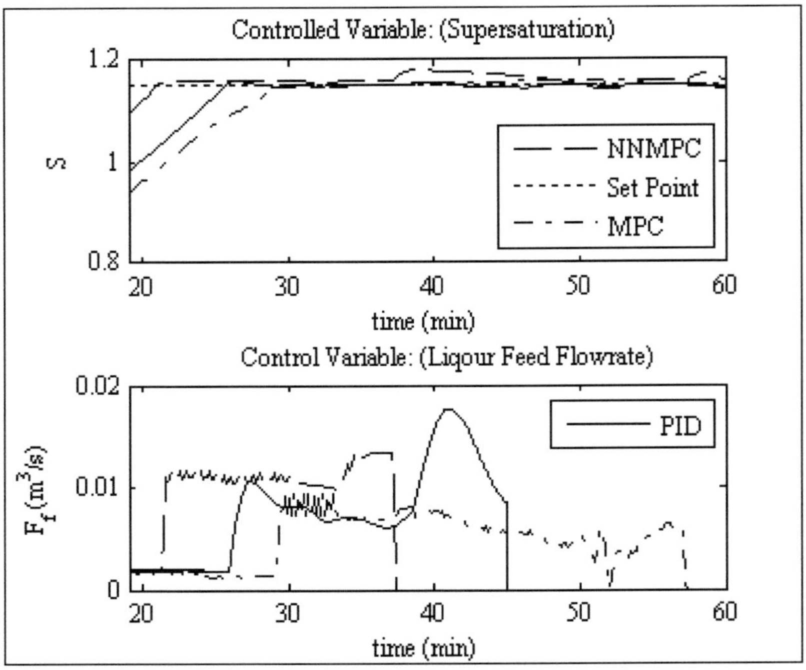

Fig. 5. Controlled (Supersaturation) and control variables (F_f- feed flowrate) over time for the 2nd control loop.

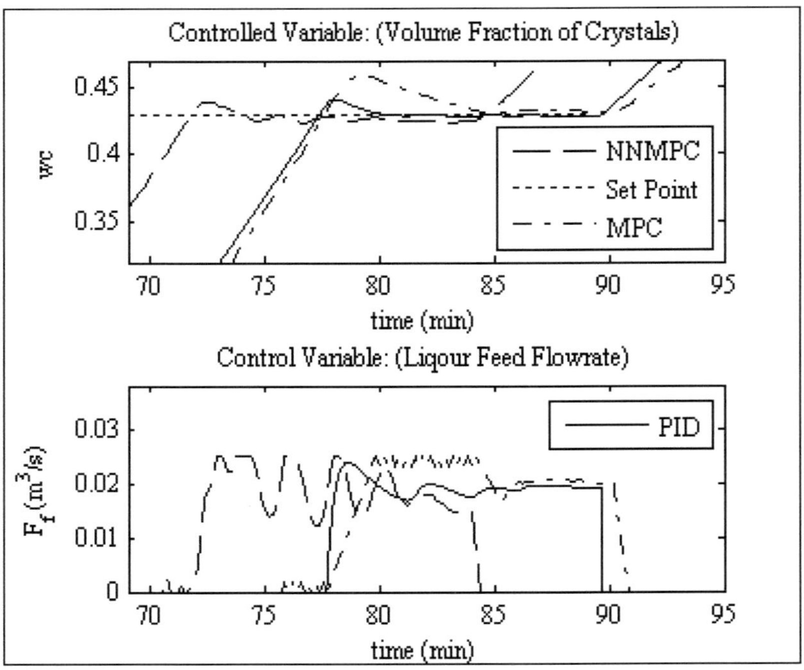

Fig. 6. Controlled (Volume fraction of crystals) and control variables (F_f- feed flowrate) over time for the 4th control loop.

Performance measures	Classical MPC	NN-MPC	PI
AM (mm) (reference 0.56)	0.586	0.584	0.590
CV (%)	32.17	31.13	32.96

Table 4-1. Batch end point performance measures (Batch - 1)

Performance measures	Classical MPC	NN-MPC	PI
AM (mm) (reference 0.56)	0.615	0.609	0.613
CV (%)	29.39	30.28	31.14

Table 4-2. Batch end point performance measures (Batch - 2)

Performance measures	Classical MPC	NN-MPC	PI
AM (mm) (reference 0.56)	0.636	0.631	0.639
CV (%)	28.74	29.42	29.23

Table 4-3. Batch end point performance measures (Batch - 3)

7. Conclusion

With the results obtained in this work it has been demonstrated that algorithm NNMPC is a viable solution to control nonlinear complexes processes, still in the case that only exists input-output information of the process.
An aspect very important to obtain successful results with NNMPC is the representative quality of the available data, which was demonstrated with the results obtained in the third control loop analyzed.
The weighting factor λ_2 has a crucial paper in the good NNMPC performance. A constrain very hard can impose that the control signal can not follow the dynamics of the process, but a very soft constrain can cause instability in the control signal, when the model is not precise.

8. Acknowledgment

Several institutions contributed for this study: 1) Foundation of Science and Technology of Portugal, which financed the scholarship of investigation of doctorate SFR/16175/2004; 2) Laboratory for Process, Environmental and Energy Engineering (LEPAE), Department of Chemical Engineering, University of Porto; 3) The Institute of Electronic Engineering and Telematics of Aveiro (IEETA); 4) Sugar refinery RAR, Portugal; The authors are thankful to all of them.

9. Appendix A. Crystallization model

Sugar crystallization occurs through the mechanisms of nucleation, growth and agglomeration. The general phenomenological model of the fed-batch crystallization process consists of mass, energy and population balances, including the relevant kinetic rates for nucleation, linear growth and agglomeration [Ilchmann, et al., 1994]. While the mass and energy balances are common expressions in many chemical process models, the population balance is related with the crystallization phenomenon, which is still an open modeling problem.

Mass balance

The mass of all participating solid and dissolved substances are included in a set of conservation mass balance equations:

$$\dot{M} = f_1(M(t),F(t),S_1(t)), \quad t_0 \le t \le t_f, \quad M(0) = M_0 \tag{A-1}$$

where $M(t) \in \Re^q$ and $F(t) \in \Re^m$ are the mass and the flow rate vectors, with q and m dimensions respectively, and t_f is the final batch time. $S_1(t) \in \Re^{r1}$ is the vector of physical time dependent parameters as density, viscosity, purity, etc. For the process in hand, the detailed form of the macro-model (A1) is as follows

$$M_{sol} = M_a + M_i + M_w \tag{A-2}$$

$$M_m = M_{sol} + M_c \tag{A-3}$$

$$\frac{dM_w}{dt} = F_f \rho_f \left(1 - B_f\right) + F_w \rho_w - J_{vap} \tag{A-4}$$

$$\frac{dM_i}{dt} = F_f \cdot \rho_f \cdot B_f \cdot \left(1 - Pur_f\right) \tag{A-5}$$

$$\frac{dM_a}{dt} = F_f \cdot \rho_f \cdot B_f \cdot Pur_f - J_{cris} \tag{A-6}$$

$$\frac{dM_c}{dt} = J_{cris} \tag{A-7}$$

$$V_m = \frac{M_c + M_{sol}}{\rho_{sol}} \tag{A-8}$$

$$J_{vap} = \frac{W + Q}{\lambda_{vap}} + K_{vap} \cdot \left(T_m - T_{w(vac)} - BPE\right) \tag{A-9}$$

Energy balance

The general energy balance model is

$$\frac{dT_m}{dt} = aJ_{cris} + bF_f + cJ_{vap} + d \tag{A-10}$$

where parameters *a, b, c* and *d* incorporate the enthalpy terms and specific heat capacities derived as time dependent functions of physical and thermodynamic properties as follows

$$a = \frac{H_{sol} - H_c + (1 - B_{sol})\dfrac{dH_{sol}}{dB_{sol}} + \dfrac{1 - Pur_{sol}}{B_{sol}} \cdot \dfrac{dH_{sol}}{dPur_{sol}}}{M_{sol} \cdot Cp_{sol} + M_c \cdot Cp_c} \tag{A-11}$$

$$b = \frac{\rho_f \left(H_f - H_{sol} + (B_f - B_{sol})\dfrac{dH_{sol}}{dB_{sol}} + \dfrac{B_f\left(Pur_f - Pur_{sol}\right)}{B_{sol}} \cdot \dfrac{dH_{sol}}{dPur_{sol}} \right)}{M_{sol} \cdot Cp_{sol} + M_c \cdot Cp_c} \tag{A-12}$$

$$c = \frac{H_{sol} - H_{vap} - B_{sol} \cdot \dfrac{dH_{sol}}{dB_{sol}}}{M_{sol} \cdot Cp_{sol} + M_c \cdot Cp_c} \tag{A-13}$$

$$d = \frac{W + Q + F_w \rho_w (H_w - H_{sol} + B_{sol})\dfrac{dH_{sol}}{dB_{sol}}}{M_{sol} \cdot Cp_{sol} + M_c \cdot Cp_c} \tag{A-14}$$

$$\frac{dH_{sol}}{dB_{sol}} = -29.7T_m + 4.6Pur_{sol}T_m + 0.075T_m^2 \tag{A-15}$$

$$\frac{dH_{sol}}{dPur_{sol}} = 4.61Bx_{sol}T_m \tag{A-16}$$

Population balance

Mathematical representation of the crystallization rate can be achieved through basic mass transfer considerations or by writing a population balance represented by its moment equations. Employing a population balance is generally preferred since it allows to take into account initial experimental distributions and, most significantly, to consider complex mechanisms such as those of size dispersion and/or particle agglomeration/aggregation. The basic moments of the number-volume distribution function are

$$\frac{d\tilde{\mu}_0}{dt} = \tilde{B}_0 - \frac{1}{2} \cdot \beta' \cdot \tilde{\mu}_0^2 \tag{A-17}$$

$$\frac{d\tilde{\mu}_1}{dt} = G_v \cdot \tilde{\mu}_0 \tag{A3-18}$$

$$\frac{d\tilde{\mu}_2}{dt} = 2 \cdot G_v \cdot \tilde{\mu}_1 + \beta' \cdot \tilde{\mu}_1^2 \tag{A3-19}$$

$$\frac{d\tilde{\mu}_3}{dt} = 3 \cdot G_v \cdot \tilde{\mu}_2 + 3 \cdot \beta' \cdot \tilde{\mu}_2^2 \tag{A3-20}$$

$$J_{cris} = \rho_c \cdot \frac{d\tilde{\mu}_1}{dt}, \tag{A3-21}$$

where $\tilde{B}_0$, G and β' are the kinetic variables nucleation rate, linear growth rate and the agglomeration kernel, respectively with the following mathematical descriptions

$$\tilde{B}_0 = K_n \cdot 2.894 \cdot 10^{12} \cdot G^{0.51} \left(\frac{\tilde{\mu}_1}{k_v \cdot V_m} \right)^{0.53} \cdot V_m \tag{A-22}$$

$$\beta' = \frac{K_{ag} \cdot G \cdot \tilde{\mu}_1}{V_m^2} \tag{A3-23}$$

$$G = K_g \cdot \exp\left(-\frac{57000}{R(T_m + 273)} \right) \cdot (S-1) \cdot \exp(-13.863(1 - P_{sol})) \cdot \left(1 + 2 \cdot \frac{v}{V_m} \right) \tag{A-24}$$

$$G_v = 3 \cdot k_v \left(\frac{v}{\tilde{\mu}_0} \right)^{2/3} \cdot G. \tag{A-25}$$

The crystallization quality is evaluated by the particle size distribution (PSD) at the end of the process which is quantified by two parameters - the final average (in mass) particle size (AM) and the final coefficient of particle variation (CV) with the following definitions:

$$AM = \overline{L} \tag{A-26}$$

$$CV = \frac{\sigma}{\overline{L}} \tag{A-28}$$

Where σ and $\overline{L}$ are computed from:

$$\overline{L} = \left(\frac{\eta_3}{1 + 3 \cdot \left(\frac{\sigma}{\overline{L}} \right)^2} \right)^{1/3} \tag{A-29}$$

$$15 \cdot \eta_3^2 \cdot \left(\frac{\sigma}{\overline{L}} \right)^6 + \left(45 \cdot \eta_3^2 - 9 \cdot \eta_6 \right) \left(\frac{\sigma}{\overline{L}} \right)^4 + \left(15 \cdot \eta_3^2 - 6 \cdot \eta_6 \right) \left(\frac{\sigma}{\overline{L}} \right)^2 + \eta_3^2 - \eta_6 = 0 \tag{A-30}$$

In (A-29, A-30), η_j represent moments of mass-size distribution functions, that are related to the moments of the number-volume distribution functions (μ_j by the following relationships:

$$\eta_3 = \frac{\mu_2}{k_v \cdot \mu_1}, \quad \text{(A-31)}$$

and

$$\eta_6 = \frac{\mu_3}{k_v^{\,2} \cdot \mu_1} \quad \text{(A3-32)}$$

Correlations for physical properties

$$Q = \alpha_s \cdot F_s \cdot \Delta H_s \quad \text{(A-33)}$$

$$\rho_f = \left(1000 + \frac{Bx_f \cdot (200 + Bx_f)}{54}\right) \cdot \left(1 - 0.036 \cdot \frac{T_f - 20}{160 - T_f}\right) \quad \text{(A-34)}$$

$$Cp_f = 4186.8 - 29.7 \cdot Bx_f + 4.61 \cdot Bx_f \cdot Pur_f + 0.075 \cdot Bx_f \cdot T_f \quad \text{(A-35)}$$

$$H_f = Cp_f \cdot T_f \quad \text{(A-36)}$$

$$\rho^*_{sol} = \left(1000 + \frac{Bx_{sol} \cdot (200 + Bx_{sol})}{54}\right) \cdot \left(1 - 0.036 \cdot \frac{T_m - 20}{160 - T_m}\right) \quad \text{(A-37)}$$

$$\rho_{sol} = \rho^*_{sol} + 1000 \cdot \left(-1 + \exp\left[\left(-6.927 \cdot 10^{-6} \cdot Bx_{sol}^{\,2} - 1.164 \cdot 10^{-4} \cdot Bx_{sol}\right) \cdot (Pur_{sol} - 1)\right]\right) \quad \text{(A-38)}$$

$$Cp_{sol} = 4186.8 - 29.7 \cdot Bx_{sol} + 4.61 \cdot Bx_{sol} \cdot Pur_{sol} + 0.075 \cdot Bx_{sol} \cdot T_m \quad \text{(A-39)}$$

$$H_{sol} = Cp_{sol} \cdot T_m \quad \text{(A-40)}$$

$$\rho_m = \frac{\rho_{sol} \cdot \rho_c}{\rho_c - w_c \cdot (\rho_c - \rho_{sol})} \quad \text{(A-41)}$$

$$Pur_{sol} = \frac{M_a}{M_a + M_i} \quad \text{(A-42)}$$

$$B_{sol} = \frac{M_a + M_i}{M_{sol}} \quad \text{(A-43)}$$

$$Bx_{sol} = 100 \cdot B_{sol} \quad \text{(A-44)}$$

$$Bx_{sat} = 64.447 + 8.222 \cdot 10^{-2} \cdot T_m + 1.66169 \cdot 10^{-3} \cdot T_m^{\,2} - 1.558 \cdot 10^{-6} \cdot T_m^{\,3} - 4.63 \cdot 10^{-8} \cdot T_m^{\,4} \quad \text{(A-45)}$$

$$S^* = 1.129 - 0.284 \cdot (1 - Pur_{sol}) + (2.333 - 0.0709 \cdot (T_m - 60)) \cdot (1 - Pur_{sol})^2 \quad \text{(A-46)}$$

$$S = \frac{\dfrac{Bx_{sol}}{100 - Bx_{sol}}}{\dfrac{Bx_{sat}}{100 - Bx_{sat}} \cdot C_{sat}} \quad \text{(A-47)}$$

$$C_{sat} = 0.1 \cdot \frac{Bx_{sol}}{100 - Bx_{sol}} \cdot (1 - Pur_{sol}) + 0.4 + 0.6 \cdot \exp\left(-0.24 \cdot \frac{Bx_{sol}}{100 - Bx_{sol}} \cdot (1 - Pur_{sol}) \right) \quad \text{(A-48)}$$

$$v = \frac{M_c}{\rho_c} \quad \text{(A-49)}$$

$$w_c = \frac{M_c}{M_c + M_{sol}} \quad \text{(A-50)}$$

$$Cp_c = 1163.2 + 3.488 \cdot T_m \quad \text{(A-51)}$$

$$H_c = Cp_c \cdot T_w \quad \text{(A-52)}$$

$$\rho_w = 1016.7 - 0.57 \cdot T_w \quad \text{(A-53)}$$

$$T_{w(vac)} = 122.551 \cdot \exp(-0.246 \cdot P_{vac}) \cdot (P_{vac})^{0.413} \quad \text{(A-54)}$$

$$T_{w(s)} = 100.884 \cdot \exp(-1.203 \cdot 10^{-2} \cdot P_s) \cdot (P_s)^{0.288} \quad \text{(A-55)}$$

$$\lambda_{w(vac)} = 2263.28 - 58.21 \cdot \ln(P_{vac}) \quad \text{(A-56)}$$

$$\lambda_s = 2257.51 - 85.95 \cdot \ln(P_s) \quad \text{(A-57)}$$

$$H_w = 2323.3 + 4106.7 \cdot T_w + T_w^{\ 2} \quad \text{(A-58)}$$

$$H_{w(s)} = 2323.3 + 4106.7 \cdot T_{w(s)} + T_{w(s)}^{\ 2} \quad \text{(A-59)}$$

$$H_s = 2491860 - 13270 \cdot P_s + (1946.5 + 37.9 \cdot P_s) \cdot T_s \quad \text{(A-60)}$$

$$H_{vac} = 2499980 - 24186 \cdot P_{vac} + (1891.1 + 106.1 \cdot P_{vac}) \cdot T_m \quad \text{(A-61)}$$

$$\Delta H_s = H_s + H_{w(s)} \quad \text{(A-62)}$$

$$BPE = (0.03 - 0.018 \ Pur_{sol}) \cdot (T_{w(vac)} + 84) \cdot \left(\frac{Bx_{sol}}{100 - Bx_{sol}} \right) \quad \text{(A-63)}$$

For more detailed presentation of the process model, refer to [Georgieva et al., 2003].

10. References

Allgöwer, F., Findeisen, R. & Nagy, Z. K. (2004). Nonlinear model predicitve control: From theory to application. Journal of Chinese Institute of Chemical Engineers, 35 (3), 299-315.

Aström, K. J., Hägglund, T. (1995). Pid controllers : theory, design, and tuning. North Carolina: Research Triangle Park, Instrument Society of America.

Balasubramhanya, L. S., Doyle, F. J (2000). Nonlinear model-based control of a batch reactive distillation column. Journal of Process Control, 10, 209-218.

Bemporad, A., Morari, M. & Ricker, N. L. (2005). User's Guide: Model predictive control toolbox for use with MatLab: The MathWorks Inc.

Camacho, E. F., Bordons, C. (2004). Model predictive control in the process industry. London: Springer-Verlag.

Chorão, J. M. N. 1995. Operação assistida por comutador dum cristalizador industrial de açúcar, Ph. D. Tesis, Faculdade de Engenharia, Departamento de Eng. Química, Universidade de Porto, Porto

Diehl, M., H. G. Booc, J. P. Schlder, R. Findeisen, A. Nagy, and F. Allgöwer. (2002). Real-time optimization and nonlinear model predictive control of processes governed by deferential algebraic equations. *Jornal of Process Control* 12:577–585.

Feyo de Azevedo, S., and M. J. Gonçalves. (1988). Dynamic Modelling of a Batch Evaporative Crystallizer. Recent Progrés en Génie de Procedés, Lavoisier, Paris: Ed. S. Domenech, X. Joulia, B. Koehnet, 199-204.

Georgieva, P., Meireles, M. J. & Feyo de Azevedo, S. (2003). Knowledge Based Hybrid Modeling of a Batch Crystallization When Accounting for Nucleation, Growth and Agglomeration Phenomena. Chemical Engineering Science, 58, 3699-3707.

Jancic, S. J., and P. A. M. Grootscholten. (1984). *Industrial Crystallization*. Delft, Holland: Delft University Press.

Morari, M. (1994). *Advances in Model-Based Predictive Control*. Oxford: Oxford University Press.

Qin, S. J., and T. A. Badgwell. (2003). A survey of model predictive control technology. *Control Engineering Practice* 11 (7):733-764.

Rawlings, J. (2000). Tutorial Overview of Model Predictive Control. *IEEE Control Systems Magazine*:38-52.

Rossiter, J. A. (2003). Model based predictive control. A practical approach. New York: CRC Press.

Seki, H., Ogawa, M., Ooyama, S., Akamatsu, K., Ohshima, M. & Yang, W. (2001). Industrial application of a nonlinear model predictive control to polymerization reactors. Control Engineering Practice, 9, 819-828.

Simoglou, A., Georgieva, P., Martin, E. B., Morris, J. & Feyo de Azevedo, S. (2005). On-line Monitoring of a Sugar Crystallization Process. Computers & Chemical Engineering, 29 (6), 1411-1422.

Soeterboek, R. (1992). Predictive control. A unified approach. New York: Prentice Hall International.

12

Predictive Control for Active Model and its Applications on Unmanned Helicopters

Dalei Song, Juntong Qi, Jianda Han and Guangjun Liu
Shenyang Institute of Automation Science, Chinese Academy of Sciences
China

1. Introduction

Unmanned helicopters are increasingly popular platforms for unmanned aerial vehicles (UAVs). With the abilities such as hovering, taking off and landing vertically, unmanned helicopters extend the potential applications of UAVs. However, due to the complex mechanism and complicated aero-flow during flight, it is almost impossible to accurately model the dynamics of an unmanned helicopter in full flight envelope, and the significant model uncertainties associated with a nominal model may degrade the performance and even stability of an onboard controller.

Due to the difficulty in obtaining a high fidelity full envelope model, the multi-mode modeling technique has been proposed for rotor aircrafts, such as tilt-rotor aircraft XV-15 [1], helicopter BO-105 [2], UH-60 [3], R-50 [4] and X-Cell [5]. The mode-dependent model, which is identified and simplified according to a specific flight mode, such as hovering, cruising, taking off and landing, can be used for control design for the corresponding flight mode. However, the mode-dependent control suffers from at least two problems: one is the difficulty in accommodating the mode transition dynamics, and the other is the compensation of the 'model drift' due to flight dynamics change within one particular mode. Up to now, for the purpose of practical implementation, the mode transition problem can be partially dealt with by limiting the mode switching conditions [6], e.g., mode change is made through hovering mode.

Robust and adaptive control techniques [7-8], on the other hand, have been used to deal with the 'model-shift' within a flight mode. However, such control schemes normally need to know the boundary of internal and external uncertainties and relative noise distribution, which are difficult to identify accurately for a helicopter in full flight envelope. Although online identification technology can be used to obtain the real-time dynamics and disturbance, it is a large burden for the flight computer to reconstruct the robust controllers and reach the requested control period (>50Hz) for sampling and actuating due to the complex calculation of the robust/adaptive optimization process [9-10] and the strict weight limits of micro flight computers.

Besides the model uncertainties, another critical problem that limits the control performance of a helicopter is the time delay between the actuator command and the generation of relative aerodynamic force/torque [11], which will be called aerodynamics-delay/time-delay in the following sections. Normally, this time delay may cause reduced feedback gain of a model-based controller and result in poor robustness [12-13], i.e., sensitive to disturbances.

In recent years, the encouraging achievement in sequential estimation makes it an important direction for online modeling and model-reference control [14]. Among stochastic estimations, the most popular one is the Kalman-type filters (KFs) [15, 16, and 17]. Although widely used, the KFs suffer from sensitivity to bias and divergence in the estimates, relying on assumptions on statistic distribution such as white noise and known mean or covariance for optimal estimation. In many cases, it is more practical to assume that the noises or uncertainties are unknown but bounded (UBB). In view of this, the set-membership filter (SMF), which computes a compact feasible set in which the true state or parameter lies only under the UBB noise assumption, provides an attractive alternative [18-19].

On the control issue, model predictive control (MPC) can compensate for the aerodynamics delay and does not require a high accuracy reference nonlinear model [20]. Among these methods, linear generalized predictive control (GPC) has become one of the most popular MPC methods in industry and academia. However, the normal GPC is sensitive to process noise and model errors [21], which are unknown but bounded for helicopters when sudden 'mode change' happen and model-drift in full flight envelope. This makes the prediction biased, and results in the non-optimal process of controller solving.

In this paper, for realizing the coupling control of unmanned helicopters in full flight envelope, an active modeling based controller is developed based on a modified generalized predictive control and adaptive set-membership filter estimation (ASMF). The time varying model error and its boundary are estimated by the adaptive set-member filter, which is first proposed in [19]. Incremental prediction process and dimension reduction method is embedded into traditional GPC, which can decreases the computation burden and maintain prediction unbiased when 'mode change' happens. Based on this active estimation and the modified GPC controller, a novel optimal strategy for on-line compensation of model error is developed. Thus, aggressive flight can be achieved only based on the hovering model with time-delay terms. Using the identified hovering dynamics model as nominal model for controller, flight experiments have been conducted to test the performance of the proposed controller in full flight envelope on our UAV platform, and experimental results have demonstrated the effectiveness of the proposed method.

2. Active model based control scheme and reference model of a helicopter

Fig. 1 illustrates the active model based control scheme. The error between the reference model and the actual dynamics of the controlled plant is estimated by an on-line modeling strategy. The control, which is designed according to the reference model, should be able to compensate the estimated model error and it in real time. In the followings of this paper, we use the ASMF as the active modeling algorithm and the modified GPC as the control.

For normal missions of an unmanned helicopter, the flight modes include hovering (velocity under 5m/s), cruising (velocity above 5m/s), taking off and landing (distance to the ground is below 3m while significant ground effect exists) and the transitions among these modes. A reference model is typically obtained by linearizing the nonlinear dynamics of a helicopter at one flying mode. The model errors from linearization, external disturbance, simplification, and un-modeled dynamics can be considered as additional process noise [22]. Thus, a linearized state-space model for helicopter dynamics in full flight envelope can be formulated as

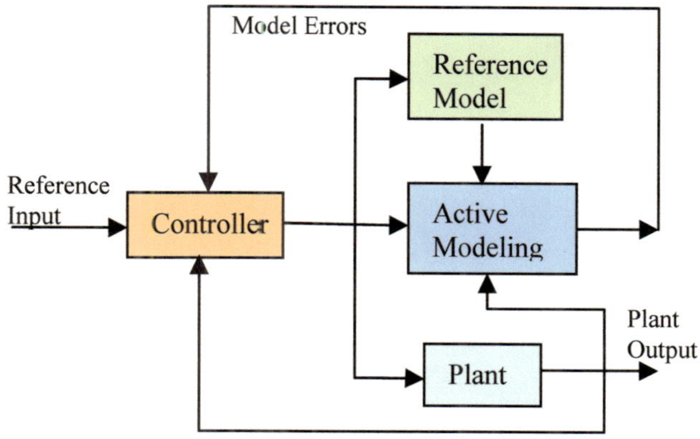

Fig. 1. The scheme of active model based control

$$\begin{cases} \dot{X}_t = A_0 X_t + B_0 U_{t-k} + B_f f(X_t, \dot{X}_t, W_t) \\ Y_t = C X_t \end{cases} \tag{1}$$

where $X \in R^{13}$ is the state, including 3-axis velocity, pitch and roll angle, 3-axis angle rate, flapping angles of main rotor and stabilizer bar, and the feedback of yaw gyro. $Y_t \in R^8$ is the output, including 3-axis velocity, pitch and roll angle and 3-axis angle rate, A_0 and B_0 contain parameters that can be identified in different flight modes, and we use them to describe the parameters in hovering mode. $U \in R^4$ is the control input vector. $C \in R^{13\times 8}$ is the output matrix, $k \in R$ is the time-delay for the driving system. The detail of building the nominal model and physical meaning of parameters is explanted in Appendix A.

To describe the dynamics change, in equation (1), here, we introduce $f(X_t, \dot{X}_t, W_t) \in R^{13}$ to represent the time varying model error in full flight envelope, and $W_t \in R^{13}$ is the process noise.

The following two sections, based on model (1) will describe the way to estimate $f(X_t, \dot{X}_t, W_t)$ and to compensate for model errors from process noise, parameters change, control delay and flight mode change in real applications.

3. ASMF based active model error estimation

As illustrated in Fig.1, adopting the active modeling process to get the model error f and system state X is the basis for elimination of the model error. Controller can only work based on nominal model and feedback of state and model error from active modeling process. In this section, the active modeling process is built based on an adaptive set-membership filter (ASMF) [19] since the UBB process noise.

First, we must obtain the reference equation for estimation. Compared with the sampling frequency (often >50Hz for flight control) of the control system, the model error $f(X, \dot{X}, W)$ can be considered as a slow-varying vector, which means

$$f_{t+1} = f_t + h_t$$

where f_t is the sampling value of $f(X,\dot{X},W)$ at sampling time t, and h_t is the assumed unknown but bounded (UBB) process noise.
Let the extended sampling state

$$X_t^a = \begin{pmatrix} X_t^T & f_t^T \end{pmatrix}^T$$

Then, we can obtain the discrete equation from Eq. (1) as

$$\begin{cases} X_{t+1}^a = A_d^a X_t^a + B_d^a U_t + W_t^a \\ Y_t = C_d^a X_t^a + V_t \end{cases} \tag{2}$$

where $A_d^a = \begin{pmatrix} A_d & B_f \\ 0_{13\times13} & I_{13\times13} \end{pmatrix}$, $B_d^a = \begin{pmatrix} B_d \\ 0_{13\times4} \end{pmatrix}$, $C_d^a = \begin{pmatrix} C_d & 0_{8\times13} \end{pmatrix}$, $W_t^a = \begin{pmatrix} W_t^T & h_t^T \end{pmatrix}^T$, $B_f = I_{13\times13}$ and f_t is a 13×1 vector for model errors. Here, *t* is the sampling time, $I_{m\times m}$ is the $m\times m$ unit matrix and $0_{m\times n}$ is the $m\times n$ zero matrix. $\{A_d, B_d, C_d\}$ is the discrete expression of system $\{A_0, B_0, C\}$. Here, time-delay *k* is ignored during the estimate process, and the compensation method will be discussed in the next part on modified GPC.
The model error i.e., *f* in Eq. (1), comes from the linearization while neglecting the coupling dynamics and uncertainties, and also the A_0 and B_0 because they are identified with respect to a specific flight mode, here hovering mode is selected as nominal flight mode since easy identification. Therefore, both the model error and the process noise W^a are vehicle dynamics and flight states dependent, and do the following assumption

Assumption:

W^a does not necessarily have a normal distribution.

Thus, the Kalman type filter cannot be applied, and adaptive set-membership filter, which is developed for UUB process noise and can get the uncertain boundaries of the states, is considered to estimate the states and model errors here.
In this section we only present the result of ASMF and please refer to [19] for the details about ASMF. With respect to Eq. (2), we can build the adaptive set-membership filter as Eq. (3), where Q^a and R^a are the initial elliptical boundary of process and measurement noise respectively, r_m is the maximum eigenvalue of R, p_m is the maximum eigenvalue of $C_d^a P_{t|t-1} C_d^{aT}$, $Tr(\bullet)$ is the trace of a matrix, δ_t and β_t are the adaptive parameters of the filter. We can also obtain the boundary of the ith element $\hat{X}_i^a$ of extended state $\hat{X}_{t|t}^a$ as $\left(\hat{X}_i^a - \sqrt{P_{ii}}, \hat{X}_i^a + \sqrt{P_{ii}}\right)$, where P_{ii} is the *i*-th diagonal element of matrix *P*.

$$
\begin{cases}
\rho_t = \dfrac{\sqrt{r_{mt}}}{\sqrt{r_{mt}} + \sqrt{p_{mt}}} \\
W_t = \dfrac{C_d^a P_{t|t-1} C_d^{aT}}{1-\rho_t} + \dfrac{R^a}{\rho_t} \\
K_t^e = \dfrac{P_{t|t-1} C_d^{aT} W_t^{-1}}{1-\rho_t} \\
\delta_t = 1 - (Y_t - C_d^a \hat{X}_{t|t-1}^a)^T W_t^{-1} (Y_t - C_d^a \hat{X}_{t|t-1}^a) \\
\hat{X}_{t|t}^a = \hat{X}_{t|t-1}^a + K_t^e (Y_t - C_d^a \hat{X}_{t|t-1}^a) \\
P_{t|t} = \delta_t \left(\dfrac{P_{t|t-1}}{1-\rho_t} - \dfrac{P_{t|t-1}}{1-\rho_t} C_d^{aT} W_t^{-1} C_d^a \dfrac{P_{t|t-1}}{1-\rho_t} \right) \\
\hat{X}_{t+1|t}^a = A_d^a \hat{X}_{t|t}^a + B_d^a U_t \\
\beta_t = \dfrac{\sqrt{Tr(Q^a)}}{\sqrt{Tr(Q^a)} + \sqrt{Tr(A_d^a P_{t|t} A_d^{aT})}} \\
P_{t+1|t} = \dfrac{A_d^a P_{t|t} A_d^{aT}}{1-\beta_t} + \dfrac{Q^a}{\beta_t}
\end{cases}
\qquad (3)
$$

4. Modified GPC for unmanned helicopters

To eliminate the negative influence of model errors and control delay in flight, besides the active estimation algorithm like ASMF that does not require a normal distribution assumption, an effective control algorithm has to be designed according to the reference model of Eq. (1) while adopting the on-line estimation of f as compensation.
We describe the normal GPC in Section 4.1, and then, the modified scheme is proposed in Section 4.2 & 4.3 to eliminate the negative influence of model errors in real applications.

4.1 Preliminary work for generalized predictive control

Generally, for a linear system with actuator time delay like,

$$
\begin{cases}
X_{t+1} = A_d X_t + B_d u_{t-k} + W_t \\
y_t = C_d X_t
\end{cases}
\qquad (4)
$$

where $X_t \in R^{n\times 1}$ is the system state vector at sampling time t, $y_t \in R^{l\times 1}$ is the output vector, $u_t \in R^{m\times 1}$ is the control input vector, k is the actuators' time-delay and W_t is process noise; traditional Generalized Predictive Control (GPC) [23] can be designed as:

Step I: Make prediction

Firstly, for the case that predictive step *i* is less than time-delay *k* (i.e., the time instant that system behavior cannot be regulated through current and future control action), prediction can be denoted as following equation.

$$\hat{X}_{t+i|t} = A_d\hat{X}_{t+i-1|t} + B_d u_{t+i-1-k} \triangleq \hat{X}^1_{t+i|t} \tag{5}$$

where $\hat{X}_{t+i} | t$ is the prediction state at time *t+i*, the superscript 1 denotes that the part of predicted variable that is independent of the current and future's control actions.
Secondly, for the case that prediction step *i* is larger than the time delay k,

$$\begin{aligned}\hat{X}_{t+k+i|t} &= A_d\hat{X}_{t+k+i-1|t} + B_d u_{t+i} \\ &= A_d\hat{X}^1_{t+k+i-1|t} + \sum_{n=0}^{i} A_d^n B_d u_{t+n} \\ &= A_d\hat{X}^1_{t+k+i|t} + \sum_{n=0}^{i} A_d^n B_d u_{t+n} \quad , \quad 1 \le i \le p\end{aligned} \tag{6}$$

where p is the prediction range; similarly, $\hat{X}^1_{t+k+i-1|t}$ denotes the sub-variable of $\hat{X}_{t+k+i-1|t}$ that is independent of the current and future's control actions.

Step II: Receding horizon optimization

After making prediction, the control vector can be obtained by minimize the following cost function:

$$J = (R^x{}_t - X_t^v)^T (R^x{}_t - X_t^v) + U_t^T \gamma U_t \tag{7}$$

And the optimal control inputs can be denoted as,

$$U_t^* = (G_0^T G_0 + \gamma)^{-1} G_0^T (R^x{}_t - X_t^1) \tag{8}$$

where G_0 is the predictive matrix, X_t^v is the predictive state vector, X_t^1 is the known vector inside X_t^v, λ is the weight of control input, and R_t^x is the reference of system states. The detailed definition of these matrixes can be referenced in [23].

Step III: Control implementation

The first element of vector U_t^* is used as the control to the real plant. After that, go back to step I at the next time instant.
However, with application to the unmanned helicopters, this kind of GPC algorithm has the following three disadvantages, which will be solved in the next two sections:
1. It cannot reject the influence of working mode changes, i.e., if

$$\begin{aligned}X_t &= x_t - x_0 > \pi(x_0, u_0) \\ U_t &= u_t - u_0 > \pi(x_0, u_0)\end{aligned} \tag{9}$$

where (x_0, u_0) is the current operation point, which cannot be ensured on-line, $\pi(x_0, u_0)$ is the valid range for model linearization and x_t is the absolute state at time t, u_t is the absolute control input at time t. The biased prediction, due to the changing operation point (x_0, u_0), will bring steady errors for velocity tracking.

2. Normal GPC is sensitive to mismatch of the nominal model, which means slow change in parameters (A_d, B_d) may result in prediction error and unstable control.
3. The transient model errors of the nominal model from external disturbance, estimated by ASMF, cannot be eliminated. And this will also result in the non-minimum variance and the instability of the closed control loop.

4.2 Stationary increment predictive control

To reject the influence of working mode change and sensitivity to nominal parameters change in real application, i.e. the problem 1) and 2) in Section 4.1, we assume that the process noise W_t's increment in Eq. (4) is a stationary random process, which means

$$W_t^C \triangleq \Delta W_t = W_t - W_{t-1} \tag{10}$$

is normal distribution. Where $\Delta = 1 - q^{-1}$ is the difference operator; q^{-1} is one-step delay factor. Thus, Eq. (4) can be rewritten as follows,

$$\Delta X_{t+1} = A_d \Delta X_t + B_d \Delta u_{t-k} + W_t^0 \tag{11}$$

Consider

$$\Delta X_t = (x_t - x_0) - (x_{t-1} - x_0) = \Delta x_t$$
$$\Delta U_t = (u_t - u_0) - (u_{t-1} - u_0) = \Delta u_t$$

if behavior prediction is made based on Eq. (11), only the absolute state x_t and control input u_t, which can be measured or estimated directly from sensors, are used and the current operation point (x_0, u_0) disappears in prediction. Thus, the problem of biased prediction due to changing of working point, i.e., problem 1), can be solved.

Otherwise, according to the process of traditional GPC, the set-point R_t^x must be obtained for every prediction step, and this is often set as current reference states. However, for helicopter system, only measurable outputs are cared, such as position, velocity and etc; and the internal states, such as rotor's pitch angle and yaw gyro's feedback and so on, are coupled with the measurable states/outputs, and cannot be set independently. Others, this reference input often comes from position track planning, which changes quickly for flight and often cause a step-like signal for tracking. To avoid the step signal reference tracking, which is dangerous for unmanned helicopter system, we use a low pass filter to calculate the set-point inputs of the output in the future *i*-th step, *i*=1, …, *p*.

Let $SP_t \in R^{l \times 1}$ be the set-point input at time *t*, then we have

$$r_{t+k+i} = SP_t + \alpha(r_{t+k+i-1} - SP_t) \quad ,1 \le i \le p \tag{12}$$

where α is the cut-off frequency of the filter, the initial value $r_{t+k} = \hat{y}_{t+k|t}$, r_{t+k+i} is the *i*-th set-point input, and $\hat{y}_{t+k|t}$ is the estimate of output at time *t+k*.

Thus, the set-point problem is solved and the output prediction can be implanted based on increment model (11) as follows:

When the prediction step i is less than time-delay k,

$$\hat{X}_{t+i|t} = \hat{X}_{t+i-1|t} + A_d \Delta \hat{X}_{t+i-1|t} + B_d \Delta u_{t+i-1-k} = \hat{X}^1_{t+i|t} \tag{13}$$

When the prediction step is larger than time-delay k, let

$$\Delta \hat{X}^1_{t+i|t} = \hat{X}^1_{t+i|t} - \hat{X}^1_{t+i-1|t}$$

Then,

$$\begin{aligned}
\hat{X}_{t+k+i|t} &= \hat{X}_{t+k+i-1|t} + A_d \Delta \hat{X}_{t+k+i-1|t} + B_d \Delta u_{t+i} \\
&= \hat{X}^1_{t+k+i-1|t} + A_d \Delta \hat{X}^1_{t+k+i-1|t} \\
&+ \sum_{m=0}^{i-1} \left\{ \{ \sum_{n=0}^{i-1-m} A_d^n B_d \} \Delta u_{t+m} \right\} \\
&= \hat{X}^1_{t+k+i|t} + \sum_{m=0}^{i-1} \left\{ \{ \sum_{n=0}^{i-1-m} A_d^n B_d \} \Delta u_{t+m} \right\}, 1 \le i \le p
\end{aligned} \tag{14}$$

Hence, the above problem 1), which comes from working mode change, is solved because x_0 disappears in predictive equation (14).

We can obtain the following prediction matrix for the output, which is often cared in helicopter tracking problem, from Eq. (12) and (13):

$$\begin{aligned}
\hat{Y}_t &= \left(\hat{y}_{t+k+1|t} \quad \hat{y}_{t+k+2|t} \quad \cdots \quad \hat{y}_{t+k+p|t} \right)^T \\
&= \left(C_d \hat{X}^1_{t+k+1|t} \quad C_d \hat{X}^1_{t+k+2|t} \quad \cdots \quad C_d \hat{X}^1_{t+k+p|t} \right)^T \\
&+ G \left(\Delta u_t^T \quad \Delta u_{t+1}^T \quad \cdots \quad \Delta u_{t+p-1}^T \right)^T \\
&= Y_t^1 + G \Delta U
\end{aligned} \tag{15}$$

where Y_t^1 is the known part of p steps' prediction, which cannot be influenced by current control input, and matrix G has the following form:

$$G = \begin{pmatrix}
C_d B_d & 0 & \cdots & 0 \\
C_d B_d + C_d A_d B_d & C_d B_d & \cdots & 0 \\
\cdots & \cdots & \cdots & \cdots \\
C_d \sum_{i=0}^{p-1} A_d^i B_d & C_d \sum_{i=0}^{p-2} A_d^i B_d & \cdots & C_d B_d
\end{pmatrix} \tag{16}$$

Compared with the normal GPC, the prediction of SIPC has better characteristics that can be described by the following theorem, which solves the above problem 2) in Section IV.A.

Theorem: for nominal model (11), when the nominal model parameters (A_d, B_d) change into (A_{dr}, B_{dr}).

1. $\exists M,N>0\in R$, let the matrix norms satisfy

$$\|A_d\|<M,\|B_d\|<M$$
$$\|A_{dr}\|<N,\|B_{dr}\|<N$$

2. Define

$R_{\max}\{\bullet\}$ is the operator for the maximum of eigenvalue of matrix •.
Thus, if

$$R_{\max}\{A_{dr}\}<0$$
$$R_{\max}\{A_d\}<0$$

Then, the state prediction obtained by Eqs. (13-14) maintains unbiased, and the characteristic is also guaranteed in traditional GPC conditions, i.e. Eq. (4), where W_t is normal distribution.

Proof: See Appendix B.

In Eq. (14), ΔU , including p control inputs, need to be optimized, while only the first one is used for control. This will occupy a great deal of computation resource and result in very low computational efficiency, especially with respect to the fast applications.

In order to reduce the computational burden of Eq. (14), we propose here a 'step plan' technique,

$$\Delta u_{t+i+1}=\beta\Delta u_{t+i} \tag{17}$$

where β is an $m\times m$ diagonal matrix presenting the length of one step, which will be a parameter to be selected. Then, we can simplify Eq. (14) by only calculating the unknown control, which has smaller dimensions.

$$\begin{aligned}\hat{Y}_t&=Y_t^1+G\left(I_{m\times m}\quad\beta\quad\ldots\quad\beta^{p-1}\right)^T\Delta u_t\\&=Y_t^1+G_2\Delta u_t\end{aligned} \tag{18}$$

where $I_{m\times m}$ is an $m\times m$ unit matrix. Thus, the number of the unknown control input vector (from current time t to the future time $t+p$-1) is reduced from p to 1, and the dimension of predictive matrix is changed from $pl\times pm$ to $pl\times m$. This reduction brings low computer memory consuming and simplifies the receding horizon optimization in the following calculation.

To complete the horizon optimization and obtain the control input, the cost function of the stationary increment predictive control is designed as:

$$J=(R_t-\hat{Y}_t)^TW(R_t-\hat{Y}_t)+\Delta u_t{}^T\lambda\Delta u_t \tag{19}$$

where $R_t=\left(r_{t+k+1}^T\quad r_{t+k+2}^T\quad\ldots\quad r_{t+k+p}^T\right)^T$, $W\in R^{lp\times lp}$ is the weight matrix for tracking error, and $\lambda\in R^{m\times m}$ is the weight matrix of the control increment.

In order to minimize the cost function of Eq. (19), we can calculate the control vector as follows:

$$\begin{aligned}\Delta u_t &= (G_2^T W G_2 + \lambda)^{-1} G_2^T W (R_t - Y_t^1) \\ &= K_f (R_t - Y_t^1)\end{aligned} \tag{20}$$

where $K_f = (G_2^T W G_2 + \lambda)^{-1} G_2^T W$ can be completed offline.
Consequently, the proposed stationary increment predictive controller (SIPC) can be designed as followings.

Step I: Make increment prediction

Based on the current and history measure value, use Eqs. (13-15) to obtain the prediction for future output $\hat{Y}_t$ and initial plan point

$$r_{t+k} = \hat{y}_{t+k|t}$$

Step II: Plan for the set-point input

Use Eq. (12) to plan the future set-points, and obtain

$$R_t = \begin{pmatrix} r_{t+k+1}^T & r_{t+k+2}^T & \cdots & r_{t+k+p}^T \end{pmatrix}^T$$

Step III: Receding horizon optimization

Calculate the control increment Δu_t, based on Eq. (20).

Step IV: Control implementation

Current control input $u_t = u_{t-1} + \Delta u_t$, which is used as the control to the real plant. After that, go back to step I at the next time instant.
Thus, for real implementation, only the prediction of Eq. (13-15), the intenerating of Eq. (12), and the control law (20) need to be calculated online, thus the real time computation load, and steady tracking error are both reduced greatly compared with GPC, and the real test in section V has shown its feasibility.
The model error, problem 3), will be compensated by an online optimal strategy, which will be described later.

4.3 Optimal strategy for model error compensation

In order to compensate the model error in Eq. (1), the control vector has to match the following equation, which can be directly obtained from Eq. (1):

$$B_d U_t + B_f f_t = B_d U_t^0 \tag{21}$$

where U_t^0 is the control vector need to be calculated by the predictive controller in section 4.2, designed based on the original model (1) without the model error f.
The control input at sampling time *t* cannot be solved directly from Eq. (21), because:

1. Eq. (21) is difficult to be implemented because the dimension of U_t is less than that of f_t. Thus, only the approximate solution can be obtained with respect to (21);
2. f_t is actually an uncertainty set, an static optimal problem must be considered.

Thus, we introduce the following cost function with quadratic form to solve the above problem 1).

$$\begin{aligned} U_t^* &= \arg\min_{U_t} J_t(U_t) \\ J_t(U_t) &\triangleq \left(B_d U_t + B_f f_t - B_d U_t^0\right)^T H \left(B_d U_t + B_f f_t - B_d U_t^0\right) \end{aligned} \tag{22}$$

where H is a weight matrix, which can be selected.

On the other hand, f_t is obtained from the ASMF algorithm introduced in section III, thus its convergence is very important for the validity of the whole controller. Actually, the convergence of ASMF algorithm is also influenced by the control action U_t. This is because the stability of the ASMF can be represented by the filter parameter δ_t, while δ_t in Eq. (3) can be rewritten as follows,

$$\begin{aligned} \delta_t &= 1 - (Y_t - C_d^a \hat{X}_{t|t-1}^a)^T W_t^{-1} (Y_t - C_d^a \hat{X}_{t|t-1}^a) \\ &= 1 - (Y_{t+1} - C_d^a (A_d^a \hat{X}_{t|t}^a + B_d^a U_t))^T \\ &\qquad W_t^{-1} (Y_{t+1} - C_d^a (A_d^a \hat{X}_{t|t}^a + B_d^a U_t))] \end{aligned} \tag{23}$$

In [19], it has been shown the stability of the ASMF can be represented by the filter parameter δ_t, i.e., the ASMF is stable when $\delta_t > 0$.

Firstly, define

$$\begin{aligned} J_t^\delta(U_t, Y_{t+1}) &\triangleq (Y_{t+1} - C_d^a (A_d^a \hat{X}_{t|t}^a + B_d^a U_t))^T \\ &\qquad W_t^{-1} (Y_{t+1} - C_d^a (A_d^a \hat{X}_{t|t}^a + B_d^a U_t))] \end{aligned} \tag{24}$$

Thus, from Eq. (23), in order to maintain $\delta_{t+1} > 0$, the maximum value of $J_t^\delta(U_t, Y_{t+1})$ with respect to $\hat{X}_{t|t}^a$ should be less than or equal to 1, i.e.,

$$\begin{aligned} J_t^{\delta *}(U_t, Y_{t+1}) &= \max_{\hat{X}_{t|t}^a} J_t^\delta(U_t, Y_{t+1}) \\ &= \max_{\hat{X}_{t|t}^a} \{ \left[Y_{t+1} - C_d^a \left(A_d^a \hat{X}_{t|t}^a + B_d^a U_t \right) \right]^T W_t^{-1} \left[Y_{t+1} - C_d^a \left(A_d^a \hat{X}_{t|t}^a + B_d^a U_t \right) \right] \} \le 1 \end{aligned} \tag{25}$$

In general, larger δ_t often means more rapid convergence of ASMF algorithm. That is, we should select an Ut to make $J_t^{\delta *}(U_t, Y_{t+1})$ small as far as possible, that is,

$$J_t^{\delta *}(Y_{t+1}) = \min_{U_t} J_t^{\delta *}(U_t, Y_{t+1}) \tag{26}$$

We introduce the following cost function $J_t(U_t)$ with consideration of both (22) and (25) at the same time:

$$\begin{aligned} {}^*U_t &= \arg\min_{U_t} \bar{J}_t(U_t) \\ \bar{J}_t(U_t) &\triangleq J_t(U_t) + \alpha J_t^{\delta *}(U_t, Y_{t+1}) \end{aligned} \tag{27}$$

where $\alpha = 1 - \delta_t \in R$ are the positive definite weight matrix. To minimize $J_t(U_t)$, considering $J_t(U_t) > 0$, the control can be obtained at $\partial J_t(U_t) / \partial U_t = 0$, i.e.,

$$\partial J_t(U_t) / \partial U_t = 2(MU_t + N) \tag{28}$$

where

$$M = B_d^T H B_d + \alpha B_d^{aT} C_d^{aT} W_t^{-1} C_d^a B_d^a$$

$$N = B_d^T H (B_f f_t - B_d U_t^0) - \alpha B_d^{aT} C_d^{aT} W_t^{-1} Y_{t+1}$$

Here H can be selected as $H = \delta_t C_d^T C_d$. Thus, we can obtain the optimal control that minimizes $J_t(U_t)$ as:

$$\begin{aligned} U_t(Y_{t+1}) &= -M^{-1}N \\ &= (B_d^T H B_d + \alpha B_d^{aT} C_d^{aT} W_t^{-1} C_d^a B_d^a)^{-1} \\ &\quad \left[\alpha B_d^{aT} C_d^{aT} W_t^{-1} Y_{t+1} - B_d^T H (B_f f_t - B_d U_t^0)\right] \end{aligned} \tag{29}$$

For the unknown measurement at time t+1 in Eq. (24), we consider that the control system is stable, so, $Y_{t+1} \in \Delta(Y_t)$. Here, $\Delta(Y_t)$ is the elliptical domain of Y_t. Because $J_t^{\delta}(U_t, Y_{t+1})$ in Eq. (25) is positive definite, its maximum value point must be on the boundary, which can be estimated by the ASMF. Thus, we first define array S_t^i to include the estimate of the i-th element's two boundary endpoints as

$$S_t^i \triangleq \left\{ \hat{Y}_{t+1}^i \,|\, \{Y_t^i + (-1)^h (\underset{j_l = \{\pm\sqrt{p_{ll}}, l=1,\ldots,13\}}{Max} |C_d Col\{j\}|_i)\} \right\} \tag{30}$$

where Y_t^i is the i-th element in the vector Y_t, $\hat{Y}_{t+1}^i$ is the corresponding output Y_{t+1}'s endpoints estimation. For set S_t^i, $i \in \{1,2,\ldots,8\}$ and h is 0 or 1 for every i, $|\bullet|_i$ is the operator for absolute value of the *i*-th element in vector $\bullet$, and the function $Col\{j\}$ is defined as follows:

$$Col\{j\} = \begin{pmatrix} j_1 & \ldots & j_{13} \end{pmatrix}^T \tag{31}$$

Then, we define a set S_t to describe all possible endpoint vector of the Y_{t+1} as

$$S_t \triangleq \left\{ \hat{Y}_{t+1}^{EP} \,|\, \begin{pmatrix} S_t^1 & \ldots & S_t^{13} \end{pmatrix} \right\} \tag{32}$$

where $\hat{Y}_{t+1}^{EP}$ is the possible endpoint (EP) for output Y_{t+1} at next sampling time t+1.

Thus, the proposed active modeling based predictive controller can be implemented by using the following steps:

Step I: Make increment prediction

Based on the current estimated state $\hat{X}^a_{t|t}$, use the stationary increment predictive controller, as in section 4.2, to obtain the nominal control input U_t^0;

Step II: Model error estimation and elimination

Based on U_t^0, compute the optimal control input ${}^*U_t$:

Estimate the values and boundaries of state X_t and model error f_t, using ASMF in (3);

Calculate the corresponding $U_t(\hat{Y}_{t+1}^{EP})$ for every $\hat{Y}_{t+1}^{EP}$ in set S_t by Eq. (29);

For every $U_t(\hat{Y}_{t+1}^{EP})$ in step 1), use Eq. (24) to obtain the maximum of function $J_t^\delta(U_t(\hat{Y}_{t+1}^{EP}), \hat{Y}_{t+1}^{EP})$, and get the ${}^*\hat{Y}_{t+1}^{EP}$ to let

$$^*\hat{Y}_{t+1}^{EP} = \arg\underset{\hat{Y}_{t+1}^{EP} \in S_t}{Max}\left\{J_t^\delta\{U_t(\hat{Y}_{t+1}^{EP}), \hat{Y}_{t+1}^{EP}\}\right\};$$

The corresponding $U_t({}^*\hat{Y}_{t+1}^{EP})$ is the optimal control ${}^*U_t$ at time t, i.e. ${}^*U_t = U_t({}^*\hat{Y}_{t+1}^{EP})$.

Step III: Receding horizon strategy

Go back to step I at the next time instant.

5. Flight test

5.1 Flight test platform

All flight tests are conducted on the Servoheli-40 setup, which was developed in the State Key Laboratory, SIACAS. It is equipped with a 3-axis gyro, a 3-axis accelerometer, a compass and a GPS. The sensory data can be sampled and stored into an SD card through an onboard DSP. Tab.1 shows the physical characteristics of SERVOHELI-40 small-size helicopter. More details of this experimental platform can be found in [24].

Fig. 2. SERVOHELI-40 small-size helicopter platform

Length	2.12m
Height	0.73m
Main rotor diameter	2.15m
Stabilizer bar diameter	0.75m
Rotor speed	1450rpm
Dry weight	20kg
Engine	2-stroke, air cooled
Flight time	45 min

Table 1. Physical characteristics of SERVOHELI-40 small-size helicopter

5.2 Experiment for the verification of model error estimate when mode-change

We use the identified hovering parameters, through frequency estimate [25], as the nominal model for hovering dynamics of the ServoHeli-40 platform. The model accuracy is verified in hovering mode (speed less than 3m/s) and cruising mode (speed more than 5m/s), the results for lateral velocity are shown in Fig.3a.

Fig. 3 further shows the model difference due to mode change, where the red lines are the results calculated by the identified model with the inputs of hovering and cruising actuations, respectively, and blue lines are the measurements of the onboard sensors. Comparison shows that the hovering model outputs match the hovering state closely, but clear differences occur while being compared to the cruising state, even though the cruising actuations are used as the model inputs. This is the model error when flight mode is changed.

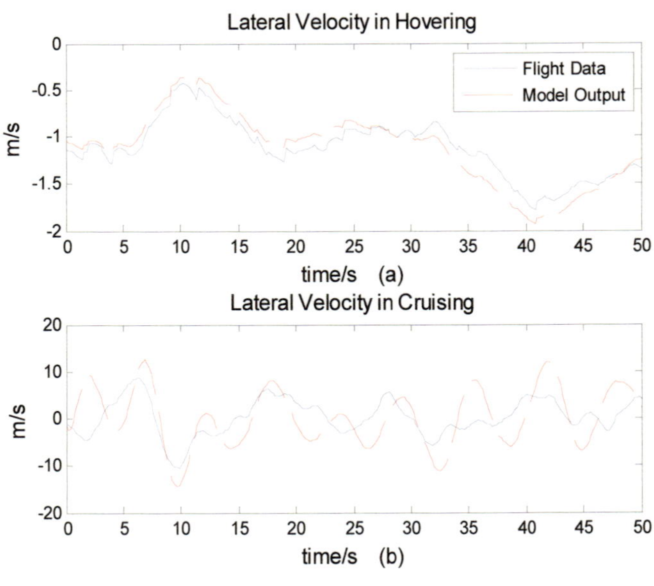

Fig. 3. Model difference due to mode change: (a) hovering conditions; (b) cruising conditions

To verify the accuracy of the estimate of the model error, described in Fig.3, the following experiment is designed:

1. Actuate the longitudinal control loop to keep the speed more than 5 meter per second;
2. Get the lateral model error value and boundaries through ASMF, and add them to the hovering model we built above;
3. Compare the model output before and after compensation for model error.

This process of experiment can be described by Fig.4, and the results are shown in Fig.5. Fig.5a shows that model output (red line) cannot describe the cruising dynamics due to the model error when 'mode-change', similar with Fig.3b; however, after compensation, shown in Fig.5b, the model output (red line) is very close with real cruising dynamics (blue line), and the uncertain boundaries can include the changing lateral speed, which mean that the proposed estimation method can obtain the model error and range accurately by ASMF when mode-change.

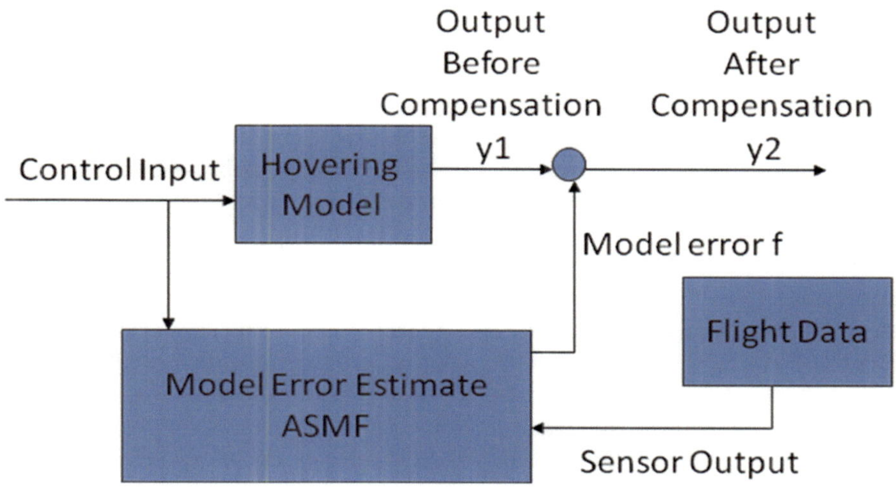

Fig. 4. The experiment process for model-error estimate

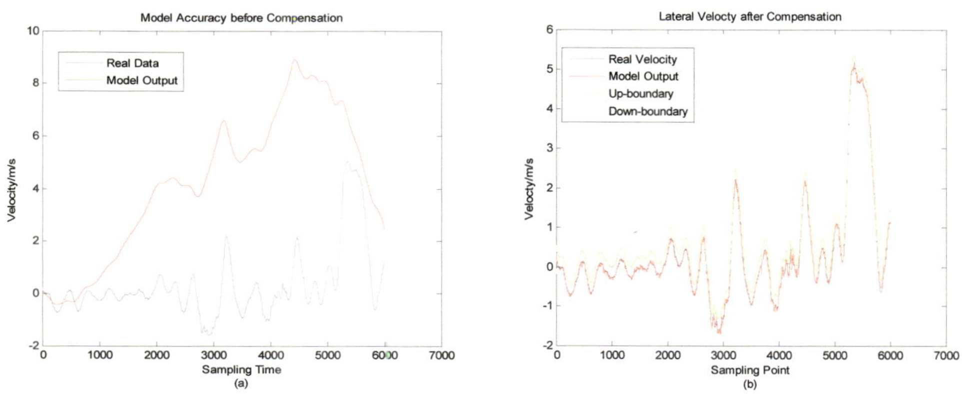

Fig. 5. Model output before/after compensation: (a) before compensation; (b) after compensation

5.3 Flight experiment for the comparison of GPC SIPC and AMSIPC when sudden mode-change

In Section 5.2, the model-error occurrence and the accuracy of the proposed method for estimation are verified. So, the next is the performance of the proposed controller in real flight. In this section, the performance of the modified GPC (Generalized Predictive Control, designed in Section 4.1), SIPC (Stationary Increment Predictive Control, designed in Section 4.2) and AMSIPC (Active Modeling Based Stationary Increment Predictive Control, designed in Section 4.3), are tested in sudden mode-change, and are compared with each other on the ServoHeli-40 test-bed. To complete this mission, the following experimental process is designed:

1. Using large and step-like reference velocity, red line in Fig.6-8, input it to longitudinal loop, lateral loop and vertical loop;
2. Based on the same inputted reference velocity, using the 3 types of control method, GPC, SIPC and AMSIPC to actuate the helicopter to change flight mode quickly;
3. Record the data of position, velocity and reference speed for the 3 control loops, and obtain reference position by integrating the reference speed;
4. Compare errors of velocity and position tracking of GPC, SIPC and AMSIPC, executively, in this sudden mode-change flight.

GPC, SIPC and AMSIPC are all tested in the same flight conditions, and the comparison results are shown in Figs. 6-8. We use the identified parameters in Section 5.2 to build the nominal model, based on the model structure in Appendix A, and parameters' selection in Appendix C for controllers

It can be seen that, when the helicopter increases its longitudinal velocity and changes flight mode from hovering to cruising, GPC (brown line) has a steady velocity error and increasing position error because of the model errors. SIPC (blue line) has a smaller velocity error because it uses increment model to reject the influence of the changing operation point and dynamics' slow change during the flight. The prediction is unbiased and obtains better tracking performance, which is verified by Theorem. However, the increment model may enlarge the model errors due to the uncertain parameters and sensor/process noises, resulting in the oscillations in the constant velocity period (clearly seen in Fig.6&7) because the error of its prediction is only unbiased, but not minimum variance. While for AMSIPC (green line), because the model error, which makes the predictive process non-minimum variance, has

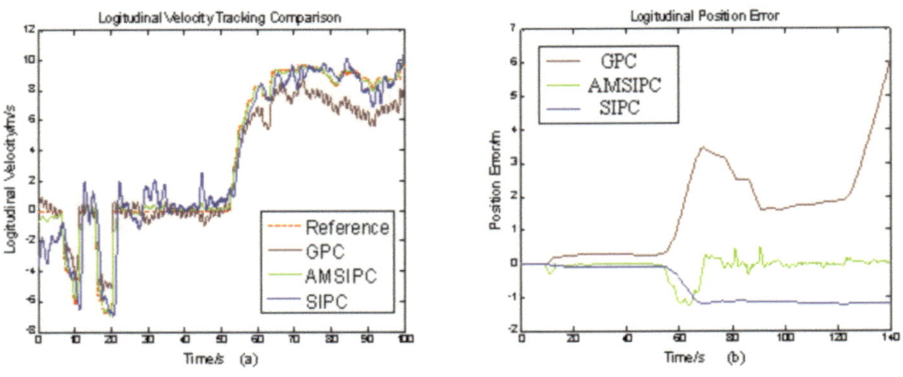

Fig. 6. Longitudinal tracking results: (a) velocity; (b) position error (<50s hovering, >50s cruising)

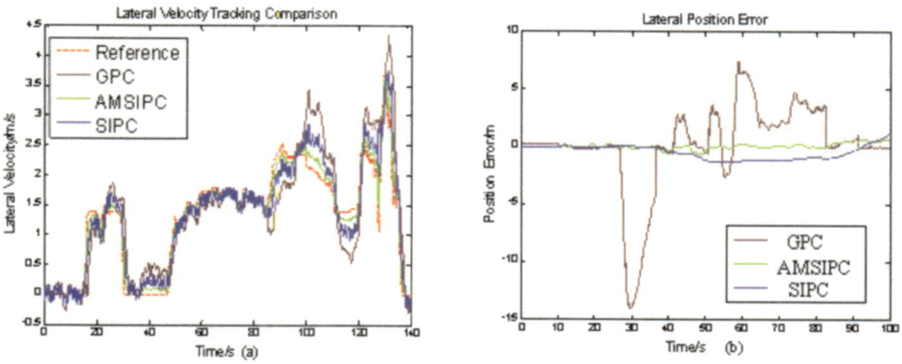

Fig. 7. Lateral tracking results: (a) velocity; (b) position error (25s~80s cruising, others hovering)

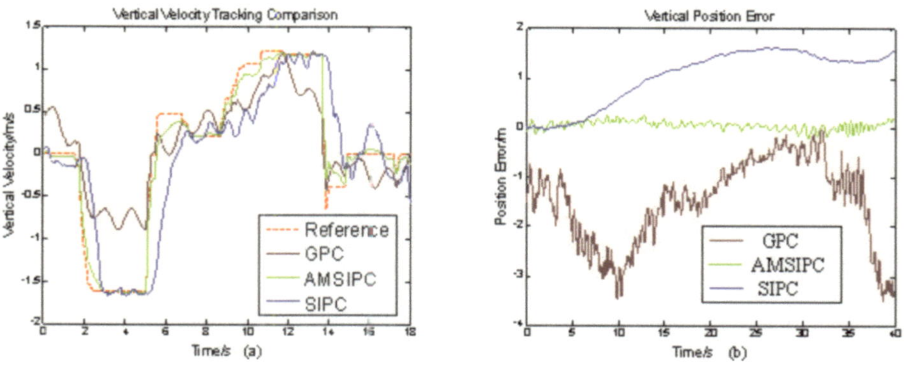

Fig. 8. Vertical tracking results: (a) velocity; (b) position error (<5s hovering; >5s cruising)

been online estimated by the ASMF and compensated by the strategy in section 4.3, the proposed AMSIPC successfully reduces velocity oscillations and tracking errors together.

6. Conclusion

An active model based predictive control scheme was proposed in this paper to compensate model error due to flight mode change and model uncertainties, and realize full flight envelope control without multi-mode models and mode-dependent controls.

The ASMF was adopted as an active modeling technique to online estimate the error between reference model and real dynamics. Experimental results have demonstrated that the ASMF successfully estimated the model error even though it is both helicopter dynamics and flight-state dependent.In order to overcome the aerodynamics time-delay, also with the active estimation for optimal compensation, an active modeling based stationary increment predictive controller was designed and analyzed.

The proposed control scheme was implemented on our developed ServoHeli-40 unmanned helicopter. Experimental results have demonstrated clear improvements over the normal GPC without active modeling enhancement when sudden mode-change happens.

It should be noted that, at present, we have only tested the control scheme with respect to the flight mode change from hovering to cruising, and vice versa. Further mode change conditions will be flight-tested in near future.

7. Appendix

A. Helicopter dynamics

A helicopter in flight is free to simultaneously rotate and translate in six degrees of freedom. Fig. A-1 shows the helicopter variables in a body-fixed frame with origin at the vehicle's center of gravity.

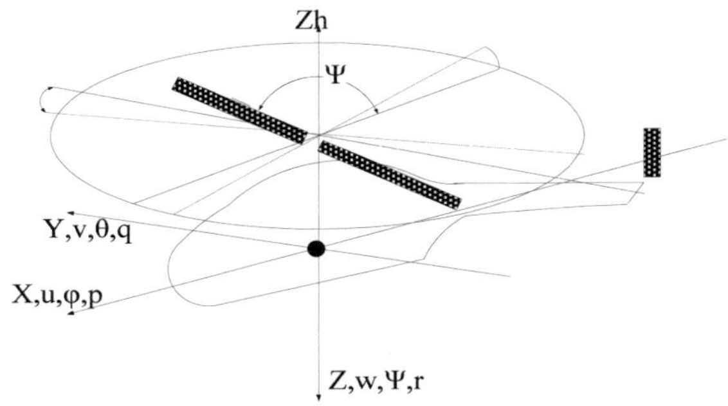

Fig. A-1. Helicopter with its body-fixed reference frame

Ref.[18] developed a semi-decoupled model for small-size helicopter, i.e.,

$$\begin{pmatrix} \delta\dot{u} \\ \delta\dot{q} \\ \delta\dot{\theta} \\ \dot{a} \\ \dot{c} \end{pmatrix} = \begin{pmatrix} X_u & 0 & -g & X_a & 0 \\ M_u & 0 & 0 & M_a & 0 \\ 0 & 1 & 0 & 0 & 0 \\ 0 & -1 & 0 & -1/\tau_f & A_c/\tau_f \\ 0 & -1 & 0 & 0 & -1/\tau_f \end{pmatrix} \begin{pmatrix} \delta u \\ \delta q \\ \delta\theta \\ a \\ c \end{pmatrix} + \begin{pmatrix} X_{lon} & X_{lat} \\ M_{lon} & M_{lat} \\ 0 & 0 \\ A_{lon} & A_{lat} \\ C_{lon} & C_{lat} \end{pmatrix} \begin{pmatrix} \delta_{lon} \\ \delta_{lat} \end{pmatrix}, \text{ i.e.,}$$

$$\begin{cases} \dot{X}_{lon} = A_{lon}\delta X_{lon} + B_{lon}\delta u_{lon} \\ y_{lon} = \begin{pmatrix} I_{3\times3} & 0_{3\times2} \end{pmatrix}\delta X_{lon} = C_{lon}\delta X_{lon} \end{cases} \tag{A-1}$$

$$\begin{pmatrix} \delta\dot{v} \\ \delta\dot{p} \\ \delta\dot{\varphi} \\ \dot{b} \\ \dot{d} \end{pmatrix} = \delta\dot{X}_{lat} = \begin{pmatrix} Y_u & 0 & g & Y_a & 0 \\ L_u & 0 & 0 & L_a & 0 \\ 0 & 1 & 0 & 0 & 0 \\ 0 & -1 & 0 & -1/\tau_f & B_d/\tau_f \\ 0 & -1 & 0 & 0 & -1/\tau_f \end{pmatrix} \begin{pmatrix} \delta v \\ \delta p \\ \delta\varphi \\ b \\ d \end{pmatrix} + \begin{pmatrix} Y_{lon} & Y_{lat} \\ L_{lon} & L_{lat} \\ 0 & 0 \\ B_{lon} & B_{lat} \\ D_{lon} & D_{lat} \end{pmatrix} \begin{pmatrix} \delta_{lon} \\ \delta_{lat} \end{pmatrix}, \text{ i.e.,}$$

$$\begin{cases} \dot{X}_{lat} = A_{lat}\delta X_{lat} + B_{lat}\delta u_{lat} \\ y_{lat} = \begin{pmatrix} I_{3\times3} & 0_{3\times2} \end{pmatrix}\delta X_{lat} = C_{lat}\delta X_{lat} \end{cases} \tag{A-2}$$

$$\begin{pmatrix} \delta\dot{w} \\ \delta\dot{r} \\ \delta\dot{r}_{fb} \end{pmatrix} = \delta\dot{X}_{yaw-heave} = \begin{pmatrix} Z_w & Z_r & 0 \\ N_w & N_r & -N_{ped} \\ 0 & K_r & -K_{rfb} \end{pmatrix} \begin{pmatrix} \delta w \\ \delta r \\ \delta r_{fb} \end{pmatrix} + \begin{pmatrix} Z_{ped} & Z_{col} \\ N_{ped} & N_{col} \\ 0 & 0 \end{pmatrix} \begin{pmatrix} \delta_{ped} \\ \delta_{col} \end{pmatrix}, \text{ i.e.}$$

$$\begin{cases} \dot{X}_{yaw-heave} = A_{yaw-heave}\delta X_{yaw-heave} + B_{yaw-heave}\delta u_{yaw-heave} \\ y_{yaw-heave} = \begin{pmatrix} I_{2\times 2} & 0_{2\times 1} \end{pmatrix}\delta X_{yaw-heave} = C_{yaw-heave}\delta X_{yaw-heave} \end{cases} \tag{A-3}$$

where δu, δv, δw are longitudinal, lateral and vertical velocity, δp, δq, δr are roll, pitch and yaw angle rates, $\delta\varphi$ and $\delta\theta$ are the angles of roll and pitch, respectively, a and b are the first harmonic flapping angle of main rotor, c and d are the first harmonic flapping angle of stabilizer bar, δr_{fb} is the feedback control value of the angular rate gyro, δ_{lat} is the lateral control input, δ_{lon} is the longitudinal control input, δ_{ped} is the yawing control input, and δ_{col} is the vertical control input. All the symbols except gravity acceleration g in A_{lon}, A_{lat}, $A_{yaw-heave}$, B_{lon}, B_{lat} and $B_{yaw-heave}$ are unknown parameters to be identified. Thus, all of the states and control inputs in (A-1), (A-2) and (A-3) are physically meaningful and defined in body-axis.

B. Proof for the predictive theorem

Proof:

Assume the real dynamics is described as:

$$X_{t+1} = A_{dr}X_t + B_{dr}U_{t-k} + W_t \tag{B-1}$$

which is different from the reference model of Eq. (11). In Eq. (B-1), X_t is system state, A_{dr} is the system matrix, B_{dr} is the control matrix, U_t is control input, W_t is process noise. The one-step prediction, according to Eq. (B-1), can be obtained by Eq. (13-14),

$$\begin{aligned} \hat{X}_{t|t+1} &= X_t + A_d\Delta X_t + B_d\Delta U_{t-k} \\ &= A_{dr}X_{t-1} + B_{dr}U_{t-1-k} + W_{t-1} \\ &\quad + A_d\Delta X_t + B_d\Delta U_{t-k} \end{aligned} \tag{B-2}$$

And

$$\begin{aligned} & E\left\{X_{t+1} - \hat{X}_{t+1|t}\right\} \\ & = E\{A_{dr}X_t + B_{dr}U_{t-k} + W_t \\ & \quad - (A_{dr}X_{t-1} + B_{dr}U_{t-1-k} + W_{t-1} + A_d\Delta X_t + B_d\Delta U_{t-k})\} \\ & = E\{(A_{dr} - A_d)\Delta X_t + (B_{dr} - B_d)\Delta U_{t-k} + \Delta W_t\} \end{aligned} \tag{B-3}$$

According to condition 1) and 2), prediction is bounded, then,

$$\left\| X_{t+1} - \hat{X}_{t+1|t} \right\| < +\infty$$

and, when the system of Eq. (B-1) works around a working point in steady state, the mean value of control inputs and states should be constant, so we can obtain:

$$\begin{aligned} & E\left\{X_{t+1} - \hat{X}_{t+1|t}\right\} \\ & = (A_{dr} - A_d)E\{\Delta X_t\} + (B_{dr} - B_d)E\{\Delta U_{t-k}\} + E\{\Delta W_t\} \\ & = (A_{dr} - A_d)\bullet 0 + (B_{dr} - B_d)\bullet 0 + 0 = 0 \end{aligned} \tag{B-4}$$

Eq. (B-4) indicates that the one step prediction of Eq. (B-2) is unbiased.
Assuming that prediction at time i-1 is unbiased, i.e..

$$E\{X_{t+i-1} - \hat{X}_{t+i-1|t}\} = 0 \tag{B-5}$$

for the prediction at time i, there is

$$\begin{aligned}
&E\{X_{t+i} - \hat{X}_{t+i|t}\} \\
&= E\{A_{dr}X_{t+i-1} + B_{dr}U_{t+i-1-k} + W_{t+i-1} \\
&-(\hat{X}_{t+i-1|t} + A_d\hat{\Delta}X_{t+i-1|t} + B_d\Delta U_{t+i-1-k})\} \\
&= E\{A_{dr}X_{t+i-1} + B_{dr}U_{t+i-1-k} + W_{t+i-1} - X_{t+i-1} \\
&+(X_{t+i-1} - \hat{X}_{t+i-1|t}) - W_{t+i-2} \\
&-A_d\Delta\hat{X}_{t+i-1|t} - B_d\Delta U_{t+i-1-k}\} \\
&= E\{A_{dr}\Delta X_{t+i-1} - A_d\hat{\Delta}X_{t+i-1|t} + \\
&(B_{dr} - B_d)\Delta U_{t+i-1-k} + \Delta W_{t+i-1}\} \\
&= (A_{dr} - A_d)E\{\Delta X_{t+i-1}\} + \\
&(B_{dr} - B_d)E\{\Delta U_{t+i-1-k}\} + E\{\Delta W_{t+i-1}\} \\
&= (A_{dr} - A_d)\bullet 0 + (B_{dr} - B_d)\bullet 0 + 0 = 0
\end{aligned} \tag{B-6}$$

Therefore, the prediction at time i is also unbiased.

C. Parameters' selection for estimate and control in flight experiment

1. For Modeling

The identification results for hovering dynamics are listed in Tab.D-1.

Longitudinal Loop		Lateral Loop		Vertical Loop	
Para.	Val.	Para.	Val.	Para.	Val.
Xu	0.2446	Yv	-0.0577	Zw	1.666
Xa	-4.962	Yb	9.812	Zr	-3.784
Xlat	-0.0686	Ylat	-1.823	Zped	2.304
Xlon	0.0896	Ylon	2.191	Zcol	-11.11
Mu	-1.258	Lv	15.84	Yaw Loop	
Ma	46.06	Lb	126.6	Para.	Val.
Mlat	-0.6269	Llat	-4.875	Nw	-0.027
Mlon	3.394	Llon	28.64	Nr	-1.087
Ac	0.1628	Bd	-1.654	Nrfb	-1.845
Alat	-0.0178	Blat	0.04732	Nped	1.845
Alon	-0.2585	Blon	-9.288	Ncol	-0.972
Clat	2.238	Dlat	-0.7798	Kr	-0.040
Clon	-4.144	Dlon	-5.726	Krfb	-2.174
tf	0.5026	ts	0.5054		

Table D-1. The parameters of hovering model

2. For ASMF

$$Q = \begin{pmatrix} 0.01I_{13\times13} & 0_{13\times13} \\ 0_{13\times13} & 0.1I_{13\times13} \end{pmatrix},\ R = 0.01I_{8\times8}$$

where $I_{m\times m}$ is the m×m unit matrix and $0_{m\times n}$ is the m×n zero matrix.

3. For GPC

$$p = 10,\ \gamma = 2.32I_{40\times40},\ k = 10$$

4. For SIPC

$$p = 10,\ \gamma = 2.32I_{4\times4},\ \alpha = 0.99I_{8\times8}$$

$$W = I_{80\times80},\ k = 10,\ \beta = 0.8I_{4\times4}$$

5. For AMSIPC

$$p = 10,\ \gamma = 2.32I_{4\times4},\ \alpha = 0.99I_{8\times8}$$

$$W = I_{80\times80},\ k = 10,\ \beta = 0.8I_{4\times4},\ H = I_{13\times13}$$

8. References

Tischler M.B., "Frequency-domain Identification of XV-15 Tilt-rotor Aircraft Dynamics in Hovering Flight," Journal of the American Helicopter Society, Vol. 30 (2), pp.38-48, 1985.

Tischler M. B. and Cauffman M. G., "Frequency-Response Method for Rotorcraft System Identification: Flight Application to BO-I05 Coupled Rotor/Fuselage Dynamics," Journal of the American Helicopter Society, Vol. 37 (3), pp.3-17, 1992.

Fletcher J. W., "Identification of UH-60 Stability Derivative Models in Hover from Flight Test Data," Journal of the American Helicopter Society, Vol. 40 (1), pp.8-20, 1995.

Mettler B., Tischler M. B. and Kanade T., "System Identification of Small-Size Unmanned Helicopter Dynamics," American Helicopter Society 55th Annual Forum Proceedings, Vol. 2, pp.1706-1717, Montreal, Quebec, Canada, May 25-27, 1999.

Gavrilets V., Metlter B. and Feron E., "Nonlinear Model for a Small-scale Acrobatic Helicopter," Proceedings of the American Institute of Aeronautics Guidance, Navigation, and Control Conference, pp.8, Montreal, Quebec, Canada, August 6-9, 2001.

Massimiliano M. and Valerio S., "A Full Envelope Small Commercial Aircraft Flight Control Design Using Multivariable Proportional-Integral Control," IEEE Transactions on Control Systems Technology, Vol. 16 (1), pp.169-176, January, 2008.

Voorsluijs M. and Mulder A., "Parameter-dependent robust control for a rotorcraft UAV," AIAA Guidance, Navigation, and Control Conference and Exhibit, pp.1-11, San Francisco, California, USA, August 15-18, 2005.

Bijnens B., Chu Q.P. and Voorsluijs M., "Adaptive feedback linearization flight control for a helicopter UAV," AIAA Guidance, Navigation, and Control Conference and Exhibit, pp.1-10, San Francisco, California, USA, August 15-18, 2005.

Kahveci N.E., Ioannou P.A., Mirmirani M.D., "Adaptive LQ Control With Anti-Windup Augmentation to Optimize UAV Performance in Autonomous Soaring

Applications," IEEE Transactions on Control Systems Technology, Vol. 16(4): pp.691 - 707, 2008

MacKunis W., Wilcox Z.D., Kaiser M.K., Dixon W.E., "Global Adaptive Output Feedback Tracking Control of an Unmanned Aerial Vehicle," IEEE Transactions on Control Systems Technology, Vol. 18(6): pp.1390-1397, 2010.

Cummings M.L., Mitchell P.J., "Predicting Controller Capacity in Supervisory Control of Multiple UAVs Systems," IEEE Transactions on Man and Cybernetics, Part A: Systems and Humans, Vol. 38(2): pp.451-460, 2008.

Jiang X., Han Q.L., "On guaranteed cost fuzzy control for nonlinear systems with interval time-varying delay," Control Theory & Applications, IET, Vol. 1(6): pp.1700-1710, 2007.

Natori K., Oboe R., Ohnishi, K., "Stability Analysis and Practical Design Procedure of Time Delayed Control Systems With Communication Disturbance Observer," IEEE Transactions on Industrial Informatics, Vol. 4(3): pp.185-197, 2008.

Haykin S. and De Freitas N., "Special Issue on Sequential State Estimation," Proceedings of the IEEE, Vol. 92(3), pp.423-574, 2004.

Lerro D. and Bar-Shalom Y. K., " Tracking with Debiased Consistent Converted Measurements vs. EKF," IEEE Transactions on Aerosp. Electron.System, AES-29, pp.1015-1022, 1993 ·

Julier S. and Uhlmann J., "Unscented filtering and nonlinear estimation," Proceedings of the IEEE, Vol. 92(3), pp. 401-422, 2004.

Song Q., Jiang Z., and Han J. D., "UKF-Based Active Model and Adaptive Inverse Dynamics Control for Mobile Robot," IEEE International Conference on Robotics and Automation, 2007.

Shamma J. S. and Tu K. Y., "Approximate set-valued observers for nonlinear systems," IEEE Transactions on Automatic Control, Vol. 42(5), pp.648–658, 1997.

Zhou B., Han J.D. and Liu G., "A UD factorization-based nonlinear adaptive set-membership filter for ellipsoidal estimation," International Journal of Robust and Nonlinear Control, Vol 18 (16), pp.1513-1531, November 10, 2007.

Scholte E., Campbell M.E., "Robust Nonlinear Model Predictive Control With Partial State Information," Control Systems Technology, IEEE Transactions on, Vol. 16(4): pp.636-651, 2008.

Ding B. C., Xi Y. G., "A Synthesis Approach of On-line Constrained Robust Model Predictive Control." Automatica. Vol. 40(1): pp. 163-167, 2004.

Crassidis J. L., "Robust Control of Nonlinear Systems Using Model-Error Control Synthesis," Journal of guidance, control and dynamics, Vol. 22 (4), pp.595-601, 1999.

Gregor K. and Igor S., "Tracking-error Model-based Predictive Control for Mobile Robots in real time." Robotics and Autonomous Systems. Vol. 55, No. 7, pp. 460 - 469, 2007.

Qi J.T., Song D.L., Dai. L., Han J.D., "The ServoHeli-20 Rotorcraft UAV Project," International Conference on Mechatronics and Machine Vision in Practice, Auckland, New Zealand, pp.92-96, 2008.

Song D.L., Qi J.T., Dai. L., Han J.D. and Liu G., "Modeling a Small-size Unmanned Helicopter Using Optimal Estimation in The Frequency Domain," International Conference on Mechatronics and Machine Vision in Practice, Auckland, New Zealand, December 2-4, pp.97-102, 2008.

Song D.L., Qi J.T. and Han J.D., "Model Identification and Active Modeling Control for Small-Size Unmanned Helicopters: Theory and Experiment," AIAA Guidance Navigation and Control, Toronto, Canada, AIAA-2010-7858, 2010.

13

Nonlinear Autoregressive with Exogenous Inputs Based Model Predictive Control for Batch Citronellyl Laurate Esterification Reactor

Siti Asyura Zulkeflee, Suhairi Abdul Sata and Norashid Aziz
School of Chemical Engineering, Engineering Campus, Universiti Sains Malaysia, Seri Ampangan, 14300 Nibong Tebal, Seberang Perai Selatan, Penang, Malaysia

1. Introduction

Esterification is a widely employed reaction in organic process industry. Organic esters are most frequently used as plasticizers, solvents, perfumery, as flavor chemicals and also as precursors in pharmaceutical products. One of the important ester is Citronellyl laurate, a versatile component in flavors and fragrances, which are widely used in the food, beverage, cosmetic and pharmaceutical industries. In industry, the most common ester productions are carried out in batch reactors because this type of reactor is quite flexible and can be adapted to accommodate small production volumes (Barbosa-Póvoa, 2007). The mode of operation for a batch esterification reactor is similar to other batch reactor processes where there is no inflow or outflow of reactants or products while the reaction is being carried out. In the batch esterification system, there are various parameters affecting the ester rate of reaction such as different catalysts, solvents, speed of agitation, catalyst loading, temperature, mole ratio, molecular sieve and water activity (Yadav and Lathi, 2005). Control of this reactor is very important in achieving high yields, rates and to reduce side products. Due to its simple structure and easy implementation, 95% of control loops in chemical industries are still using linear controllers such as the conventional Proportional, Integral & Derivative (PID) controllers. However, linear controllers yield satisfactory performance only if the process is operated close to a nominal steady-state or if the process is fairly linear (Liu & Macchietto, 1995). Conversely, batch processes are characterized by limited reaction duration and by non-stationary operating conditions, then nonlinearities may have an important impact on the control problem (Hua *et al.*, 2004). Moreover, the control system must cope with the process variables, as well as facing changing operation conditions, in the presence of unmeasured disturbances. Due to these difficulties, studies of advanced control strategy have received great interests during the past decade. Among the advanced control strategies available, the Model Predictive Control (MPC) has proved to be a good control for batch reactor processes (Foss *et al.*, 1995; Dowd *et al.*, 2001; Costa *et al.*, 2002; Bouhenchir *et al.*, 2006). MPC has influenced process control practices since late 1970s. Eaton and Rawlings (1992) defined MPC as a control scheme in which the control algorithm optimizes the manipulated variable profile over a finite future time horizon in order to maximize an objective function subjected to plant models and constraints. Due to these features, these model based control algorithms can be extended to include multivariable systems and can be formulated to handle process constraints explicitly. Most of the improvements on MPC algorithms are based on the developmental reconstruction of the MPC basic elements which include prediction model, objective function and optimization algorithm. There are several comprehensive technical surveys of theories and future exploration direction of MPC by Henson, 1998, Morari & Lee, 1999, Mayne *et al.*, 2000 and Bequette, 2007. Early development of this kind of control strategy, the Linear Model Predictive Control (LMPC) techniques such as Dynamic Matrix Control (DMC) (Gattu and Zafiriou, 1992) have been successfully implemented on a large number of processes. One limitation to the LMPC methods is that they are based on linear system theory and may not perform well on highly nonlinear system. Because of this, a Nonlinear Model Predictive Control (NMPC) which is an extension of the LMPC is very much needed.

NMPC is conceptually similar to its linear counterpart, except that nonlinear dynamic models are used for process prediction and optimization. Even though NMPC has been successfully implemented in a number of applications (Braun *et al.*, 2002; M'sahli et al., 2002; Ozkan *et al.*, 2006; Nagy *et al.*, 2007; Shafiee et al., 2008; Deshpande *et al.*, 2009), there is no common or standard controller for all processes. In other words, NMPC is a unique controller which is meant only for the particular process under consideration. Among the major issues in NMPC development are firstly, the development of a suitable model that can represent the real process and secondly, the choice of the best optimization technique. Recently a number of modeling techniques have gained prominence. In most systems, linear models such as partial least squares (PLS), Auto Regressive with Exogenous inputs (ARX) and Auto Regressive Moving Average with Exogenous inputs (ARMAX) only perform well over a small region of operations. For these reasons, a lot of attention has been directed at identifying nonlinear models such as neural networks, Volterra, Hammerstein, Wiener and NARX model. Among of these models, the NARX model can be considered as an outstanding choice to represent the batch esterification process since it is easier to check the model parameters using the rank of information matrix, covariance matrices or evaluating the model prediction error using a given final prediction error criterion. The NARX model provides a powerful representation for time series analysis, modeling and prediction due to its strength in accommodating the dynamic, complex and nonlinear nature of real time series applications (Harris & Yu, 2007; Mu *et al.*, 2005). Therefore, in this work, a NARX model has been developed and embedded in the NMPC with suitable and efficient optimization algorithm and thus currently, this model is known as NARX-MPC.

Citronellyl laurate is synthesized from DL-citronellol and Lauric acid using immobilized *Candida Rugosa* lipase (Serri *et. al.*, 2006). This process has been chosen mainly because it is a very common and important process in the industry but it has yet to embrace the advanced control system such as the MPC in their plant operation. According to Petersson *et al.* (2005), temperature has a strong influence on the enzymatic esterification process. The temperature should preferably be above the melting points of the substrates and the product, but not too high, as the enzyme's activity and stability decreases at elevated temperatures. Therefore, temperature control is important in the esterification process in order to achieve maximum ester production. In this work, the reactor's temperature is controlled by manipulating the flowrate of cooling water into the reactor jacket. The performances of the NARX-MPC were evaluated based on its set-point tracking, set-point change and load change. Furthermore, the robustness of the NARX-MPC is studied by using four tests i.e. increasing heat transfer coefficient, increasing heat of reaction, decreasing inhibition activation energy and a simultaneous change of all the mentioned parameters. Finally, the performance of NARX-MPC is compared with a PID controller that is tuned using internal model control technique (IMC-PID).

2. Batch esterification reactor

The synthesis of Citronellyl laurate involved an exothermic process where Citronellol reacted with Lauric acid to produce Citronellyl Laurate and water.

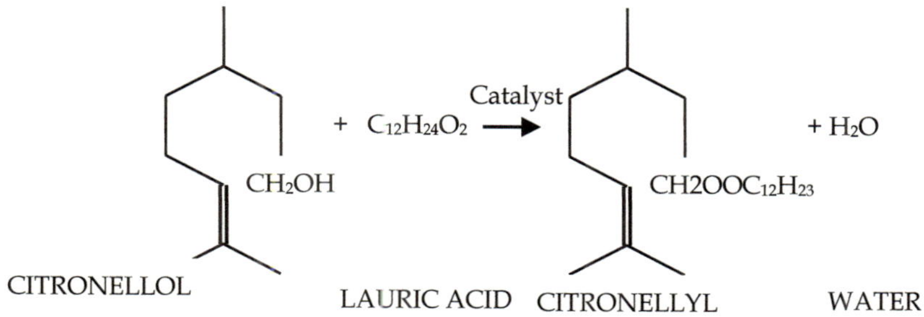

Fig. 1. Schematic represent esterification of Citronellyl laurate

The esterification process took place in a batch reactor where the immobilized lipase catalyst was mixed freely in the reactor. A layout of the batch esterification reactor with associated heating and cooling configurations is shown in Fig.2.

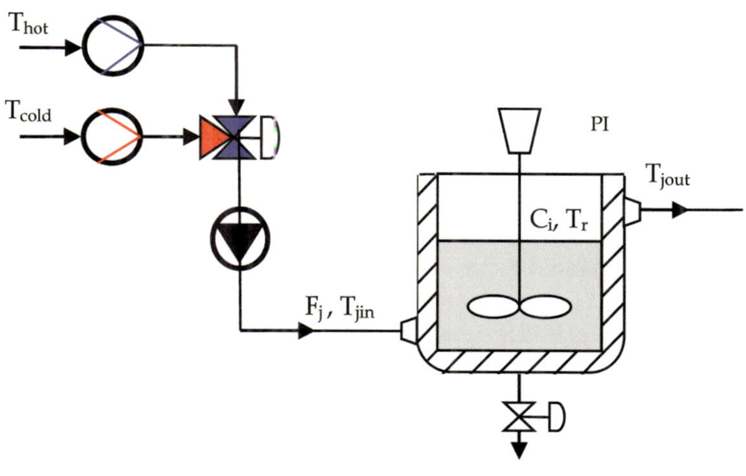

Fig. 2. Schematic diagram of the batch esterification reactor.

Typical operating conditions were 310K and 1 bar. The reactor temperature was controlled by manipulating the water flowrate within the jacket. The reactor's temperature should not exceed the maximal temperature of 320K, due to the temperature sensitivity of the catalysts (Yadav & Lathi, 2004; Serri *et. al.*, 2006; Zulkeflee & Aziz, 2007). The reactor's temperature control can be achieved by treating the limitation of the jacket's flowrate, Fj, which can be viewed as a state of the process and as the constraint control problem. The control strategy proposed in this paper was designed to meet the specifications of the laboratory scale batch reactor at the Control Laboratory of School of Chemical Engineering, University Sains Malaysia, which has a maximum of 0.2 L/min limitation on the jacket's flowrate. Therefore, the constraint of the jacket's flowrate will be denoted as F_{jmax} = 0.2 L/min.

The fundamental equations of the mass and energy balances of the process are needed to generate data for empirical model identification. The equations are valid for all $t \in [0, \infty]$. The reaction rate and kinetics are given by (Yadav & Lathi, 2004; Serri *et. al.*, 2006; Zulkeflee & Aziz, 2007):

$$\frac{dC_{Al}}{dt} = \frac{[C_{Al}]r_{max}}{\alpha K_{Al}\left(1+\left(\frac{K_{Ac}}{[C_{Ac}]}\right)+\left(\frac{K_{Ac}}{K_i}\right)+\left(\frac{[C_{Ac}]}{\beta K_i}\right)\right)} + [C_{Al}]\left(1+\left(\frac{\alpha K_{Ac}}{[C_{Ac}]}\right)\right) \tag{1}$$

$$\frac{dC_{Ac}}{dt} = \frac{[C_{Ac}]r_{max}}{\alpha K_{Ac}\left(1+\left(\frac{K_{Ac}}{[C_{Ac}]}\right)\right)} + [C_{Ac}]\left(1+\left(\frac{\alpha K_{Al}}{[C_{Al}]}\right)\right) + \frac{\alpha K_{Al}K_{Ac}}{K_i[C_{Al}]} + \frac{\alpha K_{Al}[C_{Ac}]}{\beta K_i[C_{Al}]} \tag{2}$$

$$\frac{dC_{Es}}{dt} = -\frac{[C_{Al}]r_{max}}{\alpha K_{Al}\left(1+\left(\frac{K_{Ac}}{[C_{Ac}]}\right)+\left(\frac{K_{Ac}}{K_i}\right)+\left(\frac{[C_{Ac}]}{\beta K_i}\right)\right)} + [Al]\left(1+\left(\frac{\alpha K_{Ac}}{[C_{Ac}]}\right)\right) \tag{3}$$

$$\frac{dC_W}{dt} = -\frac{[C_{Ac}]r_{max}}{\alpha K_{Ac}\left(1+\left(\frac{K_{Ac}}{[C_{Ac}]}\right)\right)} + [C_{Ac}]\left(1+\left(\frac{\alpha K_{Al}}{[C_{Al}]}\right)\right) + \frac{\alpha K_{Al}K_{Ac}}{K_i[C_{Al}]} + \frac{\alpha K_{Al}[C_{Ac}]}{\beta K_i[C_{Al}]} \tag{4}$$

$$K_i = A_i exp^{-E_i/RT_r} \tag{5}$$

$$K_{Ac} = A_{Ac} exp^{-E_{Ac}/RT_r} \tag{6}$$

$$K_{Al} = A_{Al} exp^{-E_{Al}/RT_r} \tag{7}$$

where C_{Ac}, C_{Al}, C_{Es} and C_W are concentrations (mol/L) of Lauric acid, Citronellol, Citronellyl laurate and water respectively; r_{max} (mol l^{-1} min^{-1} g^{-1} of enzyme) is the maximum rate of reaction, K_{Ac} (mol l^{-1} g^{-1} of enzyme), K_{Al} (mol l^{-1} g^{-1} of enzyme) and K_i (mol l^{-1} g^{-1} of enzyme) are the Michealis constant for Lauric acid, Citronellol and inhibition respectively; A_i, A_{Ac} and A_{Al} are the pre-exponential factors (L mol/s) for inhibition, Lauric acid and Citronellol respectively; E_i, E_{Ac} and E_{Al} are the activation energy (J mol/K) for inhibition, acid lauric and Citronellol respectively; R is the gas constant (J/mol K).

The reactor can be described by the following thermal balances (Aziz *et al.*, 2000):

$$\frac{dT_r}{dt} = \Delta H_{rxn} r_{Ac} V + \frac{\dot{Q}}{[V(C_{Ac}Cp_{Ac}+C_{Al}Cp_{Al}+C_{Es}Cp_{Es}+C_W Cp_W)]} \tag{8}$$

$$\frac{dT_j}{dt} = \frac{(F_j Cp_W \rho_W (T_{jin}-T_j)-\dot{Q})}{V_j Cp_W \rho_W} \tag{9}$$

$$\dot{Q} = UA\left(T_j - T_r\right) \tag{10}$$

where T_r (K) , T_j (K) and T_{jin} is reactor, jacket and inlet jacket temperature respectively; ΔH_{rxn} (kJ/mol) is heat of reaction; V(l) and V_j(l) is the volume of the reactor and jacket respectively; $Cp_{Ac}, Cp_{Al}, Cp_{Es}$ and Cp_W are specific heats (J/mol K) of Lauric acid, Citronellol, Citronellyl laurate and water respectively; ρ_W is the water density (g/L) in the jacket; F_j is the flowrate of the jacket (L/min); $\dot{Q}$ (kW) is the heat transfer through the jacket wall; A and U are the heat exchange area (m^2) and the heat exchange coefficient (W/m^2/K) respectively. Eq. 1 - Eq. 10 were simulated using a 4th/5th order of the Runge Kutta method in MATLAB® environment. The model of the batch esterification process was derived under the assumption that the process is perfectly mixed where the concentrations of $[Ac]$, $[Al]$, $[Es]$, $[w]$ and temperature of the fluid in the tank is uniform. Table 1 shows all the value of the parameters for the batch esterification process under consideration. The validations of corresponding dynamic models have been reported in Zulkeflee & Aziz (2007).

Parameters	Units	Values	Parameters	Units	Values
A_{Ac}	L mol/s	18.20871	Cp_w	J/mol K	75.40
A_{Al}	L mol/s	24.04675	V	L	1.5
A_i	L mol/s	0.319947	V_j	L	0.8
E_{Ac}	J mol/K	-105.405	$\dot{Q}$	J/m^3	11.648
E_{Al}	J mol/K	-66.093	ΔH_{rxn}	kJ	16.73
E_i	J mol/K	-249.944	a	-	1
T_{ji}	K	294	β	-	1
Cp_{Ac}	J/mol K	420.53	U	J/s m^2 K	2.857
Cp_{Al}	J/mol K	235.27	A	m^2	0.077
Cp_{Es}	J/mol K	617.79	R	J/mol K	8.314

Table 1. Operating Conditions and Calculated Parameters

3. NARX model

The Nonlinear Autoregressive with Exogenous inputs (NARX) model is characterized by the non-linear relations between the past inputs, past outputs and the predicted process output and can be delineated by the high order difference equation, as follows:

$$y(t) = f\{y(t-1), \dots y(t-n_y), u(t-1) \dots u(t-n_u)\} + e(t) \tag{11}$$

where $u(t)$ and $y(t)$ represents the input and output of the model at time t in which the current output $y(t) \in \Re$ depends entirely on the current input $u(t) \in \Re$. Here $n_u$and n_y are the input and output orders of the dynamical model which are $n_u \geq 0, n_y \geq 1$. The function f is a nonlinear function. $\bar{X} = [y(t-1) \dots y(t-n_y)\; u(t-1) \dots u(t-n_{u})]^T$ denotes the system input vector with a known dimension $n = n_y + n_u$. Since the function f is unknown, it is approximated by the regression model of the form:

$$y(t) = \sum_{i=0}^{n_u} a(i).u(t-i) + \sum_{j=1}^{n_y} b(j).y(t-j) + \sum_{i=0}^{n_u}\sum_{j=i}^{n_u} a(i,j).u(t-i).u(t-j) + \sum_{i=1}^{n_y}\sum_{j=i}^{n_y} b(i,j).y(t-i).y(t-j) + \sum_{i=o}^{n_u}\sum_{j=1}^{n_y} c(i,j).u(t-i).y(t-j) + e(t) \tag{12}$$

where $a(i)$ and $a(i,j)$ are the coefficients of linear and nonlinear for originating exogenous terms; $b(i) and\, b(i,j)$ are the coefficients of the linear and nonlinear autoregressive terms; $c(i,j)$ are the coefficients of the nonlinear cross terms. Eq. 12 can be written in matrix form:

$$\begin{bmatrix} y(t) \\ y(t+1) \\ \vdots \\ y(t+n_y) \end{bmatrix} = a.u^T + b.y^T + A.[U]^T + B.[Y]^T + C.[X]^T \tag{13}$$

where

$$a = [a(0)\ a(1) \dots a(n_u)]^T \tag{14}$$

$$b = [b(1)\ b(2) \dots b(n_y)]^T \tag{15}$$

$$A = [a(0,0)\ a(0,1) \dots a(0,n_u)\ a(1,1) \dots a(n_u,n_u)]^T \tag{16}$$

$$B = [b(1,1)\ b(1,2) \dots b(1,n_y)\ b(2,2) \dots b(n_y,n_y)]^T \tag{17}$$

$$C = [c(0,1)\ c(0,2) \dots c(0,n_y)\ c(1,1) \dots c(n_u,n_y)]^T \tag{18}$$

$$u = [u(t)\ u(t-1) \dots u(n_u)] \tag{19}$$

$$y = [y(t-1)\ u(t-2) \dots u(n_y)] \tag{20}$$

$$U = [u(t).u(t)\ u(t).u(t-1) \dots u(t).u(t-n_u)\ u(t-1).u(t-1) \dots u(t-n_u).u(t-n_u)] \tag{21}$$

$$Y = [y(t-1).y(t-1)\ y(t-1).y(t-2) \dots y(t-1).y(t-n_y)\ y(t-2).y(t-2) \dots y(t-n_y).y(t-n_y)] \tag{22}$$

$$X = [u(t).y(t-1)u\ .y\ (t-2)..u(t).y(t-n_y)u(t-1).y(t-1)..u(t-n_u).u(t-n_y)] \tag{23}$$

The Eq. 13 can alternatively be expressed as:

$$y(t) = [u^T\ y^T\ U^T\ Y^T\ X^T] \begin{bmatrix} a \\ b \\ A \\ B \\ C \end{bmatrix} \tag{24}$$

and can be simplified as:

$$\bar{Y} = \bar{U}.\bar{C} \tag{25}$$

where

$$\bar{Y} = y(t) \tag{26}$$

$$\bar{U} = [u^T\ y^T\ U^T\ Y^T\ X^T] \tag{27}$$

$$\bar{C} = [a\ b\ A\ B\ C]^T \tag{28}$$

Finally, the solution of the above identification problem is represented by

$$\bar{C} = \bar{U} \backslash \bar{Y} \tag{29}$$

The procedures for a NARX model identification is shown in Fig. 3. This model identification process includes:

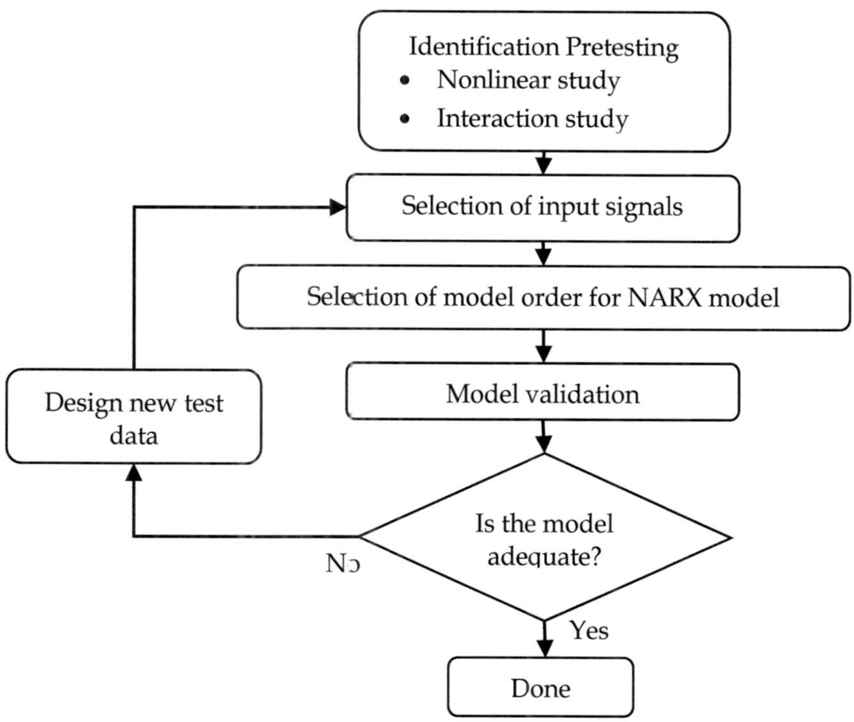

Fig. 3. NARX model identification procedure

- Identification pre-testing: This study is very important in order to choose the important controlled, manipulated and disturbance variables. A preliminary study of the response plots can also gives an idea of the response time and the process gain.
- Selection of input signal: The study of input range has to be done, to calculate the maximal possible values of all the input signals so that both inputs and outputs will be within the desired operating conditions range. The selection of input signal would allow the incorporation of additional objectives and constraints, i.e. minimum or maximum input event separations which are desirable for the input signals and the resulting process behavior.
- Selection of model order: The important step in estimating NARX models is to choose the model order. The model performance was evaluated by the Means Squared Error (MSE) and Sum Squared Error (SSE).
- Model validation: Finally, the model was validated with two sets of validation data which were unseen independent data sets that are not used in NARX model parameter estimation.

The details of the identification of NARX model for the batch esterification can be found at Zulkeflee & Aziz (2008).

4. MPC algorithm

The conceptual structure of MPC is depicted in Fig. 4. The conception of MPC is to obtain the current control action by solving, at each sampling instant, a finite horizon open-loop optimal control problem, using the current state of the plant as the initial state. The desired objective function is minimized within the optimization method and related to an error function based on the differences between the desired and actual output responses. The first optimal input was actually applied to the plant at time t and the remaining optimal inputs were discarded. Meanwhile, at time *t+1*, a new measurement of optimal control problem was resolved and the receding horizon mechanism provided the controller with the desired feedback mechanism (Morari & Lee, 1999; Qin & Badgwell, 2003; Allgower, Findeisen & Nagy, 2004).

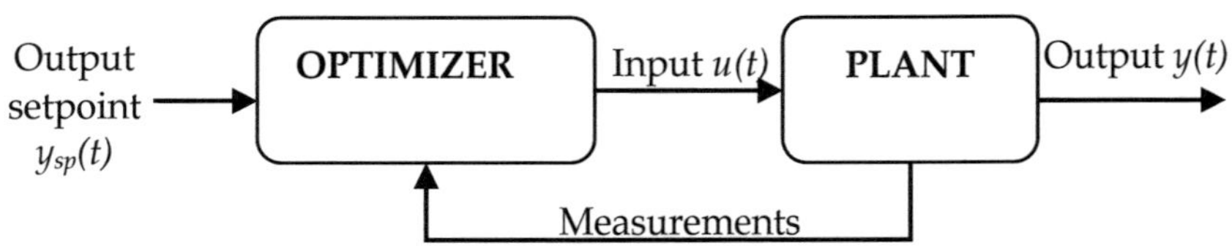

Fig. 4. Basic structure of Model Predictive Control

A formulation of the MPC on-line optimization can be as follows:

$$\min_{u[t|t],\ldots u[m+p|t]_{i=0}} J(y(t), u(t)) \tag{30}$$

$$\min_{u[t|t],\ldots u[m+p|t]_{k=0}} \sum_{k=1}^{P} w_k (y[t+k|t] - y_{sp})^2 + \sum_{k=1}^{M} r_k \Delta u[t+k|t]^2 \qquad (31)$$

Where P and M is the length of the process output prediction and manipulated process input horizons respectively with $P \leq M$. $u[t+k|t]_{k=0,\ldots P}$ is the set of future process input values. The vector w_k is the weight vector .

The above on-line optimization problem could also include certain constraints. There can be bounds on the input and output variables:

$$u_{max} \geq u[t+k|t] \geq u_{min} \qquad (32)$$

$$\Delta u_{max} \geq \Delta u[t+k|t] \geq -\Delta u_{min} \qquad (33)$$

$$y_{max} \geq y[t+k|t] \geq y_{min} \qquad (34)$$

It is clear that the above problem formulation necessitates the prediction of future outputs

$$y[t+k|t]$$

In this NARX model, for k step ahead:

The error *e(t)*:

$$\begin{aligned} e[t|t] = y(t) - \sum_{i=0}^{n_u} a(i).u(t-i) - \sum_{j=1}^{r_y} b(j).y(t-j) - \sum_{i=0}^{n_u}\sum_{j=i}^{n_u} a(i,j).u(t-i).u(t-j) \\ - \sum_{i=1}^{n_y}\sum_{j=i}^{n_y} b(i,j).y(t-i).y(t-j) - \sum_{i=o}^{n_u}\sum_{j=1}^{n_y} c(i,j).u(t-i).y(t-j) \end{aligned} \qquad (35)$$

The prediction of future outputs:

$$\begin{aligned} y(t+k) = \sum_{i=0}^{n_u} a(i).u(t-i+k) + \sum_{j=1}^{n_y} b(j).y(t-j+k) \\ + \sum_{i=0}^{n_u}\sum_{j=i}^{n_u} a(i,j).u(t-i+k).u(t-j+k) \\ + \sum_{i=1}^{n_y}\sum_{j=i}^{n_y} b(i,j).y(t-i+k).y(t-j+k) \\ + \sum_{i=o}^{n_u}\sum_{j=1}^{n_y} c(i,j).u(t-i+k).y(t-j+k) + e(t+k) \end{aligned} \qquad (36)$$

Substitution of Eq. 35 and Eq. 36 into Eq 31 yields:

$$\min_{u[t|t],\ldots u[m+p|t]_{k=0}} \sum_{k=1}^{P} w_k \left(\left(\sum_{i=0}^{n_u} a(i).u(t-i+k \right) + \sum_{j=1}^{n_y} b(j).y(t-j+k \right)$$
$$+ \sum_{i=0}^{n_u}\sum_{j=i}^{n_u} a(i,j).u(t-i+k).u(t-j+k)$$
$$+ \sum_{i=1}^{n_y}\sum_{j=i}^{n_y} b(i,j).y(t-i+k).y(t-j+k)$$
$$+ \sum_{i=o}^{n_u}\sum_{j=1}^{n_y} c(i,j).u(t-i+k).y(t-j+k) + y(t) - \sum_{i=0}^{n_u} a(i).u(t-i) \tag{37}$$
$$- \sum_{j=1}^{n_y} b(j).y(t-j) - \sum_{i=0}^{n_u}\sum_{j=i}^{n_u} a(i,j).u(t-i).u(t-j)$$
$$- \sum_{i=1}^{n_y}\sum_{j=i}^{n_y} b(i,j).y(t-i).y(t-j) - \sum_{i=o}^{n_u}\sum_{j=1}^{n_y} c(i,j).u(t-i).y(t-j))$$
$$- y_{sp})^2 + \sum_{k=1}^{M} r_k \Delta u[t+i|t]^2$$

Where

$$y_{sp}(t) = [y_{sp}(t+1)\; y_{sp}(t+2) \ldots .\, y_{sp}(t+P)]^T \tag{38}$$

$$\Delta u(t) = [\Delta u[t|t]\; \Delta u[t+1|t] \ldots \Delta u[t+M-1|t]]^T \tag{39}$$

The above optimization problem is a nonlinear programming (NLP) which can be solved at each time *t*. Even though the input trajectory was calculated until *M-1* sampling times into the future, only the first computed move was implemented for one sampling interval and the above optimization was repeated at the next sampling time. The structure of the proposed NARX-MPC is shown in Fig. 5.

In this work, the optimization problem was solved using constrained nonlinear optimization programming (*fmincon*) function in the MATLAB. A lower flowrate limit of 0 L/min and an upper limit of 0.2 L/min and a lower temperature limit of 300K and upper limit of 320K were chosen for the input and output variables respectively. In order to evaluate the performance of NARX-MPC controller, the NARX-MPC has been used to track the temperature set-point at 310K. For the set-point change, a step change from 310K to 315K was introduced to the process at t=25 min. For load change, a disturbance was implemented with a step change (+10%) for the jacket temperature from 294K to 309K. Finally, the performance of NARX-MPC is compared with the performance of PID controller. The parameters of PID controller have been estimated using the internal model based controller. The details of the implementation of IMC-PID controller can be found in Zulkeflee & Aziz (2009).

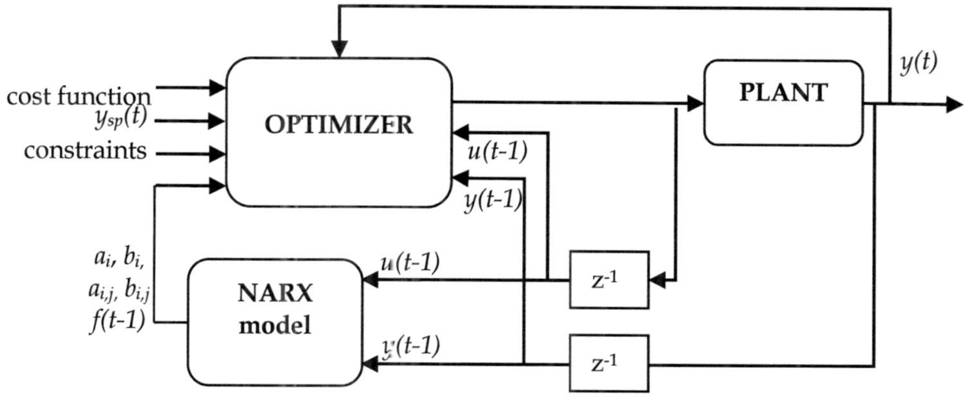

Fig. 5. The structure of the NARX-MPC

5. Results

5.1 NARX model identification

The input and output data for the identification of a NARX model have been generated from the validated first principle model. The input and output data used for nonlinear identification are shown in Fig. 6. The minimum-maximum range input (0 to 0.2 L/min) under the amplitude constraint was selected in order to achieve the most accurate parameter to determine the ratio of the output parameter. For training data, the inputs signal for jacket flowrate was chosen as multilevel signal. Different orders of NARX models which was a mapping of past inputs (n_u) and output (n_y) terms to future outputs were tested and the best one was selected according to the MSE and SSE criterion. Results have been summarized in Table 2. From the results, the MSE and SSE value decreased by increasing the model order until the NARX model with n_u =1 and n_y= 2. Therefore, the NARX model with n_u =1 and n_y= 2 was selected as the optimum model with MSE and SSE equal to 0.0025 and 0.7152 respectively. The respective graphical error of identification for training and validation of estimated NARX model is depicted in Fig. 7.

5.2 NARX-MPC

The identified NARX model of the process has been implemented in the MPC algorithm. Agachi *et al.*, (2007) proposed some criteria to select the significant tuning parameters (prediction horizon, P; control horizon, M; penalty weight matrices w_k and r_k) for the MPC controller. In many cases, the prediction (P) and control horizons (M) are introduced as P>M>1 due to the fact that it allows consequent control over the variables for the next future cycles. The value of weighting (w_k and r_k) of the controlled variables must be large enough to minimize the constraint violations in objective function. Tuning parameters and SSE values of the NARX-MPC controller are shown in Table 3. Based on these results, the effect of changing the control horizon, M for M: 2, 3, 4 and 5 indicated that M=2 gave the smallest error of output response with SSE value=424.04. From the influence of prediction horizon, P results, the SSE value was found to decrease by increasing the number of prediction horizon until P=11 with the smallest SSE value = 404.94. SSE values shown in Table 3 demonstrate that adjusting the elements of the w_k and r_k weighting matrix can improve the control performance. The value of w_k= 0.1 and r_k = 1 had resulted in the smallest error with SSE=386.45. Therefore, the best tuning parameters for the NARX-MPC controller were P=11; M=2; w_k = 0.1 and r_k= 1.

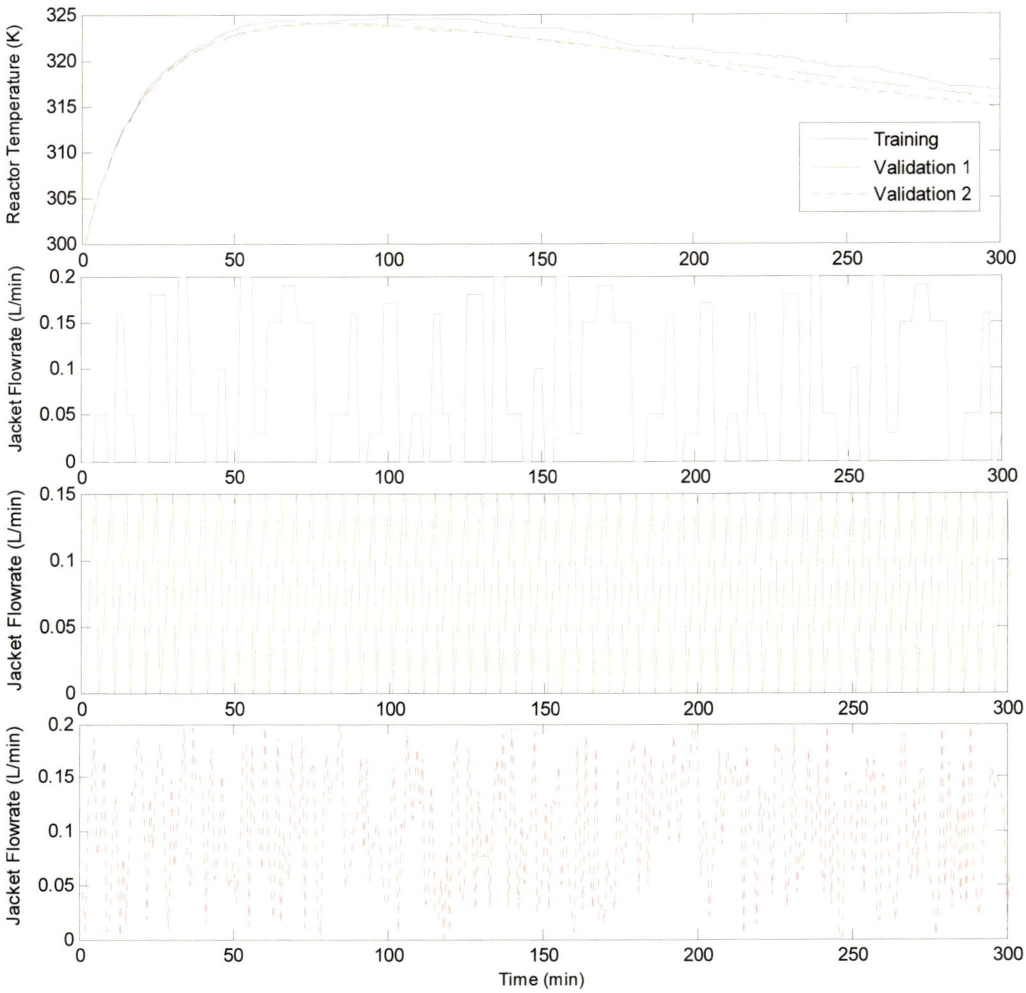

Fig. 6. Input output data for NARX model identification

Model	Training		Validation1		Validation2	
(n_u, n_y)	mse	sse	mse	sse	mse	sse
0,1	0.0205	6.1654	0.0285	8.5909	0.0254	7.6357
1,1	0.0202	6.0663	0.0307	9.2556	0.0251	7.5405
2,1	0.0194	5.8419	0.0392	11.8036	0.0266	8.0157
1,2	0.0025	0.7512	0.0034	1.0114	0.0059	1.7780
2,2	0.0026	0.7759	0.0029	0.8639	0.0038	1.1566
3,2	0.0024	0.7289	0.0035	1.0625	0.0097	2.9141
2,3	0.0024	0.7143	0.0033	0.9930	0.0064	1.9212

Table 2. MSE and SSE values of NARX model for different number of n_u and n_y

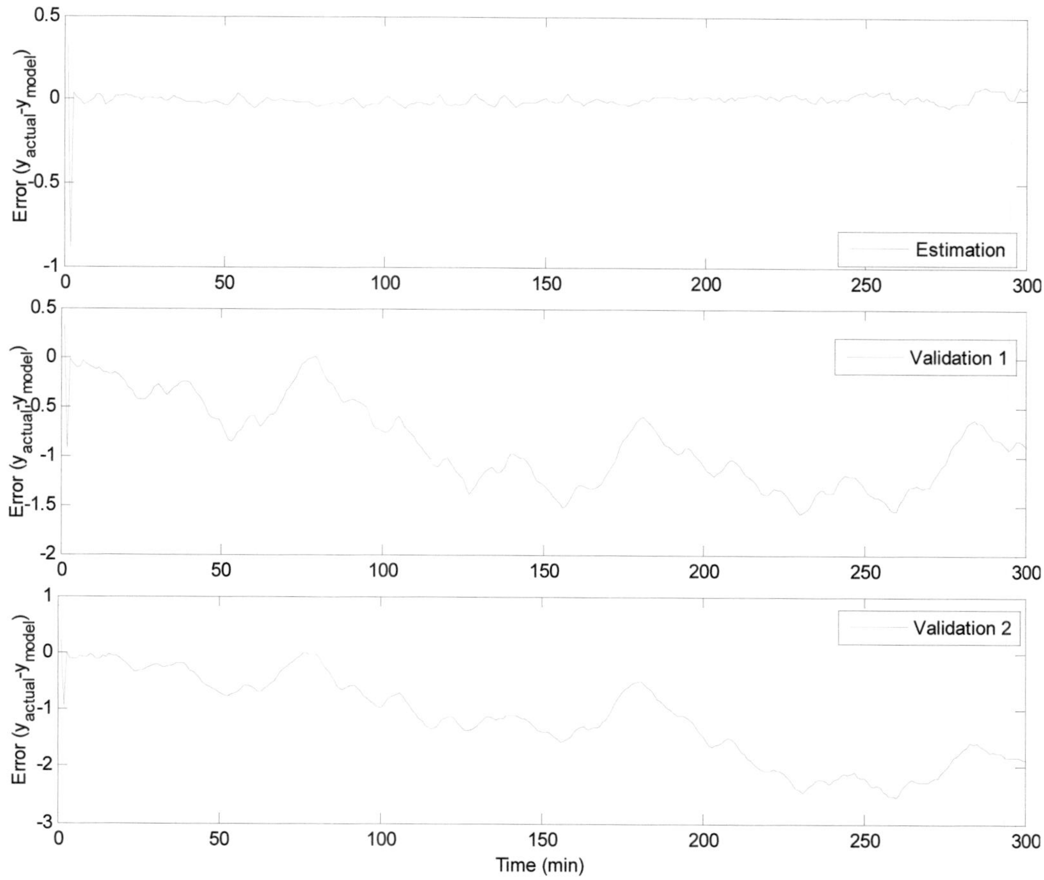

Fig. 7. Graphical error of identification for the training and validation of estimated NARX model

Tuning Parameter	SSE	Tuning Parameter	SSE
M=2	424.04	w_k= 10	410.13
M=3	511.35	w_k= 1	404.94
M =4	505.26	w_k= 0.1	386.45
M=5	509.95	w_k= 0.01	439.37
with P= 7; w_k= 1; r_k= 1		with P= 11; M= 2; r_k = 1	
P =7	424.04	r_k = 10	439.23
P =10	405.31	r_k = 1	386.45
P =11	404.94	r_k= 0.1	407.18
P =12	406.06	r_k = 0.01	410.02
with M= 2; w_k= 1; r_k= 1		with P =11; M=2; w_k=0.1	

Table 3. Tuning parameters and SSE criteria for applied controllers in set-point tracking

The responses obtained from the NARX-MPC and the IMC-PID controllers with parameter tuning, K_c=8.3; T_I=10.2; T_D=2.55 (Zulkeflee & Aziz, 2009) during the set-point tracking are shown in Fig. 8. The results show that the NARX-MPC controller had driven the process output to the desired set-point with a fast response time (10 minutes) and no overshoot or oscillatory response with SSE value = 386.45. In comparison, the output response for the unconstrained IMC-PID controller only reached the set-point after 25 minutes and had shown smooth and no overshoot response with SSE value = 402.24. However, in terms of input variable, the output response for the IMC-PID controller has shown large deviations as compared to the NARX-MPC. Normally, *actuator saturation is among the most conventional and notable problem in control system designs and the IMC-PID controller did not take this into consideration.* Concerning to this matter, an alternative to set a constraint value for the IMC-PID manipulated variable has been developed. As a result, the new IMC-PID control variable with constraint had resulted in higher overshoot with a settling time of about 18 minutes with SSE=457.12.

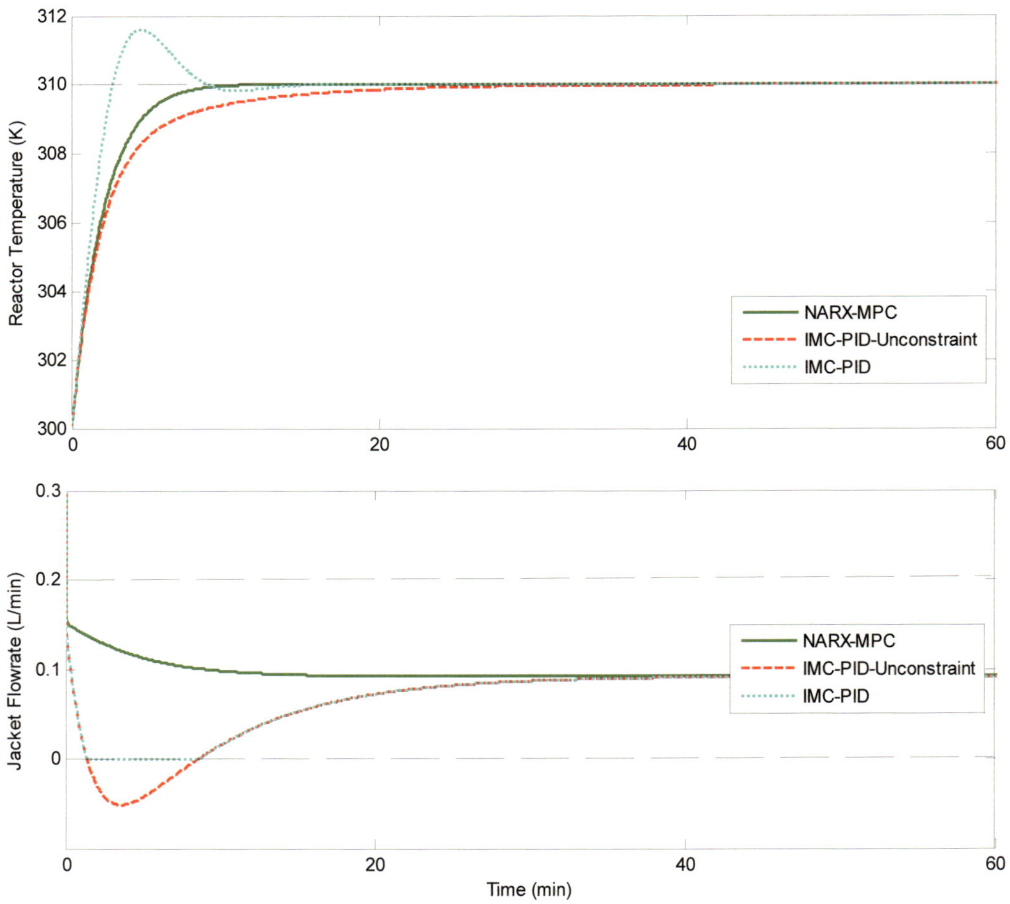

Fig. 8. Control response of NARX-MPC and IMC-PID controllers for set-point tracking with their respective manipulated variable action.

With respect to the conversion of ester, the implementation of the NARX-MPC controller led to a higher conversion of Citronellyl laurate (95% conversion) as compared to the IMC-PID, with 90% at time=150min (see Fig. 9). Hence, it has been proven that the NARX-MPC is far better than the IMC-PID control scheme.

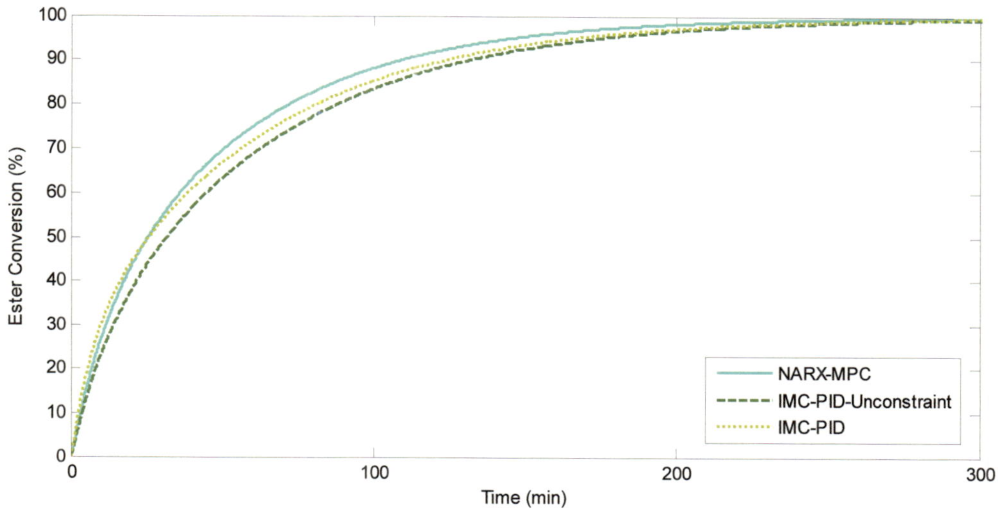

Fig. 9. Profile of ester conversion for NARX-MPC, IMC-PID-Unconstraint and IMC-PIC controllers.

With a view to set-point changing (see Fig. 10), the responses of the NARX-MPC and IMC-PID for set-point change have been varied from 310K to 315K at t=25min. The NARX-MPC was found to drive the output response faster than the IMC-PID controller with settling time, t= 45min and had shown no overshoot response with SSE value = 352.17. On the other hand, the limitation of input constraints for IMC-PID was evidenced in the poor output response with some overshoot and longer settling time, t= 60min (SSE=391.78). These results showed that NARX-MPC response controller had managed to cope with the set-point change better than the IMC-PID controllers.

Fig. 11 shows the NARX-MPC and the IMC-PID responses for 10% load change (jacket temperature) from the nominal value at t=25min. The NARX-MPC was found to drive the output response faster than the IMC-PID controller. As can be seen in the lower axes of Fig 9, the input variable response for the IMC-PID had varied extremely as compared to the input variable from NARX-MPC. From the results, it was concluded that the NARX-MPC controller with SSE=10.80 was able to reject the effect of disturbance better than the IMC-PID with SSE=32.94.

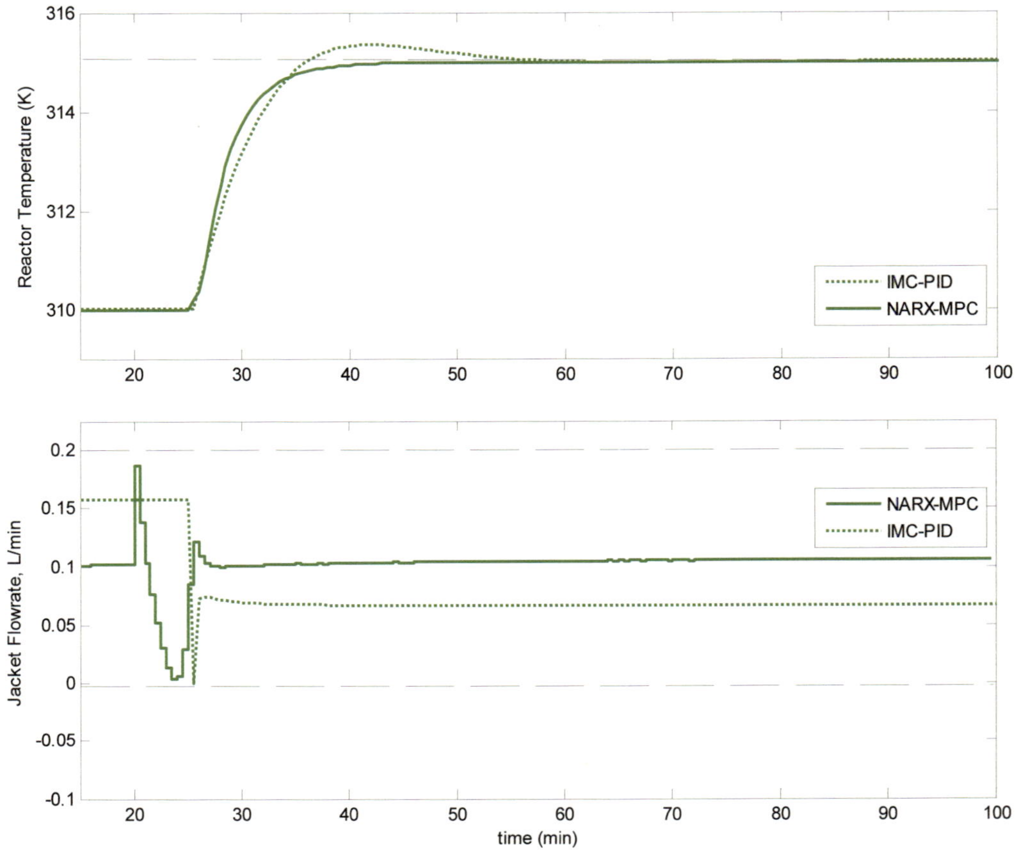

Fig. 10. Control response of NARX-MPC and IMC-PID controllers for set-point changing with their respective manipulated variable action.

The performance of the NARX-MPC and the IMC-PID controllers was also evaluated under a robustness test associated with a model parameter mismatch condition. The tests were;

- **Test 1:** A 30% increase for the heat of reaction, from 16.73 KJ to 21.75 KJ. It represented a change in the operating conditions that could be caused by a behavioral phase of the system.
- **Test 2:** Reduction of heat transfer coefficient from 2.857 J/s m^2 K to 2.143 J/s m^2 K, which was a 25 % decrease. This test simulated a change in heat transfer that could be expected due to the fouling of the heat transfer surfaces.
- **Test 3**: A 50% decrease of the inhibition activation energy, from 249.94 J mol/K to 124.97 J mol/K. This test represented a change in the rate of reaction that could be expected due to the deactivation of catalyst.
- **Test 4**: Simultaneous changes in heat of reaction, heat transfer coefficient and inhibition activation energy based on previous tests. This test represented the realistic operation of an actual reactive batch reactor process which would involve more than one input variable changes at one time.

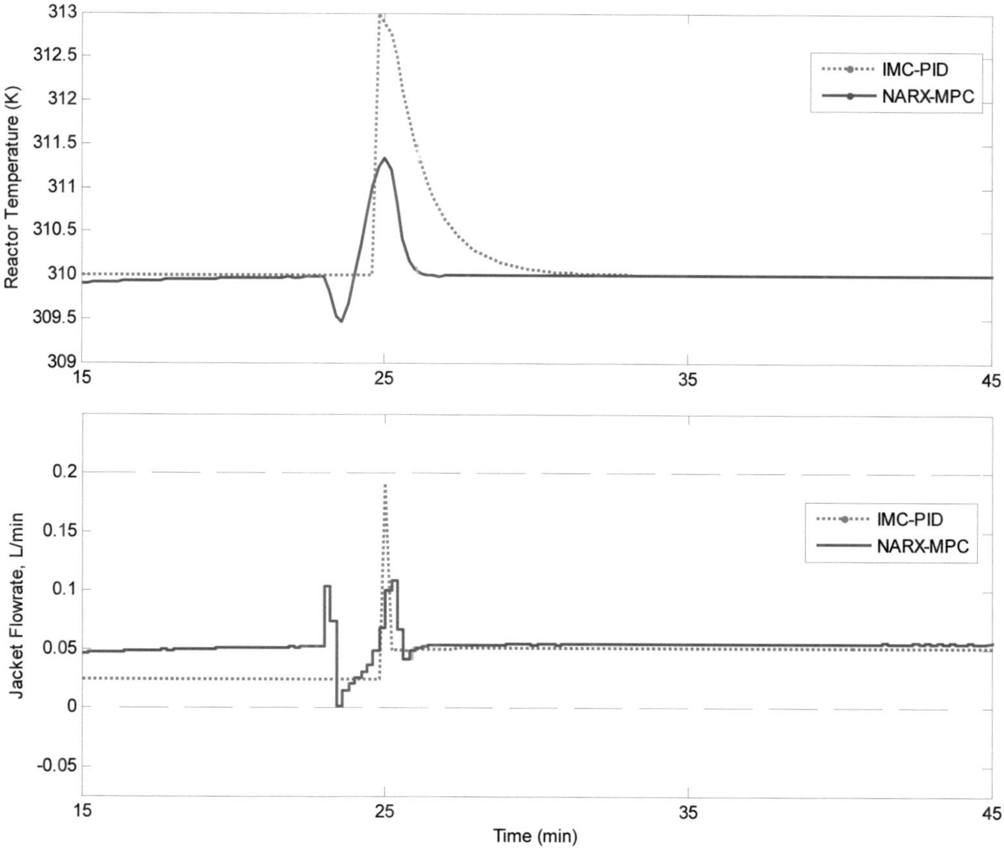

Fig. 11. Control response of NARX-MPC and IMC-PID controllers for load change with their respective manipulated variable action.

Fig.12- Fig.15 have shown the comparison of both IMC-PID and NARX-MPC control scheme's response for the reactor temperature and their respective manipulated variable action for robustness test 1 to test 4 severally. As can be seen in Fig. 12- Fig. 15, in all tests, the time required for the IMC-PID controllers to track the set-point is greater compared to the NARX-MPC controller. Nevertheless, NARX-MPC still shows good profile of manipulated variable, maintaining its good performance. The SSE values for the entire robustness test are summarized in Table 4. These SSE values shows that both controllers manage to compensate with the robustness. However, the error values indicated that the NARX-MPC still gives better performance compared to the both IMC-PID controllers.

Controller	Test 1	Test 2	Test 3	Test 4
NARX-MPC	415.89	405.37	457.21	481.72
IMC-PID	546.64	521.47	547.13	593.46

Table 4. SSE value of NARX-MPC and IMC-PID for robustness test

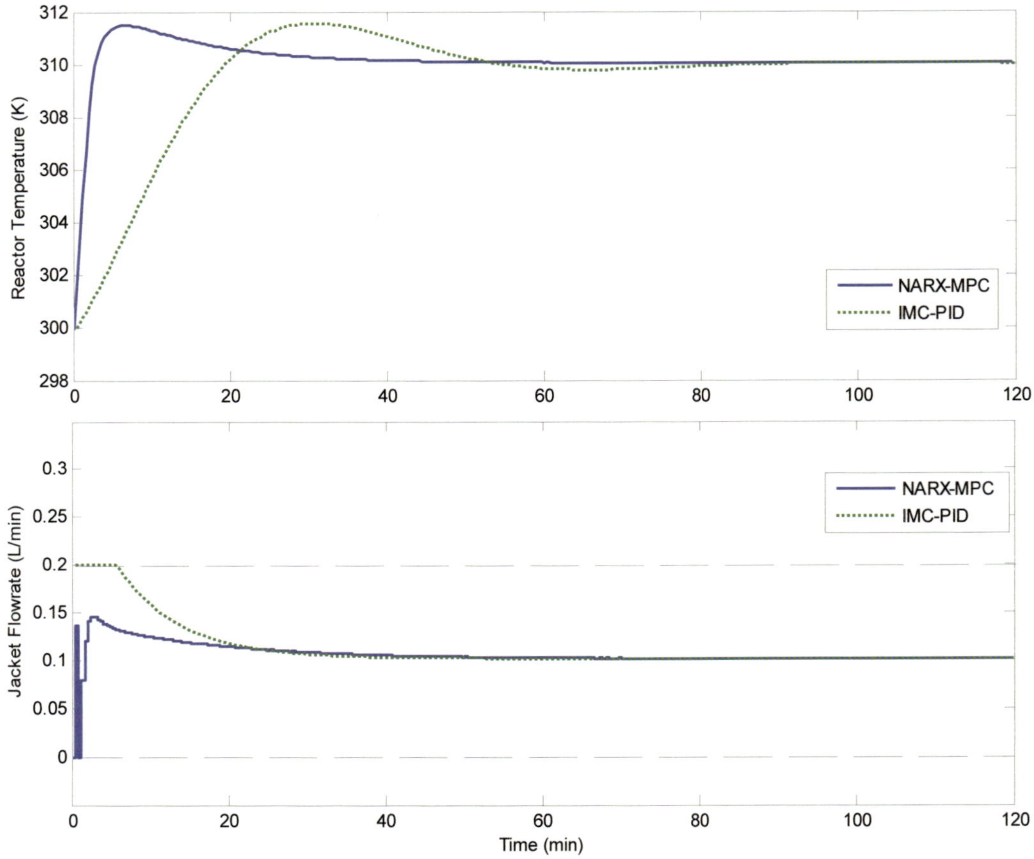

Fig. 12. Control response of NARX-MPC and IMC-PID controllers for robustness Test 1 with their respective manipulated variable action.

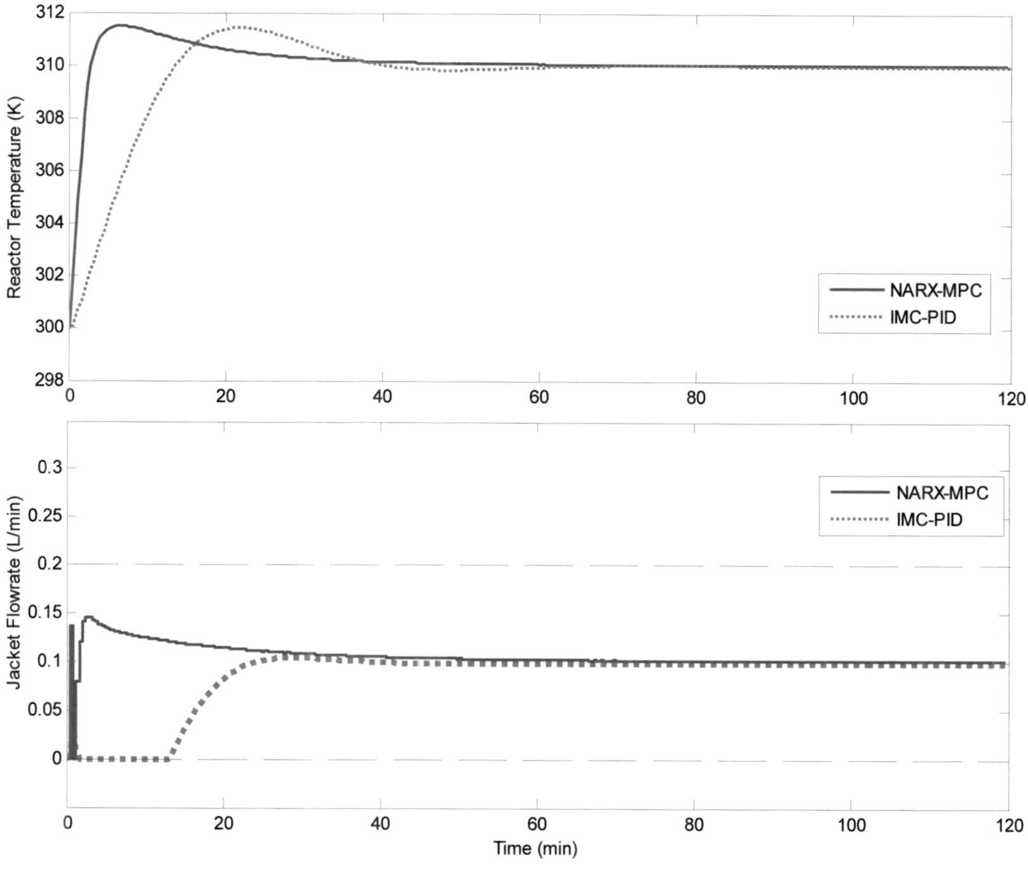

Fig. 13. Control response of NARX-MPC and IMC-PID controllers for robustness Test 2 with their respective manipulated variable action.

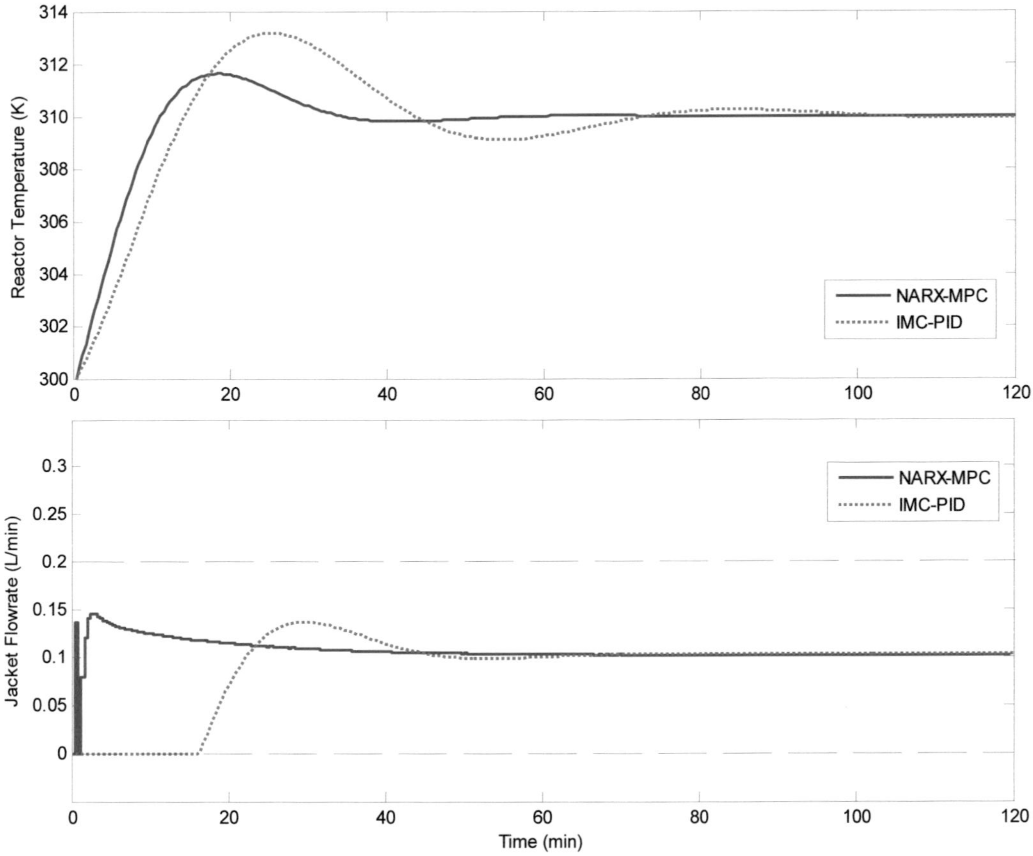

Fig. 14. Control response of NARX-MPC and IMC-PID controllers for robustness Test 3 with their respective manipulated variable action.

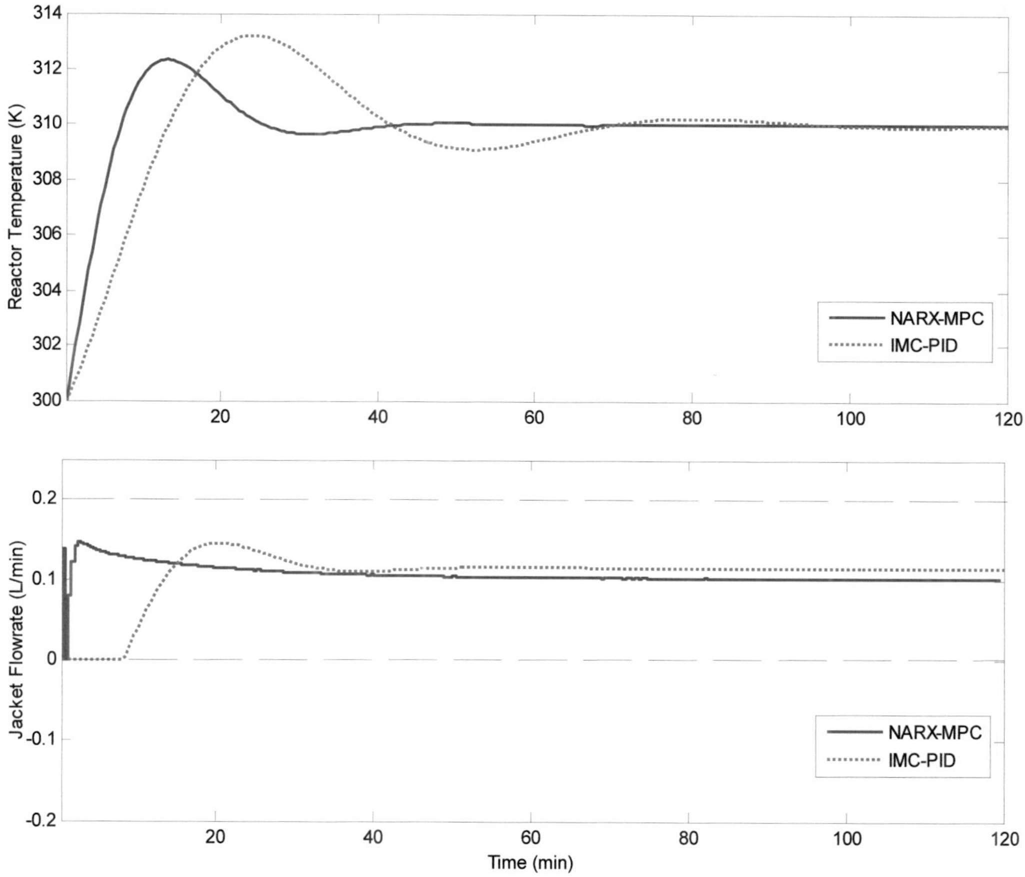

Fig. 15. Control response of NARX-MPC and IMC-PID controllers for robustness Test 4 with their respective manipulated variable action.

6. Conclusion

In this work, the NARX-MPC controller for the Batch Citronellyl Laurate Esterification Reactor has been developed. The validated first principle model was used as a process model to generate data required for NARX model identification. The NARX model with n_u =1 and n_y= 2 was chosen since it gave the best performance with MSE and SSE equal to 0.0025 and 0.7152 respectively. Finally, the performances of the NARX-MPC controller were evaluated for set-point tracking, set-point change, load change and robustness test. For all cases, the developed controller strategy (NARX-MPC) has been proven to perform well in controlling the temperature of the batch esterification reactor, as compared to the IMC-PID controllers.

7. Acknowledgement

The authors wish to acknowledge the financial support by Universiti Sains Malaysia (USM), for the research funding through the Research University Grant scholarship and Ministry of Science, Technology and Innovation, Malaysia (MOSTI) for the scholarship for the first author.

8. References

Agachi, P.S., Nagy, Z.K. Cristea, M.V. and Imre-Lucaci, A. (2007) *Model Based Control: Case studies in process Engineering*. Wiley-VCH Germany, 12-13.

Allgöwer, F., Findeisen, R. and Nagy, Z.: 2000, Nonlinear model predictive control: From theory to application, *Journal of the Chinese Institute of Chemical Engineers,* 35(3), 299–315.

Aziz,N., Hussain, M.A. and Mujtabba, I.M. (2000) Performance of different types of controllers on tracking optimal temperature profiles in batch reactors. *Computers and Chemical Engineering*, 24, 1069-1075.

Barbosa-Póvoa, A.P. (2007) A critical review on the design and retrofit of batch plants. *Computers and Chemical Engineering*, 31, 833-855

Bequette, B.W. (2007) Nonlinear model predictive control: A personal retrospective. *The Canadian Journal of Chemical Engineering*, 85, 408-415

Bouhenchir, H., Le Lann, M.V., and Cabassud, M. (2006). Predictive functional control for the temperature control of a chemical batch. *Computers and Chemical Engineering*. 30, 1141–1154

Braun, M.W., Ortiz-Mojica, R. and Rivera, D.E. (2002). Application of minimum crest factor multisinusoidal signals for "plant-friendly" identification of nonlinear process systems. *Control Engineering Practice* 10, 301–313

Costa, A.C., Meleiro, L.A.C., and Maciel Filho, R. (2001). Non-linear predictive control of an extractive alcoholic fermentation process. *Process Biochemistry*. 38, 743-750

Deshpande, A.P., Patwardhan, S.C., Narasimhan, S.S. (2009). Intelligent state estimation for fault tolerant nonlinear predictive control. *Journal of Process Control*. 19, 187-204

Dowd, J.E., Kwok, K.E., and Piret, J.M. (2001) Predictive modeling and loose-loop control for perfusion bioreactors. *Biochemical Engineering Journal*. 9, 1-9

Eaton, J.W. and Rawlings, J.B. (1992) Model predictive control of chemical processes. *Chemical Engineering Science*. 47, 705-720.

Foss, B.A., Johansen, T.A. and Sorensen, A.V. (1995). Nonlinear predictive control using local models-applied to a batch fermentation process. *Control Engineering Practice*. 3 (3), 389-396.

Garcia, C.E. and Morari, M., (1982) Internal model control. 1. A Unifying review and some new results, *Ind Eng Chem Process Des Dev*, 21: 308-323.

Garcia, T., Coteron, A., Martinez, M. and Aracil, J. (1996) Kinetic modelling for the esterification reactions catalysed by immobilized lipases" Chemical Engineering Science. 51, 2846-2846

Garcia, T., Sanchez, N., Martinez, M. and Aracil, J. (1999) Enzymatic synthesis of fatty esters-Part I: Kinetic approach. Enzyme and Microbial Technology. 25, 584-590.

Garcia, T., Coteron, A., Martinez, M. and Aracil, J. (2000) Kinetic model for the esterification of oleic acid and cetyl alcohol using immobilized lipase as catalyst. Chemical Engineering Science. 55, 1411-1423

Garcia, R.M., Wornle, F., Stewart, B.G. and Harrison D.K. (2002) Real-Time remote network control of an inverted pendulum using ST-RTL. *Proceeding for 32nd ASEE/IEEE Frontiers in Education Conference Boston*. 18-23

Gattu, G. and Zafiriou, E. (1992) Nonlinear quadratic dynamic matrix control with state estimation. *Ind. Eng. Chem. Res.*, 31 (4), 1096-1104.

Harris, T.J. and Yu, W. (2007) Controller assessment for a class of nonlinear systems. *Journal of Process Control*, 17: 607-619

Henson, M.A. (1998) Nonlinear model predictive control: Current status and future directions. *Computers and Chemical Engineering*, 23, 187-202

Hua, X., Rohani, S., and Jutan, A. (2004). Cascade closed-loop optimization and control of batch reactors. *Chemical Engineering Sciencel 59*, 5695-5708

Li, S., Lim, K.Y. and Fisher, D.G. (1989) A State Space Formulation for Model Predictive Control. *AIChe Journal*. 35(2), 241-249.

Liu, Z.H. and Macchietto, S. (1995). Model based control of a multipurpose batch reactor. *Computers Chem. Engng 19*, 477-482.

Mayne, D.Q., Rawlings, J.B., Rao, C.V. & Scokaert, P.O.M. (2000). Constrained model predictive control: Stability and optimality. *Automatica*, 36, 789-814.

M'sahli, F., Abdennour, R. B. and Ksouri, M. (2002) Nonlinear Model-Based Predictive Control Using a Generalised Hammerstein Model and its Application to a Semi-Batch Reactor. International Journal *Advanced Manufacturing Technology* 20, 844–852.

Morari, M. & Lee, J.H. (1999) Model predictive control: Past, present and future, Computers and Chemical Engineering, 23, 667-682

Mu, J., Rees, D. and Liu, G.P., 2005. Advanced controller design for aircraft gas turbine engines. *Control Engineering Practice*, 13: 1001–1015

Nagy, Z.K., Mahn, B., Franke, R. and Allgower, F. (2007). Evaluation study of an efficient output feedback nonlinear model predictive control for temperature tracking in an industrial batch reactor. *Control Engineering Practice* 15, 839–850.

Ozkan, G., Hapoglu, H. and Alpbaz, M. (2006). Non-linear generalised predictive control of a jacketed well mixed tank as applied to a batch process- A polymerisation reaction. *Applied Thermal Engineering* 26, 720–726.

Petersson, A.E.V., Gustafsson, L.M., Nordblad, M,, Boerjesson, P., Mattiasson, B., & Adlercreutz, P. (2005) Wax esters produced by solvent-free energy efficient enzymatic synthesis and their applicability as wood coatings. *Green Chem.* 7(12), 837–843.

Qin, S.J., T.A. Badgwell (2003). A survey of industrial model predictive control technology, *Control Engineering Practice,* 11(7), 733-764.

Serri N.A., Kamaruddin A.H., and Long W.S. (2006). Studies of reaction parameters on synthesis of Citronellyl laurate ester via immobilized *Candida rugosa* lipase in organic media. *Bioprocess and Biosystems Engineering* 29, 253-260.

Shafiee, G., Arefi, A.A., Jahed-Motlagh, M.R. and Jalali, A.A. (2008) Nonlinear predictive control of a polymerization reactor based on piecewise linear Wiener model. *Chemical Engineering Journal.* 143, 282-292

Valivety, R.H., Halling, P.J., Macrae, A.R. (1992) Reaction rate with suspended lipase catalyst shows similar dependence on water activity in different organic solvents. *Biochim Biophys Acta 1118.* 3, 218–222.

Yadav, G.D., Lathi, P.S. (2003). Kinetics and mechanism of synthesis of butyl isobutyrate over immobilized lipases. Biochemical Engineering Journal. 16, 245-252.

Yadav G.D., Lathi P.S. (2004) Synthesis of citronellol laurate in organic media catalyzed by immobilizied lipases: kinetic studies, *Journal of Molecular Catalysis B:Enzymatic, 27,* 113-119.

Yadav, G.D., Lathi, P.S. (2005) Lipase catalyzed transeterification of methyl acetoacetate with n-butanol. *Journal of Molecular Catalysis B: Enzymatic,* 32, 107-113

Zulkeflee, S.A. & Aziz, N. (2007) Kinetic Modelling of Citronellyl Laurate Esterification Catalysed by Immobilized Lipases. *Curtin University of Technology Sarawak Engineering Conference, Miri, Sarawak, Malaysia,* Oct 26-27.

Zulkeflee, S.A. & Aziz, N. (2008) Modeling and Simulation of Citronellyl Laurate Esterification in A batch Reactor. *15th Regional Symposium on Chemical Engineering in conjunction with the 22nd Symposium of Malaysian Chemical Engineering (RSCE-SOMCHE), Kuala Lumpur, Malaysia,* Dec 2-3.

Zulkeflee, S.A. & Aziz, N. (2009) IMC based PID controller for batch esterification reactor. *International Conference on Robotics, Vision, Signal Processing and Power Applications (RoViSP'09), Langkawi Kedah, Malaysia,* Dec 19-20.

14

Using Model Predictive Control for Local Navigation of Mobile Robots

Lluís Pacheco, Xavier Cufí and Ningsu Luo
University of Girona
Spain

1. Introduction

Model predictive control, MPC, has many interesting features for its application to mobile robot control. It is a more effective advanced control technique, as compared to the standard PID control, and has made a significant impact on industrial process control (Maciejowski, 2002). MPC usually contains the following three ideas:

- The model of the process is used to predict the future outputs along a horizon time.
- An index of performance is optimized by a control sequence computation.
- It is used a receding horizon idea, so at each instant of time the horizon is moved towards the future. It involves the application of the first control signal of the sequence computed at each step.

The majority of the research developed using MPC techniques and their application to WMR (wheeled mobile robots) is based on the fact that the reference trajectory is known beforehand (Klancar & Skrjanc, 2007). The use of mobile robot kinematics to predict future system outputs has been proposed in most of the different research developed (Kühne et al., 2005; Gupta et al., 2005). The use of kinematics have to include velocity and acceleration constraints to prevent WMR of unfeasible trajectory-tracking objectives. MPC applicability to vehicle guidance has been mainly addressed at path-tracking using different on-field fixed trajectories and using kinematics models. However, when dynamic environments or obstacle avoidance policies are considered, the navigation path planning must be constrained to the robot neighborhood where reactive behaviors are expected (Fox et al., 1997; Ögren & Leonard, 2005). Due to the unknown environment uncertainties, short prediction horizons have been proposed (Pacheco et al., 2008). In this context, the use of dynamic input-output models is proposed as a way to include the dynamic constraints within the system model for controller design. In order to do this, a set of dynamic models obtained from experimental robot system identification are used for predicting the horizon of available coordinates. Knowledge of different models can provide information about the dynamics of the robot, and consequently about the reactive parameters, as well as the safe stop distances. This work extends the use of on-line MPC as a suitable local path-tracking methodology by using a set of linear time-varying descriptions of the system dynamics when short prediction horizons are used. In the approach presented, the trajectory is dynamically updated by giving a straight line to be tracked. In this way, the control law considers the local point to be achieved and the WMR coordinates. The cost function is formulated with parameters that involve the capacity of turning and going straight. In the

case considered, the Euclidean distance between the robot and the desired trajectory can be used as a potential function. Such functions are CLF (control Lyapunov function), and consequently asymptotic stability with respect to the desired trajectory can be achieved. On-line MPC is tested by using the available WMR. A set of trajectories is used for analyzing the path-tracking performance. In this context, the different parameter weights of the cost function are studied. The experiments are developed by considering five different kinds of trajectories. Therefore, straight, wide left turning, less left turning, wide right turning, and less right turning are tested. Experiments are conducted by using factorial design with two levels of quantitative factors (Box et al., 2005). Studies are used as a way of inferring the weight of the different parameters used in the cost function. Factor tuning is achieved by considering aspects, such as the time taken, or trajectory deviation, within different local trajectories. Factor tuning depicts that flexible cost function as an event of the path to be followed, can improve control performance when compared with fixed cost functions. It is proposed to use local artificial potential attraction field coordinates as a way to attract WMR towards a local desired goal. Experiments are conducted by using a monocular perception system and local MPC path-tracking. On-line MPC is reported as a suitable navigation strategy for dynamics environments.
This chapter is organized as follows: Section 1 gives a brief presentation about the aim of the present work. In the Section 2, the WMR dynamic models are presented. This section also describes the MPC formulation, algorithms and simulated results for achieving local path-tracking. Section 3 presents the MPC implemented strategies and the experimental results developed in order to adjust the cost function parameters. The use of visual data is presented as a horizon where trajectories can be planned by using MPC strategies. In this context local MPC is tested as a suitable navigation strategy. Finally, in Section 4 some conclusions are made.

2. The control system identification and the MPC formulation

This section introduces the necessary previous background used for obtaining the control laws that are tested in this work as a suitable methodology for performing local navigation. The WMR PRIM, available in our lab, has been used in order to test and orient the research (Pacheco et al., 2009). Fig. 1 shows the robot PRIM and sensorial and system blocs used in

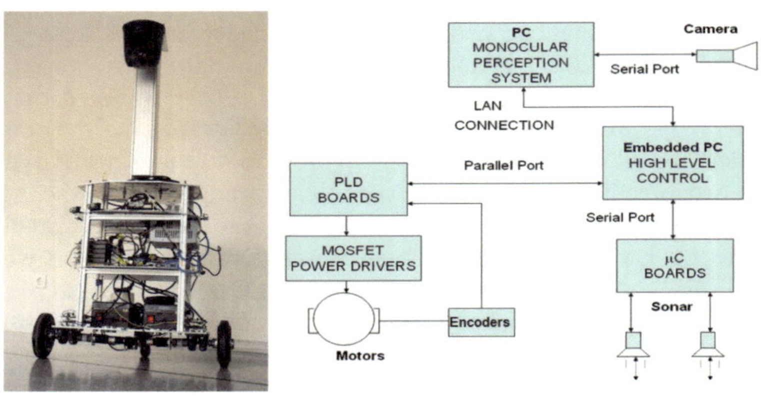

(a) (b)

Fig. 1. (a) The robot PRIM used in this work; (b) The sensorial and electronic system blocs the research work. The mobile robot consists of a differential driven one, with two independent wheels of 16cm diameters actuated by two DC motors. A third spherical omni-directional wheel is used to guarantee the system stability. Next subsection deals with the problem of modeling the dynamics of the WMR system. Furthermore, dynamic MPC techniques for local trajectory tracking and some simulated results are introduced in the remaining subsections. A detailed explanation of the methods introduced in this section can be found in (Pacheco et al., 2008).

2.1 Experimental model and system identification

The model is obtained through the approach of a set of lineal transfer functions that include the nonlinearities of the whole system. The parametric identification process is based on black box models (Norton, 1986; Ljung, 1989). The nonholonomic system dealt with in this work is considered initially to be a MIMO (multiple input multiple output) system, as shown in Fig. 2, due to the dynamic influence between two DC motors. This MIMO system is composed of a set of SISO (single input single output) subsystems with coupled connection.

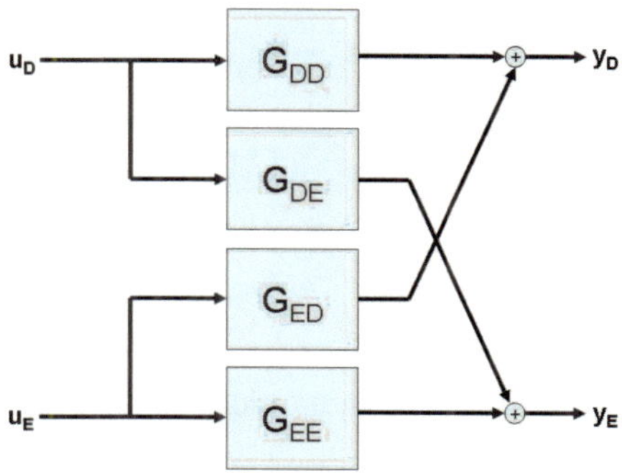

Fig. 2. The MIMO system structure

The parameter estimation is done by using a PRBS (Pseudo Random Binary Signal) such as excitation input signal. It guarantees the correct excitation of all dynamic sensible modes of the system along the whole spectral range and thus results in an accurate precision of parameter estimation. The experiments to be realized consist in exciting the two DC motors in different (low, medium and high) ranges of speed. The ARX (auto-regressive with external input) structure has been used to identify the parameters of the system. The problem consists in finding a model that minimizes the error between the real and estimated data. By expressing the ARX equation as a lineal regression, the estimated output can be written as:

$$\hat{y} = \theta\varphi \tag{1}$$

with $\hat{y}$ being the estimated output vector, θ the vector of estimated parameters and φ the vector of measured input and output variables. By using the coupled system structure, the transfer function of the robot can be expressed as follows:

$$\begin{pmatrix} Y_R \\ Y_L \end{pmatrix} = \begin{pmatrix} G_{RR} & G_{LR} \\ G_{RL} & G_{LL} \end{pmatrix} \begin{pmatrix} U_R \\ U_L \end{pmatrix} \tag{2}$$

where Y_R and Y_L represent the speeds of right and left wheels, and U_R and U_L the corresponding speed commands, respectively. In order to know the dynamics of robot system, the matrix of transfer function should be identified. In this way, speed responses to PBRS input signals are analyzed. The filtered data, which represent the average value of five different experiments with the same input signal, are used for identification. The system is identified by using the identification toolbox "ident" of Matlab for the second order models. Table 1 shows the continuous transfer functions obtained for the three different lineal speed models.

Linear Transfer Function	High velocities	Medium velocities	Low velocities
G_{DD}	$\frac{0.20s^2-3.15s+9.42}{s^2+6.55s+9.88}$	$\frac{0.20s^2+3.10s+8.44}{s^2+6.17s+9.14}$	$\frac{0.16s^2+2.26s+5.42}{s^2+5.21s+6.57}$
G_{ED}	$\frac{-0.04s^2-0.60s-0.32}{s^2+6.55s+9.88}$	$\frac{-0.02s^2-0.31s-0.03}{s^2+6.17s+9.14}$	$\frac{-0.02s^2-0.20s+0.41}{s^2+5.21s+6.57}$
G_{DE}	$\frac{-0.01s^2-0.08s-0.36}{s^2+6.55s+9.88}$	$\frac{0.01s^2+0.13s+0.20}{s^2+6.17s+9.14}$	$\frac{-0.01s^2-0.08s-0.17}{s^2+5.21s+6.57}$
G_{EE}	$\frac{0.31s^2+4.47s+8.97}{s^2+6.55s+9.88}$	$\frac{0.29s^2+4.11s+8.40}{s^2+6.17s+9.14}$	$\frac{0.25s^2+3.50s+6.31}{s^2+5.21s+6.57}$

Table 1. The second order WMR models

The coupling effects should be studied as a way of obtaining a reduced-order dynamic model. It can be seen from Table 1 that the dynamics of two DC motors are different and the steady gains of coupling terms are relatively small (less than 20% of the gains of main diagonal terms). Thus, it is reasonable to neglect the coupling dynamics so as to obtain a simplified model. In order to verify the above facts from real results, a set of experiments have been done by sending a zero speed command to one motor and different non-zero speed commands to the other motor. The experimental result confirms that the coupled dynamics can be neglected. The existence of different gains in steady state is also verified experimentally. Finally, the order reduction of the system model is carried out through the analysis of pole positions by using the root locus method. It reveals the existence of a dominant pole and consequently the model order can be reduced from second order to first order. Table 2 shows the first order transfer functions obtained. Afterwards, the system models are validated through the experimental data by using the PBRS input signal.

Linear Transfer Function	High velocities	Medium velocities	Low velocities
G_{DD}	$\frac{0.95}{0.42s+1}$	$\frac{0.92}{0.41s+1}$	$\frac{0.82}{0.46s+1}$
G_{EE}	$\frac{0.91}{0.24s+1}$	$\frac{0.92}{0.27s+1}$	$\frac{0.96}{0.33s+1}$

Table 2. The reduced WMR model

2.2 Dynamic MPC techniques for local trajectory tracking

The minimization of path tracking error is considered to be a challenging subject in mobile robotics. In this subsection the LMPC (local model predictive control) techniques based on the dynamics models obtained in the previous subsection are presented. The use of dynamic models avoids the use of velocity and acceleration constraints used in other MPC research based on kinematic models. Moreover, contractive constraints are proposed as a way of guaranteeing convergence towards the desired coordinates. In addition, real-time implementations are easily implemented due to the fact that short prediction horizons are used. By using LMPC, the idea of a receding horizon can deal with local on-robot sensor information. The LMPC and contractive constraint formulations as well as the algorithms and simulations implemented are introduced in the next subsections.

2.2.1 The LMPC formulation

The main objective of highly precise motion tracking consists in minimizing the error between the robot and the desired path. Global path-planning becomes unfeasible since the sensorial system of some robots is just local. In this way, LMPC is proposed in order to use the available local perception data in the navigation strategies. Concretely, LMPC is based on minimizing a cost function related to the objectives for generating the optimal WMR inputs. Define the cost function as follows:

$$J(n,m) = \min_{\left\{U(k+i|k)_{i=0}^{i=m-1}\right\}} \left\{ \begin{array}{l} \left[X(k+n|k) - X_{ld}\right]^T P \left[X(k+n|k) - X_{ld}\right] \\ +\sum_{i=1}^{n-1} \left[X(k+i|k) - \overline{X_{ld}X_{l0}}\right]^T Q \left[X(k+i|k) - \overline{X_{ld}X_{l0}}\right] \\ +\sum_{i=1}^{n-1} \left[\theta(k+i|k) - \theta_{ld}\right]^T R \left[\theta(k+i|k) - \theta_{ld}\right] \\ +\sum_{i=0}^{m-1} U^T(k+i|k) S U(k+i|k) \end{array} \right\} \tag{3}$$

The first term of (3) refers to the attainment of the local desired coordinates, $X_{ld}=(x_d, y_d)$, where (x_d, y_d) denote the desired Cartesian coordinates. $X(k+n/k)$ represents the terminal value of the predicted output after the horizon of prediction n. The second one can be considered as an orientation term and is related to the distance between the predicted robot positions and the trajectory segment given by a straight line between the initial robot Cartesian coordinates $X_{l0}=(x_{l0}, y_{l0})$ from where the last perception was done and the desired local position, X_{ld}, to be achieved within the perceived field of view. This line orientation is denoted by θ_{ld} and denotes the desired orientation towards the local objective. $X(k+i/k)$ and $\theta(k+i/k)$ $(i=1,\dots n-1)$ represents the predicted Cartesian and orientation values within the prediction horizon. The third term is the predicted orientation error. The last one is related to the power signals assigned to each DC motor and are denoted as U. The parameters P, Q, R and S are weighting parameters that express the importance of each term. The control horizon is designed by the parameter m. The system constraints are also considered:

$$\left\{ \begin{array}{ll} G_0 < |U(k)| \le G_1 & \alpha \in (0,1] \\ |X(K+n/k) - X_{ld}| \le \alpha |X(k) - X_{ld}| & \\ \text{or } |\theta(k+n/k) - \theta_{ld}| \le \alpha |\theta(k) - \theta_{ld}| & \end{array} \right\} \tag{4}$$

where $X(k)$ and $\theta(k)$ denote the current WMR coordinates and orientation, $X(k+n/k)$ and $\theta(k+n/k)$ denote the final predicted coordinates and orientation, respectively. The limitation of the input signal is taken into account in the first constraint, where G_0 and G_1 respectively denote the dead zone and saturation of the DC motors. The second and third terms are contractive constraints (Wang, 2007), which result in the convergence of coordinates or orientation to the objective, and should be accomplished at each control step.

2.2.2 The algorithms and simulated results

By using the basic ideas introduced in the previous subsection, the LMPC algorithms have the following steps:

1. Read the current position
2. Minimize the cost function and to obtain a series of optimal input signals
3. Choose the first obtained input signal as the command signal.
4. Go back to the step 1 in the next sampling period.

The minimization of the cost function is a nonlinear problem in which the following equation should be verified:

$$f(\alpha x+\beta y)\leq\alpha f(x)+\beta f(y) \tag{5}$$

The use of interior point methods can solve the above problem (Nesterov et al., 1994; Boyd & Vandenberghe, 2004). Gradient descent method and complete input search can be used for obtaining the optimal input. In order to reduce the set of possibilities, when optimal solution is searched for, some constraints over the DC motor inputs are taken into account:

- The signal increment is kept fixed within the prediction horizon.
- The input signals remain constant during the remaining interval of time.

The above considerations will result in the reduction of the computation time and the smooth behavior of the robot during the prediction horizon (Maciejowski, 2002). Thus, the set of available input is reduced to one value, as it is shown in Fig. 3.

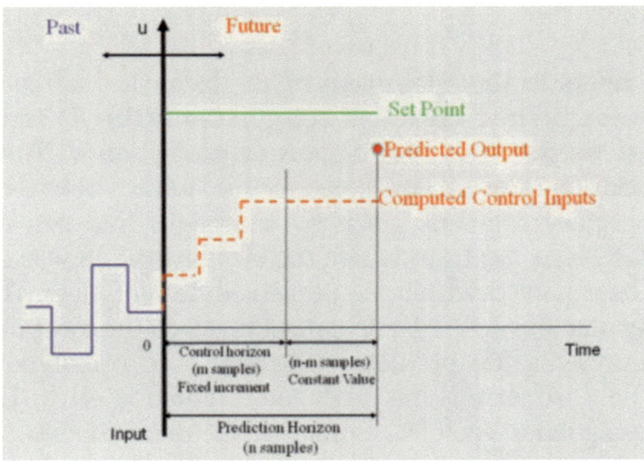

Fig. 3. LMPC strategy with fixed increment of the input during the control horizon and constant value for the remaining time

Both search methods perform accurate path-tracking. Optimal input search has better time performance and subinterval gradient descent method does not usually give the optimal solution. Due to these facts obtained from simulations, complete input search is selected for the on-robot experiences presented in the next section.

The evaluation of the LMPC performance is made by using different parametric values in the proposed cost function (3). In this way, when only the desired coordinates are considered, (*P=1, Q=0, R=0, S=0*), the trajectory-tracking is done with the inputs that can minimize the cost function by shifting the robot position to the left. The reason can be found in Table 2, where the right motor has more gain than the left one for high speeds. This problem can be solved, (*P=1, Q=1, R=0, S=0*) or (*P=1, Q=0, R=1, S=0*) by considering either the straight-line trajectory from the point where the last perception was done to the final desired point belonging to the local field of perception or the predicted orientations. Simulated results by testing both strategies provide similar satisfactory results. Thus, the straight line path or orientation should be considered in the LMPC cost function. Fig. 4 shows a simulated result of LMPC for WMR by using the orientation error, the trajectory distance and the final desired point for the cost function optimization (*P=1, Q=1, R=1, S=0*). Obtained results show the need of *R* parameter when meaningful orientation errors are produced.
The prediction horizon magnitude is also analyzed. The possible coordinates available for prediction when the horizon is larger (*n=10, m=5*), depict a less dense possibility of coordinates when compared with shorter horizons of prediction. Short prediction horizon strategy is more time effective and performs path-tracking with better accuracy. For these reasons, a short horizon strategy (*n=5, m=3*) is proposed for implementing experimental results.

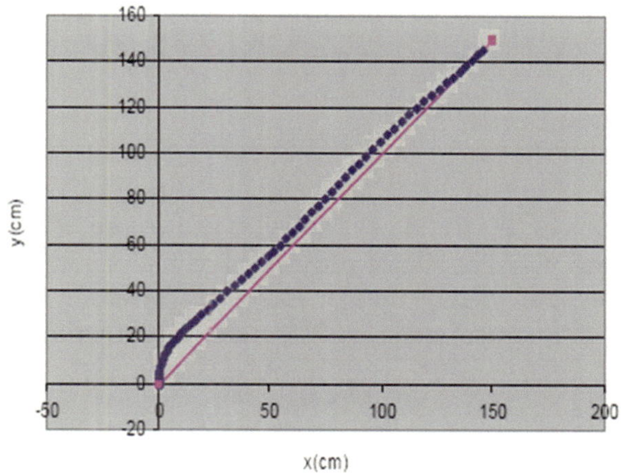

Fig. 4. Trajectory tracking simulated result by using the orientation error, trajectory distance and the final desired point for the optimization.

The sampling time for each LMPC step was set to 100ms. Simulation time performance of complete input search and gradient descent methods is computed. For short prediction horizon (*n=5, m=3*), the simulation processing time is less than 3ms for the complete input search strategy and less than 1ms for the gradient descent method when algorithms are running in a standard 2.7 GHz PC. Real on-robot algorithm time performance is also compared for different prediction horizons by using the embedded 700 Mhz PC and additional hardware system. Table 3 shows the LMPC processing time for different horizons of prediction when complete optimal values search or the gradient descent method are used. Surprisingly, when the horizon is increased the computing time is decreased. It is due to the fact that the control horizon is also incremented, and consequently less range of signal increments are possible because the signal increment is kept fixed within the control horizon. Thus, the maximum input value possibilities decrease with larger horizons. Hence for n=5 there are 1764 possibilities (42x42), and for n=10 there are 625 (25x25).

Horizon of prediction (n)	Complete search method	Gradient descent method
n=5	45ms	16ms
n=8	34ms	10ms
n=10	25ms	7ms

Table 3. LMPC processing times

3. Tuning the control law parameters by using path-tracking experimental results

In this section, path-tracking problem and the cost function parameter weights are analyzed, within a constrained field of perception provided by the on-robot sensor system. The main objective is to obtain further control law analysis by experimenting different kind of trajectories. The importance of the cost function parameter weights is analyzed by developing the factorial design of experiments for a representative set of local trajectories. Statistical results are compared and control law performance is analyzed as a function of the path to be followed. Experimental LMPC results are conducted by considering a constrained horizon of perception provided by a monocular camera where artificial potential fields are used in order to obtain the desired coordinates within the field of view of the robot.

3.1 The local field of perception

In order to test the LMPC by using constrained local perception, the field of view obtained by a monocular camera has been used. Ground available scene coordinates appear as an image, in which the camera setup and pose knowledge are used, and projective perspective is assumed to make each pixel coordinate correspond to a 3D scene coordinate (Horn, 1998). Fig. 5 shows a local map provided by the camera, which corresponds to a field of view with a horizontal angle of 48°, a vertical angle of 37°, *H* set to 109cm and a tilt angle of 32°.

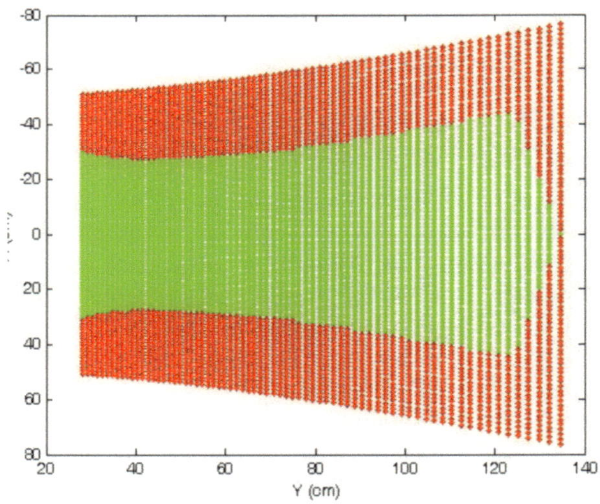

Fig. 5. Available local map coordinates (in green), the necessary coordinates free of obstacles and the necessary wide-path (in red).

It is pointed out that the available floor coordinates are reduced due to the WP (wide-path) of the robot (Schilling, 1990). It should also be noted that for each column position corresponding to scene coordinates Y_j, there are R row coordinates X_i. Once perception is introduced, the problem is formulated as finding the optimal cell that brings the WMR close to the desired coordinates (X_d, Y_d) by searching for the closest local desired coordinates (X_{ld}, Y_{ld}) within the available local coordinates (X_i, Y_j). In this sense, perception is considered to be a local receding horizon on which the trajectory is planned. The local desired cell is obtained by minimizing a cost function J that should act as a potential field corridor. Thus, the cost function is minimized by attracting the robot to the desired objective through the free available local cell coordinates. It is noted that from local perception analysis and attraction potential fields a local on field path can be obtained. The subsequent subsections infer control law parameter analysis by considering a set of path possibilities obtained within the perception field mentioned in this section.

3.2 The path-tracking experimental approach by using LMPC methods

The path tracking performance is improved by the adequate choice of a cost function that is derived from (3) and consists of a quadratic expression containing some of the following four parameters to be minimized:

- The squared Euclidean *approaching point distance* (APD) between the local desired coordinates, provided by the on-robot perception system, and the actual robot position. It corresponds with the parameter "*P*" of the LMPC cost function given by (3).
- The squared *trajectory deviation distance* (TDD) between the actual robot coordinate and a straight line that goes from the robot coordinates, when the local frame perception was acquired, and the local desired coordinates belonging to the referred frame of perception. It corresponds with the parameter "*Q*" of the cost function shown by (3).
- The third parameter consists of the squared *orientation deviation* (OD); it is expressed by the difference between the robot desired and real orientations. It corresponds with the parameter "*R*" of the LMPC cost function depicted by (3).
- The last parameter refers to changes allowed to the input signal. It corresponds with the parameter "*S*" of the LMPC cost function given by (3).

One consideration that should be taken into account is the different distance magnitudes. In general, the approaching distance could be more than one meter. However, the magnitude of the deviation distance is normally in the order of cm, which becomes effective only when the robot is approaching the final desired point. Hence, when reducing the deviation distance further to less than 1cm is attempted, an increase, in the weight value for the deviation distance in the cost function, is proposed.

The subsequent subsections use statistical knowledge for inferring APD (P) and TDD (Q) or APD (P) and OD (R) factor performances as a function of the kind of paths to be tracked. Other cost function parameters are assumed to be equal to zero.

3.3 Experimental tuning of APD and TDD factors

This subsection presents the results achieved by using factorial design in order to study the LMPC cost function tuning when APD and TDD factors are used. Path-tracking performance is analyzed by the mean of the different factor weights. The experiments are developed by considering five different kinds of trajectories within the reduced field of view as shown in Fig. 5. Therefore, straight, wide left turning, less left turning, wide right turning and less right turning trajectories are tested. Experiments are conducted by using factorial design with two levels of quantitative factors (Box et al, 2005). Referred to the cost function, let us assume that high value (H) is equal to "1" and low value (L) is equal to "0.5". For each combination of factors three different runs are experimented. The averaged value of the three runs allows statistical analysis for each factor combination. From these standard deviations, the importance of the factor effects can be determined by using a rough rule that considers the effects when the value differences are similar or greater than 2 or 3 times their standard deviations. In this context, the main effects and lateral effects, related to APD and TDD, are analyzed. Fig. 6 shows the four factor combinations (APD, TDD) obtained by both factors with two level values.

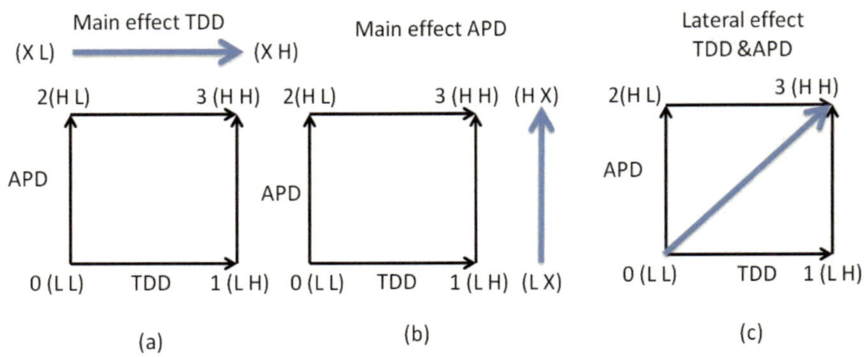

Fig. 6. The different factor combinations and the influence directions, in which the performances should be analyzed.

The combinations used for detecting lateral and main effect combinations are highlighted by blue arrows. Thus, the main effect of APD factor, ME_{APD}, can be computed by the following expression:

$$ME_{APD} = \frac{Y_3 + Y_2}{2} - \frac{Y_1 + Y_0}{2} \tag{6}$$

Path-tracking statistical performances to be analyzed in this research are represented by Y. The subscripts depict the different factor combinations. The main effect for TDD factor, ME_{TDD}, is computed by:

$$ME_{TDD} = \frac{Y_3 + Y_1}{2} - \frac{Y_2 + Y_0}{2} \tag{7}$$

The lateral effects are computed by using the following expression:

$$LE_{APD\_TDD} = Y_3 - Y_0 \tag{8}$$

The detailed measured statistics with parameters such as time (T), trajectory error (TE), and averaged speeds (AS) are presented in (Pacheco & Luo, 2011). The results were tested for straight trajectories, wide and less left turnings, and wide and less right turnings. The main and lateral effects are represented in Table 4.

The performance is analyzed for the different trajectories:

- The factorial analysis for straight line trajectories, (σ_T = 0.16s, σ_{TE} = 0.13cm, σ_{AS} = 2.15cm/s), depicts a main time APD effect of -0.45s, and an important lateral effect of -0.6s and -0.32cm. Speed lateral effect of only 1.9cm/s is not considered as meaningful. Considering lateral effects that improve time and accuracy, high values (APD, TDD) are proposed for both factors.
- The analysis for wide left turning trajectories, (σ_T = 0.26s, σ_{TE} = 0.09cm, σ_{AS} = 0.54cm/s) show negative APD main effect of 0.53s, and 0.15cm. However, the TDD factor tends to decrease the time and trajectory deviation. The 0.3cm/s speed TDD main factor is irrelevant. In this case, low value for APD factor and high value for the TDD factor is proposed.
- The factor analysis for less left turning, (σ_T = 0.29s, σ_{TE} = 0.36cm, σ_{AS} = 0.84cm/s), depicts a considerable lateral effect of -0.46s and -0.31cm. Speed -0.2cm/s lateral effect is not important. In this sense high values are proposed for APD and TDD factors.
- The analysis for wide right turning, (σ_T = 0.18s, σ_{TE} = 0.15cm, σ_{AS} = 1.04cm/s) does not provide relevant clues, but small time improvement seems to appear when TDD factor is set to a low value. Low values are proposed for APD and TDD factors.
- Finally, the factorial analysis for less right turning trajectories, (σ_T = 0.12s, σ_{TE} = 0.18cm, σ_{AS} = 1.94cm/s), depicts APD and lateral effects that increase the trajectory time with 0.32s and 0.44s. Main or lateral effects related to the speed have not been detected. Low values are proposed for APD and TDD factors.

Straight line trajectory			
Parameter Performance	Main Effect TDD factor	Main Effect APD factor	Lateral Effect TDD & APD factors
Time	-0.05s	-0.45s	-0.6s
Trajectory accuracy	-0.18cm	-0.14cm	-0.32cm
Averaged speed	1.25cm/s	0.6cm/s	1.9cm/s
Wide left turn trajectory			
Time	-0.34s	0.53s	0.16s
Trajectory accuracy	-0.17cm	0.15cm	-0.01cm
Averaged speed	0.3cm/s	0.4cm/s	0.7cm/s
Slight left turn trajectory			
Time	-0.24s	0.02s	-0.46s
Trajectory accuracy	-0.14cm	-0.17cm	-0.31cm
Averaged speed	0.8cm/s	-1cm/s	-0.2cm/s
Wide right turn trajectory			
Time	0.27s	-0.10s	0.17s
Trajectory accuracy	-0.22cm	0.1cm	-0.12cm
Averaged speed	0.7cm/s	0.2cm/s	0.9cm/s
Slight right turn trajectory			
Time	0.12s	0.32s	0.44s
Trajectory accuracy	-0.18cm	-0.06cm	-0.25cm
Averaged speed	-1.3cm/s	2.8cm/s	1.5cm/s

Table 4. Main and lateral effects

3.4 Experimental performance by using fixed or flexible APD & TDD factors

Once factorial analysis is carried out, this subsection compares path-tracking performance by using different control strategies. The experiments developed consist in analyzing the performance when a fixed factor cost function or a flexible factor cost function is used. The trajectories to be analyzed are formed by straight lines, less right or left turnings, and wide right or left turnings. The fixed factor cost function maintains the high values for APD and TDD factors, while the flexible factor cost function is tested as function of the path to be tracked.

Different experiments are done; see (Pacheco & Luo, 2011). As instance one experiment consists in tracking a trajectory that is composed of four points ((0, 0), (-25, 40), (-25, 120), (0, 160)) given as (x, y) coordinates in cm. It consists of wide left turning, straight line and wide right turning trajectories. The results obtained by using fixed and flexible factor cost function are depicted in Table 5. Three runs are obtained for each control strategy and consequently path-tracking performance analysis can be done.

Results show that flexible factor strategy improves an 8% the total time performance of the fixed factor strategy. The turning trajectories are done near 50% of the path performed. Remaining path consists of a straight line trajectory that is performed with same cost

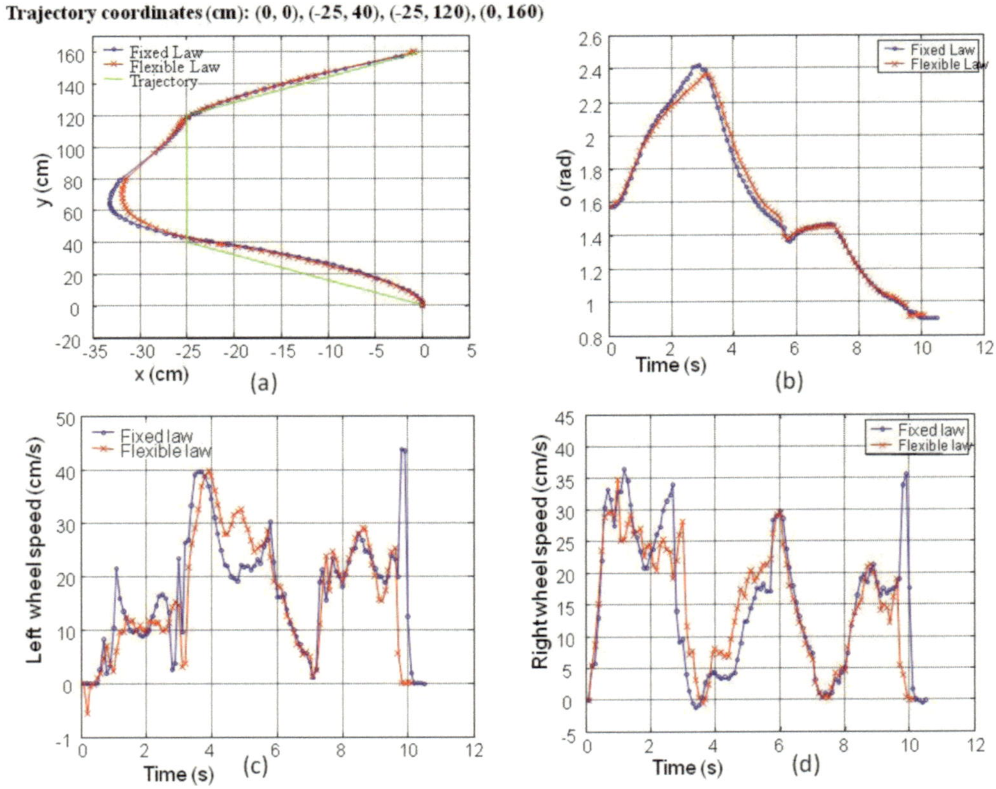

Fig. 7. (a) Trajectory-tracking experimental results by using flexible or fixed cost function. (b) WMR orientation experimental results by using flexible or fixed cost function. (c) Left wheel speed results by using flexible or fixed cost function. (d) Right wheel speed results by using flexible or fixed cost function.

function values for fixed and flexible control laws. It is during the turning actions, where the two control laws have differences, when time improvement is nearly 16%. Fig. 7 shows an example of some results achieved. Path-tracking coordinates, angular position, and speed for the fixed and flexible cost function strategies are shown.

It can be seen that flexible cost function, when wide left turning is performed approximately during the first three seconds, produces less maximum speed values when compared with fixed one. However, a major number of local maximum and minimum are obtained. It results in less trajectory deviation when straight line trajectory is commanded. In general flexible cost function produces less trajectory error with less orientation changes and improves time performance.

	Trajectory points: (0,0), (-25,40), (-25,120), (0,160) ((x,y) in cm)					
	Time (s)		Trajectory error (cm)		Averaged Speed (cm/s)	
Experiment	Fixed Law	Flexible Law	Fixed Law	Flexible Law	Fixed Law	Flexible Law
Run 1	10,5	10,3	3,243	3,653	18,209	16,140
Run 2	10,9	9,8	3,194	2,838	16,770	16,632
Mean	10,70	10,05	3,219	3,245	17,489	16,386
Variance	0,0800	0,1250	0,0012	0,3322	1,0354	0,1210
Standart deviation	0,2828	0,3536	0,0346	0,5764	1,0175	0,3479

Table 5. Results obtained by using fixed or flexible cost function

Developed experiences with our WMR platform show that flexible LMPC cost function related with the path to be tracked can improve the control system performance.

3.5 Experimental tuning using APD and OD factors

In a similar way APD and OD factors can be used. This subsection compares path-tracking performance by using different control strategies. The experiments developed consist in analyzing the performance when a fixed factor cost function or a flexible factor cost function is used. The trajectories to be analyzed are formed by straight lines, less right or left turnings, and wide right or left turnings. The fixed factor cost function maintains the high values for APD and OD factors, while the flexible factor cost function is tested as function of the path to be tracked. The experiments developed show the measured performance statistics, time, trajectory accuracy, and averaged speeds, for straight trajectories, wide and less left turnings, and wide and less right turnings. The standard deviation obtained as well as the main and lateral effects are represented in Table 6. The time, trajectory error and averaged speed standard deviations are respectively denoted by σ_T, σ_{TE}, and σ_{AS}. Table 6 represents the experimental statistic results obtained for the set of proposed trajectories. The standard deviations computed for each kind of trajectory by testing the different factor weights under different runs are also depicted.

The main and lateral effects were calculated by using (6), (7), (8), and the mean values obtained for the different factor combinations. Therefore, in Table 6 are highlighted the significant results achieved using experimental factorial analysis. The inferred results obtained can be tested using different trajectories.

Straight trajectory				
Parameters		OD	APD	APD & OD
Time (s)	$\sigma_T = 0.06s$	0,02	-0,13	-0,10
Trajectory error (cm)	$\sigma_{TE} = 0.69cm$	-0,24	1,34	1,10
Speed (cm/s)	$\sigma_{AS} = 0.88cm/s$	0,87	0,70	1,57
Wide left turning				
Parameters		OD	APD	APD & OD
Time (s)	$\sigma_T = 0.06s$	-0,10	0,20	0,10
Trajectory error (cm)	$\sigma_{TE} = 0.18cm$	0,36	0,38	0,02
Speed (cm/s)	$\sigma_{AS} = 0.59cm/s$	0,36	-0,87	-0,52
Less left turning				
Parameters		OD	APD	APD & OD
Time (s)	$\sigma_T = 0.09s$	-0,12	0,07	-0,05
Trajectory error (cm)	$\sigma_{TE} = 0.11cm$	0,58	1,08	0,50
Speed (cm/s)	$\sigma_{AS} = 0.92cm/s$	0,60	-0,13	0,47
Wide right turning				
Parameters		OD	APD	APD & OD
Time (s)	$\sigma_T = 0.11s$	0,10	0,35	0,45
Trajectory error (cm)	$\sigma_{TE} = 0.08cm$	0,44	0,45	0,01
Speed (cm/s)	$\sigma_{AS} = 0.67cm/s$	-0,58	-1,67	-2,25
Less right turning				
Parameters		OD	APD	APD & OD
Time (s)	$\sigma_T = 0.26s$	-0,07	0,07	0,00
Trajectory error (cm)	$\sigma_{TE} = 0.20cm$	1,38	0,65	-0,73
Speed (cm/s)	$\sigma_{AS} = 0.13cm/s$	-0,33	-0,14	-0,48

Table 6. Main and lateral effects

The experiments developed consist in analyzing the time performance when a fixed factor cost function or a flexible factor cost function is used. The trajectories to be analyzed are formed by straight lines, less right or left turnings, and wide right or left turnings. The fixed factor cost function maintains the high values for APD and OD factors, while the flexible factor cost function is tested as function of the trajectory to be tracked. The experiments presented consist in tracking a trajectory that is composed of three points ((0, 0), (-25, 40), (-25, 120)) given as (x, y) coordinates in cm. The results obtained by using fixed and flexible factor cost function are depicted in Table 7.

Trajectory (x,y) in cm: (0,0), (-25,40), (-25,120)						
Features	Time (s)		Error (cm)		Aver. speed (cm/s)	
Experiment	Fixed	Flexible	Fixed	Flexible	Fixed	Flexible
Run 1	7,2	7,0	3,8	3,0	19,4	17,5
Run 2	7,4	6,6	2,2	3,5	16,5	20,1
Mean	7,3	6,8	3,0	3,2	18,0	18,8
Variance	0,02	0,1	1,3	0,1	4,2	3,4
Stand. dev.	0,14	0,3	1,1	0,3	2,0	1,9

Table 7. Experimental performances

Two runs are obtained for each strategy and consequently time performance analysis can be done. The averaged standard deviation between the two cost function systems is of 0.22s, and the difference of means are 0.5s. Thus, flexible factor strategy improves a 6.85% the time performance of the fixed factor strategy. However, left turning is done only a 33% of the trajectory. Thus, time improvement during the left turning is of near 20%. Fig. 8 shows an example of some results achieved. Path-tracking coordinates, angular position, and speed for the fixed and flexible cost function strategies are shown. Trajectory error and averaged speed statistical results are not significant, due to the fact that the differences of means between fixed and flexible laws are less than two times the standard deviations.

4. Conclusion

This research can be used on dynamic environments in the neighborhood of the robot. On-line LMPC is a suitable solution for low level path-tracking. LMPC is more time expensive when compared with traditional PID controllers. However, instead of PID speed control approaches, LMPC is based on a horizon of available coordinates within short prediction horizons that act as a reactive horizon. Therefore, path planning and convergence to coordinates can be more easily implemented by using LMPC methods. In this way, contractive constraints are used for guaranteeing the convergence towards the desired coordinates. The use of different dynamic models avoids the need of kinematical constraints that are inherent to other MPC techniques applied to WMR. In this context the control law is based on the consideration of two factors that consist of going straight or turning. Therefore, orientation deviation or trajectory deviation distance can be used as turning factors. The methodology used for performing the experiments is shown. From on-robot depicted experiences, the use of flexible cost functions with relationships to the path to be tracked can be considered as an important result. Thus, control system performance can be improved by considering different factor weights as a function of path to be followed.

The necessary horizon of perception is constrained to just few seconds of trajectory planning. The short horizons allow real time implementations and accuracy trajectory tracking. The experimental LMPC processing time was 45ms, (m=3, n=5), running in the WMR embedded PC of 700MHz. The algorithms simplicity is another relevant result obtained. The factorial design, with two levels of quantitative factors, is presented as an easy way to infer experimental statistical data that allow testing feature performances as function

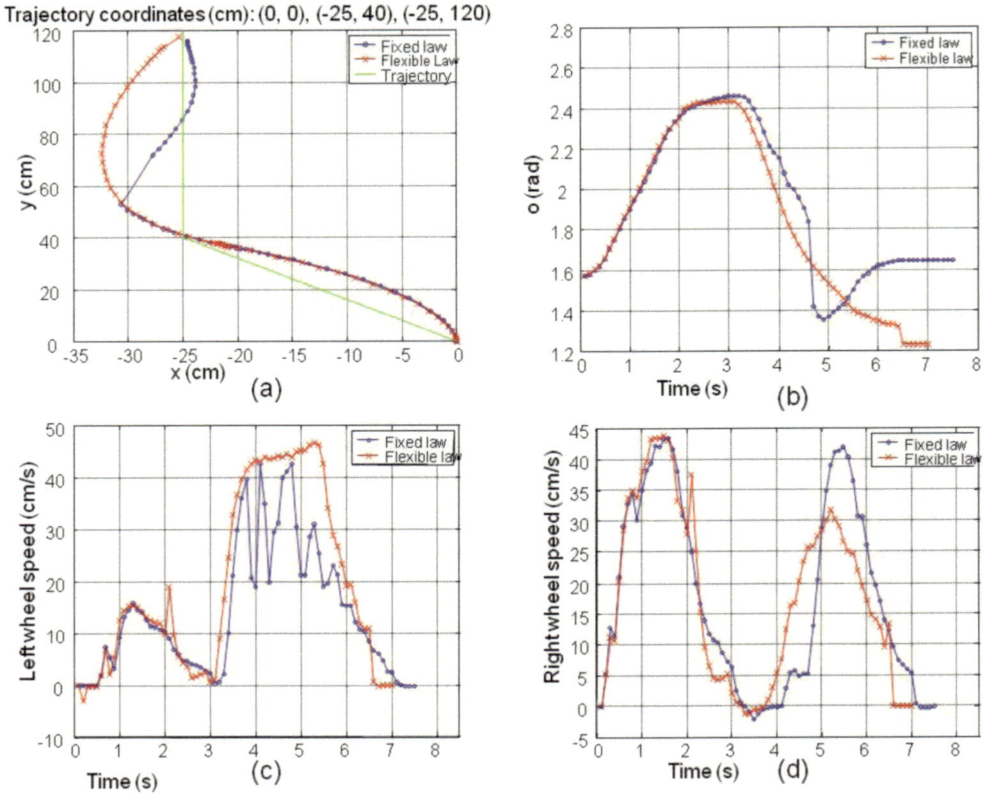

Fig. 8. (a) Trajectory-tracking experimental results by using flexible or fixed cost function. (b) WMR orientation experimental results by using flexible or fixed cost function. (c) Left wheel speed results by using flexible or fixed cost function. (d) Right wheel speed results by using flexible or fixed cost function.

of the different factor combinations. Further studies on LMPC should be done in order to analyze its relative performance with respect to other control laws or to test the cost function performance when other factors are used. The influence of the motor dead zones is also an interesting aspect that should make further efforts to deal with it.

5. Acknowledgement

This work has been partially funded by the Commission of Science and Technology of Spain (CICYT) through the coordinated projects DPI2007-66796-C03-02 and DPI 2008-06699-C02-01.

6. References

Box, G. E. P., Hunter, J. S., Hunter W. G. (2005). Statistics for Experimenters, Ed. John Wiley & Sons, ISBN 0-471-71813-0, New Jersey (USA).

Boyd, S. & Vandenberghe, L. (2004). Convex Optimization, Cambridge University Press, ISBN-13: 9780521833783, New York (USA).

Fox, D.; Burgard, W. & Thrun, S. (1997). The dynamic window approach to collision avoidance. IEEE Robotics & Automation Magazine, Vol. 4, No. 1, (Mar 1997) 23-33, ISSN 1070-9932

Gupta, G.S.; Messom, C.H. & Demidenko, S. (2005). Real-time identification and predictive control of fast mobile robots using global vision sensor. IEEE Trans. On Instr. and Measurement, Vol. 54, No. 1, (February 2005) 200-214, ISSN 1557-9662

Horn, B. K. P. (1998). Robot Vision. MIT press, Ed. McGraw–Hill, ISBN 0-262-08159-8, London (England)

Klancar, G & Skrjanc, I. (2007). Tracking-error model-based predictive control for mobile robots in real time, Robotics and Autonomous Systems, Vol. 55, No. 6, (June 2007), pp. 460-469, ISSN: 0921-8890.

Kühne, F.; Lages, W. F. & Gomes da Silva Jr., J. M., (2005). Model predictive control of a mobile robot using input-output linearization, Proceedings of Mechatronics and Robotics, ISBN 0-7803-9044-X, Niagara Falls Canada, July 2005

Ljung, L. (1991). Issues in System Identification. IEEE Control Systems Magazine, Vol. 11, No.1, (January 1991) 25-29, ISSN 0272-1708

Maciejowski, J.M. (2002). Predictive Control with Constraints, Ed. Prentice Hall, ISBN 0-201-39823-0, Essex (England)

Nesterov, I. E.; Nemirovskii, A. & Nesterov, Y. (1994). Interior_Point Polynomial Methods in Convex Programming. Siam Studies in Applied Mathematics, Vol 13, Publications, ISBN 0898713196

Norton, J. (1986). An Introduction to Identification. Academic Press, ISBN 0125217307, London and New York, 1986

Ögren, P. & Leonard, N. E. (2005). A convergent dynamic window approach to obstacle avoidance. IEEE Transaction on Robotics, Vol. 21, No. 2., (April 2005) 188-195, ISSN: 1552-3098

Pacheco, L., Luo, N., Cufí, X. (2008). Predictive Control with Local Visual Data, In: Robotics, Automation and Control, Percherková, P., Flídr, M., Duník, J., pp. 289-306, Publisher I-TECH, ISBN 978-953-7619-18-4, Printed in Croatia.

Pacheco, L., Luo, N.; Ferrer, I. and Cufí, X. (2009). Interdisciplinary Knowledge Integration Through an Applied Mobile Robotics Course, The International Journal of Engineering Education, Vol. 25, No. 4, (July, 2009), pp. 830-840, ISSN: 0949-149X

Pacheco, L., Luo, N. (2011) Mobile robot local trajectory tracking with dynamic model predictive control techniques, International Journal of Innovative Computing, Information and Control, Vol.7, No.6, (June 2011), in press, ISSN 1349-4198

Schilling, R.J. (1990). Fundamental of Robotics. Prentice-Hall (Ed.), New Jersey (USA) 1990, ISBN 0-13-334376-6

Wan, J. (2007) Computational reliable approaches of contractive MPC for discrete-time systems, PhD Thesis, University of Girona.

15

Model Predictive Control and Optimization for Papermaking Processes

Danlei Chu, Michael Forbes, Johan Backström,
Cristian Gheorghe and Stephen Chu
Honeywell,
Canada

1. Introduction

Papermaking is a large-scale two-dimensional process. It has to be monitored and controlled continuously in order to ensure that the qualities of paper products stay within their specifications. There are two types of control problems involved in papermaking processes: machine directional (MD) control and cross directional (CD) control. Machine direction refers to the direction in which paper sheet travels and cross direction refers to the direction perpendicular to machine direction. The objectives of MD control and CD control are to minimize the variation of the sheet quality measurements in machine direction and cross direction, respectively. This chapter considers the design and applications of model predictive control (MPC) for papermaking MD and CD processes.

MPC, also known as moving horizon control (MHC), originated in the late seventies and has developed considerably in the past two decades (Bemporad and Morari 2004; Froisy 1994; Garcia et al. 1998; Morari & Lee 1999; Rawlings 1999; Chu 2006). It can explicitly incorporate the process' physical constraints in the controller design and formulate the controller design problem into an optimization problem. MPC has become the most widely accepted advanced control scheme in industries. There are over 3000 commercial MPC implementations in different areas, including petro-chemicals, food processing, automotives, aerospace, and pulp and paper (Qin and Badgwell 2000; Qin and Badgwell 2003).

Honeywell introduced MPC for MD controls in 1994; this is likely the first time MPC technology was applied to MD controls (Backström and Baker, 2008). Increasingly, paper producers are adopting MPC as a standard approach for advanced MD controls.

MD control of paper machines requires regulation of a number of quality variables, such as paper dry weight, moisture, ash content, caliper, etc. All of these variables may be coupled to the process manipulated variables (MV's), including thick stock flow, steam section pressures, filler flow, machine speed, and disturbance variables (DV's) such as slice lip adjustments, thick stock consistency, broke recycle, and others. Paper machine MD control is truly a multivariable control problem.

In addition to regulation of the quality variables during normal operation, a modern advanced control system for a paper machine may be expected to provide dynamic economic optimization on the machine to reduce energy costs and eliminate waste of raw materials. For machines that produce more than one grade of paper, it is desired to have an automatic grade change feature that will create and track controlled variable (CV) and MV trajectories to quickly and safely transfer production from one grade to the next. Basic MD-MPC, economic optimization, and automatic grade change are discussed in this chapter.

MPC for CD control was introduced by Honeywell in 2001 (Backström et al. 2001). Today, MPC has become the trend of advanced CD control applications. Some successful MPC applications for CD control have been reported in (Backström et al. 2001, Backström et al. 2002; Chu 2010a; Gheorghe 2009).

In papermaking processes, it is desired to control the CD profile of quality variables such as dry weight, moisture, thickness, etc. These properties are measured by scanning sensors that traverse back and forth across the paper sheet, taking as many as 2000 or more samples per sheet property across the machine. There may be several scanners installed at different points along the paper machine and so there may be multiple CD profiles for each quality variable.

The CD profiles are controlled using a number of CD actuator arrays. These arrays span the paper machine width and may contain up to 300 individual actuators. Common CD actuators arrays allow for local adjustment, across the machine, of: slice lip opening, headbox dilution, rewet water sprays, and induction heating of the rolls. As with the CD measurements, there may be multiple CD actuator arrays of each type available for control. By changing the setpoints of the individual CD actuators within an array, one can adjust the local profile of the CD measurements.

The CD process is a multiple-input-multiple-output (MIMO) system. It shows strong input and output off-diagonal coupling properties. One CD actuator array can have impact on multiple downstream CD measurement profiles. Conversely, one CD measurement profile can be affected by multiple upstream CD actuator arrays. Therefore, the CD control problem consists of attempting to minimize the variation of multiple CD measurement profiles by simultaneously optimizing the setpoints of all individual CD actuators (Duncan 1989).

MPC is a natural choice for paper machine CD control because it can systematically handle the coupling between multiple actuator and multiple measurement arrays, and also incorporate actuator physical constraints into the controller design. However, different from standard MPC problems, the most challenging part of the cross directional MPC (CD-MPC) is the size of the problem. The CD-MPC problem can involve up to 600 MVs, 6000 CVs, and 3000 hard constraints. Also, the new setpoints of MVs are required as often as every 10 to 20 seconds. This chapter discusses the details of the design for an efficient large-scale CD-MPC controller.

This chapter has 5 sections. Section 2 provides an overview of the papermaking process highlighting both the MD and CD aspects. Section 3 focuses on modelling, control and optimization for MD processes. Section 4 focuses on modelling, control and optimization for CD processes. Both Sections 3 and 4 give industrial examples of MPC applications. Finally, Section 5 draws conclusions and provides some perspective on the future of MD-MPC and CD-MPC.

2. Overview of papermaking processes

A flat sheet of paper is a network consisting of cellulose fibres bound to one another. A paper machine transforms a slurry of water and wood cellulose fibres into this type of network. The whole papermaking process can be regarded as a water-removal system: the consistency of fibre solutions, called stock by papermakers, increases from around 1% at the beginning of a paper machine (the headbox) to around 95% at the end (the reel).

2.1 Brief description of papermaking processes

In general a paper machine can be divided into four sections: forming section, press section, drying section, and calendering section. In the forming section, the stock flow enters the headbox to be distributed evenly across a continuously running fabric felt called the wire. The newly formed sheet is carried by the wire along the Fourdrinier table, which has a set of drainage elements that promote water removal by various gravity and suction mechanisms. These elements include suction boxes, couch rolls, foils, etc. The solid consistency of the paper web can reach 20% by the time the web leaves the forming section and enters the press section. Figure 1 illustrates the configuration of a Fourdrinier-type paper machine.

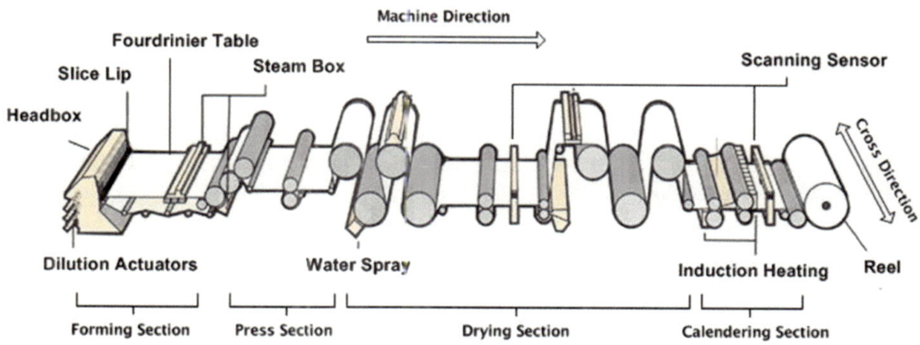

Fig. 1. The configuration of a Fourdrinier-type paper machine

The press section may be considered as an extension of the water-removal process that was started on the wire in the forming section. Typically, it consists of 1 - 3 rolling press nips. When the paper web passes through these nips, the pressing roll squeezes water out and consolidates the web formation at the same time. In the press section, both the surface smoothness and the web strength are improved. As higher web strength is achieved in the press section, better runability will be observed in the drying section. A paper machine is typically operated at a very high speed. The fastest machine speed may be as high as 2,200 meters per minute.

The drying section includes multiple drying cylinders which are heated by high temperature and high pressure steam. The heat is transferred from steam onto the paper surface through these rotating steel cylinders. The heat flow increases the paper surface temperature to the point where water starts evaporating and escaping from the paper web. The drying section is the most energy consuming part of paper manufacturing. Before the paper enters the drying section, the solid consistency is around 50%. After the drying section, the consistency can reach 95%, which corresponds to a finished product moisture specification.

The last section of the paper machine is called the calendering section. Calendering is a terminology referring to pressing with a roll. The surface and the interior properties of the paper web are modified when it passes through one or more calendering nips. Typically the calendering nip consists of one or multiple soft/hard or hard/hard roll pairs. The hard roll presses the paper web against the other roll, and deforms the paper web plastically. By this means, the calender roll surface is replicated onto the paper web. Depending on the type of paper being produced, the primary objective of calendering may be to produce a smooth paper surface (for printing), or to improve the uniformity of CD properties, such as paper caliper (thickness).

More details of paper machine design and operation are given in (Smook 2002; Gavelin 1998).

2.2 Paper quality measurement

A paper machine can have one or more measurement scanners. The quality measurement sensors are mounted on the scanner head which travels back and forth across the paper web to provide online quality measurements. The most common paper machine quality measurements include dry weight, moisture, and caliper. Dry weight indicates the solid weight per unit area of a sheet of paper. For different types of products, the value of dry weight can vary from 10 grams per square meter (gsm), in the case of paper tissue, to 400 gsm, in the case of heavy paper board. Moisture content is another critical quality property of the finished paper product. It indicates the mass percentage of water contained in a sheet of paper. Moisture content is a key factor determining the strength of the finished product. Typical moisture targets range from 5% to 9%. Caliper is the measure of the thickness of a sheet of paper. It is a key factor determining the gloss and printability of the finished product. The caliper targets are in the range from 70 μm to 300 μm depending on the production grade. In general the online measurements for dry weight, moisture and caliper are available and used for both the MD and CD feedback controller designs.

As the scanners travel across a moving sheet, the real data collected actually comes from a zig-zag trajectory (See Figure 2). These data contain both CD and MD variation. A reliable MD/CD separation scheme is the prerequisite for MD and CD control designs. Since the MD/CD separation is a separate topic, the rest of this chapter assumes that the pure MD/CD measurements have been obtained prior to the MD/CD controller development. The scanner measurements are denoted by $x(i, t)$, $i = 1, \cdots, n$ indexes the n measurements taken across the sheet each scan (CD measurement index), and t is the time stamp of each scan (MD measurement index). $\bar{x}(t)$ is the MD measurement given by

$$\bar{x}(t)=\frac{1}{n}\sum_{i=1}^{n} x(i, t). \tag{1}$$

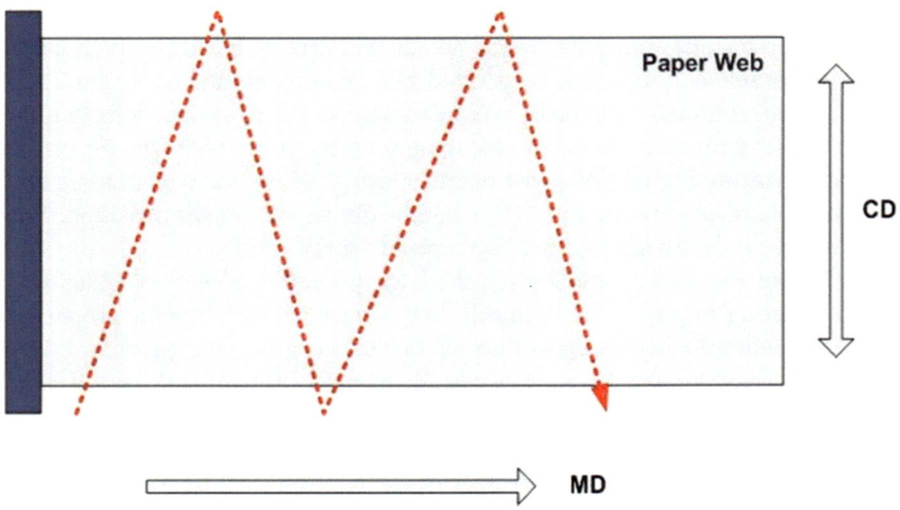

Fig. 2. The zig-zag scanner trajectories

2.3 Brief description of MD control

The objective of MD control is to minimize the variation of the sheet quality measurements in machine direction.

A number of actuators are available for control of the MD variables. Stock flow to the headbox is regulated by the stock flow valve or variable speed pump. As stock flow increases, the amount of fibre flowing into the forming section increases and dry weight and caliper increase. At the same time, there is also more water coming through the machine and moisture will increase. So, changes in the stock flow affect dry weight, moisture and caliper. The steam pressure in the cylinders of the drying section may be adjusted. As the steam pressure in the cylinders increases, so does the temperature in the cylinders, and more heat is transferred to the paper. In this way, steam pressure affects moisture. Typically, the dryer cylinders are divided into groups, and the steam pressure for each of these dryer sections may be adjusted independently. Machine speed affects dry weight and caliper, as increasing the machine speed stretches the paper web thinner, giving it less mass per unit area. Machine speed also affects moisture as both the drying properties of the paper and the residence time in the dryer change. Clearly, paper machine MD control is a multivariable control problem.

2.4 Brief description of CD control

The objective of CD control is to achieve uniform paper qualities in the cross direction, i.e., to minimize the variation of CD profiles. The CD variation, can be formulated as two times of the standard deviation of the CD profile,

$$2\sigma_{CD}(t)=2^*\left(\frac{1}{n-1}\sum_{i=1}^{n}\left(x(i,t)-\bar{x}(t)\right)^2\right)^{\frac{1}{2}} \tag{2}$$

Often the term 'CD spread' is used interchangeably with $2\sigma_{CD}$.

CD actuators are used to regulate CD profiles and improve the uniformity of paper quality properties in the cross direction i.e., reduce the value of $2\sigma_{CD}$.

The most common dry weight CD actuators are both located at the headbox. The headbox slice opening is a full-width orifice or nozzle that can be adjusted at points across the width of the paper machine. This allows for differences in the local stock flow onto the wire across the machine. The consistency profiler changes the consistency of local stock flow by injecting dilution water and altering the local concentration of pulp fibre across the headbox. Headbox slice and consistency profiler are primarily designed for dry weight control, but they have the effects on both moisture and caliper profiles. Figure 1 indicates the location of headbox dry weight actuators.

The most common moisture actuators are the steam box and water spray. The steam box applies high temperature steam to the surface of the moving paper web. As the latent heat in the steam is released and heats up the paper web, it lowers the web viscosity and eases dewatering in the press section. The water spray regulates the moisture profiles according to a different mechanism. It deploys a fine water spray to the paper surface through a set of nozzles across the machine width to re-moisturize the paper web. Similar to the dry weight actuators, moisture actuators are designed for moisture profile regulation but they may have effects on the caliper profile. The steam box is typically installed in the press section and the water spray is located in the drying section. Figure 1 indicates the physical locations of moisture actuators.

The most common caliper actuators are hot air showers and induction heaters. Both types of actuators provide surface heating for calendering rolls. The hot shower uses the high temperature steam or air; the induction heater uses the high frequency alternating current. By heating up the calender roll, caliper actuators alter the local diameter of the calender roll and subsequently increase the local pressure applied to the paper web. The physical location of the caliper actuators can be also found in Figure 1.

3. Modelling, control and optimization of papermaking MD processes

Control of the MD process is typically a regulation problem where the paper quality variables need to be held within specified quality limits. At the same time, the constant pressure to increase operational efficiency demands that the paper production uses the least amount of raw materials and energy required to meet quality goals.
When multiple grades of paper are produced on a single paper machine, control must also be able to provide quick transitions of the quality variables along smooth trajectories. When the differences between the paper grades are large, it may be necessary to enhance the control algorithm to account for process nonlinearities that become apparent over a larger span of operating points.

3.1 Modeling of papermaking MD processes

In this section, modelling of the MD process for MPC controller design is discussed. The additional modelling required for paper grade change control is discussed in section 3.4.
For effective MPC control of paper MD quality variables, it is necessary to build a matrix of linear models relating the process MV's to the quality variables (the CV's). A basic paper machine model matrix most often includes stock flow, steam pressure of multiple dryer sections, and machine speed as MVs, and paper weight (basis weight or dry weight), and moisture as CV's. Many other MV's and CV's can be included in the model matrix depending on the complexity of the paper machine and the paper quality requirements. An example model matrix is given in Figure 3.

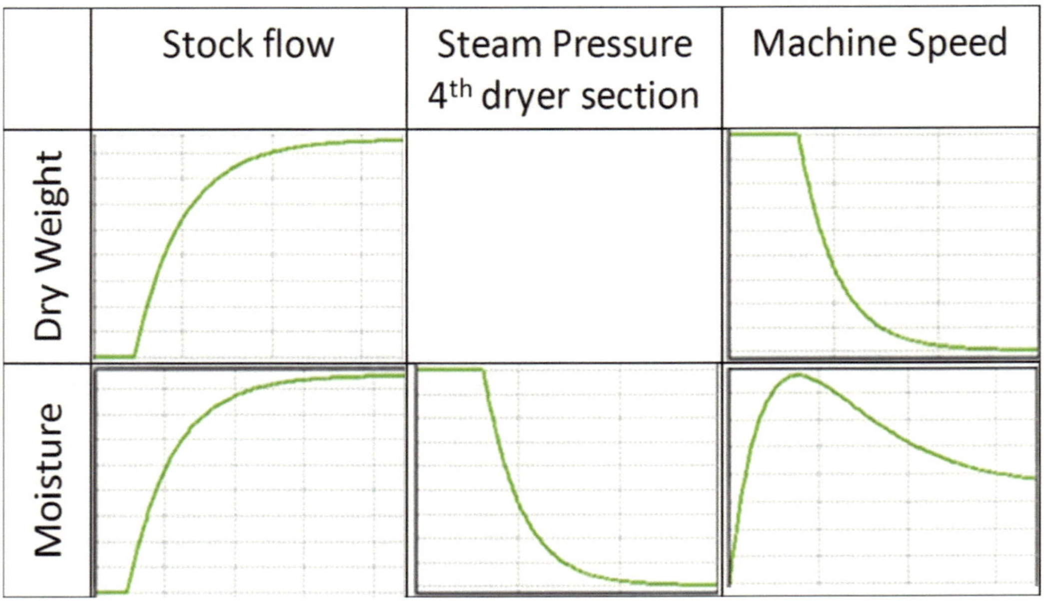

Fig. 3. A basic model matrix for CD-MPC. The models are step responses.

The process models used for MPC control are often developed in transfer function form, such as:

$$y(s) = g(s)u(s) + d(s) \tag{3}$$

and

$$y(s)=\begin{bmatrix} y_1(s) \\ \vdots \\ y_{N_y}(s) \end{bmatrix},\ g(s) = \begin{bmatrix} g_{11}(s) & \cdots & g_{1N_u}(s) \\ \vdots & \ddots & \vdots \\ g_{N_y1}(s) & \cdots & g_{N_yN_u}(s) \end{bmatrix},\ u(s)=\begin{bmatrix} u_1(s) \\ \vdots \\ u_{N_u}(s) \end{bmatrix}, \tag{4}$$

where $y(s) \in \mathbb{C}^{N_y}$ is the Laplace transformation of the N_y MD quality measurements (such as dry weight, moisture, caliper, etc). $u(s) \in \mathbb{C}^{N_u}$ is the Laplace transformation of the actuator setpoints (such as thick stock flow, dryer section steam pressure(s), filler flows, machine speed, etc). $d(s) \in \mathbb{C}^{N_y}$ is the Laplace transformation of the augmented process disturbance array. $g_{ij}(s) \in \mathbb{C}$ ($i = 1 \ldots N_y$ and $j = 1 \ldots N_u$) are the transfer functions from the jth actuator u_j to the ith MD quality measurement y_i.

3.1.1 Model identification

Typically the MD process models that are used as the basis for the MD-MPC controller are identified from data obtained during simple process experiments. A series of step changes is made for each MV. There is a delay after each step long enough so that the full responses of all of CV's can be observed. That is, the CV responses reach steady state before the next step change in the MV is made. These types of process experiments are known as bump tests.

Once a set of identification data has been obtained, various techniques may be used to generate a process model from this data. In the simplest case, plots of the bump tests are reviewed to graphically estimate a process gain, k_{ij}, dead time, $T_{d_{ij}}$, and time constant, $T_{p_{ij}}$, yielding the process model:

$$y_i(s) = \frac{k_{ij}e^{-T_{d_{ij}}s}}{1+T_{p_{ij}}s} u_j(s) \tag{5}$$

More complex methods involve use of regression and search techniques to find both the optimum model parameters, and the optimum model structure (transfer function numerator and denominator orders). These techniques typically use minimization of squared model prediction errors as the objective:

$$J = \sum_{k=1}^{N} \left(\hat{y}_i(k) - y_i(k)\right)^2 \tag{6}$$

Where $\hat{y}_i(k)$ and $y_i(k)$ are respectively the predicted and actual values of the ith CV at time k. (Ljung 1998) is the classic reference on system identification, and there are commercial software packages available that automate much of the system identification work.

3.2 MD-MPC design

Once all of the bump test, and system identification activities have been performed, the complete process model (3) is used directly in the model predictive controller. MPC solves an optimization problem at each control execution. One robust MPC problem formulation is:

$$\min_{\Delta u, y_r} \frac{1}{2} \|W(y_r - S\Delta u)\|_2^2, \tag{7}$$

Subject to:

$$y_l \leq y_r \leq y_h$$

$$u_l \leq u \leq u_h$$

$$\Delta u_l \leq \Delta u \leq \Delta u_h$$

The values of y_r are the CV targets and SΔU are the predicted future values of the CV's. S is the prediction matrix, containing all the information from the process model (3). W is a weighting matrix, and $\|\cdot\|_2^2$ is the two-norm squared operator. y_l and y_h are the low and high CV quality limits, u_l and u_h are the low and high MV limits, and Δu_l and Δu_h are the low and high limits for MV moves. As discussed in the section below, this problem formulation, combined with techniques employed in its solution implicitly provide robustness characteristics in the controller design.
The technical details of the solution of the problem (7) are given in (Ward 1996); however, some notable aspects of the solution methodology and beneficial characteristics of the solution are given in the section below.

3.2.1 Model scaling and controller robustness

Robust numerical solution of the optimization problem (7) depends on condition number of the system gain matrix, G. Prior to performing the controller design, the gain matrix condition number is minimized by solving the problem:

$$\min_{D_r, D_c} \{\kappa(D_r g D_c)\}, \tag{8}$$

Where κ is condition number, and D_r and D_c are diagonal transformation matrices. The scaled system gain matrix g_s is then:

$$g_s = D_r g D_c, \tag{9}$$

g_s is then used for all MPC computations.
It should be noted that the objective (7) does not explicitly penalize the MV moves Δu as a method to promote controller robustness. Instead, controller robustness is provided by the CV range formulation and singular value thresholding.
First, the CV range formulation refers to the inequalities given in the problem formulation (7). Under this formulation, if a CV is predicted to be within it range in the future, no MV action is taken. Since MV moves are not made unless absolutely necessary, this is a very robust policy.
Second, the solution of the problem involves an active set method that allows the constrained optimization problem to be converted into an unconstrained problem. A URV orthogonal decomposition (see Ward 1996) of the matrix characterizing the unconstrained problem is then employed to solve the unconstrained problem. Prior to the decomposition, singular values of the problem matrix that are less than a certain threshold are dropped, reducing the dimension of the problem, and ensuring that the controller does not attempt to control weakly controllable directions of the process.

3.3 Economic optimization

Energy consumption is a big concern for papermakers. Increasing profits by minimizing operating costs without sacrificing paper quality and runability is always a goal for them. In theory, if the number of MVs of a process is greater than the number of CVs plus the number of active constraints, the process has degrees of freedom allowing for steady-state optimization. Product value optimization can be systematically integrated with the MD-MPC control. One can then take the feed, product, and utility costs into account with the MD controller design.

For the economic optimization of a process, the following objective is to be minimized:

$$J = \sum_i \left(a_{y_i}(y_i - y_{i0})^2 + b_{y_i}y_i\right) + \sum_j \left(a_{u_j}\left(u_j - u_{j0}\right)^2 + b_{u_j}u_j\right) \tag{10}$$

Here y_{i0} and u_{j0} are the desired steady state values of the process CV's and MV's, a_{y_i} and a_{u_j} are the costs of quadratic deviation from the desired values, and b_{y_i} and b_{u_j} are the linear costs of the CV's and MV's. This objective is useful for paper machines, for example, by placing costs on the different energy sources used in drying.

The economic objective is combined with the MPC control problem objective to give an augmented problem formulation. The augmented problem is then solved using the same solution method as described above.

Economic optimization is a lower priority for paper machines than quality control. If the paper does not meet quality specifications, it cannot be sold, and any savings made from economic optimization are more than lost. Therefore, economic optimization only occurs when there are extra degrees of freedom for the controller. Economic optimization is not attempted unless all of the CV's are predicted to remain within their quality specifications over the whole of the controller's prediction horizon.

3.3.1 Mill implementation results

MPC including an economic optimization layer was implemented for a tissue machine. A diagram of the tissue machine is given in Figure 4. As can be seen in the diagram, tissue dry weight and moisture are measured at the reel; moisture is measured between the second through-air dryer (TAD2) and the Yankee dryer, and TAD1 exhaust pressure must also be monitored and controlled. These four variables are the CV's in this example. A large number of MV's are available to control this machine. Stock flow, TAD1 supply temperature, TAD1 dry end differential pressure, TAD1 gap pressure, TAD2 exhaust temperature, TAD2 dry end differential pressure, TAD2 gap pressure, Yankee hood temperature, and Yankee supply fan speed are all used as MV's in the MPC. Machine speed and tickler refiner were added as DV's. The MPC model matrix is shown in Figure 5.

MV	Energy Fuel	Units	Linear Obj Coef Cost / eng unit
TAD1 Supply Temp	Gas	deg F	0.680
TAD1 DE DP	Electricity	inch H2O	47.267
TAD1 Gap Pres	Electricity	inch H2O	-0.030
TAD2 Exh Temp	Gas	deg F	5.858
TAD2 DE DP	Electricity	inch H2O	40.249
TAD2 Gap Pres	Electricity	inch H2O	-16.415

Table 1. Tissue machine MV's with linear objective coefficients

A large number of the MV's in this control problem have an impact on the paper moisture, both after TAD2, and at the reel; however each MV uses a different energy source and has different drying efficiency. Overall, since there are more MV's than CV's, and significant cost differences between the MV's, there is an opportunity for economic optimization in this system. Table 1 shows the energy sources and different energy cost efficiencies (Linear Obj Coeff Cost/eng unit) associated with each MV. Economic optimization can be accomplished by including these variables in the linear part of the economic objective function given by (10).

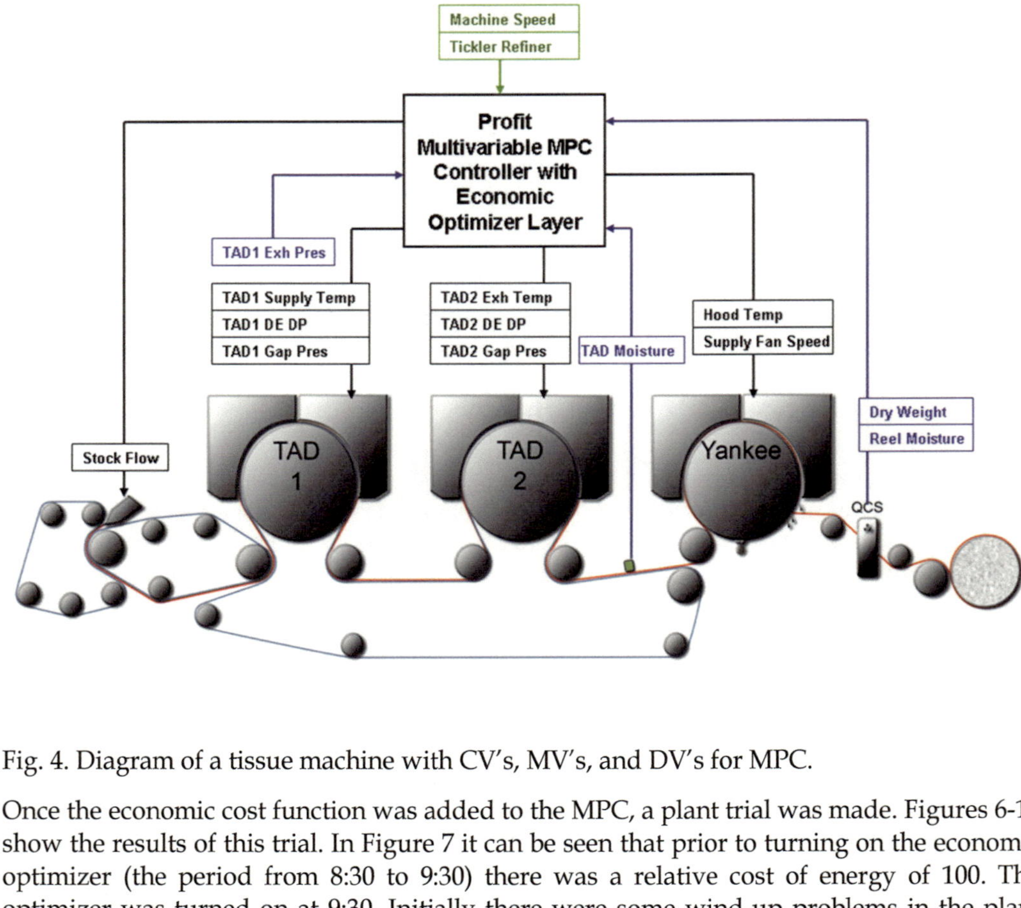

Fig. 4. Diagram of a tissue machine with CV's, MV's, and DV's for MPC.

Once the economic cost function was added to the MPC, a plant trial was made. Figures 6-10 show the results of this trial. In Figure 7 it can be seen that prior to turning on the economic optimizer (the period from 8:30 to 9:30) there was a relative cost of energy of 100. The optimizer was turned on at 9:30. Initially there were some wind-up problems in the plant DCS which were preventing the MPC from optimizing. Once these were cleared, at 10:44, the controller drove the process to the low cost operating point (from 10:44 to 12:30). The relative cost of energy at this operating point was 98.8. In order to better interpret these results, it is necessary to rank the costs of each MV on the common basis of Cost/% Moi. This is accomplished by dividing the linear objective coefficients given in Table 1 by their respective process gains. These are shown in Table 2, along with the MV high and low limits, and the optimization behaviour. Looking again at the Figures 8 and 9, it can be seen that the highest costing MV's are driven to their minimum operating points, and the lowest costing MV's are driven to their maximum operating points. The TAD1 dry end differential pressure is left as the MV that is within limits and actively controlling the paper moistures. Figure 10 shows that throughout this trial, the MV's are optimized without causing any disturbance to the CV's.

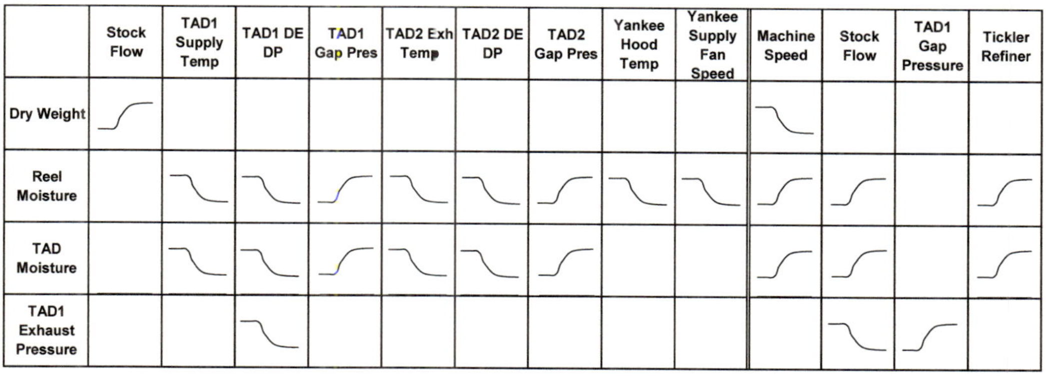

Fig. 5. The MPC model matrix for the tissue machine control and optimization example

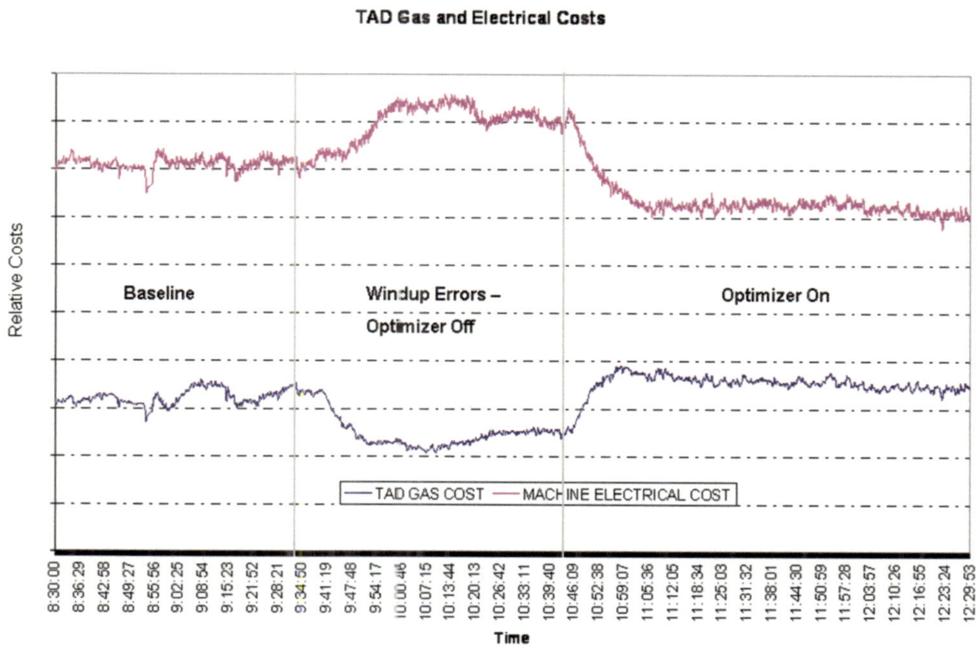

Fig. 6. Natural gas costs and electricity costs during the trial

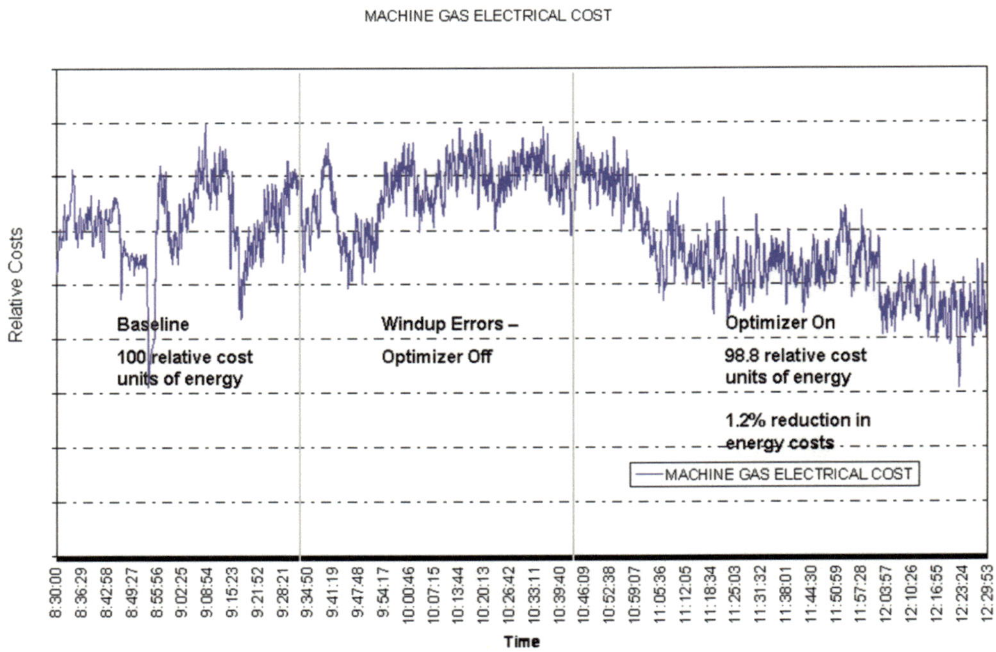

Fig. 7. Total costs during the trial

MV	eng unit	Low Limit	High Limit	Linear Obj Coef (Cost/eng unit)	Process Gain (%Moi/eng unit)	Cost (Cost / % Moi)	Rank	Optimization Behavior
TAD1 Supply Temp	deg F	300.0	450.0	0.68	-0.12	5.48	4	450 (max)
TAD1 DE DP	inch H2O	1.0	3.9	47.30	-5.12	9.24	3	controlling Moi
TAD1 Gap Prs	inch H2O	0.4	1.5	-0.03	1.95	0.02	6	0.4 (max)
TAD2 Exh Temp	deg F	175.0	250.0	5.86	-0.45	13.02	1	175 (min)
TAD2 DE DP	inch H2O	1.0	3.5	40.26	-3.14	12.82	2	1 (min)
TAD2 Gap Prs	inch H2O	0.2	1.5	-16.40	4.25	3.86	5	0.2 (max)

Table 2. The MV cost rankings.

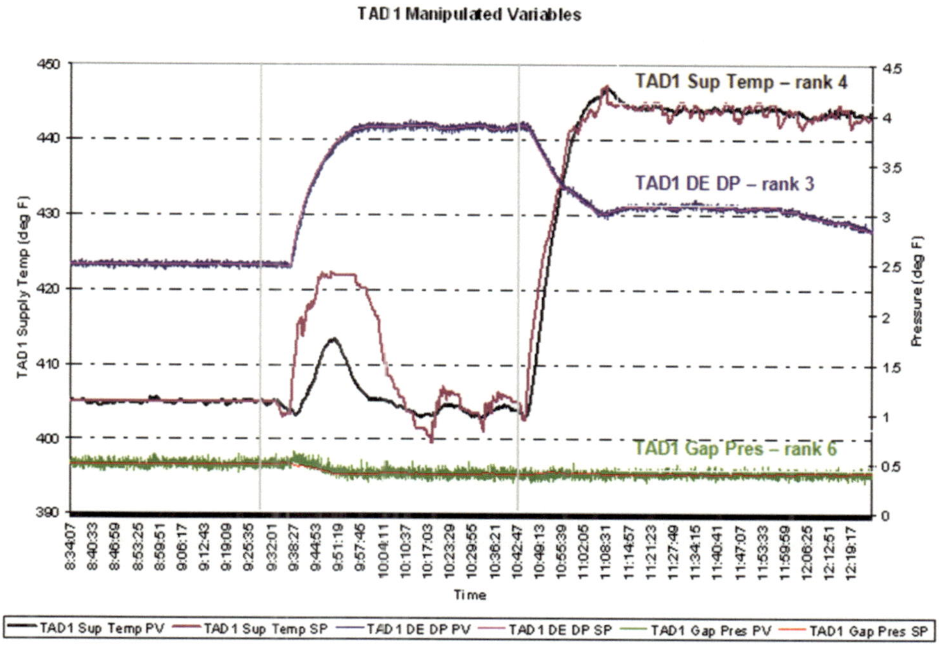

Fig. 8. Manipulated variables during the trial

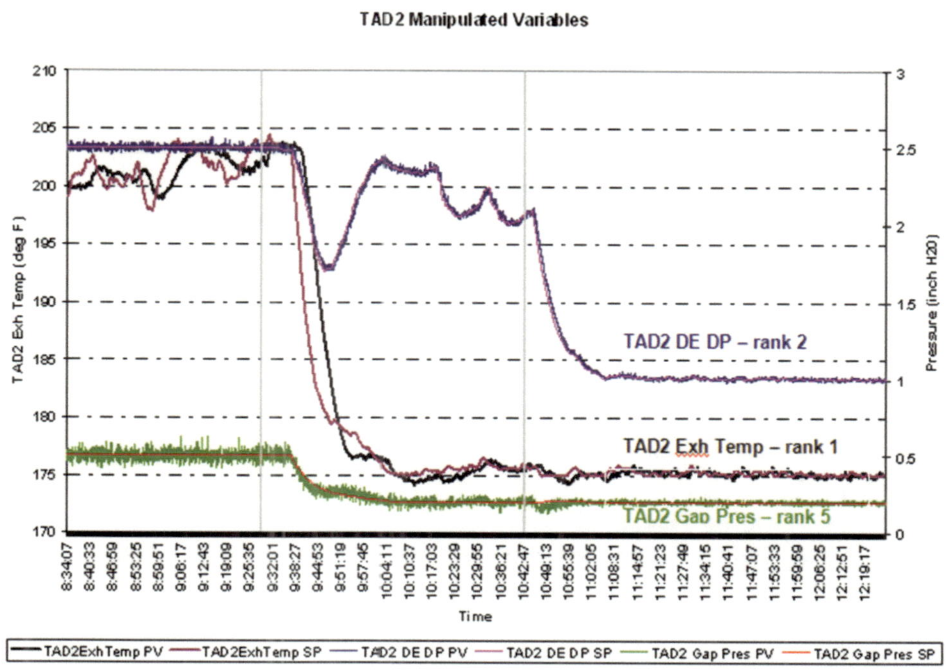

Fig. 9. Manipulated variables during the trial

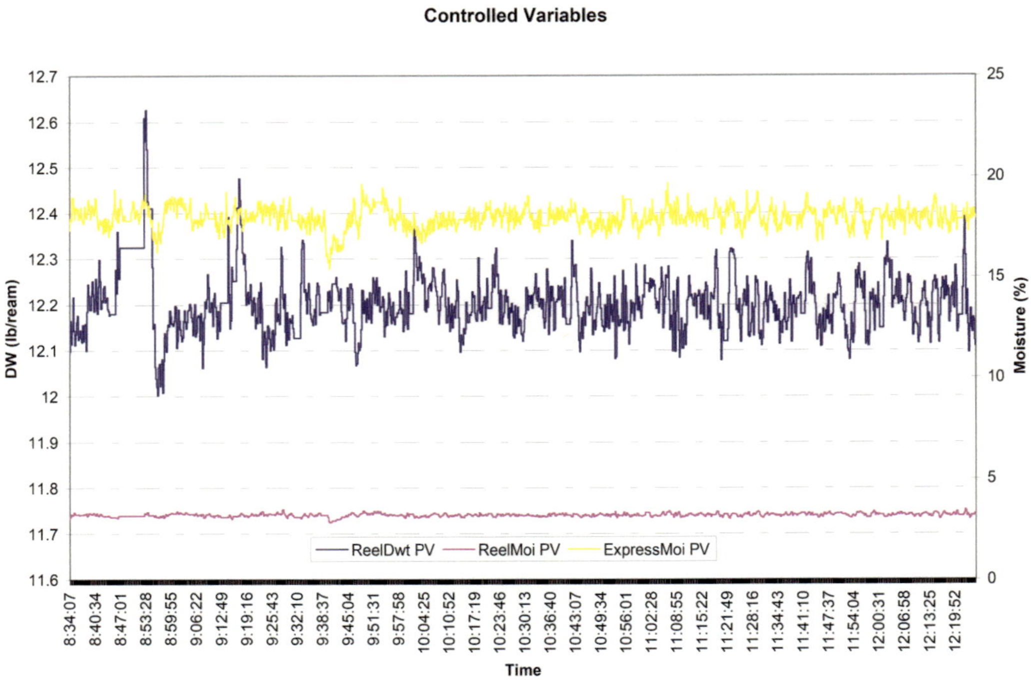

Fig. 10. Controlled variables during the trial

3.4 Grade change strategies

Grade change is a terminology in MD control. It refers to the process of transitioning a paper machine from producing one grade of paper product to another. One can achieve a grade change by gradually ramping up a set of MVs to drive the setpoints of CVs from one operating point to another. During a grade change, the paper product is often off-specification and not sellable. It is important to develop an automatic control scheme to coordinate the MV trajectories and minimize the grade change transition times and the off-spec product. An offline model predictive controller can be designed to produce CV and MV trajectories to meet these grade change criteria. MPC is well-suited to this problem because it explicitly considers MV and CV trajectories over a finite horizon. By coordinating the offline grade change controller (linear or nonlinear) and an online MD-MPC, one can derive a fast grade change that minimizes off-spec production. This section discusses the design of MPC controllers for linear and nonlinear grade changes.

Figure 11 gives a block diagram of the grade change controller incorporated into an MD control system. The grade change controller calculates the MV and CV trajectories to meet the grade change criteria. This occurs as a separate MPC calculation performed offline so that grade change specific process models can be used, and so that the MPC weightings can be adjusted until the MV and CV trajectories meet the design criteria. The MV trajectories are sent to the regulatory loop as a series of MV setpoint changes. The CV trajectories are sent as setpoint changes to the MD controller. If the grade change is performed with the MD controller in closed-loop, additional corrections to the MV setpoints are made to eliminate any deviation of the CV from its target trajectory.

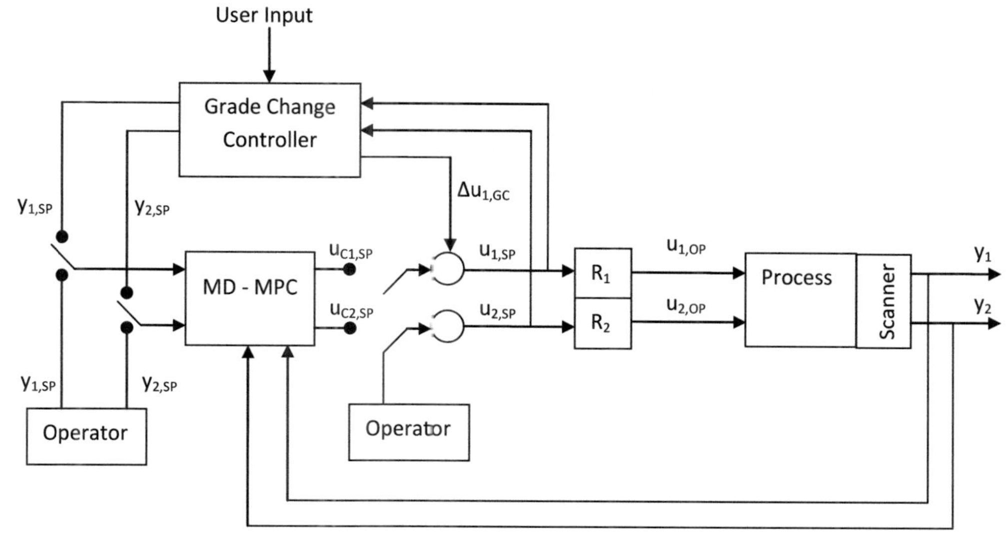

Fig. 11. Block diagram of MD-MPC control enhanced with grade change capability

The MV and CV trajectories are generated in a two step procedure. First there is a target calculation step that generates the MV setpoints required to bring the CV's to their target values for the new grade. Once the MV setpoints are generated, then there is a trajectory generation step where the MV and CV trajectories are designed to meet the specifications of the grade change.
The MV targets are generated from solving a set of nonlinear equations:

$$\begin{aligned} &y_{dw}^1 - f_{dw}^1(u_1, u_2, u_3, \ldots, C_1, C_2, C_3, \ldots) = 0, \\ &y_{dw}^2 - f_{dw}^2(u_1, u_2, u_3, \ldots, C_1, C_2, C_3, \ldots) = 0, \\ &\vdots \\ &y_{moi}^1 - f_{moi}^1(u_1, u_2, u_3, \ldots, C_1, C_2, C_3, \ldots) = 0, \\ &y_{moi}^2 - f_{moi}^2(u_1, u_2, u_3, \ldots, C_1, C_2, C_3, \ldots) = 0, \\ &\vdots \end{aligned} \tag{11}$$

Here y_{dw}/y_{moi} represents the CV target for the new grade. The functions $f(\cdot)$ are the models of dry weight and moisture. The process MV's are denoted u_i and model constants are denoted C_i. The superscripts indicate the same paper properties measured by different scanners. Since the number of MV's and the number of CV's is not necessarily equal, these equations may have one, multiple or no solutions. To allow for all of these cases, the problem is recast as:

$$\min_{u_i} F(u_1, u_2, \ldots), \tag{12}$$

Subject to:

$$G(u_1, u_2, \ldots) \leq 0,$$

$$H(u_1, u_2, \ldots) = 0,$$

Where $F(\cdot)$ is a quadratic objective function formulated to find the minimum travel solution. $H(\cdot)$ represents the equality constraints given above, and $G(\cdot)$ represents the physical limitations of the CVs and MVs (high, low, and rate of change limits).
Once the MV targets have been generated, the MV and CV trajectories are then designed. Figure 12 gives a schematic representation of the trajectory generation algorithm. The process models are linearized (if necessary) and then scaled and normalized for application in an MPC controller. Process constraints such as the MV and CV targets, and the MV high and low limits are also given to the MPC controller. Internal controller tuning parameters are then used to adjust the MV and CV trajectories to meet the grade change requirements.

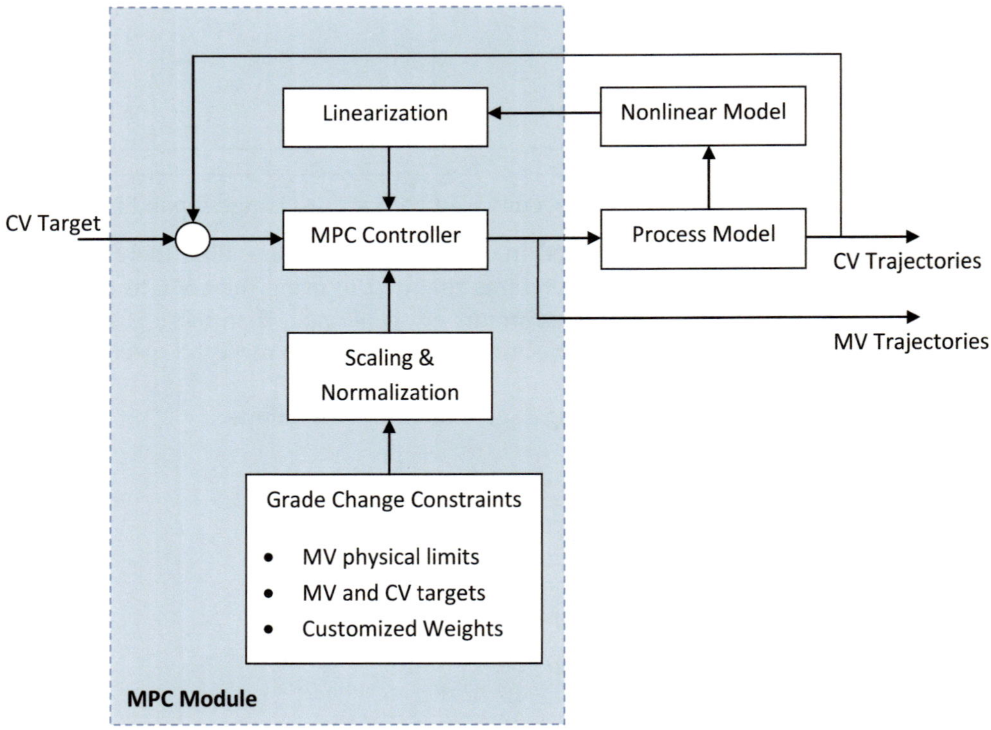

Fig. 12. Diagram of MPC-based grade change trajectory generation.

3.4.1 Linear grade change

In a linear grade change, the MD process models that are used in the MD-MPC controller are also used as the models for determining the MD targets, and for designing the MD grade change trajectories.

3.4.2 Nonlinear grade change

In a nonlinear grade change, a first principles model may be used for the target and trajectory generation. For example, a simple dry weight model is:

$$m_{dry} = K\frac{q_{stock}}{v}, \tag{13}$$

Where m_{dry} is the paper dry weight, q_{stock} is the thick stock flow, and v is machine speed. K is the expression of a number of process constants and values including fibre retention, consistency, and fibre density. (Chu et al. 2008) gives a more detailed treatment of this dry weight model.
(Persson 1998, Slätteke 2006, and Wilhelmsson 1995) are examples of first principles moisture models that may be used.

3.4.3 Mill implementation results

In this section, some results of MPC-based grade changes for a fine paper machine are given. The grade change is from a paper with a dry weight of 53 lb/3000ft^2 (86 g/m^2) to a paper with a dry weight of 44 lb/3000ft^2 (72 g/m^2). Both paper grades have the same reel moisture setpoint of 4.8%. For the grade change, stock flow, 6th section steam pressure, and machine speed are manipulated.
Figures 13 and 14 show a grade change performed on the paper machine using linear process models, and keeping the regular MPC in closed-loop during the grade change. The grade change was completed in 10 minutes, which is a significant improvement over the 22 minutes required by the grade change package of the plant's previous control system. In Figure 13, the CV trajectories are shown. Here it can be seen that although there is initially a small gap between the actual dry weight and the planned trajectory, the regular MPC takes action with the thick stock valve (as shown in Figure 14) to quickly bring dry weight back on target. The deviation in the reel moisture is more obvious. This might be expected as the moisture dynamics of the paper machine display more nonlinear behaviour for this range of operations. The steam trajectory in Figure 14 is ramping up at its maximum rate and yet the paper still becomes too wet during the initial part of the grade change. This indicates that the grade change package is aggressively pushing the system to achieve short grade change times.

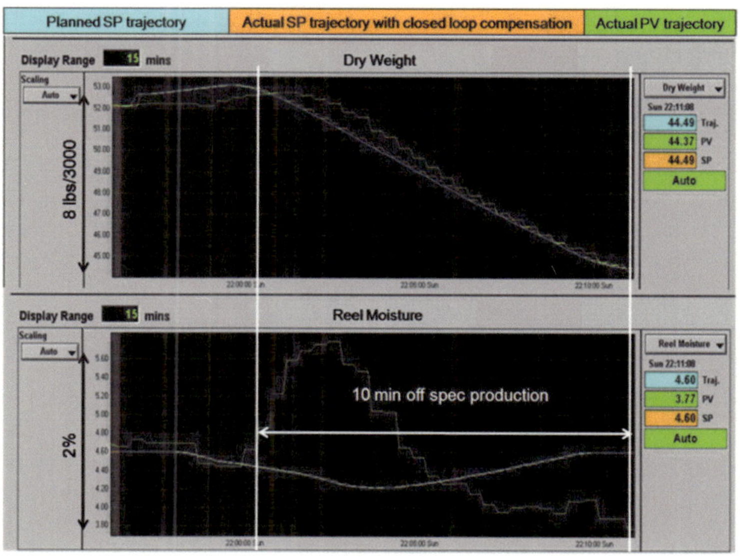

Fig. 13. CV trajectories under closed-loop GC with linear models

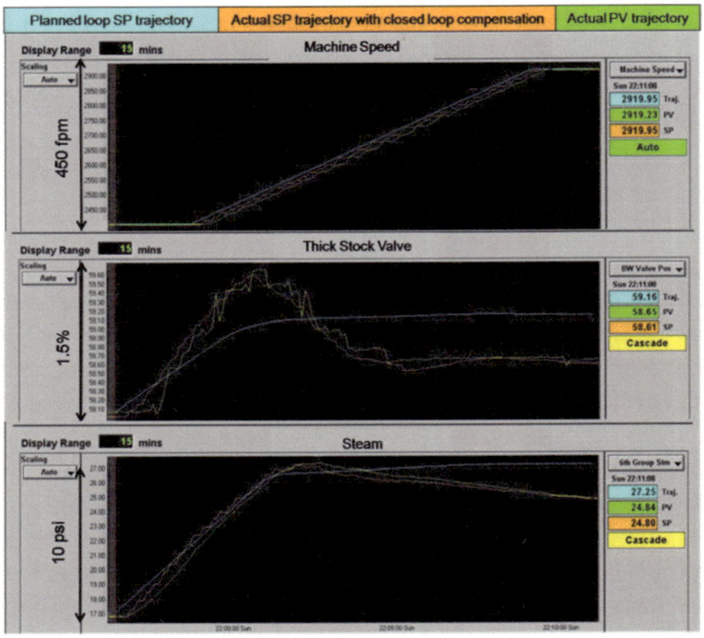

Fig. 14. MV trajectories under closed-loop GC with linear models

Figures 15 and 16 show a grade change performed on a high fidelity simulation of the fine paper machine. This grade change uses a nonlinear process model, and the regular MPC is kept in closed-loop during the grade change. Here it can be seen that the duration of the grade change is reduced to 8 minutes. Part of the improvement comes from using stock flow setpoint instead of stock valve position, allowing improved dry weight control. Another improvement is that the planned trajectories allow for some deviation in the reel moisture that cannot be eliminated. Both dry weight and reel moisture follow their trajectories more closely. At the end of the grade change, the nonlinear grade change package is able to anticipate the need to reduce steam preventing the sheet from becoming dry.

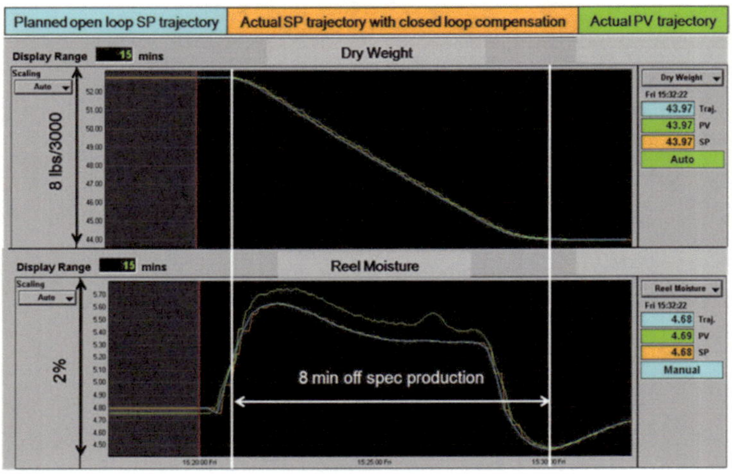

Fig. 15. CV trajectories under closed-loop GC with nonlinear models

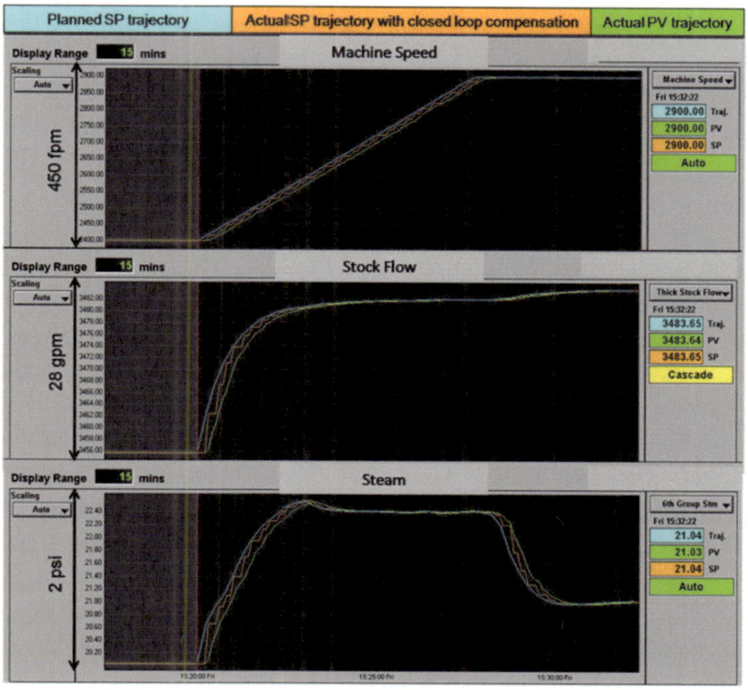

Fig. 16. MV trajectories under closed-loop GC with nonlinear models

4. Modelling, control and optimization of papermaking CD processes

To produce quality paper it is not enough that the average value of paper weight, moisture, caliper, etc across the width of the sheet remains on target. Paper properties must be uniform across the sheet. This is the purpose of CD control.

4.1 Modelling of papermaking CD processes

The papermaking CD process is a large scaled two-dimensional process. It involves multiple actuator arrays and multiple quality measurement arrays. The process shows very strong input-output off-diagonal coupling properties. An accurate CD model is the prerequisite for an effective CD-MPC controller. We begin by discussing a model structure for the CD process and the details of the model identification.

4.1.1 A two-dimensional linear system

The CD process can be modelled as a linear multiple actuator arrays and multiple measurement arrays system,

$$Y(s) = G(s)U(s) + D(s), \tag{14}$$

and

$$Y(s)=\begin{bmatrix} y_1(s) \\ \vdots \\ y_{N_y}(s) \end{bmatrix},\quad G(s)=\begin{bmatrix} G_{11}(s) & \dots & G_{1N_u}(s) \\ \vdots & \ddots & \vdots \\ G_{N_y 1}(s) & \dots & G_{N_y N_u}(s) \end{bmatrix},\quad U(s)=\begin{bmatrix} u_1(s) \\ \vdots \\ u_{N_u}(s) \end{bmatrix}, \tag{15}$$

where $Y(s) \in \mathbb{C}^{(N_y \cdot m)}$ is the Laplace transformation of the augmented CD measurement array. The element $y_i(s) \in \mathbb{C}^m$ $(i = 1, \dots, N_y)$ is the Laplace transformation of the ith individual CD measurement profile, such as dry weight, moisture, or caliper. N_y is the total number of the quality measurements, and m is the number of elements of individual measurement arrays. $U(s) \in \mathbb{C}^{(\sum_{j=1}^{N_u} n_j)}$ is the Laplace transformation of the augmented actuator setpoint array. The element $u_j(s) \in \mathbb{C}^{n_j}$ $(j = 1 \dots N_u)$ is the Laplace transformation of the jth individual CD actuator setpoint profile, such as the headbox slice, water spray, steam box, or induction heater. N_u is the total number of actuator beams available as MV's, and n_j is the number of individual zones of the jth actuator beam. In general a CD system has the same number of elements for all CD measurement profiles, but different numbers of actuator beam setpoints. $D(s) \in \mathbb{C}^{(N_y \cdot m)}$ is the Laplace transformation of the augmented process disturbance array. It represents process output disturbances.
$G_{ij}(s) \in \mathbb{C}^{m \times n_j}$ $(i = 1 \dots N_y$ and $j = 1 \dots N_u)$ in (15) is the transfer matrix of the sub-system from the jth actuator beam u_j to the ith CD quality measurement y_i. The model of this sub-system can be represented by a spatial static matrix $P_{ij} \in \mathbb{R}^{m \times n_j}$ with a temporal dynamic transfer function $h_{ij}(s)$. In practice, $h_{ij}(s)$ is simplified as a first-order plus dead time system. Therefore, $G_{ij}(s)$ is given by

$$G_{ij}(s) = P_{ij} h_{ij}(s) = P_{ij} \frac{1}{1+T_p s} e^{-T_d s} \tag{16}$$

where T_p is the time constant and T_d is the time delay. The static spatial matrix P_{ij} is a matrix with n_j columns, i.e., $P_{ij} = [p^1 \ \ p^2 \ \ \cdots \ \ p^{n_j}]$ and its kth column p^k represents the spatial response of the kth individual actuator zone of the jth actuator beam. As proposed in (Gorinevsky & Gheorghe 2003), p^k can be formulated by,

$$p^k = \frac{g}{2}\{e^{-\frac{\alpha((x-x_k)-\beta\omega)^2}{\omega^2}} \cos(\frac{\pi}{\omega}((x-x_k)-\beta\omega)) + e^{-\frac{\alpha((x-x_k)+\beta\omega)^2}{\omega^2}} \cos(\frac{\pi}{\omega}((x-x_k)+\beta\omega))\} \tag{17}$$

where x is the coordinate of CD measurements (CD bins), g is the process gain, ω is the response width, α is the attenuation and β is divergence. x_k is the CD alignment that indicates the spatial relationship between the centre of the kth individual CD actuator and the center of the corresponding measurement responses. A fuzzy function may be used to model the CD alignment. Refer to (Gorinevsky & Gheorghe 2003) for the technical details.
Figure 17 illustrates the structure of the spatial response matrix P_{ij}. The colour map on the left shows the band-diagonal property of P_{ij}; and the plot in the right shows the spatial response of the individual spatial actuator p^k. It can be seen that each individual actuator affects not only its own spatial zone area, but also adjacent zone areas.

4.1.2 Model identification

Model identification of the papermaking CD process is the procedure to determine the values of the parameters in (16, 17), i.e., the dynamic model parameters $\theta_T = \{T_p, T_d\}$, the spatial model parameters $\theta_{CD} = \{g, \omega, \alpha, \beta\}$, and the alignment x_k. An iterative identification algorithm has been proposed in (Gorinevsky & Gheorghe 2003). As with MD model identification, this algorithm is an open-loop model identification approach. Identification experiment data are first collected by performing open-loop bump tests.

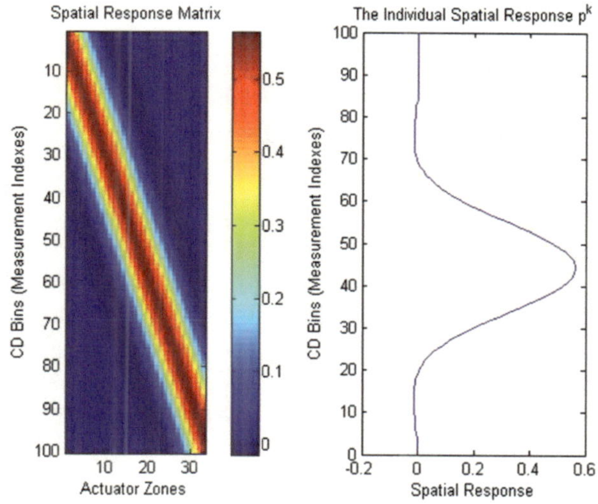

Fig. 17. The illustration to spatial response matrix P_{ij}.

Figure 18 illustrates the logic flow of this algorithm. This nontrivial system identification approach first estimates the overall dynamic response and spatial response, and subsequently identifies the dynamic model parameter θ_T and the spatial model parameter θ_{CD}. $\hat{h}$ in Figure 18 is the estimated finite impulse response (FIR) of the dynamic model h(s) in (16). $\hat{p}$ in Figure 18 is the estimated steady state measurement profile, i.e., overall spatial response. For easier notation, we omit the indexes i and j here. The key concept of the algorithm is to optimize the model parameters iteratively. Refer to (Gorinevsky & Gheorghe 2003) for technical details of this algorithm, and (Gorinevsky & Heaven, 2001) for the theoretical proof of the algorithm convergence.

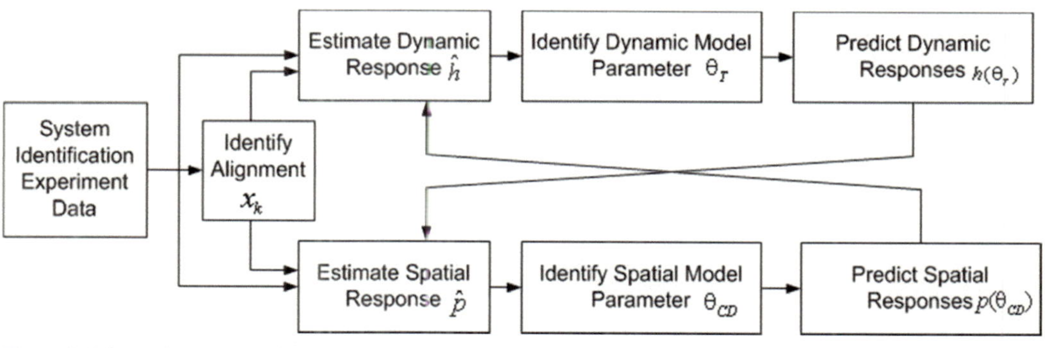

Fig. 18. The schematic of the iterative CD system identification algorithm

The algorithm described above has been implemented in a software package, named IntelliMap™, which has been widely used in pulp and paper industries. The tool executes the open-loop bump tests automatically and, at the end of the experiments, provides a continuous-time transfer matrix model (defined in (14)). For convenience, the MPC controller design discussed in the next section will use the state space model representation. Conversion of the continuous-time transfer matrix model into the discrete-time state space model is trivial (Chen 1999) and is omitted here.

4.2 CD-MPC design

In this section, a state space realization of (14) is used for the MPC controller development,

$$\begin{aligned} X(k+1) &= AX(k) + B\Delta U(k) \\ Y(k) &= CX(k) + D(k) \end{aligned} \quad . \tag{18}$$

$X(k) \in \mathbb{R}^{(2\cdot N_y\cdot m)}$, $Y(k) \in \mathbb{R}^{(N_y\cdot m)}$, $\Delta U(k) \in \mathbb{R}^{(\sum_{j=1}^{N_u} n_j)}$, and$D(k) \in \mathbb{R}^{(N_y\cdot m)}$ are the augmented state, output, actuator move, and output disturbance arrays of the papermaking CD process with multiple CD actuator beams and multiple quality measurement arrays. {A, B, C} are the model matrices with compatible dimensions. Assume (A, B) is controllable and (A, C) is observable. In this section, the objective function of CD-MPC is developed first. Then the CD actuator constraints are incorporated in the objective function. Finally a fast QP solver is presented for solving the large scale constrained CD-MPC optimization problem. How to tune a CD-MPC controller is also covered in this section

4.2.1 Objective function of CD-MPC

The first step of MPC development is performing the system output prediction over a certain length of prediction horizon. From the state space model defined in (18), we can predict the future states,

$$\mathcal{X}(k) = \mathcal{P}_A X(k) + \mathcal{P}_B \Delta\mathcal{U}(k), \tag{19}$$

where $\mathcal{X}(k) \in \mathbb{R}^{(2\cdot N_y\cdot m\cdot H_p)}$ is the state prediction, $\Delta\mathcal{U}(k) \in \mathbb{R}^{(H_u\cdot \sum_{j=1}^{N_u} n_j)}$ is the augmented actuator moves. $\mathcal{P}_A$ and $\mathcal{P}_B$ are the state and input prediction matrices with the compatible dimensions. H_p and H_u are the output and input prediction horizons, respectively.
The explicit expressions of the parameters in (19) are

$$\mathcal{X}(k) = \begin{bmatrix} X(k+1|k) \\ X(k+2|k) \\ \vdots \\ X(k+H_p|k) \end{bmatrix}, \quad \mathcal{P}_A = \begin{bmatrix} A \\ A^2 \\ \vdots \\ A^{H_p} \end{bmatrix}, \quad \mathcal{P}_B = \begin{bmatrix} B & \cdots & 0 \\ AB & \cdots & 0 \\ \vdots & \ddots & \vdots \\ A^{H_p-1}B & \cdots & A^{H_p-H_u}B \end{bmatrix},$$
$$\text{and} \qquad \Delta\mathcal{U}(k) = \begin{bmatrix} \Delta U(k|k) \\ \Delta U(k+1|k) \\ \vdots \\ \Delta U(k+H_u-1|k) \end{bmatrix} \tag{20}$$

The initial state $\hat{X}^0(k|k-1)$ at instant k can be estimated from the previous state estimation $\hat{X}(k-1)$ and the previous actuator move $\Delta U(k-1)$, i.e.,

$$\hat{X}^0(k|k-1) = A\hat{X}(k-1) + B\Delta U(k-1). \tag{21}$$

The measurement information at instant k can be used to improve the estimation,

$$\hat{X}(k) = \hat{X}^0(k|k-1) + L(Y(k) - C\hat{X}^0(k|k-1)), \tag{22}$$

where $L \in \mathbb{R}^{(2\cdot N_y\cdot m)\times(N_y\cdot m)}$ is the state observer matrix.
Replace the state X(k) by its estimation $\hat{X}(k)$, and perform the output prediction $\mathcal{Y}(k)$,

$$\mathcal{Y}(k) \;= \mathcal{P}_c\mathcal{P}_A\hat{X}(k) + \mathcal{P}_c\mathcal{P}_B\Delta\mathcal{U}(k), \tag{23}$$

where $\mathcal{P}_c \in \mathbb{R}^{(N_y\cdot m\cdot H_p)\times(2\cdot N_y\cdot m\cdot H_p)}$ is the output prediction matrix, given by

$$\mathcal{P}_c = \text{diag}(C,\cdots C) = \begin{bmatrix} C & 0 & \cdots & 0 \\ 0 & C & \vdots & 0 \\ \vdots & \vdots & \ddots & \vdots \\ 0 & 0 & \cdots & C \end{bmatrix}. \text{ Also}, \mathcal{Y}(k) = \begin{bmatrix} Y(k+1|k) \\ Y(k+2|k) \\ \vdots \\ Y(k+H_p|k) \end{bmatrix}. \tag{24}$$

From the expression in (24), one can define the objective function of a CD-MPC problem,

$$\min_{\Delta\mathcal{U}(k)}||\mathcal{Y}(k) - \mathcal{Y}_{tgt}||^2_{\mathcal{Q}_1} + |\ \Delta\mathcal{U}(k)||^2_{\mathcal{Q}_2} + ||\mathcal{U}(k) - \mathcal{U}_{tgt}||^2_{\mathcal{Q}_3} + ||\mathcal{F}_b\mathcal{U}(k)||^2_{\mathcal{Q}_4}. \tag{25}$$

$\mathcal{Y}_{tgt} = [Y^T_{tgt}, Y^T_{tgt}, \cdots, Y^T_{tgt}]^T$ defines the measurement targets over the prediction horizon H_p. Similarly, $\mathcal{U}_{tgt} = [U^T_{tgt}, U^T_{tgt}, \cdots, U^T_{tgt}]^T$ defines the input actuator setpoint targets over the control horizon H_u. ($\mathcal{Q}_1$, $\mathcal{Q}_2$, $\mathcal{Q}_3$, $\mathcal{Q}_4$) are the diagonal weighting matrices. $\mathcal{Q}_1$ defines the relative importance of the individual quality measurements. $\mathcal{Q}_2$ defines the relative aggressiveness of the individual CD actuators. $\mathcal{Q}_3$ defines the relative deviation from the targets of the individual CD actuators. $\mathcal{Q}_4$ defines the relative picketing penalty of the individual CD actuators. The matrix $\mathcal{F}_b = \text{diag}(F_b, \cdots, F_b)$ is the augmented actuator bending matrix. The detailed definition of F_b will be covered in Section 4.2.2. $||\cdot||^2_{\mathcal{R}_i}$ is the square of weighted 2-norm, i.e., $||\cdot||^2_{\mathcal{R}_i} = (\cdot)^T\mathcal{R}_i(\cdot)$. In general, ($\mathcal{Q}_1$, $\mathcal{Q}_2$, $\mathcal{Q}_3$, $\mathcal{Q}_4$) are used as the tuning parameters for CD-MPC.
$\mathcal{U}(k)$ is the future input prediction. It can be expressed by

$$\mathcal{U}(k) = \begin{bmatrix} U(k|k) \\ U(k+1|k) \\ \vdots \\ U(k+H_{u-1})|k) \end{bmatrix} = \underbrace{\begin{bmatrix} I \\ I \\ \vdots \\ I \end{bmatrix}}_{\mathcal{S}_I} U(k-1) + \underbrace{\begin{bmatrix} I & 0 & \cdots & 0 \\ I & I & \vdots & 0 \\ \vdots & \cdots & \ddots & \vdots \\ I & I & I & I \end{bmatrix}}_{\mathcal{S}_p} \Delta\mathcal{U}(k), \tag{26}$$

where $I \in \mathbb{R}^{(\sum_{j=1}^{N_u} n_j)\times(\sum_{j=1}^{N_u} n_j)}$ is the identity matrix. Inserting (26) into (25) and replacing $\mathcal{U}(k)$ by $\Delta\mathcal{U}(k)$, the QP problem can be recast into

$$\min_{\Delta\mathcal{U}(k)} \frac{1}{2}\Delta\mathcal{U}^T(k)\Phi\Delta\mathcal{U}(k) + \varphi^T\Delta\mathcal{U}(k), \tag{27}$$

where Φ is the Hessian matrix and φ is the gradient matrix. Both can be derived from the prediction matrices ($\mathcal{P}_A, \mathcal{P}_B, \mathcal{P}_C$) and weighting matrices ($\mathcal{Q}_1, \mathcal{Q}_2, \mathcal{Q}_3, \mathcal{Q}_4$). Refer to (Fan 2003) for the detailed expressions of Φ and φ.
By solving the QP problem in (27), one can derive the predicted optimal array $\Delta\mathcal{U}(k)$. Only the first component of $\Delta\mathcal{U}(k)$, i.e., $\Delta U(k)$, is sent to the real process and the rest are rejected. By repetition of this procedure, the optimal MV moves at any instant are derived for unconstrained CD-MPC problems.

4.2.2 Constraints

In Section 4.2.1 the CD-MPC controller is formulated as an unconstrained QP problem. In practice the new actuator setpoints given by the CD-MPC controller in (27) should always respect the actuator's physical limits. In other words, the hard constraints on $\Delta\mathcal{U}(k)$ should be added into the problem in (27).

The CD actuator constraints include:

- First and second order bend limits;
- Average actuator setpoint maintenance;
- Maximum actuator setpoints;
- Minimum actuator setpoints; and
- Maximum change of actuator setpoints between consecutive CD-MPC iterations.

Of these five types of actuator constraints, most of them are very common for the typical MPC controllers, except for the bend limits which are special for papermaking CD processes. The first and second bend limits define the allowable first and second order difference between the adjacent actuator setpoints of the actuator beam. It typically applies to slice lips and induction heaters to prevent the actuator beams from being overly bent or locally over-heated. The bending matrix of the jth actuator beam, $F_{b,j}$ ($j = 1, \cdots, N_u$) can be defined by

$$-\underbrace{\begin{bmatrix} \delta_{1,j} \\ \delta_{2,j} \\ \vdots \\ \delta_{2,j} \\ \delta_{1,j} \end{bmatrix}}_{\gamma_{b,j}} \leq \underbrace{\begin{bmatrix} -1 & 1 & 0 & \cdots 0 & \cdots & 0 \\ 0.5 & -1 & 0.5 & 0 & \cdots & 0 \\ 0 & 0.5 & -1 & 0.5 & \cdots & 0 \\ \vdots & \cdots & \cdots & \cdots & \ddots & \vdots \\ 0 & 0 & \cdots & 0.5 & -1 & 0 \\ 0 & 0 & \cdots & 0 & 1 & -1 \end{bmatrix}}_{F_{b,j}} \underbrace{\begin{bmatrix} u_{1,j} \\ u_{2,j} \\ \vdots \\ u_{n_j-1,j} \\ u_{n_j,j} \end{bmatrix}}_{u_j} \leq \underbrace{\begin{bmatrix} \delta_{1,j} \\ \delta_{2,j} \\ \vdots \\ \delta_{2,j} \\ \delta_{1,j} \end{bmatrix}}_{\gamma_{b,j}}, \tag{28}$$

where $\delta_{1,j}$ and $\delta_{2,j}$ are the first order and the second order bend limit of the jth actuator beam u_j. γ_b and $F_{b,j}$ define the bend limit vector and the bend limit matrix of the jth actuator u_j, respectively. The bend limit matrix $F_{b,j}$ is not only part of the constraints, but also the objective function in (27). In (27), $\mathcal{F}_b = \mathrm{diag}(F_b, \cdots, F_b)$and $F_b = \mathrm{diag}(F_{b,1}, \cdots, F_{b,N_u})$.

The individual bend limit constraint on the jth actuator beam u_j in (28) can be extended to the overall bend limit matrix F_b for the augmented actuator setpoint array U, i.e.,

$$\begin{bmatrix} F_b \\ -F_b \end{bmatrix} U \leq \begin{bmatrix} \gamma_b \\ \gamma_b \end{bmatrix} \tag{29}$$

where γ_b is the overall bend limit vector, and $\gamma_b = [\gamma_{b,1}^T, \cdots \gamma_{b,N_u}^T]^T$.

Similar to the bend limits, other types of actuator physical constraints can be formulated as the matrix inequalities,

$$\begin{bmatrix} F_{max} \\ -F_{min} \\ F_{avg} \\ -F_{avg} \\ F_{\Delta U} \\ -F_{\Delta U} \end{bmatrix} U \leq \begin{bmatrix} \gamma_{max} \\ \gamma_{min} \\ \gamma_{avg} \\ \gamma_{avg} \\ \gamma_{\Delta U} \\ \gamma_{\Delta U} \end{bmatrix}, \tag{30}$$

where the subscripts "max", "min", "avg", and "ΔU" stand for the maximum, minimum, average limit, and maximum setpoint changes between two consecutive CD-MPC iterations of the augmented actuator setpoint array, U. It is straightforward to derive the expressions of F_{max}, F_{min}, F_{avg}, $F_{\Delta U}$. Therefore the detailed discussion is omitted.

From (29) and (30), one can see that the constraints on the augmented actuator setpoint array U can be represented by a linear matrix inequality, i.e.,

$$FU \leq \gamma, \tag{31}$$

where F and γ are constant coefficients used to combine the inequalities in (29) and (30) together.
(26) is inserted into (31). The constraint in (31) is then added to the objective function in (27). Finally the CD-MPC controller is formulated as a constrained QP problem,

$$\min_{\Delta\mathcal{U}(k)} \frac{1}{2}\Delta\mathcal{U}^{T}(k)\Phi\Delta\mathcal{U}(k) + \phi^{T}\Delta\mathcal{U}(k)$$
$$\text{subject to,} \qquad , \qquad (32)$$
$$\mathcal{F}(\mathcal{S}_{I}U(k-1) + \mathcal{S}_{p}\Delta\mathcal{U}(k)) \leq \Gamma$$

where $\mathcal{F} = \mathrm{diag}(F, F, \cdots, F)$ and $\Gamma = \mathrm{diag}(\gamma, \gamma, \cdots, \gamma)$. By solving the QP problem in (32), the optimal actuator move at instant k can be achieved.

4.2.3 CD-MPC tuning

Figure 19 illustrates the implementation of the CD-MPC controller. First, the process model is identified offline from input/output process data. Then the CD-MPC tuning algorithm is executed to generate optimal tuning parameters. Subsequently these tuning parameters are deployed to the CD-MPC controller. The controller generates the optimal actuator setpoints continuously based on the feedback measurements.

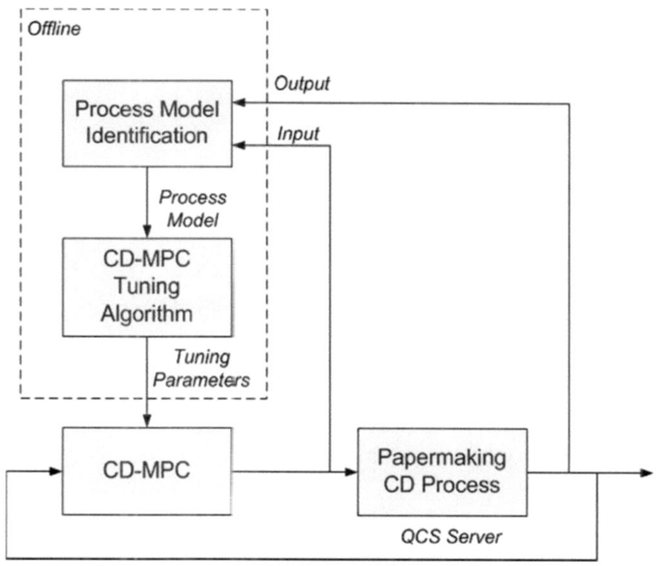

Fig. 19. The implementation of the CD-MPC controller

The objective for CD-MPC tuning algorithm in Figure 19 is to determine the values of $\mathcal{Q}_1$, $\mathcal{Q}_2$, $\mathcal{Q}_3$, and $\mathcal{Q}_4$ in (25). It has been proven that $\mathcal{Q}_1$ defines the relative importance of quality measurements, $\mathcal{Q}_2$ defines the dynamic characteristics of the closed-loop CD-MPC system, and $\mathcal{Q}_3$ and $\mathcal{Q}_4$ define the spatial frequency characteristics of the closed-loop CD-MPC system. $\mathcal{Q}_3$ is for the high spatial frequency behaviours and $\mathcal{Q}_4$ is for the low spatial frequencies (Fan 2004).
Strictly speaking, the CD-MPC tuning problem requires analyzing the robust stability of a closed-loop control system with nonlinear optimization. An analytic solution to the QP

problem in (32) is the prerequisite for the CD-MPC tuning algorithm. However, in practice it is very challenging; almost impossible to derive the explicit solution to (32) due to the large size of CD-MPC problems. A novel two-dimensional loop shaping approach is proposed in (Fan 2004) to overcome limitations for large scaled MPC systems. The algorithm consists of four steps:

Step 1. Ignore the inequality constraint in (32) such that the closed-loop system given by (27) is linear.

Step 2. Compute the closed-loop transfer function of the unconstrained CD-MPC system given by (27).

Step 3. By performing two-dimensional loop shaping, optimize the weighting matrices to get the best trade off between the performance and robustness of the unconstrained CD-MPC system.

Step 4. Finally, re-introduce the constraint in (32) for implementation.

Figure 20 shows the closed-loop diagram of the unconstrained CD-MPC system with unstructured model uncertainties. The derivation of the pre-filtering matrix K_r and feedback controller K is standard and can be found in (Fan 2003).

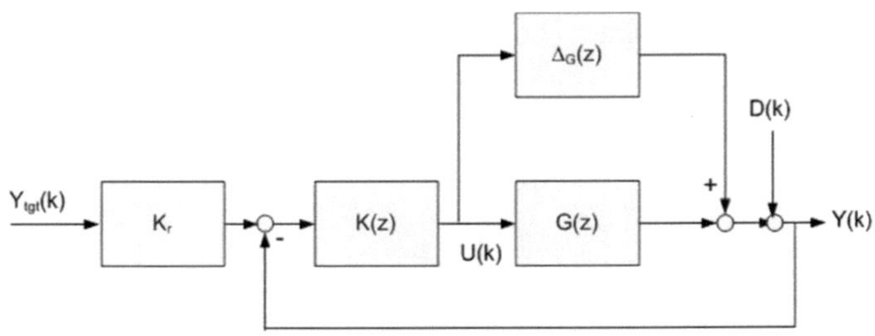

Fig. 20. Closed-loop diagram of unconstrained CD-MPC system with unstructured model uncertainties

From the small gain theory (Khalil 2001), the linear closed-loop system in Figure 20 is robustly stable if the closed-loop in (32) is nominally stable and,

$$||G_{ud}(z)\ \Delta_G\ (z)||_\infty < 1 \Leftarrow \bar{\sigma}(G_{ud}(e^{j\omega})) < \frac{1}{\bar{\sigma}(\Delta_G(e^{j\omega}))}, \forall\omega. \tag{33}$$

Here $G_{ud}(z)$ is the control sensitivity function which defines the linear transfer function from the output disturbance D(k) to the actuator setpoint U(k),

$$G_{ud}(z) = K(z)[I - G(z)K(z)]^{-1}. \tag{34}$$

The sensitivity function of the system in Figure 20 defines the linear transfer function from the output disturbance D(k) to the output Y(k),

$$G_{yd}(z) = [I - G(z)K(z)]^{-1}.$$

By properly choosing the weighting matrices Q_1 to Q_4, both the control sensitivity function $G_{ud}(z)$ and the sensitivity function $G_{yd}(z)$ can be guaranteed stable, and also the small gain condition in (33) can be satisfied. The two-dimensional loop shaping approach uses $G_{ud}(z)$ and $G_{yd}(z)$ to analyze the behaviour of the closed-loop system in Figure 20.

It has been shown that both $G_{ud}(z)$ and $G_{yd}(z)$ can be approximated as rectangular circulant matrices. One important property of circulant matrixes is that the circulant matrix can be block-diagonalized by left- and right-multiplying Fourier matrices. Fourier matrices multiplication is equivalent to performing the standard discrete Fourier transformation. Therefore, the two-dimensional frequency representation of $G_{ud}(z)$ and $G_{yd}(z)$ can be obtained by,

$$\hat{g}_{ud}(\nu,\ e^{j\omega}) = F_n G_{ud}(e^{j\omega}) F_m^H,\ \text{and}\ \hat{g}_{yd}(\nu,\ e^{j\omega}) = F_m G_{yd}(e^{j\omega}) F_m^H, \tag{35}$$

where ν represents the spatial frequency. F_m and F_n are m-points and n-points Fourier matrices, respectively. The detailed definitions of Fourier matrices can be found in (Fan 2004). The two-dimensional frequent representation $\hat{g}_{ud}(\nu,\ e^{j\omega})$ and $\hat{g}_{yd}(\nu,\ e^{j\omega})$ are block diagonal matrices. The singular values of $\hat{g}_{ud}(\nu,\ e^{j\omega})$ and $\hat{g}_{yd}(\nu,\ e^{j\omega})$ are directly linked to the spatial frequencies.

Instead of tune $\hat{g}_{ud}(\nu,\ e^{j\omega})$ and $\hat{g}_{yd}(\nu,\ e^{j\omega})$ in full ν and ω frequency ranges, two dimensional loop shaping approach decouples the spatial tuning and dynamic tuning by firstly tuning the controller at zero spatial frequency, i.e., setting $\nu = 0$, and then tuning the controller at zero dynamic frequency, i.e., setting $\omega = 0$. The theoretical proof of this strategy can be founded in (Fan 2004).

From spatial tuning, the value of the weighting matrices Q_3 and Q_4 can be determined, and from the dynamic tuning, the value of Q_2 are determined. Q_1, as mentioned above, defines the relative importance of quality measurements and its value is defined by a CD-MPC user. In practice, the process gain matrix P_i in (16) is ill-conditioned. Similar to MD-MPC tuning, the scaling matrices have to be applied before tuning the controller. A scaling approach discussed in (Lu 1996) is used by CD-MPC to reduce the condition number of the gain matrices.

4.2.4 Fast QP solver

The technical challenge of the CD-MPC optimization is how to solve the problem in (32) efficiently and accurately. The typical scanning rate of the paper machine is 10 - 30 seconds. Also considering the time cost of software implementation and data acquisition, the computation time of the problem in (32) is typically limited to 5 to 10 seconds.

Different optimization techniques have been developed to solve QP problems efficiently, such as the active set method, interior point method, QR factorization, etc. This section presents a fast QP solver, called QPSchur, which is specifically designed to solve a large scaled CD-MPC problem. QPSchur is a dual space algorithm, where an unconstrained optimal solution is found first and violated constraints are added until the solution is feasible (Bartlett 2002).

Let's consider the Lagrangian of the constrained QP in (32)

$$\Lambda(\Delta\mathcal{U},\lambda) = \frac{1}{2}\Delta\mathcal{U}^T(k)\Phi\Delta\mathcal{U}(k) + \phi^T\Delta\mathcal{U}(k) + \lambda^T(\xi^T\Delta\mathcal{U}(k) - \psi), \tag{36}$$

where $\xi = \mathcal{S}_p^T\,\mathcal{F}^T$ and $\psi = \Gamma - \mathcal{F}\mathcal{S}_I U(k-1)$. In (36), $\Delta\mathcal{U}(k)$ is called the primary variable and $\lambda \leq 0$ is called as the dual variable (also known as the Lagrangian variable).

At the starting point, QPSchur ignores all the constraints in (32) and solves unconstrained QP problem. This is equivalent to set the dual variable $\lambda = 0$. By this means, the initial optimal solution $\Delta\mathcal{U}^*(k)$ is determined,

$$\Delta\mathcal{U}^*(k) = -\Phi^{-1}\phi. \tag{37}$$

If $\Delta\mathcal{U}^*(k)$ satisfies all the inequality constraints, i.e., $\xi^T\Delta\mathcal{U}^*(k) \le \psi$, then $\Delta\mathcal{U}^*(k)$ is the optimal solution, i.e. $\Delta\mathcal{U}^o(k) = \Delta\mathcal{U}^*(k)$. The first elements of $\Delta U^o(k)$ are sent to the real process, and the CD-MPC optimization stops the search iteration.
If $\Delta\mathcal{U}^*(k)$ violates one or more of the inequality constraints in (32), all the violation inequalities are noted, such that

$$\xi_{sub}^T\Delta\mathcal{U}^*(k) \ge \psi_{sub}, \tag{38}$$

where $(\xi_{sub}^T, \psi_{sub})$ is the violating subset of the inequality constraints in (32), and called the active set matrix and the active set vector, respectively. The Lagrangian in (36) is redefined by using $(\xi_{sub}^T, \psi_{sub})$. The Karush-Kuhn-Tucker (KKT) condition of the updated Lagrangian is,

$$\begin{bmatrix} \Phi & \xi_{sub} \\ \xi_{sub}^T & \Xi \end{bmatrix}\begin{bmatrix} \Delta\mathcal{U}(k) \\ \lambda_{sub} \end{bmatrix} = \begin{bmatrix} -\phi \\ \psi_{sub} \end{bmatrix}, \tag{39}$$

Here $\Xi = 0$ for the first searching iteration. Since Φ is non-singular (refer to Fan 2003), the problem in (39) can be solved by using Gaussian elimination. The Schur complement of the block Ξ is given by

$$\mathbb{S} = \Xi - \xi_{sub}^T\Phi^{-1}\xi_{sub}. \tag{40}$$

The Schur complement theorem guarantees that $\mathbb{S}$ is non-singular if the Hessian matrix Φ is non-singular. From $\mathbb{S}$, (39) can be solved by

$$\begin{aligned} \lambda_{sub} &= \mathbb{S}^{-1}(\psi_{sub} + \xi_{sub}^T\Phi^{-1}\phi) \\ &= \mathbb{S}^{-1}(\psi_{sub} + \xi_{sub}^T\Delta\mathcal{U}^*(k)) \\ \Delta\mathcal{U}(k) &= \Phi^{-1}(-\phi - \xi_{sub}\lambda_{sub}) \end{aligned} \tag{41}$$

The inequality constraints in (32) are re-evaluated, and the new active constraints (violated constraints) and the positive dual variables inequalities are added into the subset pair $(\xi_{sub}^T, \psi_{sub})$. The KKT condition of (39) is updated to derive

$$\begin{bmatrix} \Phi & \hat{\xi}_{sub} \\ \hat{\xi}_{sub}^T & \hat{\Xi} \end{bmatrix}\begin{bmatrix} \Delta\mathcal{U}(k) \\ \hat{\lambda}_{sub} \end{bmatrix} = \begin{bmatrix} -\phi \\ \hat{\psi}_{sub} \end{bmatrix}, \tag{42}$$

where

$$\hat{\xi}_{sub} = [\xi_{sub}, \xi_{new}],\ \hat{\Xi} = \begin{bmatrix} \Xi & \rho \\ \rho^T & \chi \end{bmatrix},\ \hat{\lambda}_{sub} = \begin{bmatrix} \lambda_{sub} \\ \lambda_{new} \end{bmatrix}, \text{and } \hat{\psi}_{sub} = \begin{bmatrix} \psi_{sub} \\ \psi_{new} \end{bmatrix}. \tag{43}$$

In the same fashion, the Schur complement of the block $\hat{\Xi}$ can be represented by,

$$\begin{aligned} \hat{\mathbb{S}} &= \hat{\Xi} - \hat{\xi}_{sub}^T\Phi^{-1}\hat{\xi}_{sub} \\ &= \begin{bmatrix} \Xi & \rho \\ \rho^T & \chi \end{bmatrix} - \begin{bmatrix} \xi_{sub}^T \\ \xi_{new}^T \end{bmatrix}\Phi^{-1}[\xi_{sub}, \xi_{new}] \\ &= \begin{bmatrix} \mathbb{S} & \rho - \xi_{sub}^T\Phi^{-1}\xi_{new} \\ \rho^T - \xi_{new}\Phi^{-1}\xi_{sub}^T & \chi - \xi_{new}^T\Phi^{-1}\xi_{new} \end{bmatrix} \end{aligned} \tag{44}$$

From (44), the new Schur complement $\widehat{\mathbb{S}}$ can be easily derived from $\mathbb{S}$. The Schur complement update requires only multiplication with Φ^{-1} that is calculated in the initial search step and stored for reuse. This feature makes the SchurQP much faster than a standard QP solver. Removing the non-active constraints (zero dual variables) of each search step is achieved easily: the columns of the Schur complement $\widehat{\mathbb{S}}$ corresponding to the non-active constraints is removed before pursuing the next search iteration.

At the current search iteration, if all the inequality constraints in (32) and the sign of dual variables are satisfied, the solution to (42) will be the final optimal solution of the CD-MPC controller, i.e.,

$$\begin{aligned} \widehat{\lambda}_{sub} &= \widehat{\mathbb{S}}^{-1}(\widehat{\psi}_{sub} + \hat{\xi}_{sub}^{T}\Delta\mathcal{U}^{*}(k)) \\ \Delta\mathcal{U}^{o}(k) &= \Phi^{-1}(-\phi - \hat{\xi}_{sub}\widehat{\lambda}_{sub}) \end{aligned} \tag{45}$$

$\Delta U^{o}(k)$ (the first component of the optimal solution $\Delta\mathcal{U}^{o}(k)$) is sent to the real process, and a new constrained QP problem is formed at the end of the next scan.

4.3 Mill implementation results

CD-MPC has been implemented in Honeywell's quality control system (QCS) and widely deployed on different types of paper mills including fine paper, newsprint, liner board, and tissue, etc. In this chapter, a CD-MPC application for a fine paper machine will be used as an example to demonstrate the effectiveness of the CD-MPC controller.

4.3.1 Paper machine configuration

The paper machine discussed here is a fine paper machine, equipped with three CD actuator beams and two measurement scanner frames. The CD actuators include headbox slice lip (63 zones), infrared dryer (40 zones), and induction heater (79 zones). The two scanner frames hold the paper quality gauges for dry weight, moisture, and caliper. Each measurement profile includes 250 measurement points with the measurement interval equal to 25.4 mm (CD bin width). The production range of this machine is from 26 gsm (gram per square meter) to 85 gsm. The machine speed varies from 2650 feet per minute (13.5 meter/second) to 3100 feet per minute (15.7 meter/second). The scanning rates of the two scanners are 32 and 34 seconds, respectively. In order to capture the nonlinearity of the process, three model groups are setup to represent the products of light weight paper, medium weight paper, and heavy weight paper, respectively. All three CD actuator beams and three quality measurement profiles are included into the CD-MPC controller. In this section, the medium weight scenario is used to illustrate the control performance of the CD-MPC controller.

4.3.2 Multiple actuator beams and multiple quality measurements model

Figure 21 shows the two-dimensional process models from the slice lip actuators (Autoslice) to the measurements of dry weight, moisture and caliper profiles. The system identification algorithm discussed in Section 4.1.2 is used to derive these models. The plots on the left are the spatial responses, and the plots on the right are the dynamic responses. The purple profiles are the average of the real process data, and the white profiles are the estimated profiles based on identified process model. It can be seen from comparison to the model for Autoslice to caliper that the models for Autoslice to dry weight and to moisture have high model fit. In general, the bump test with a larger bump magnitude and longer bump duration will lead to a more accurate process model (better model fit). However, the open-loop bump tests degrade the quality of the finished product and excessive bump tests are always prevented. The criterion of the CD model identification is to provide a process model accurate enough for a CD-MPC controller.

From the model identification results in Figure 21, we can see the strong input-output coupling properties of papermaking CD processes. The response width from slice lip to dry weight equals to 226.8mm. This is equivalent to 2.3 times the zone width of the slice lip CD actuator. Therefore, each individual zone of the slice lip affects not only its own spatial zone but also adjacent zones. As we discussed above, a CD-MPC process has two-fold process couplings: one is the coupling between different actuator beams; and the other is the coupling between the different zones of the same actuator beams. Considering these strong coupling characteristics, MPC strategy is a good candidate for CD control design.

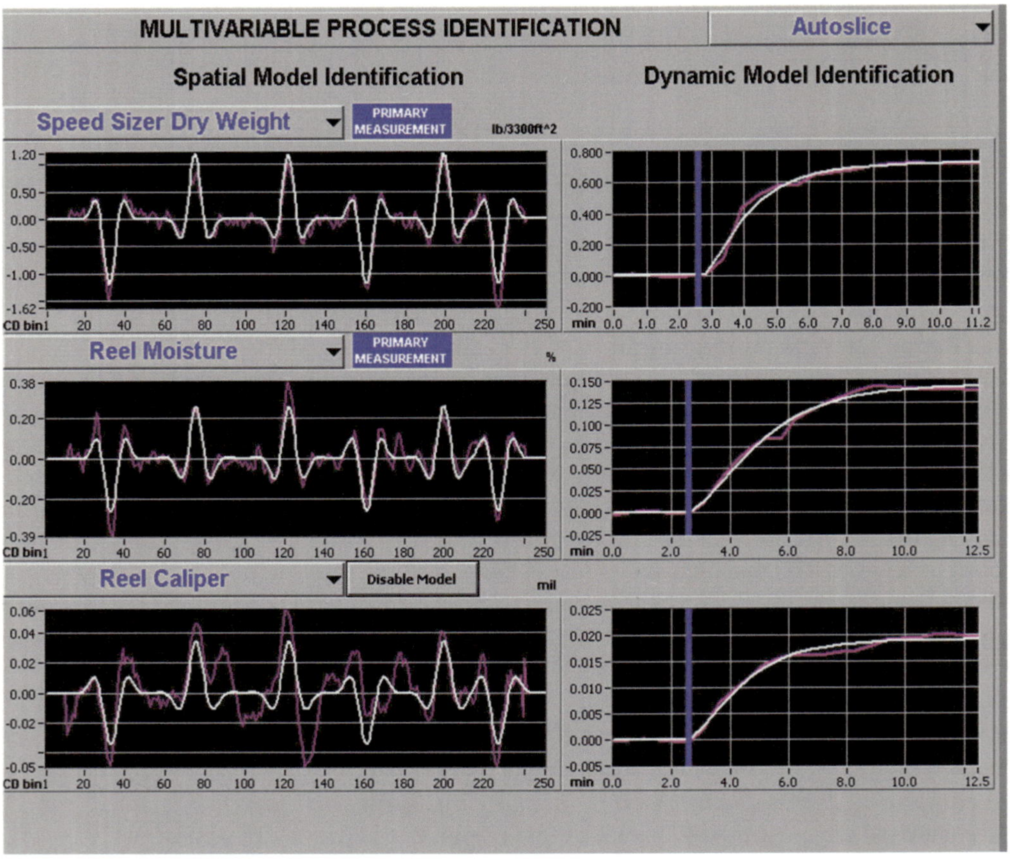

Fig. 21. The multiple CD actuator beams and quality measurement model display

4.3.3 Control performance of the CD-MPC controller

Table 3 summarizes the performance comparison between the CD-MPC controller and the traditional single-input-single-output (SISO) CD controller (a Dahlin controller). Although traditional CD control is still quite common in paper mills, CD-MPC is becoming more and more popular. The significant performance improvement can be observed after switching CD control into multivariable CD-MPC.

Paper Properties	Traditional CD Control 2σ	Multivariable CD-MPC Control 2σ	Improvement (%)
Dry Weight (gsm)	0.40	0.24	40%
Moisture (%)	0.31	0.19	39%
Caliper (mil)	0.032	0.025	22%

Table 3. Traditional CD versus CD-MPC

Figures 22–24 provide a visual performance comparison for the different quality measurements in both spatial domain and spatial frequency domain. It can be seen that the peak-to-peak values (the proxy of $2\sigma_{CD}$ indexes) are smaller when using the CD-MPC controller. Also the controllable disturbances (the disturbances with the spatial frequency less than Xc) are effectively rejected by the CD-MPC controller. Here X3db represents the spatial frequency where the spatial process power drops to 50% of the maximum spatial power over the full spatial frequency band, Xc represents the frequency where the spatial power drops to 4% of the maximum power, and 1/2Xa represents the Nyquist frequency.

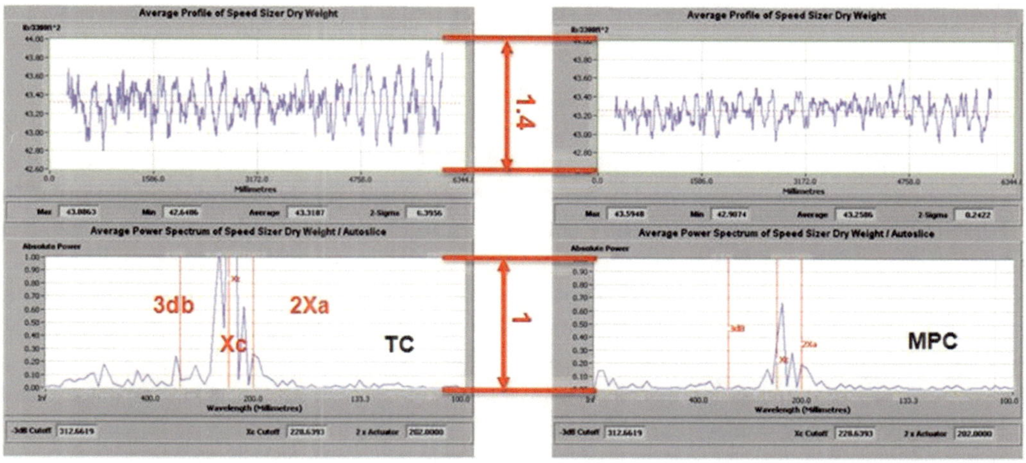

Fig. 22. Performance comparison of dry weight profiles

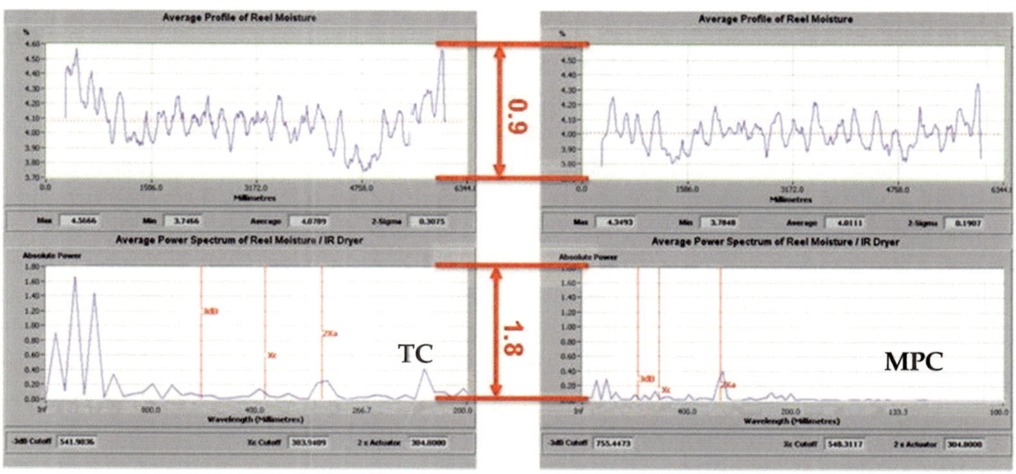

Fig. 23. Performance comparison of moisture profiles

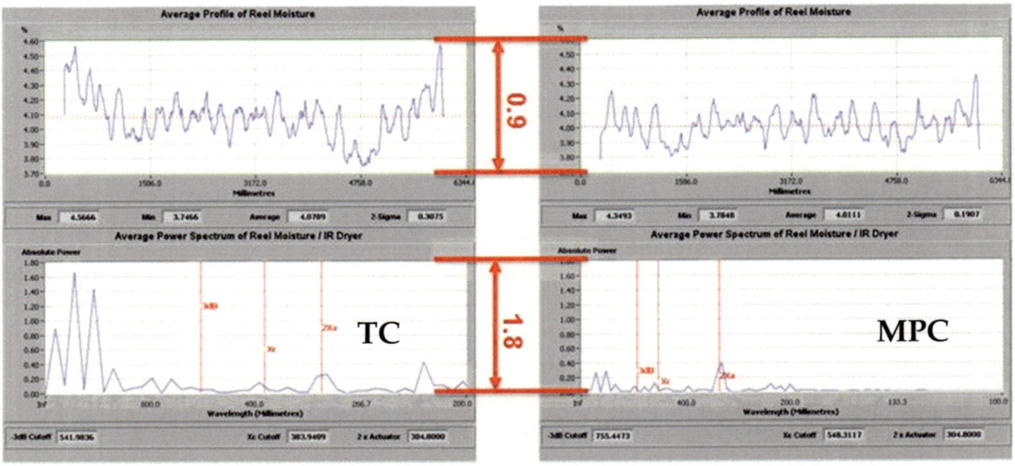

Fig. 24. Performance comparison of caliper profiles

5. Conclusion and perspective

We have seen that MPC has a number of applications in paper machine control. MPC performs basic MD control, and allows for enhanced MD-MPC control that incorporates economic optimization, and orchestrates transitions between paper grades. MPC can also be used for CD controls, using a carefully chosen solution technique to handle the large scale nature of the problem within the required time scale.

While MD-MPC provides robust and responsive control, and also easily scales to demanding paper machine applications with larger numbers of CV's and MV's. The MD-MPC formulation may also be augmented with an economic objective function so that paper machine operational efficiency can be optimized (maximum production, minimum energy costs, maximizing filler to fibre ratio etc.) while all quality variables continue to be regulated.

In the future as new online sensors, such as the extensional stiffness sensor, gain acceptance additional quality variables can be adding to MD-MPC. In the case of extensional stiffness, this online strength measurement could allow economic optimization to minimize fibre use while maintaining paper strength.

The papermaking CD process is a large scaled two-dimensional system. It shows strong input-output coupling properties. MPC is a standard technique in controlling multivariable systems, and has become a standard advanced control strategy in papermaking systems. However, there are several barriers for the acceptance of CD-MPC by mill personnel: one is the novel multivariable control concept and the other is the non-trivial tuning technique. Commercial offline tools, such as IntelliMap, facilitate the acceptance of CD-MPC by providing automatic model identification and easy-to-use offline CD-MPC tuning. Such packages enable the CD-MPC users to review the predicted CD steady states before they update their CD control to CD-MPC (Fan et al. 2005). CD-MPC has been successfully deployed in over 70 paper mills and applied to practically all types of existing CD processes from fine paper, to board, to newsprints, to tissues, etc. Without doubt, CD-MPC will have a significant impact in papermaking CD control applications over the next decade.

CD-MPC offers the significant capability to include multiple CD actuator arrays and multiple CD measurement arrays into one single CD controller. The next generation CD-

MPC applications are most likely to include non-standard CD measurement, such as fibre orientation, gloss, web formation, and web porosity into the existing CD-MPC framework. A successful CD-MPC application for fibre orientation control has been reported in (Chu et al. 2010a). However there still exist technical challenges of controlling non-standard paper properties by using CD-MPC; for example, the derivation of accurate parametric models and the effectiveness of CD-MPC tuners for non-standard CD measurements.

In the current CD-MPC framework, system identification and controller design are clearly separated. The efforts towards integrating system identification and controller design may bring significant benefits to CD control. Online CD model identification has drawn extensive attention in both academia and industries. A closed-loop CD alignment identification algorithm is presented in (Chu et al. 2010b). Closed loop identification of the entire CD model remains an open problem.

6. References

Backström, J., & Baker, P. (2008). A Benefit Analysis of Model Predictive Machine Directional Control of Paper Machines, in Proc Control Systems 2008, Vancouver, Canada, June 2008.

Backström, J., Gheorghe, C., Stewart, G., & Vyse, R. (2001). Constrained model predictive control for cross directional multi-array processes. In Pulp & Paper Canada, May 2001, pp. T128– T102.

Backström, J., Henderson, B., & Stewart G. (2002). Identification and Multivariable Control of Supercalenders, in Proc. Control Systems 2002, Stockholm, Sweden, pp 85-91.

Bartlett, R., Biegler L., Backström, J., & Gopal, V. (2002). Quadratic programming algorithm for large-scale model predictive control, in Journal of Process Control, Vol. 12, pp. 775 - 795.

Bemporad, A., & Morari, M. (2004). Robust model predictive control: A survey. In Proc. of European Control Conference, pp. 939–944, Porto, Portugal.

Chen, C. (1999), Linear Systems Theory and Design. Oxford University Press, 3rd Edition.

Chu, D. (2006). Explicit Robust Model Predictive Control and Its Applications. Ph.D. Thesis, University of Alberta, Canada.

Chu, D., Backström J., Gheorghe C., Lahouaoula, A., & Chung, C. (2010a). Intelligent Closed Loop CD Alignment, in Proc Control System 2010, pp. 161-166, Stockholm, Sweden, 2010.

Chu, D., Choi, J., Backström, J., & Baker, P. (2008). Optimal Nonlinear Multivariable Grade Change in Closed-Loop Operations, in Proc Control Systems 2008, Vancouver, Canada, June 2008.

Chu, D., Gheorghe C., Backström J., Naslund, H., & Shakespeare, J. (2010b). Fiber Orientation Model and Control, pp.202 - 207, in Proc Control System 2010, Stockholm, Sweden, 2010.

Chu, S., MacHattie, R., & Backström, J. (2010). Multivariable Control and Energy Optimization of Tissue Machines, In Proc. Control System 2010, Stockholm, Sweden, 2010.

Duncan, S. (1989). The Cross-Directional Control of Web Forming Process. Ph.D. Thesis, University of London, UK.

Fan, J. (2003). Model Predictive Control For Multiple Cross-directional Processes : Analysis, Tuning, and Implementation, Ph.D. Thesis, University of British Columbia, Canada.

Fan, J., Stewart, G., Dumont G., Backström J., & He P. (2005) Approximate steady-state performance prediction of large-scale constrained model predictive control systems, in IEEE Trans on Control System Technology, Vol. 13, pp.884 - 895.

Fan, J., Stewart, G., & Dumont G. (2004) Two-dimensional frequency analysis for unconstrained model predictive control of cross-directional processes, in Automatica, Vol. 40, pp. 1891 - 1903.

Froisy, J. (1994). Model predictive control: Past, present and future. In ISA Transactions, Vol. 33, pp. 235–243.

Garcia, C., Prett, D., & Morari, M. (1989). Model predictive control: Theory and practice - a survey. In Automatica, Vol. 25, pp. 335–348.

Gavelin, G. (1998). Paper Machine Design and Operation. Angus Wilde Publications, Vancouver, Canada

Gheorghe, C., Lahouaoula, A, Backström J., & Baker P. (2009). Multivariable CD control of a large linerboard machine utilizing multiple multivariable MPC controllers, in Proc. PaperCon '09 Conference, May, SL, USA.

Gorinevsky, D., & Gheorghe C. (2003). Identification tool for cross-directional processes, in IEEE Trans on Control Systems Technology, Vol. 11, 2003

Gorinevsky, D., & Heaven, M. (2001). Performance-optimized applied identification of separable distributed-parameter processes," in IEEE Trans on Automatic Control, Vol. 46, pp. 1584 -1589

Khalil H. (2001). Nonlinear Systems, Prentice Hall, 3rd Edition.

Ljung, L. (1999). System Identification Theory for the User (2nd edition), Prentice Hall PTR, 0-13-656695-2, USA.

Lu, Z. (1996). Method of optimal scaling of variables in a multivariable controller utilizing range control, U.S. Patent 5,574,638.

MacArthur, J.W. (1996). RMPCT : A New Approach To Multivariable Predictive Control For The Process Industries, in Proc Control Systems 1996, Halifax, Canada, April 1996.

Morari, M., & Lee, J. (1999). Model predictive control: Past, present and future. In Computer and Chemical Engineering, Vol. 23, pp. 667–682.

Persson, H. (1998). Dynamic modelling and simulation of multicylinder paper dryers, Licentiate thesis, Lund Institute of Technology, Sweden.

Qin, S., & Badgwell, T. (2000). An overview of nonlinear model predictive control applications, in F. Allgöwer and A. Zheng (eds), Nonlinear Predictive Control, Birkhäuser, pp. 369–393.

Qin, S., & Badgwell, T. (2003). A Survey of industrial model prediction control technology, in Control Engineering Practice, Vol. 11, pp. 733–764.

Rawlings, J. (1999). Tutorial: Model predictive control technology. In Proc. of the American Control Conference, pp. 662 –676, San Diego, California.

Slätteke, O. (2006). Modelling and Control of the Paper Machine Drying Section, Ph.D. thesis, Lund University, Sweden.

Smook, G. (2002). Handbook for Pulp and Paper Technologists (Third Edition), Angus Wilde Publications, Vancouver, Canada

Wilhelmsson, B. (1995). An experimental and theoretical study of multi-cylinder paper drying, Lund Institute of Technology, Sweden.

16

Gust Alleviation Control Using Robust MPC

Masayuki Sato[1], Nobuhiro Yokoyama[2] and Atsushi Satoh[3]
[1]*Japan Aerospace Exploration Agency*
[2]*National Defense Academy*
[3]*Iwate University*
Japan

1. Introduction

Disturbance suppression is one of the very important objectives for controller design. Thus, many papers on this topic have been reported, e.g. (Xie & de Souza, 1992; Xie et al., 1992). This kind of problem can be described as an H_∞ controller design problem using a fictitious performance block (Zhou et al., 1996). Therefore, disturbance suppression controllers can be easily designed by applying the standard H_∞ controller design method (Fujita et al., 1993).
Disturbance suppression is also important in aircraft motions (*Military Specification: Flight Control Systems - Design, Installation and Test of Piloted Aircraft, General Specification For*, 1975), and the design problem of flight controllers which suppress aircraft motions driven by wind gust, i.e. Gust Alleviation (GA) flight controller design problem (in short GA problem), has been addressed (Botez et al., 2001; Hess, 1971; 1972). In those papers, only the state information related to aircraft motions (such as, pitch angle, airspeed, etc.) is exploited for the control of aircraft motions. However, if turbulence information is obtained *a priori* and can also be exploited for the control, it is inferred that GA performance will be improved. This idea has already been adopted by several researchers (Abdelmoula, 1999; Phillips, 1971; Rynaski, 1979a;b; Santo & Paim, 2008).
Roughly speaking, GA problem is to design flight controllers which suppress the vertical acceleration driven by turbulence. In the 1970s, turbulence was measured at the nose of aircraft (Phillips, 1971); however, the lead time from the measurement of turbulence to its acting on aircraft motions becomes very short as aircraft speed increases. Thus, the turbulence data which were measured at the nose of aircraft could not be effectively used. On this issue, as electronic and optic technologies have advanced in the last two decades, nowadays, turbulence can be measured several seconds ahead using LIght Detection And Ranging (LIDAR) system (Ando et al., 2008; Inokuchi et al., 2009; Jenaro et al., 2007; Schmitt et al., 2007). This consequently means that GA control exploiting turbulence data which are measured *a priori* now becomes more practical than before. Thus, this paper addresses the design problem of such GA flight controllers.
If disturbance data are supposed to be given *a priori* and the current state of plant is also available, then controllers using Model Predictive Control (MPC) scheme work well, as illustrated for active suspension control for automobiles (Mehra et al., 1997; Tomizuka, 1976). However, in those papers, it is supposed that the plant dynamics are exactly modeled; that is, robustness of controllers against the plant uncertainties (such as, modeling errors, neglected

nonlinearities, etc.) is not considered. From a practical standpoint, it is very difficult to obtain the exact plant model. If controllers are designed without the consideration of plant uncertainties then the controlled system might achieve very poor control performance, or even worse the controlled system might be unstable. Thus, it is very important to ensure the robustness of controllers against plant uncertainties.

There have been a lot of papers which propose the design methods of MPC ensuring robustness against plant uncertainties, e.g. (Badgwell, 1997; Bemporad & Morari, 1999; Kothare et al., 1996; Kwon & Han, 2005; Löfberg, 2003; Takaba, 2000). Generally speaking, MPC design for uncertain plant leads to a design problem with infinitely many conditions. However, it is intrinsically very hard to solve this kind of problems. In the above papers, the difficulty of solving infinitely many conditions is successfully circumvented by introducing some conservatism. For example, the controllers are designed by solving Linear Matrix Inequalities (LMIs) associated with H_∞ performance or H_2 performance using common Lyapunov functions (Kothare et al., 1996; Takaba, 2000), or invariant ellipsoids being encompassed by the original invariant sets are used to ensure robust performance (Löfberg, 2003). Common Lyapunov functions as well as invariant ellipsoids generally introduce conservatism, which should be reduced.

The plant model for real systems usually includes various types of uncertainties, e.g. parametric uncertainties, neglected nonlinearities, uncertain dead time, etc. If the operating range of aircraft is relatively small and the nominal aircraft motion model is well known, then the uncertainty to be considered most is the unmodeled dynamics, which usually lie in high frequency range. One of the effective representations of this kind of uncertainties is bounded uncertain delays at the control input channels (Miyazawa, 1995; Ohno et al., 1999; Sato & Satoh, 2008), in which the effectiveness of this model is demonstrated with applications to real aircraft. Since the delay at the control input generally augments phase lag in the high frequency range, the controller designed using this type of uncertainty would have sufficiently large stability margin in the high frequency range. Therefore, this paper supposes that plant uncertainties are expressed as bounded time-invariant uncertain delays at the control input channels.

In this paper, turbulence is supposed to be measured *a priori*. The measured data always have measurement errors, such as, calibration error, position error, etc. Therefore, when exploiting the measured turbulence data for controller design, the measurement errors should be considered.

Considering these backgrounds, this paper addresses the following controller design problem: GA flight controllers exploiting *a priori* measured turbulence data including some measurement errors for aircraft motions with bounded time-invariant uncertain delays at the control input. We show that this problem is reduced to a robust MPC with finitely many conditions with neither conservatism nor approximations being introduced. The proposed MPC is formulated in terms of a Second-Order Cone Programming (SOCP) problem (Boyd & Vandenberghe, 2004), which is easily solved by using some of the generally available software, e.g. (Sturm, 1999).

Hereafter, 0_n, $0_{n,m}$, $\mathbf{0}$ and I_n respectively denote an $n \times n$ dimensional zero matrix, an $n \times m$ dimensional zero matrix, an appropriately dimensioned zero matrix and an $n \times n$ dimensional identity matrix, $\mathbf{1}_n$ denotes an n-dimensional vector with all elements being unities, $\mathcal{Z}$, $\mathcal{R}^n$ and $\mathcal{R}^{n\times m}$ respectively denote the set of integers, the set of n-dimensional real vectors and the set of $n \times m$ dimensional real matrices, $\otimes$ denotes Kronecker product, and $\lceil p \rceil$ denotes

$\min\{n \in \mathcal{Z} | p \leq n\}$. For n-dimensional vectors $p = [p_1 \cdots p_n]^T$ and $r = [r_1 \cdots r_n]^T$, $p \leq r$ denotes that $p_i \leq r_i$ holds for all i, that is, the inequality holds element-wise.

2. Preliminaries

In this section, the supposed aircraft motion model with uncertain delays at the control input is first defined and a family of models representing the aircraft motions is derived, then *a priori* measured turbulence data with some measurement errors are defined, and finally the addressed problem is given.

2.1 Uncertain plant system

Let us define the nominal continuous-time linearized aircraft motion model including actuator dynamics as P_c.

$$P_c : \begin{cases} \dot{x}(t) = A_c x(t) + B_{1_c} w(t) + B_{2_c} u(t) \\ z(t) = C_c x(t) + D_{1_c} w(t) + D_{2_c} u(t) \end{cases}, \tag{1}$$

where $x(t) \in \mathcal{R}^n$, $w(t) \in \mathcal{R}^{n_w}$, $u(t) \in \mathcal{R}^{n_u}$, and $z(t) \in \mathcal{R}^{n_z}$ respectively denote the state which includes the variables related to the aircraft motions (e.g. velocity, pitch angle, etc.) and the variables related to the control actuators, the turbulence input, the control input (e.g. elevator command), and the performance output which characterizes the motion to be suppressed (e.g. vertical acceleration). All states are supposed to be measurable and available.
The uncertainties of the plant are supposed to be represented as delays with bounded time-invariant uncertain delay time at the control input. Thus, the control input $u(t)$ is given as

$$u(t) = v(t - T_d), \tag{2}$$

where $v(t) \in \mathcal{R}^{n_u}$ denotes the control input command created by the onboard flight computer and T_d [s] denotes the uncertain delay time which is assumed to lie in the interval between $T_{d_{\min}}$ and $T_{d_{\max}}$.

$$T_d \in \left[T_{d_{\min}}, T_{d_{\max}}\right] \tag{3}$$

Considering the delay due to the calculation of the control input command with the onboard computer, the minimum delay time $T_{d_{\min}}$ is assumed to be larger than or equal to the sampling period of the onboard computer which is given as T_s [s].

$$T_s \leq T_{d_{\min}} \tag{4}$$

Similarly to usual MPC in the literature (Bemporad & Morari, 1999; Kothare et al., 1996; Löfberg, 2003), the discretized plant of P_c is considered for controller design. Suppose that the discretized plant of P_c is given as P_d using a zero-order hold which is a common method for the discretization.

$$P_d : \begin{cases} x_{k+1} = A x_k + B_1 w_k + B_2 u_k \\ z_k = C x_k + D_1 w_k + D_2 u_k \end{cases}, \tag{5}$$

where $x_k \in \mathcal{R}^n$, $w_k \in \mathcal{R}^{n_w}$, $u_k \in \mathcal{R}^{n_u}$, and $z_k \in \mathcal{R}^{n_z}$ respectively denote the state, the turbulence, the control input, and the performance output of P_d at step k. The sampling period for the discretization is assumed to be the same as that of the onboard computer T_s [s]. As all states of P_c are supposed to be measurable and available, all states of P_d are also supposed to be measurable and available.

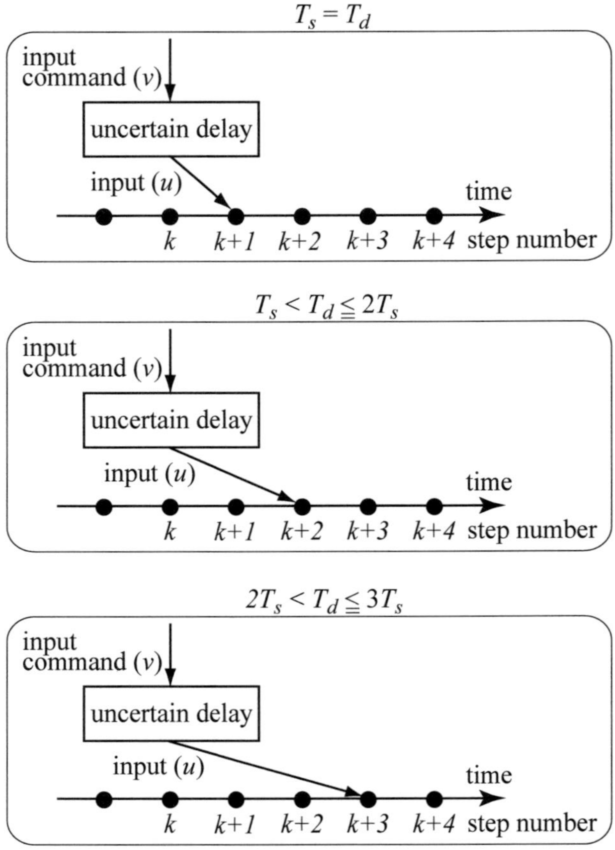

Fig. 1. Effect of uncertain delay

Remark 1. *In equation (5), the turbulence w_k is supposed be constant during the sampling period T_s. Strictly speaking, this does not hold true. However, if the sampling period is sufficiently small compared to the period corresponding to the typical frequency of turbulence then the turbulence can be regarded as constant during the sampling period.*

Similarly, the delay system (2) must be discretized. As the control input is applied not continuously but discretely, the delay effect of the control input command to the actual control input appears only at the sampling steps. That is, in a sharp contrast to a continuous-time case, in a discrete-time case, the whole elements of delay step set $\mathcal{T}_d$, which is defined as (6), are only to be considered as the uncertain delay T_d lying in the interval $\left[T_{d_{\min}}, T_{d_{\max}}\right]$.

$$\mathcal{T}_d = \left\{\underline{d}, \underline{d}+1, \cdots, \overline{d}\right\}, \tag{6}$$

where $\underline{d}$ and $\overline{d}$ are respectively defined as $\left\lceil \frac{T_{d_{\min}}}{T_s} \right\rceil$ and $\left\lceil \frac{T_{d_{\max}}}{T_s} \right\rceil$ (see Fig. 1). The number of the elements in set $\mathcal{T}_d$ is denoted by $\hat{d}(= \overline{d} - \underline{d} + 1)$. Thus, the delay steps belonging to set $\mathcal{T}_d$ are only to be considered in the discrete-time systems, while all possible delays lying in $\left[T_{d_{\min}}, T_{d_{\max}}\right]$ must be considered in the continuous-time systems. Then, the control input u_k is given as one of the elements of the following set:

$$\left\{v_{k-\underline{d}}, v_{k-\underline{d}-1}, \cdots, v_{k-\overline{d}}\right\}, \tag{7}$$

where v_m denotes the control input command of P_d created at step m; however, the real input u_k which is applied to P_d is unknown.
Now, the control input command v_k is factorized into the previous control input command v_{k-1} and the deviation between these commands to consider the rate limit of the control input command, i.e.

$$v_k = v_{k-1} + \Delta v_k, \tag{8}$$

where the previous control input command at step 0, v_{-1}, is set $\mathbf{0}$.
Under these preliminaries, the supposed plant P_u, which is composed of P_d and the uncertain delays at the control input, is described as follows.

$$P_u \in \left\{P_u^{\underline{d}}, P_u^{\underline{d}+1}, \cdots, P_u^{\overline{d}}\right\}, \tag{9}$$

where P_u^i $(i = \underline{d}, \underline{d}+1, \cdots, \overline{d})$ is defined as

$$P_u^i : \begin{cases} \hat{x}_{k+1}^i &= \hat{A}^i \hat{x}_k^i + \hat{B}_1^i w_k + \hat{B}_2^i \Delta v_k \\ z_k^i &= \hat{C}^i \hat{x}_k^i + \hat{D}_1^i w_k + \hat{D}_2^i \Delta v_k \end{cases}, \tag{10}$$

where $\hat{x}_k^i$ denotes the augmented state of i-th plant model at step k and is defined as $\hat{x}_k^i := \begin{bmatrix} x_k^{i\,T} & v_{k-\overline{d}}^T & \cdots & v_{k-1}^T \end{bmatrix}^T$ with x_k^i which denotes the state of i-th plant model at step k. The matrices $\hat{A}^i$, etc. are defined in (11).

$$\left[\begin{array}{c|c|c} \hat{A}^i & \hat{B}_1^i & \hat{B}_2^i \\ \hline \hat{C}^i & \hat{D}_1^i & \hat{D}_2^i \end{array}\right] := \begin{cases} \left[\begin{array}{ccc|c|c} A & 0_{n,n_u(\overline{d}-1)} & B_2 & B_1 & 0_{n,n_u} \\ 0_{n_u(\overline{d}-1),n} & 0_{n_u(\overline{d}-1),n_u(\overline{d}-1)} & 0_{n_u(\overline{d}-1),n_u} & 0_{n_u(\overline{d}-1),n_w} & 0_{n_u(\overline{d}-1),n_u} \\ 0_{n_u,n} & 0_{n_u,n_u(\overline{d}-1)} & 0_{n_u,n_u} & 0_{n_u,n_w} & I_{n_u} \\ \hline C & 0_{n_z,n_u(\overline{d}-1)} & 0_{n_z,n_u} & D_1 & D_2 \end{array}\right] & (i=1) \\ \left[\begin{array}{cccc|c|c} A & 0_{n,n_u(\overline{d}-i)} & B_2 & 0_{n,n_u(i-1)} & B_1 & 0_{n,n_u} \\ 0_{n_u(\overline{d}-1),n} & 0_{n_u(\overline{d}-1),n_u(\overline{d}-i)} & 0_{n_u(\overline{d}-1),n_u} & \begin{bmatrix} 0_{n_u(\overline{d}-i),n_u(i-1)} \\ I_{n_u(i-1)} \end{bmatrix} & 0_{n_u(\overline{d}-1),n_w} & 0_{n_u(\overline{d}-1),n_u} \\ 0_{n_u,n} & 0_{n_u,n_u(\overline{d}-i)} & 0_{n_u,n_u} & \begin{bmatrix} 0_{n_u,n_u(i-2)} & I_{n_u} \end{bmatrix} & 0_{n_u,n_w} & I_{n_u} \\ \hline C & 0_{n_z,n_u(\overline{d}-i)} & 0_{n_z,n_u} & \begin{bmatrix} 0_{n_z,n_u(i-2)} & D_2 \end{bmatrix} & D_1 & D_2 \end{array}\right] & (i \neq 1) \end{cases} \tag{11}$$

Remark 2. *Although the uncertainty model using the delay (2) with bounded time-invariant uncertain delay (3) generally introduces some approximations from the real uncertainties of aircraft motions, the derivation of a family of plant models (9) from the supposed uncertainties, i.e. the delay (2),* *introduces neither assumptions nor approximations. Thus, the formulation above introduces no further conservatism from the supposed uncertain plant model.*

Remark 3. *It is stressed that matrices $\hat{A}^i$, etc. in (11) have no uncertainties.*

As the current state x_k of P_d is supposed to be available, and previously created control input commands, $v_{k-\bar{d}}, \cdots, v_{k-1}$, can be memorized in the onboard computer which produces the control input command, the augmented state of i-th plant model, $\hat{x}_k^i$, is supposed to be available and given as $\hat{x}_k$. Then, the following holds.

$$\hat{x}_k = \hat{x}_k^d = \hat{x}_k^{d+1} = \cdots = \hat{x}_k^{\bar{d}} \tag{12}$$

Hereafter, the plant model for designing controllers for P_c with $u(t)$ being given as (2) is P_u.

2.2 Uncertain turbulence data

Using some system, such as, LIDAR system, it is supposed that turbulence w is measured before the turbulence affects the aircraft motions. Generally speaking, the measured data have measurement errors even if the calibration was conducted before its use. Thus, it is supposed that the j step ahead real turbulence at step k, which is denoted by w_{k+j}, satisfies the following relation with the j step ahead turbulence that is measured at step k, which is denoted by $w_{k+j|k}$.

$$\tilde{w} := \begin{bmatrix} w_k \\ w_{k+1} \\ \vdots \\ w_{k+N-1} \end{bmatrix} = \begin{bmatrix} w_{k|k} \\ w_{k+1|k} \\ \vdots \\ w_{k+N-1|k} \end{bmatrix} + \begin{bmatrix} X_0 \\ X_1 \\ \vdots \\ X_{N-1} \end{bmatrix} \Delta_w, \tag{13}$$

where N, which is given as a constant positive integer, denotes the maximum step number for measuring turbulence *a priori*, $X_j \in \mathcal{R}^{n_w \times n_w}$ $(j = 0, 1, \cdots, N-1)$ denote the constant given matrices which define the measurement errors in the measured turbulence data with uncertain constant vector $\Delta_w \in \mathcal{R}^{n_w}$ satisfying (14) (see Fig. 2, which is at the top of the next page).

$$-\mathbf{1}_{n_w} \leq \Delta_w \leq \mathbf{1}_{n_w} \tag{14}$$

Let us define set Ω as the existence region of $\tilde{w}$.

$$\Omega = \left\{ \tilde{w} \in \mathcal{R}^{n_w N} : \tilde{w} \text{ given as (13) with } \Delta_w \text{ satisfying (14)} \right\} \tag{15}$$

Remark 4. *Note that matrices X_j might be different for each j; that is, it is possible that $X_0 \neq X_1 \neq \cdots\cdots \neq X_{N-1}$ holds. Furthermore, note that matrices X_j might be different for each step k; that is, matrices $X_0, \cdots, X_{N-1}$ at step k might be different from the corresponding matrices $X_0, \cdots, X_{N-1}$ at step $k-1$. This corresponds to time-varying measurement error case.*

Remark 5. *Note that $\tilde{w}$ is affine with respect to the each element of Δ_w.*

Hereafter, it is supposed that at each step the real turbulence data w_{k+j} are not available *a priori*, but instead, the measured data $w_{k+j|k}$ which satisfy (13) with measurement errors defined as $X_j\Delta_w$ are available.

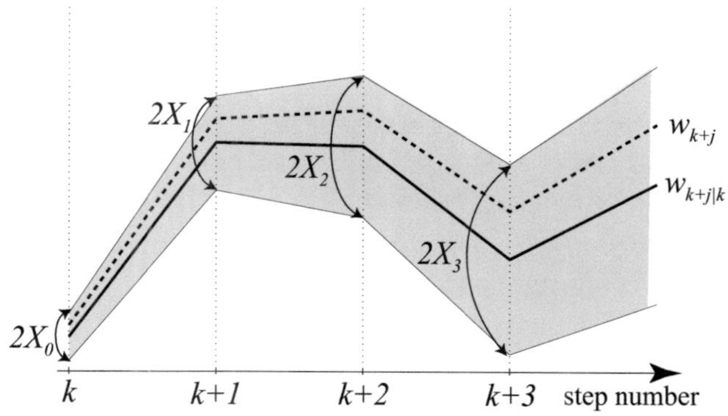

Fig. 2. Measurement error at step k in case for $n_w = 1$ and $X_j(\in \mathcal{R}) > 0$

2.3 Problem definition

Using the uncertain plant model P_u given as (9) using (10) and the *a priori* measured turbulence data $w_{k+j|k}$ satisfying (13), the addressed problem, i.e. GA problem exploiting the *a priori* measured turbulence data, is defined below.

Considering that the turbulence are measured for N steps ahead, the horizon step number in MPC, which denotes the step number during which control performance is to be optimized, is also set N. For i-th plant model P_u^i, the performance index for turbulence suppression, i.e. GA performance, is defined as

$$J_i(\hat{x}^i, \Delta v, z^i) = \sum_{j=0}^{N-1} \left((\hat{x}^i_{k+j+1|k})^T Q \hat{x}^i_{k+j+1|k} + \Delta v^T_{k+j|k} R \Delta v_{k+j|k} + (z^i_{k+j|k})^T S z^i_{k+j|k} \right), \tag{16}$$

where matrices Q and S are appropriately defined positive semidefinite matrices, matrix R is an appropriately defined positive definite matrix, $\hat{x}^i_{k+j|k}$ denotes the i-th plant's augmented state at step $k+j$ predicted at step k, $\Delta v_{k+j|k}$ denotes the control input command deviation at step $k+j$ created at step k, and $z^i_{k+j|k}$ denotes the performance output at step $k+j$ predicted at step k.

There usually exist preferable or prohibitive regions for the state, the performance output, and the control input command deviation. For the consideration of these regions, constraints for the augmented state $\hat{x}^i$, the control input command deviation Δv and the performance output z^i are introduced. That is, they should satisfy the following constraints.

$$\gamma_{\min} \leq \hat{x}^i_{k+j+1|k} \leq \gamma_{\max},\ j = 0, \cdots, N-1, \tag{17}$$

$$\delta_{\min} \leq \Delta v_{k+j|k} \leq \delta_{\max},\ j = 0, \cdots, N-1, \tag{18}$$

$$\xi_{\min} \leq z^i_{k+j|k} \leq \xi_{\max},\ j = 0, \cdots, N-1, \tag{19}$$

where $\gamma_{\min}, \gamma_{\max} \in \mathcal{R}^{n+n_u\bar{d}}$, $\delta_{\min}, \delta_{\max} \in \mathcal{R}^{n_u}$ and $\xi_{\min}, \xi_{\max} \in \mathcal{R}^{n_z}$ are given constant vectors.

If the worst performance of $J_i(\hat{x}^i, \Delta v, z^i)$ $(i = \underline{d}, \cdots, \bar{d})$ is minimized then all the other performance of $J_i(\hat{x}^i, \Delta v, z^i)$ is no more than the worst; that is, all the possible plant models P_u^i have no more worse performance than the worst. Considering this, the design objective is to obtain $\Delta v_{k+j|k}$ which minimizes the maximum of $J_i(\hat{x}^i, \Delta v, z^i)$. Thus, the addressed problem is defined as follows.

Problem 1. *Suppose that uncertain aircraft motion model is given as P_u defined as in (9) using P_u^i in (10), that the current augmented state $\hat{x}_k$ is available, and that j $(j = 0, \cdots, N-1)$ step ahead turbulence at step k is measured as $w_{k+j|k}$ which satisfies (13) for the real turbulence w_{k+j}.*
Under these assumptions, find $\Delta v_{k+j|k}$ $(j = 0, \cdots, N-1)$ which minimize $\max_{\tilde{w}\in\Omega} \max_{i\in\{\underline{d},\cdots,\overline{d}\}} J_i(\hat{x}^i, \Delta v, z^i)$ under the constraints (17), (18) and (19).

If Problem 1 is solved online, then the control input command v at step k, v_k, is calculated as $v_{k-1} + \Delta v_{k|k}$ using the previous control input command v_{k-1}. The control strategy of this paper is to obtain the optimal control input command by solving an optimization problem online using a family of plant models. That is, the proposed control strategy is MPC.
It is easily confirmed that solving Problem 1 is equivalent to solving the following problem.

Problem 2. *Find $\Delta v_{k+j|k}$ $(j = 0, \cdots, N-1)$ which minimize the following performance index.*

$$\max_{\tilde{w}\in\Omega} \max_{i\in\{\underline{d},\cdots,\overline{d}\}} J_i(\hat{x}^i, \Delta v, z^i) \text{ subject to } (17), (18), (19), (10) \text{ with } (12) \text{ and } (13)$$

Remark 6. *Note that Problem 2 seeks the common control input command deviation for all i and for all possible $\tilde{w} \in \Omega$. Therefore, solving Problem 2 produces control input command deviation $\Delta v_{k|k}$ which is robust against the uncertain delays at the control input satisfying (3) and all possible turbulence $\tilde{w} \in \Omega$.*

In the next section, the proposed method to solve Problem 2 is shown.

3. Proposed method

In this section, the proposed method to solve Problem 2 is shown. For simplicity, let us first consider the case in which the measured turbulence data have no measurement errors, i.e. $w_{k+j} = w_{k+j|k}$, next consider the case in which the measured turbulence data have the measurement errors.

3.1 No measurement error case

Let all X_j in (13) be set as $\mathbf{0}$. Then, w_{k+j} is given as $w_{k+j|k}$. That is, the following holds.

$$\tilde{w} = \begin{bmatrix} w_{k|k}^T & \cdots & w_{k+N-1|k}^T \end{bmatrix}^T$$

Define the following vectors.

$$\tilde{v} = \begin{bmatrix} \Delta v_{k|k}^T & \cdots & \Delta v_{k+N-1|k}^T \end{bmatrix}^T$$

$$\tilde{x}^i = \begin{bmatrix} (\hat{x}_{k+1|k}^i)^T & \cdots & (\hat{x}_{k+N|k}^i)^T \end{bmatrix}^T, \; i = \underline{d}, \cdots, \overline{d}$$

$$\tilde{z}^i = \begin{bmatrix} (z_{k|k}^i)^T & \cdots & (z_{k+N-1|k}^i)^T \end{bmatrix}^T, \; i = \underline{d}, \cdots, \overline{d}$$

Then, the state equation and the performance output equation of P_u^i are respectively given as follows:

$$\tilde{x}^i = \left[I_N \otimes \hat{A}^i \; 0_{(n+n_u\bar{d})N, n+n_u\bar{d}} \right] \begin{bmatrix} \hat{x}^i_{k|k} \\ \tilde{x}^i \end{bmatrix} + \left(I_N \otimes \hat{B}^i_1 \right) \tilde{w} + \left(I_N \otimes \hat{B}^i_2 \right) \tilde{v}, \tag{20}$$

$$\tilde{z}^i = \left[I_N \otimes \hat{C}^i \; 0_{n_z N, n+n_u\bar{d}} \right] \begin{bmatrix} \hat{x}^i_{k|k} \\ \tilde{x}^i \end{bmatrix} + \left(I_N \otimes \hat{D}^i_1 \right) \tilde{w} + \left(I_N \otimes \hat{D}^i_2 \right) \tilde{v}. \tag{21}$$

Define the following matrices and vectors:

$$\tilde{Q} := I_N \otimes Q, \; \tilde{R} := I_N \otimes R, \; \tilde{S} := I_N \otimes S,$$

$$\begin{aligned} \tilde{\gamma}_{\min} &:= \mathbf{1}_N \otimes \gamma_{\min}, \; \tilde{\gamma}_{\max} := \mathbf{1}_N \otimes \gamma_{\max}, \\ \tilde{\delta}_{\min} &:= \mathbf{1}_N \otimes \delta_{\min}, \; \tilde{\delta}_{\max} := \mathbf{1}_N \otimes \delta_{\max}, \\ \tilde{\zeta}_{\min} &:= \mathbf{1}_N \otimes \zeta_{\min}, \; \tilde{\zeta}_{\max} := \mathbf{1}_N \otimes \zeta_{\max}. \end{aligned}$$

Using these definitions, the following proposition, which is equivalent to Problem 2, is directly obtained.

Proposition 1. *Find $\tilde{v}$ which minimizes q subject to (22), (23), and (24).*

$$q \geq \begin{bmatrix} \tilde{x}^i \\ \tilde{v} \\ \tilde{z}^i \end{bmatrix}^T \begin{bmatrix} \tilde{Q} & \mathbf{0} & \mathbf{0} \\ \mathbf{0} & \tilde{R} & \mathbf{0} \\ \mathbf{0} & \mathbf{0} & \tilde{S} \end{bmatrix} \begin{bmatrix} \tilde{x}^i \\ \tilde{v} \\ \tilde{z}^i \end{bmatrix}, \; i = \underline{d}, \cdots, \bar{d} \tag{22}$$

$$\begin{bmatrix} \tilde{\gamma}_{\min} \\ \tilde{\zeta}_{\min} \end{bmatrix} \leq \begin{bmatrix} \tilde{x}^i \\ \tilde{z}^i \end{bmatrix} \leq \begin{bmatrix} \tilde{\gamma}_{\max} \\ \tilde{\zeta}_{\max} \end{bmatrix}, \; i = \underline{d}, \cdots, \bar{d} \tag{23}$$

$$\tilde{\delta}_{\min} \leq \tilde{v} \leq \tilde{\delta}_{\max} \tag{24}$$

As Proposition 1 is an SOCP problem (Boyd & Vandenberghe, 2004), its global optimum is easily solved by using some software, e.g. (Sturm, 1999).
Thus, if measured turbulence data have no measurement errors then the addressed problem, i.e. Problem 1, is solved by virtue of Proposition 1 without introducing any conservatism (see Remark 2).

Remark 7. *If Proposition 1 is solved, then the state is bounded by $\gamma_{\min}$ and $\gamma_{\max}$; that is, the boundedness of the state is assured.*

3.2 Measurement error case

Let us suppose that the real turbulence w_{k+j} cannot be measured and the measured turbulence $w_{k+j|k}$ satisfies (13).
First conduct full rank decompositions for matrices $\tilde{Q}$, $\tilde{R}$, and $\tilde{S}$

$$\tilde{Q} = \hat{Q}\hat{Q}^T, \tilde{R} = \hat{R}\hat{R}^T, \tilde{S} = \hat{S}\hat{S}^T.$$

Then, inequality (22) is equivalently transformed to the following inequality by applying the Schur complement (Boyd & Vandenberghe, 2004).

$$\begin{bmatrix} q & (\tilde{x}^i)^T\hat{Q} & \tilde{v}^T\hat{R} & (\tilde{z}^i)^T\hat{S} \\ \hat{Q}^T\tilde{x}^i & \mathbf{I} & \mathbf{0} & \mathbf{0} \\ \hat{R}^T\tilde{v} & \mathbf{0} & \mathbf{I} & \mathbf{0} \\ \hat{S}^T\tilde{z}^i & \mathbf{0} & \mathbf{0} & \mathbf{I} \end{bmatrix} \geq 0 \tag{25}$$

If some of matrices $\tilde{Q}, \tilde{R}, \tilde{S}$ are set zero matrices, then the corresponding rows and columns in (25) are ignored.
The state $\hat{x}^i_{k+N|k}$ and the performance output $z^i_{k+N-1|k}$ are respectively described as in (26) and (27).

$$\hat{x}^i_{k+N|k} = (\hat{A}^i)^N\hat{x}^i_{k|k} + \left[(\hat{A}^i)^{N-1}\hat{B}^i_1 \ \cdots \ \hat{A}^i\hat{B}^i_1 \ \hat{B}^i_1\right]\tilde{w} + \left[(\hat{A}^i)^{N-1}\hat{B}^i_2 \ \cdots \ \hat{A}^i\hat{B}^i_2 \ \hat{B}^i_2\right]\tilde{v} \tag{26}$$

$$z^i_{k+N-1|k} = \hat{C}^i(\hat{A}^i)^{N-1}\hat{x}^i_{k|k} + \left[\hat{C}^i(\hat{A}^i)^{N-2}\hat{B}^i_1 \ \cdots \ \hat{C}^i\hat{B}^i_1 \ \hat{D}^i_1\right]\tilde{w} + \left[\hat{C}^i(\hat{A}^i)^{N-2}\hat{B}^i_2 \ \cdots \ \hat{C}^i\hat{B}^i_2 \ \hat{D}^i_2\right]\tilde{v} \tag{27}$$

Note that both $\hat{x}^i_{k+N|k}$ and $z^i_{k+N-1|k}$ are affine with respect to each element of Δ_w, because $\tilde{w}$ is affine with respect to each element of Δ_w. Similarly, $\hat{x}^i_{k+m|k}$ $(m = 1, \cdots, N-1)$ and $z^i_{k+m|k}$ $(m = 0, \cdots, N-2)$ are also affine with respect to each element of Δ_w. Considering these and that (25) is affine with respect to $\tilde{x}^i$ and $\tilde{z}^i$, checking whether or not (25) holds for all possible Δ_w is equivalent to checking the feasibility at all vertices of Δ_w.
Now let Φ be defined as the set composed of all the vertices of Δ_w; that is,

$$\Phi = \left\{p = [p_1 \ \cdots \ p_{n_w}]^T \in \mathcal{R}^{n_w} : p_i = \pm 1, \ i = 1, \cdots, n_w\right\}. \tag{28}$$

The number of the elements belonging to Φ is 2^{n_w}.
Under these preliminaries, the following proposition, which is equivalent to solving Problem 2, is directly obtained.

Proposition 2. *Find $\tilde{v}$ which minimizes q subject to (22), (23) and (24) for all $\Delta_w \in \Phi$.*

Similarly to Proposition 1, as Proposition 2 is also an SOCP problem, its global optimum is easily obtained with the aid of some software, e.g. (Sturm, 1999).
Thus, if the measured turbulence data have measurement errors expressed as $X_j\Delta_w$ and satisfy (13) for the real turbulence, then the addressed problem, i.e. Problem 1, is solved by virtue of Proposition 2 without introducing any conservatism (see Remarks 2 and 5).
Similarly to Remark 7 for Problem 1, if Problem 2 is solved, then the state is bounded by $\gamma_{\min}$ and $\gamma_{\max}$.

Remark 8. *The increases of the numbers N, n_w and i lead to a huge numerical complexity for solving Proposition 2. Thus, obtaining the delay time bounds precisely is very important to reduce i. On the other hand, in general, n_w cannot be reduced, because this number represents the number of channels of turbulence input. The remaining number N has a great impact on controller performance, which will be shown in the next section with numerical simulation results.*

4. Numerical example

Several numerical examples are shown to demonstrate that the proposed method works well for GA problem under the condition that there exist bounded uncertain delays at the control input and the measurement errors in *a priori* measured turbulence data.

4.1 Small aircraft example

Let us first consider the linearized longitudinal aircraft motions of JAXA's research aircraft MuPAL-α (Sato & Satoh, 2008) at an altitude of 1524 [m] and a true air speed of 66.5 [m/s]. This aircraft is based on Dornier Do-228, which is a twin turbo-prop commuter aircraft.

4.1.1 Simulation setting

It is supposed that only the elevator is used for aircraft motion control. The transfer function of its actuator dynamics is modeled as $1/(0.1s+1)$. Then, the continuous-time system representing the linearized longitudinal motions with the modeled actuator dynamics is given as (1), where the state is $[u_i\ w_i\ q\ \theta\ \delta_e]^T$, the turbulence is w_g, the control input is δ_{e_c}, and the performance output is Δa_z. Here, u_i [m/s], w_i [m/s], q [rad/s], θ [rad], δ_e [rad], w_g [m/s], δ_{e_c} [rad/s] and Δa_z [m/s^2] respectively denote inertial forward-backward velocity in body axes, inertial vertical velocity in body axes, pitch rate, pitch angle, elevator deflection, vertical turbulence in inertial axes, elevator command, and vertical acceleration deviation in inertial axes.

After the discretization of (1) with sampling period T_s [s] being set as 0.1, the discrete-time system (5) is given as (29).

$$
\left[\begin{array}{c|c|c} A & B_1 & B_2 \\ \hline C & D_1 & D_2 \end{array}\right]
= \left[\begin{array}{ccccc|c|c}
0.99799 & 0.018181 & -0.54564 & -0.97647 & 0.11430 & 0.018289 & 0.048318 \\
-0.014894 & 0.87690 & 5.5175 & -0.076329 & -1.1947 & -0.12116 & -0.50903 \\
7.7845\times10^{-4} & -5.9106\times10^{-3} & 0.80765 & 5.0506\times10^{-4} & -0.23770 & -5.9576\times10^{-3} & -0.14529 \\
3.9313\times10^{-5} & -3.1213\times10^{-4} & 0.090399 & 1.0000 & -0.014529 & -3.1444\times10^{-4} & -5.3277\times10^{-3} \\
0 & 0 & 0 & 0 & 0.36788 & 0 & 0.63212 \\ \hline
-0.18089 & -1.1043 & -1.6792 & 5.8933\times10^{-3} & -4.9603 & -1.0825 & 0
\end{array}\right] \tag{29}
$$

The bounded time-invariant uncertain delay T_d [s] for elevator command is supposed to be in the interval [0.1, 0.4]. As the delay time is set as [0.1, 0.4] and the sampling period T_s is 0.1, $\underline{d}$ and $\overline{d}$ are respectively given as 1 and 4. Next the state-space matrices of P_u^i $(i=1,\cdots,4)$ are calculated. (The state-space matrices are omitted for space problem.) The augmented state $\hat{x}_k^i$ is given as $\left[u_i\ w_i\ q\ \theta\ \delta_e\ \delta_{e_c(-4)}\ \delta_{e_c(-3)}\ \delta_{e_c(-2)}\ \delta_{e_c(-1)}\right]^T$, where $\delta_{e_c(-l)}$ denotes the elevator command created at l step before. The objective is to obtain the elevator input command,

$\delta_{e_c(0)}$, which minimizes the effect of vertical turbulence to vertical acceleration for all possible delays.
The constraints for the augmented state $\hat{x}_k^i$ and the control input command deviation are given as follows:

$$\gamma_{\max} = \left[10\ 10\ \tfrac{10\pi}{180}\ \tfrac{10\pi}{180}\ \tfrac{5\pi}{180} \times \mathbf{1}_5\right]^T,\ \gamma_{\min} = -\gamma_{\max},$$
$$\delta_{\max} = \tfrac{\pi}{180},\ \delta_{\min} = -\delta_{\max}.$$

This means that the rate limit of elevator command is set as $\pm 10\,[\mathrm{deg/s}]$. The constraints for performance output $\zeta_{\min}$ and $\zeta_{\max}$ are respectively set as $-\infty$ and ∞; that is, performance output has no constraints.
Matrices Q and S in (16) are set as $Q = 0_9$ and $S = 1$ respectively. Matrix R will be set later.
The turbulence w_g is supposed to be given as

$$w_g(t) = \sin(\omega t), \tag{30}$$

where t denotes the simulation time starting from 0, and ω, which will be set later, denotes the frequency of the turbulence.

4.1.2 Simulation results without measurement errors in turbulence data

Let us first show the results of simulations in which turbulence is supposed to be exactly measured.
Numerical simulations using continuous-time system (1) composed of MuPAL-α's linearized longitudinal motions and the first-order elevator actuator model, and the proposed MPC in which Proposition 1 is solved on line are carried out for 20 [s]. In the simulations, various constant delay steps at the control input $\hat{t}_d$, various constant turbulence frequencies ω [rad/s], various constant weighting matrices R, and various constant receding horizon step numbers N are used from the following sets:

$$\begin{aligned} \hat{t}_d &\in \{1,\ 2,\ 3,\ 4\}, \\ \omega &\in \{0.1,\ 0.5,\ 1.0,\ 5.0,\ 7.0,\ 8.0,\ 10.0\}, \\ R &\in \{10^{-1},\ 10^0,\ 10^1,\ 10^2,\ 10^3,\ 10^4\}, \\ N &\in \{10,\ 20,\ 30,\ 40,\ 50\}. \end{aligned} \tag{31}$$

For comparison, the following scenarios are simultaneously carried out.

Scenario A: MPC in which Proposition 1 is solved online is applied,

Scenario B: no control is applied,

Scenario C: MPC in which Proposition 1 is solved online but with the measured turbulence data being set as zeros, i.e. MPC without prior turbulence data, is applied.

Fig. 3 shows the performance comparison for scenarios A, B and C. In this figure, J_A, J_B and J_C denote the following performance indices for the corresponding scenarios, which are obtained from the simulations:

$$\max_{\hat{t}_d \in \{1,\ 2,\ 3,\ 4\}} \int_0^{20} |\Delta a_z|^2 dt. \tag{32}$$

For comparison, mesh planes at $J_A/J_B = 1$ and $J_A/J_C = 1$ are drawn. $J_A/J_B < 1$ means that gust alleviation is effectively achieved by the proposed method, and $J_A/J_C < 1$ means that the *a priori* measured turbulence data are useful for the improvement of GA performance. The following are concluded from Fig. 3.

- It is very difficult for MuPAL-α to suppress high frequency turbulence effect, such as, over 8 [rad/s].
- MuPAL-α has no need to measure turbulence *a priori* for more than 20 steps. In other words, it is sufficient for MuPAL-α to measure turbulence for 20 steps ahead.
- Using an appropriately chosen R (e.g. $R = 10^2$), the proposed GA flight controller in which Proposition 1 is solved online improves GA performance for low and middle frequency turbulence, such as, below 5 [rad/s].

The first item is reasonable because aircraft motion model has a direct term from the vertical turbulence to the vertical acceleration and it is supposed that there exists uncertain delay at its control input. The second item is interesting, because there is a limit for the improvement of GA performance even when *a priori* measured turbulence data are available.
For reference, several time histories with $R = 10^2$ and $N = 20$ are shown in Fig. 4. For space problem, only actual elevator deflection command (δ_{e_c}) and its created command by flight computer ($\delta_{e_c(0)}$), and performance output are shown. δ_{e_c} and $\delta_{e_c(0)}$ almost overlap in some cases. These figures illustrate the usefulness of the *a priori* measured turbulence data.

4.1.3 Simulation results with measurement errors in turbulence data

Let us next show the results of simulations in which measured turbulence data have measurement errors.
Numerical simulations using continuous-time system (1) composed of MuPAL-α's linearized longitudinal motions and the first-order elevator actuator model, and the proposed MPC in which Proposition 2 is solved on line are carried out for 20 [s]. In the simulations, various constant delay steps at the control input $\hat{t}_d$, various constant turbulence frequencies ω [rad/s], various constant weighting matrices R, and various constant receding horizon step numbers N are used from the following sets:

$$\begin{aligned} &\hat{t}_d \in \{1,\ 2,\ 3,\ 4\}, \\ &\omega \in \{0.1,\ 0.5,\ 1.0,\ 3.0,\ 4.0,\ 5.0,\ 6.0,\ 7.0\}, \\ &R \in \{10^1,\ 10^2,\ 10^3,\ 10^4,\ 10^5\}, \\ &N \in \{10,\ 12,\ 14,\ 16,\ 18,\ 20,\ 22,\ 24\}. \end{aligned} \tag{33}$$

Matrices X_j in the measurement error are set as

$$X_j = 0.2 + 0.1 \times (66.5/100 \times T_s)\, j. \tag{34}$$

This means that the measurement error for w_g is composed of a constant bias error 0.2 [m/s] and a measurement error which is proportional to distance, the latter has 0.1 [m/s] measurement error at 100 [m] ahead.
Three possibilities are considered in the simulations; that is, (i) the real turbulence is the same as the measured turbulence, i.e. $w_{k+j} = w_{k+j|k}$, (ii) the real turbulence is the upper bound of the supposed turbulence, i.e. $w_{k+j} = w_{k+j|k} + X_j$ using (34), and (iii) the real turbulence is the lower bound of the supposed turbulence, i.e. $w_{k+j} = w_{k+j|k} - X_j$ using (34).
For comparison, the following scenarios are simultaneously carried out.

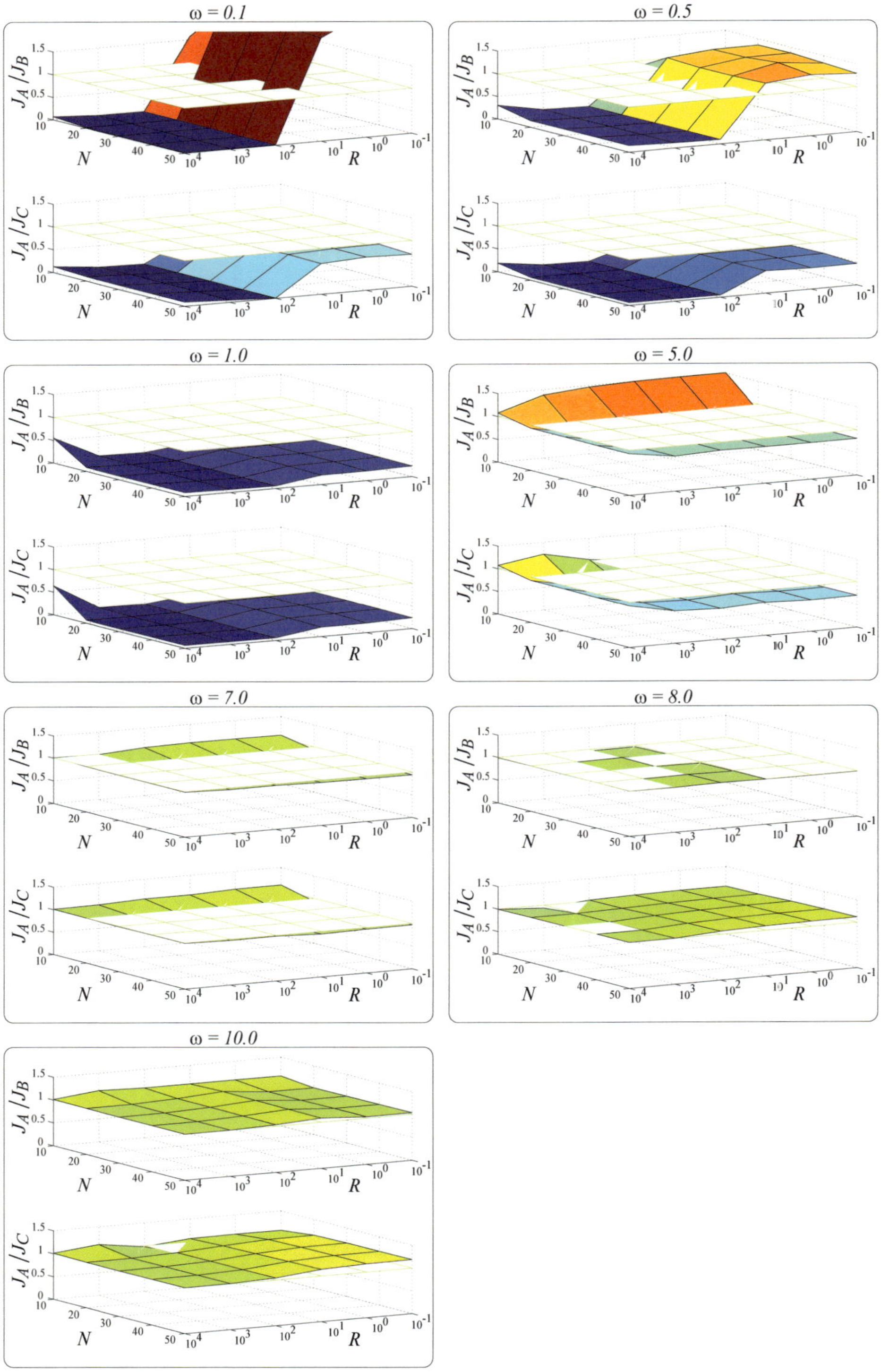

Fig. 3. GA performance comparison for MuPAL-α under no measurement errors in turbulence data

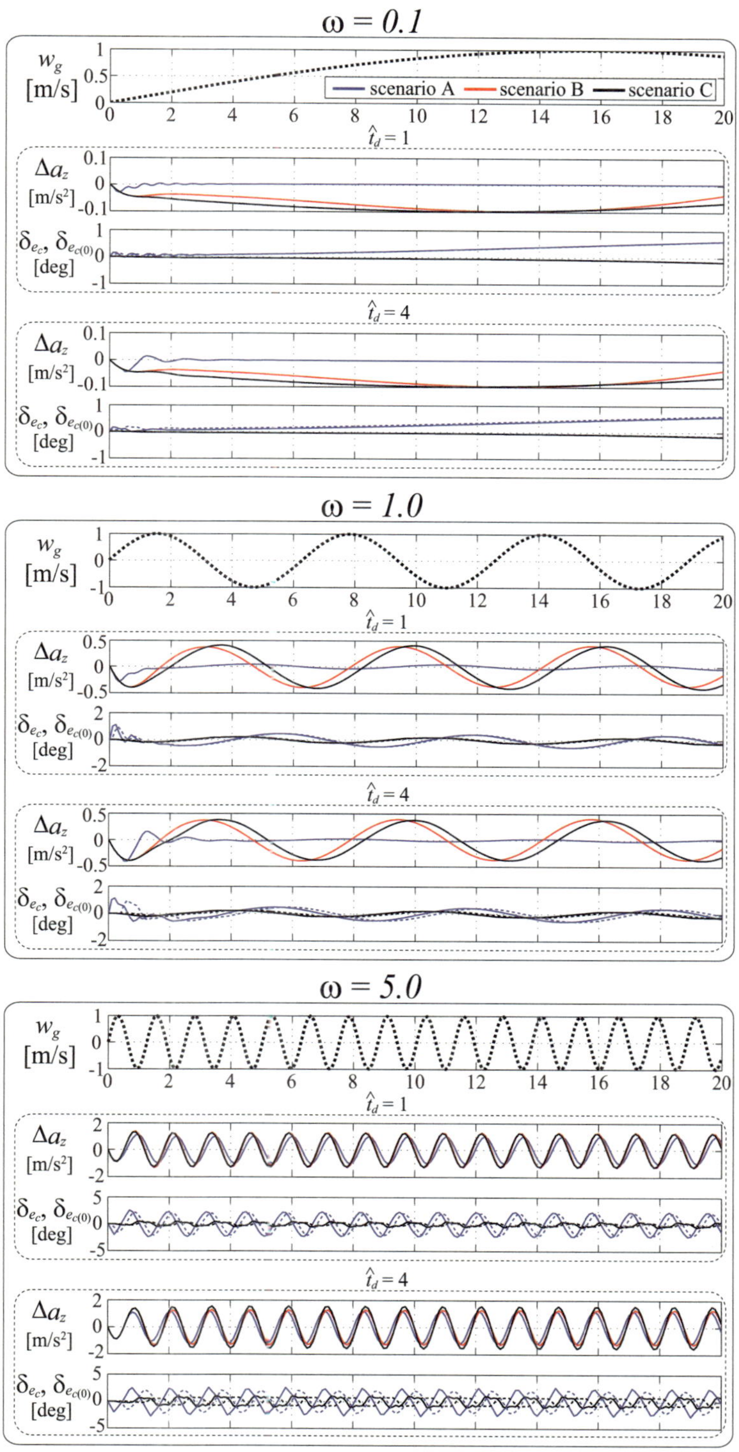

Fig. 4. Time histories under no measurement errors in turbulence data with $R = 10^2$ and $N = 20$ (δ_{e_c} is shown as dotted lines and $\delta_{e_c(0)}$ is shown as solid lines)

Scenario A: MPC in which Proposition 2 is solved online is applied,

Scenario B: no control is applied,

Scenario C: MPC in which Proposition 2 is solved online but with the measured turbulence data being set as zeros, i.e. MPC without prior turbulence data, is applied.

Fig. 5 shows the performance comparison for scenarios A, B and C. In this figure, J_A, J_B and J_C denote the following performance indices for the corresponding scenarios, which are obtained from the simulations:

$$\max_{w_{k+j}=\{w_{k+j|k},\, w_{k+j|k}\pm X_j\}} \max_{\hat{t}_d\in\{1,2,3,4\}} \int_0^{20} |\Delta a_z|^2 dt. \tag{35}$$

For comparison, mesh planes at $J_A/J_B = 1$ and $J_A/J_C = 1$ are drawn.
The following are concluded from Fig. 5.

- For turbulence, whose frequencies are no more than 0.5 [rad/s], GA performance using the proposed method is larger than the uncontrolled case.
- For turbulence, whose frequencies are more than 6 [rad/s], vertical acceleration is hardly reduced even if prior turbulence data are obtained.
- It is sufficient for MuPAL-α to measure turbulence for 20 steps ahead.
- Using an appropriately chosen R (e.g. $R = 10^3$), the proposed GA flight controller in which Proposition 2 is solved online improves GA performance for middle frequency turbulence, such as, $1 \sim 5$ [rad/s].

The first item does not hold true for no measurement error case (see also Fig. 3). Thus, GA performance deterioration for low frequency turbulence is caused by the measurement errors in the measured turbulence data. The second item is reasonable for considering that it is difficult to suppress turbulence effect on aircraft motions caused by high frequency turbulence even when the turbulence is exactly measured (see also Fig. 3). The fourth item illustrates that the *a priori* measured turbulence data improve GA performance even when there exist measurement errors in the measured turbulence data.
For reference, several time histories with $R = 10^3$, $N = 18$ and $\hat{t}_d = 4$ are shown in Fig. 6. For space problem, only actual elevator deflection command (δ_{e_c}) and its created command by flight computer ($\delta_{e_c(0)}$), and performance output are shown. These figures illustrate the usefulness of the *a priori* measured turbulence data for middle frequency turbulence (e.g. 1.0 and 5.0 [rad/s]). However, as the top figure in Fig. 6 indicates, measurement errors in turbulence data deteriorate GA performance; that is, if the real turbulence is smaller than the measured one, i.e. the case for $w_{k+j} = w_{k+j|k} - X_j$, then the proposed MPC produces surplus elevator deflections and this causes extra downward accelerations. The converse, i.e. the case for $w_{k+j} = w_{k+j|k} + X_j$, also holds true. Thus, it is very important for achieving good GA performance to measure turbulence exactly.
To evaluate the impact of the rate limit for elevator command on GA performance, the same simulations but with only $\delta_{\max}$ and $\delta_{\min}$ being doubled, i.e. $\delta_{\max} = 2\frac{\pi}{180}$ and $\delta_{\min} = -2\frac{\pi}{180}$, are carried out. The results for (35) are shown in Fig. 7.
Comparison between Figs. 5 and 7 concludes the following.

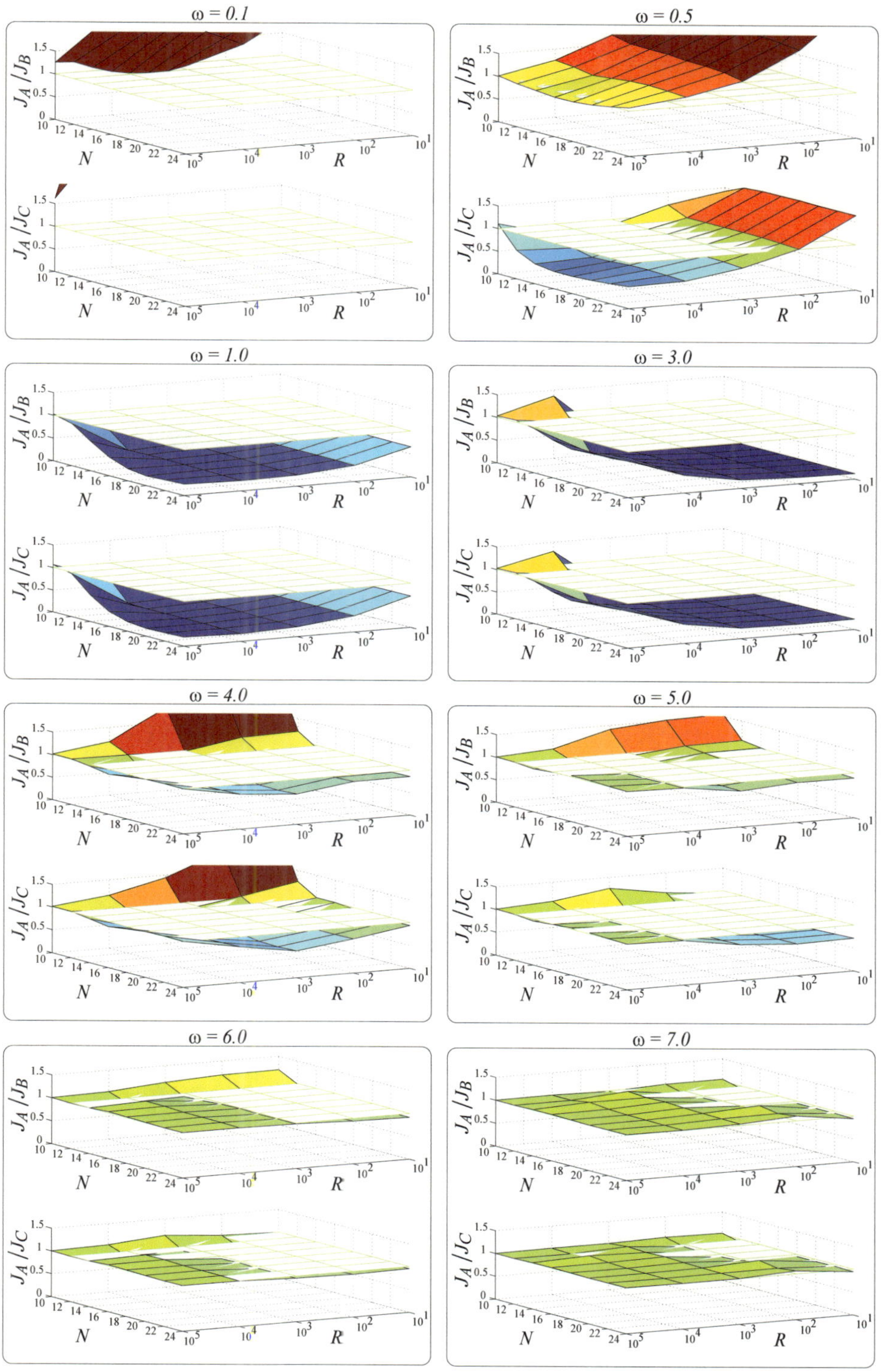

Fig. 5. GA performance comparison for MuPAL-α under measurement errors in turbulence data

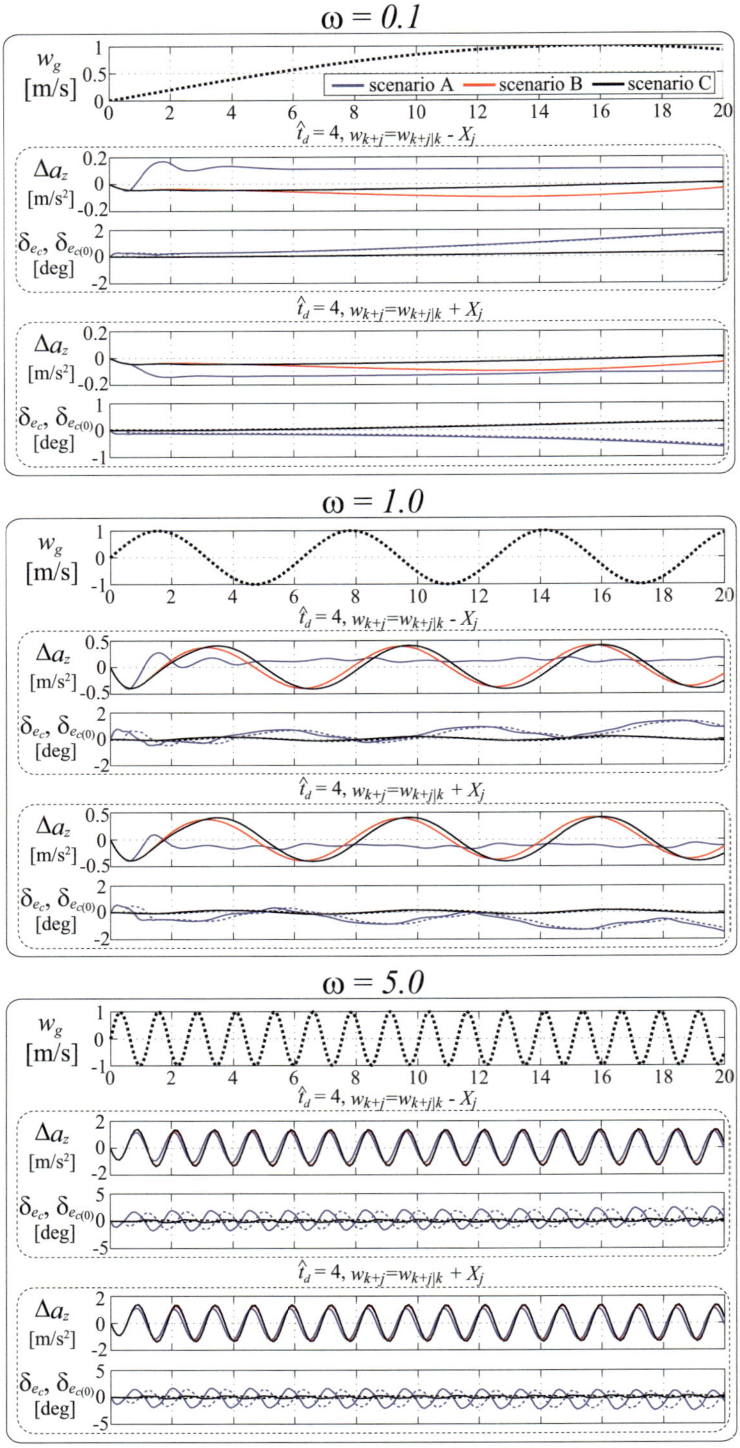

Fig. 6. Time histories under measurement errors in turbulence data with $R = 10^3$ and $N = 18$ (δ_{e_c} is shown as dotted lines and $\delta_{e_c(0)}$ is shown as solid lines)

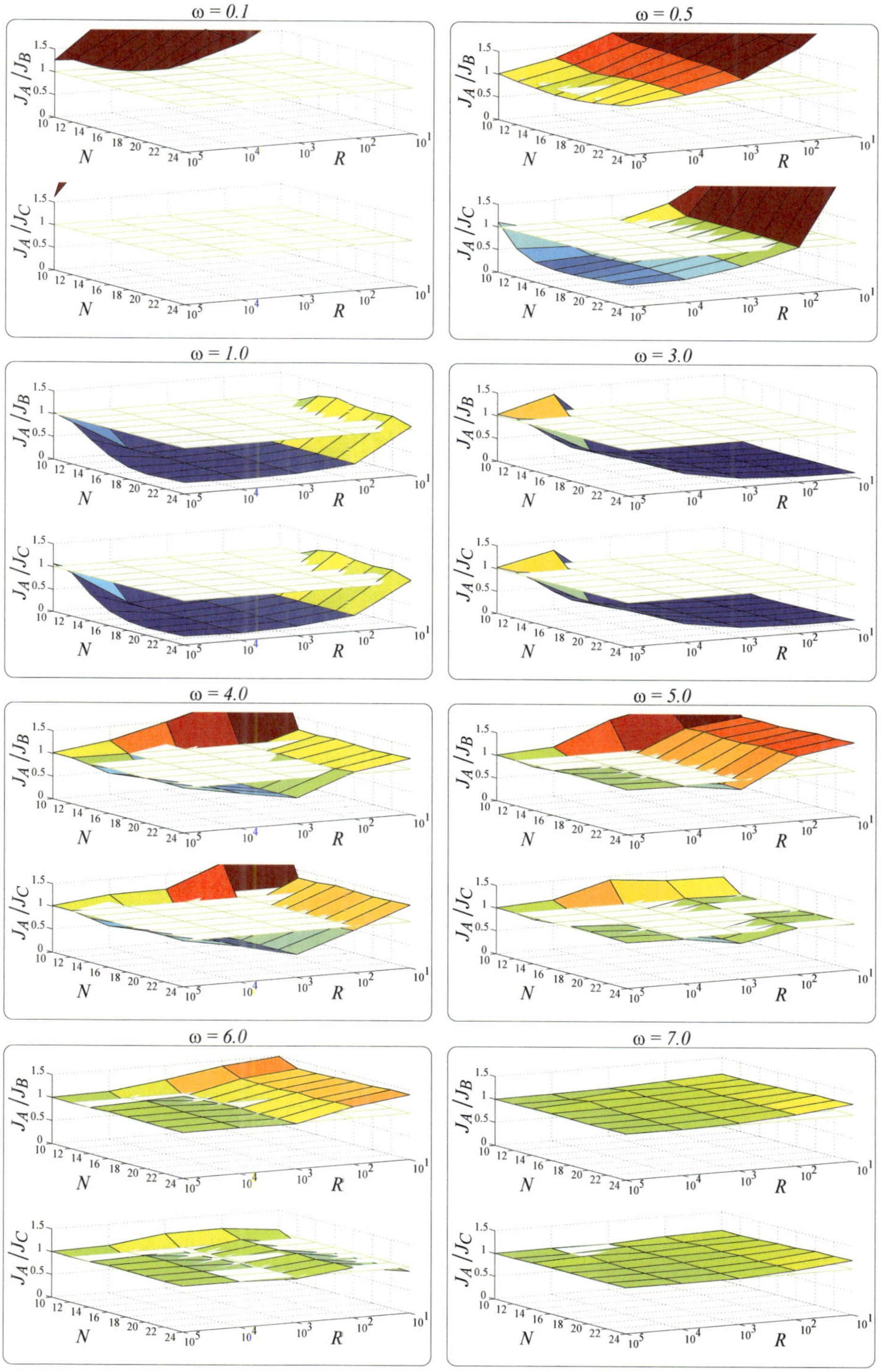

Fig. 7. GA performance comparison for MuPAL-α with relaxed rate limit for elevator command under measurement errors in turbulence data

ω [rad/s]	R	N	PT7400	P650
0.1	10^1	10	0.30 – 0.52(0.38)	0.88 – 1.45(1.18)
		24	0.47 – 1.36(0.74)	1.78 – 3.94(2.52)
	10^5	10	0.27 – 0.49(0.34)	0.84 – 1.25(1.03)
		24	0.40 – 1.00(0.59)	1.14 – 2.78(2.01)
1.0	10^1	10	0.29 – 0.53(0.37)	0.97 – 1.48(1.18)
		24	0.45 – 3.62(0.76)	1.64 – 3.94(2.52)
	10^5	10	0.26 – 0.43(0.32)	0.80 – 1.44(0.99)
		24	0.39 – 1.02(0.59)	1.36 – 2.64(1.98)
6.0	10^1	10	0.30 – 0.53(0.38)	0.77 – 1.41(1.17)
		24	0.51 – 0.96(0.69)	2.08 – 2.88(2.38)
	10^5	10	0.26 – 0.51(0.34)	0.81 – 1.28(1.00)
		24	0.43 – 0.82(0.58)	1.77 – 2.50(2.04)

Table 1. CPU time [s] (max – min (average))

- The rate limit for elevator command does not have so large impact on GA performance except for the cases using small R.

This fact is reasonable because it is difficult to suppress high frequency turbulence effect even when there are no measurement errors in the turbulence data (see Fig. 3), and if R is set relatively large, e.g. $R = 10^3$, then MuPAL-α has no need to use high rate elevator commands for suppressing middle frequency turbulence effect, such as, 1.0 [rad/s] (see Fig. 6). However, if R is set small then the proposed GA flight controller allows high rate elevator commands, which lead to severe oscillatory accelerations. Thus, GA performance deteriorates.

Finally, CPU time to solve Proposition 2 is shown in Table 1. The simulation setting is the same as for obtaining the results in Fig. 5. The simulations are conducted with Matlab® using SeDuMi (Sturm, 1999) along with a parser YALMIP (Löfberg, 2004) with a PC (Dell Precision T7400, Xeon®3.4 GHz, 32 GB RAM; PT7400) and a PC (Dell Precision 650, Xeon®3.2 GHz, 2 GB RAM; P650). Although CPU time with PT7400 is just about 30 % of P650, at the present moment, solving Proposition 2 online is impossible with these PCs even when N is set as 10. Thus, the reduction of numerical complexity for solving Proposition 2 is to be investigated.

4.2 Large aircraft example

Let us next consider the linearized longitudinal aircraft motions of large aircraft Boeing 747 (B747) (Heffley & Jewell, 1972) at an altitude of 12192 [m] and a true air speed of 236 [m/s].

4.2.1 Simulation setting

Similarly to MuPAL-α, it is supposed that only the elevator is used for aircraft motion control. The transfer function of its actuator dynamics is also supposed to be modeled as $1/(0.1s + 1)$. Then, the continuous-time system representing the linearized longitudinal motions with the modeled actuator dynamics is given as (1), where the state, the turbulence, the control input, and the performance output are the same as MuPAL-α.

After the discretization of (1) with sampling period T_s [s] being set as 0.1 using a zero-order hold, the discrete-time system (5) is given as (36).

$$\left[\begin{array}{c|c|c} A & B_1 & B_2 \\ \hline C & D_1 & D_2 \end{array}\right]$$
$$= \left[\begin{array}{ccccc|c|c} 0.99965 & 4.1350 \times 10^{-3} & -1.8544 & -0.97754 & 0.10639 & 4.1498 \times 10^{-3} & 0.044513 \\ -5.6762 \times 10^{-3} & 0.96481 & 22.643 & -0.074565 & -1.3157 & -0.034623 & -0.55260 \\ 6.5386 \times 10^{-5} & -3.1960 \times 10^{-4} & 0.95423 & 1.0244 \times 10^{-5} & -0.071212 & -3.2381 \times 10^{-4} & -0.041894 \\ 3.2771 \times 10^{-6} & -1.6194 \times 10^{-5} & 0.097759 & 1.0000 & -4.1894 \times 10^{-3} & -1.6405 \times 10^{-5} & -1.5116 \times 10^{-3} \\ 0 & 0 & 0 & 0 & 0.36788 & 0 & 0.63212 \\ \hline -0.065004 & -0.32122 & -5.3617 \times 10^{-3} & -5.2561 \times 10^{-3} & -5.5100 & -0.31497 & 0 \end{array}\right] \tag{36}$$

The delay T_d for elevator command, the constraints for the augmented state, the control input command deviation and the performance output, matrices Q and S in (16), the turbulence, and the measurement errors are all set the same as for MuPAL-α.

4.2.2 Simulation results

The same numerical simulations in section 4.1.3 but the aircraft motion model being replaced by the B747 model are carried out for the following parameter setting.

$$\begin{aligned} \hat{t}_d &\in \{1,\ 2,\ 3,\ 4\}, \\ \omega &\in \{0.1,\ 0.5,\ 1.0,\ 3.0,\ 4.0,\ 5.0,\ 6.0\}, \\ R &\in \{10^{-1},\ 10^{0},\ 10^{1},\ 10^{2},\ 10^{3},\ 10^{4},\ 10^{5},\ 10^{6},\ 10^{7}\}, \\ N &\in \{10,\ 12,\ 14,\ 16,\ 18,\ 20,\ 22,\ 24\}. \end{aligned} \tag{37}$$

Fig. 8 shows the performance comparison for scenarios A, B and C. In this figure, J_A, J_B and J_C are similarly calculated as in Fig. 5.
The following are concluded from Fig. 8.

- For turbulence, whose frequencies are no more than 0.1 [rad/s], GA performance using the proposed method is larger than the uncontrolled case.
- For turbulence, whose frequencies are more than 5 [rad/s], vertical acceleration is hardly reduced even if prior turbulence data are obtained.
- It is sufficient for B747 to measure turbulence for 20 steps ahead.
- Using an appropriately chosen R (e.g. $R = 10^3$), the proposed GA flight controller in which Proposition 2 is solved online improves GA performance for middle frequency turbulence, such as, $0.5 \sim 4$ [rad/s].

Differently from MuPAL-α, low frequency turbulence effect to B747, e.g. 0.5 [rad/s], can be reduced by the proposed GA controller; however, middle frequency turbulence effect, e.g. 5 [rad/s], cannot be reduced. The third item is interesting in a sense that the step number

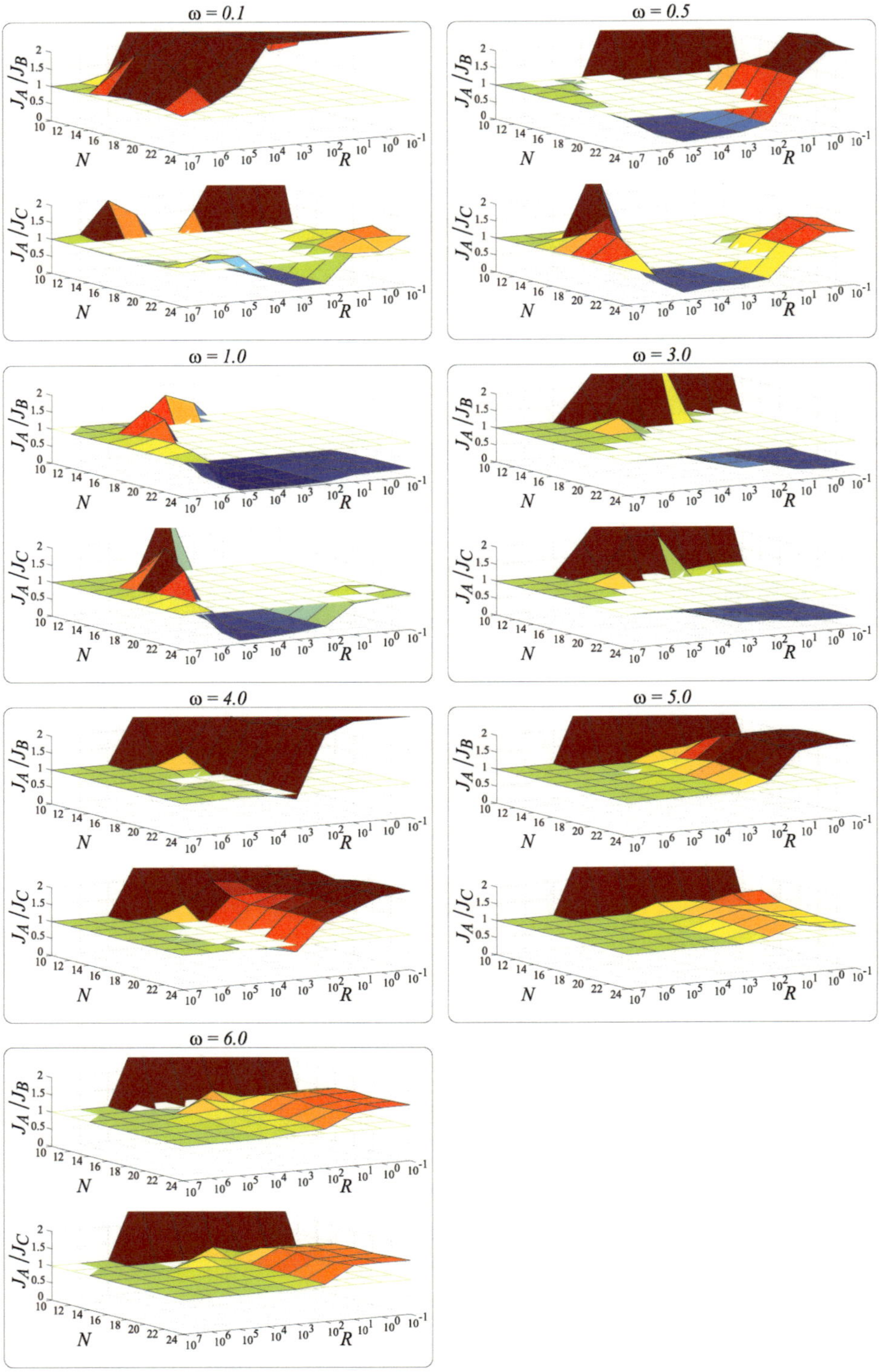

Fig. 8. GA performance comparison for B747 under measurement errors in turbulence data

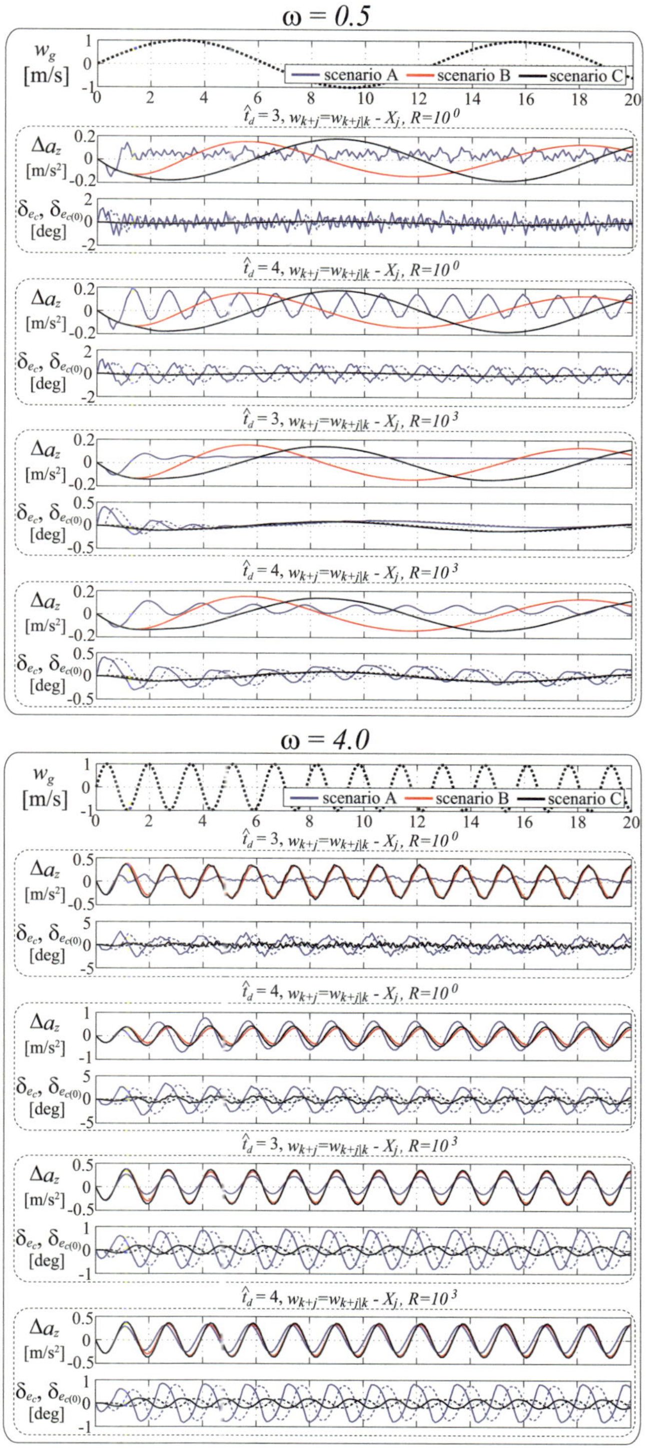

Fig. 9. Time histories for B747 under measurement errors in turbulence data with $N = 18$ (δ_{e_c} is shown as dotted lines and $\delta_{e_c(0)}$ is shown as solid lines)

for turbulence measurement does not depend on aircraft models. However, it is not sure that this fact indeed holds for other aircraft, which is to be investigated.
For reference, several time histories with $N = 18$, $R = 10^0$ or 10^3, and $\hat{t}_d = 3$ or 4 are shown in Fig. 9. For space problem, only actual elevator deflection command (δ_{e_c}) and its created command by flight computer ($\delta_{e_c(0)}$), and performance output are shown. These figures confirm that small R, i.e. $R = 10^0$, allows large elevator deviation commands and this consequently leads to severely oscillatory vertical accelerations. However, the proposed controller with an appropriately chosen R, i.e. $R = 10^3$, well suppresses turbulence effect on aircraft motions.

5. Conclusions

This paper tackles the design problem of Gust Alleviation (GA) flight controllers exploiting *a priori* measured turbulence data for suppressing aircraft motions driven by turbulence. For this problem, a robust Model Predictive Control (MPC) considering the plant uncertainties and the measurement errors in the turbulence data is proposed. In the usual setting, MPC for uncertain plant requires to solve an optimization problem with infinitely many conditions if conservatism is avoided. However, it is shown that if the plant uncertainties are represented as the bounded time-invariant uncertain delays at the control input, then the associated problem for the robust MPC is equivalently transformed to an optimization problem for finitely many plant models, which consequently means that the optimization problem has finitely many conditions.
In our problem setting, the measurement errors in the *a priori* measured turbulence data are represented as affine with respect to a constant uncertain vector, whose elements are all bounded. Using this property, it is shown that it is necessary and sufficient to evaluate the performance index in MPC at the maxima and minima of the uncertain vector. This consequently means that the robust MPC has finitely many conditions even when the measurement errors are considered.
Several numerical examples illustrate that the proposed GA flight controller with appropriately chosen controller parameters effectively suppresses turbulence effect on aircraft motions, and reveal that it is very difficult to suppress high frequency turbulence effect even when the *a priori* measured turbulence data are exploited.
To guarantee the feasibility of the proposed MPC at every step is an important issue for the implementation of the proposed method to real systems. Thus, this topic is now under investigation.

6. References

Abdelmoula, F. (1999). Design of an open-loop gust alleviation control system for airborne gravimetry, *Aerospace Science and Technology* Vol. 3(No. 6): 379–389.
Ando, T., Kameyama, S. & Hirano, Y. (2008). All-fiber coherent doppler LIDAR technologies at mitsubishi electric corporation, *14th Int. Sympo. for Advancement of Boundary Layer Remote Sensing*.
Badgwell, T. A. (1997). Robust model predictive control of stable linear systems, *Int. J. Control* Vol. 68(No. 4): 797–818.

Bemporad, A. & Morari, M. (1999). *Robustness in Identification and Control*, Springer Verlag, Berlin, chapter Robust Model Predictive Control: A Survey. Lecture Notes in Control and Information Sciences 245.

Botez, R. M., Boustani, I., Vayani, N., Bigras, P. & Wong, T. (2001). Optimal control laws for gust alleviation, *Canadian Aeronautics and Space Journal* Vol. 47(No. 1): 1–6.

Boyd, S. & Vandenberghe, L. (2004). *Convex Optimization*, Cambridge University Press, Cambridge.

Fujita, M., Hatake, K. & Matsumura, F. (1993). Loop shaping based robust control of a magnetic bearing, *IEEE Control Systems Magazine* Vol. 13(No. 4): 57–65.

Heffley, R. K. & Jewell, W. F. (1972). *Aircraft Handling Qualities Data*, National Aeronautics and Space Administration, Washington, D.C.

Hess, R. A. (1971). Optimal stochastic control and aircraft gust alleviation, *Journal of Aircraft* Vol. 8(No. 4): 284–286.

Hess, R. A. (1972). Some results of suboptimal gust alleviation, *Journal of Aircraft* Vol. 9(No. 5): 380–381.

Inokuchi, H., Tanaka, H. & Ando, T. (2009). Development of an onboard doppler lidar for flight safety, *J. Aircraft* Vol. 46(No. 4): 1411–1415.

Jenaro, G., Mirand, P., Raymond, M., Schmitt, N., Pistner, T. & Rehm, W. (2007). Airborne forward looking lidar system, *Int. Forum on Aeroelasticity and Structural Dynamics*. IF-088.

Kothare, M. V., Balakrishnan, V. & Morari, M. (1996). Robust constrained model predictive control using linear matrix inequalities, *Automatica* Vol. 32: 1361–1379.

Kwon, W. H. & Han, S. (2005). *Receding Horizon Control: model predictive control for state models*, Springer-Verlag, London.

Löfberg, J. (2003). *Minimax Approaches to Robust Model Predictive Control*, PhD thesis, Linköping University, Linköping, Sweden.

Löfberg, J. (2004). YALMIP: A toolbox for modeling and optimization in MATLAB, *Proc. the CACSD Conference*, Taipei, Taiwan.
URL: *http://control.ee.ethz.ch/ joloef/yalmip.php*

Mehra, R. K., Amin, J. N., Hedrick, K. J., Osorio, C. & Gopalasamy, S. (1997). Active suspension using preview information and model predictive control, *Proc. CCA*, pp. 860–865.

Military Specification: Flight Control Systems - Design, Installation and Test of Piloted Aircraft, General Specification For (1975). MIL-F-9490D.

Miyazawa, Y. (1995). Design with multiple-delay-model and multiple-design-point approach, *J. Guidance, Control, and Dynamics* Vol. 18(No. 3): 508–515.

Ohno, M., Yamaguchi, Y., Hata, T, Takahama, M., Miyazawa, Y. & Izumi, T. (1999). Robust flight control law design for an automatic landing flight experiment, *Control Engineering Practice* Vol. 7: 1143–1151.

Phillips, W. H. (1971). *Gust Alleviation*, National Aeronautics and Space Administration, Washington, D.C., pp. 505–553. NASA SP-258.

Rynaski, E. G. (1979a). Gust alleviation - criteria and control laws, AIAA. AIAA Paper 1979-1676.

Rynaski, E. G. (1979b). Gust alleviation using direct turbulence measurements, AIAA. AIAA Paper 1979-1674.

Santo, X. D. & Paim, P. K. (2008). Multi-objective and predictive control - application to the clear air turbulence issues, *AIAA Guidance, Navigation and Control Conference*, AIAA. AIAA Paper 2008-7141.

Sato, M. & Satoh, A. (2008). Simultaneous realization of handling and gust responses: In-flight simulator controller design, *J. Guidance, Control, and Dynamics* Vol. 37(No. 6): 1545–1560.

Schmitt, N. P., Rehm, W., Pistner, T., Zeller, P., Diehl, H. & Navé, P. (2007). The AWIATOR airborne LIDAR turbulence sensor, *Aerospace Science and Technology* Vol. 11: 546–552.

Sturm, J. S. (1999). Using SeDuMi 1.02, a MATLAB toolbox for optimization over symmetric cones, *Optimization Methods and Software* Vols. 11-12: 625–653.

Takaba, K. (2000). Robust servomechanism with preview action for polytopic uncertain systems, *Int. J. Robust and Nonlinear Control* Vol. 10: 101–111.

Tomizuka, M. (1976). Optimum linear preview control with application to vehicle suspension–revisited, *J. Dynamic Systems, Measurement, and Control* Vol. 98(No. 3): 309–315.

Xie, L. & de Souza, C. E. (1992). Robust H_∞ control for linear systems with norm-bounded time-varying uncertainty, *IEEE Trans. Automatic Control* Vol. 37(No. 8): 1188–1191.

Xie, L., Fu, M. & de Souza, C. E. (1992). H_∞ control and quadratic stabilization of systems with parameter uncertainty via output feedback, *IEEE Trans. Automatic Control* Vol. 37(No. 8): 1253–1256.

Zhou, K., Doyle, J. C. & Glover, K. (1996). *Robust and Optimal Control*, Prentice-Hall, Upper Saddle River, NJ.

MBPC – Theoretical Development for Measurable Disturbances and Practical Example of Air-path in a Diesel Engine

Jose Vicente García-Ortiz
University Jaume I and Vinci Energía
Spain

1. Introduction

Model based predictive controller is a family of control algorithms suitable for some applications. They are useful for most of the real applications. But they are not used very often due to the complexity of them. So this kind of books should be closer to the real application in order to be more used. One of the main objectives, in this chapter, is to clarify the use of the MBPC for real applications.
My experience is based on real application of MBPC. Additionally they are also suitable for small modification. One of the adaptations explained in this chapter is the one related with disturbance measurement. All real systems have disturbance. Many times the disturbance can be measured and identify. The disturbance measurable can be identified but you can not use this variable as a control variable. You can include this information in the optimization process. One the examples explained in this chapter is the management of this information. In the air management loop in a internal combustion engine use to manage VNT (Variable Nozzle turbine), EGR (Exhaust gas recirculation), SWIRL valve and others, but mass fuel injected is managed by air management loop. But this mass fuel has important influences in the air management process, so it should be considered.
The philosophy considering disturbance measurable has influences on different parts of the control algorithm development. MBPC has 4 main topics: model, control cost, optimization process and receding horizon. In this chapter the development will be explained.
Due to the high quantity of control parameters considered in the MBPC, it use to be rejected for using. In this chapter we will try to explain the influence of the most important parameters in order to show the way for tuning.

2. State of art

Control Engineering Science is a very brand new Science branch compared with other sciences. In fact the first servo system of the history was developed in 1788 by James Watt. But the strong theoretical development was done during 20th century. Additionally the control engineering science has been developed as a auxiliary tool for other real applications. One example is the development was SMC (Sliding Mode Control), which was developed to control artificial satellites.

Probably this is the main reason because, control engineering, is mainly considered as a secondary science. In this chapter MBPC will be explained, mainly the GPC (Generalized Predictive Control).

2.1 Control engineering

As explained in 1788, James Watt developed the first servo system of the history but the real development of this science was during 20th century. The first important mathematical development were done by Nyquist[27], this work studied stability of the system by means of closing the loop. One of the problems when closing the loop is the possibility of instability, that is why Nyquist developed the theory in 1932. During forties some techniques of frequency design were developed, thus control engineering were widely implemented in the Industry. During fifties Evans[14] proposed the root locus for designing algorithms with stability. From sixties the digital hardware were getting importance so digital or disserted algorithms were implemented. With discrete systems more flexible algorithms and structure were used, even thus the first approached were done by discrimination of classical controllers.

2.2 Control algorithms MBPC history

MBPC has its origin in Kalman's work. He developed an observer in which design an optimization cost is used. Additionally the optimal control based on the optimization of a quadratic cost. This control was the LQG. In comparison with Evans[14] theory, the behaviour of the controlled system is not base on system response, but it is based on the minimum cost of a quadratic cost function.

The first MBPC published was in 1976 by Richalet[33]. He formalized the development two years later in the Paper []. This algorithms is known as MPHC (Model Predictive Heuristic Control), which was commercially developed as IDCOM, this commercial software included identification, predictive controller and impulsion model. One year later Cutler and Ramaker[12] developed de DMC (Dynamic Matrix Controller) based also in the impulsion system response. Both algrothm were known as first generation MPC (Model Predictive Controllers), but neither IDCOM nor DMC were based in LQG models or state space models.

During first half of eighties the second generation of algorithms was developed. DMC with constraints were presented by García[16] by using Quadratic Programming QP in the optimization process. Connoisseur from Invensys Corporation is also an algorithm of this family.

The most important advance in MBPC were developed during second half of eighties and nineties were third generation appeared. In 1988 Keyser[20] compared some MBPC adaptative. Shell company worked in the first MBPC based in state space, its name is SMOC. An overview of MBPC was summarized by Qin[31,32], not only theoretical controllers but most of the commercial solutions were analyzed. In 1986 Clarke[8,9,10] presented the GPC or Generalized Predictive Controller. Clarke tried to develop a MBPC unifying all controllers in one algorithm. The way for getting this target is to use a general model, as general as possible, thus he decided to use the CARIMA model, this is a standard ARX plus a derivative white noise. Another proposal was done by Soeterbock[37] in which adaptative controller as general as possible. In 1990 Zafirion [42] exposed the possibility of destabilize the system by including constraints in the optimization process. This fact should be considered, because a stable system and stable controller can be destabilized by mean of including the constraints.

Galanti[15] proposed, two years later, an interesting solution for constrained by dead zone. But Zheng[43] proposed an intelligent development for solving that problem. It consists on developing soft constraints in place of constraints. The system and controller must be under the constraints, I mean, it is something compulsory. But the soft constraint is not a constraint anymore, this is a recommendation, so id you are out the soft constraint for a while, this is not an alarm, this is a warning. Thus the stability of the system is stronger, this kind of soft constraints are suitable for outout.

In 1997 Rositer and Kouvaritakis[22] proposed the terminal constraint in order to improve the stability when constraints in MBPC. A similar work was presented in Automatica by Havlena [18] where stability of optimum control when constraints were analyzed. That year some people[36] presented studies about tuning in MBPC for SISO systems.

Two years later in 1999 Bordons and Camacho[2] summarized the state of the art in their book. They mainly focus the study in a wide development of GPC as a General proposal of MBPC.

During century XXI MIMO approaches of MBPC were developed. Nuñez[26] proposed Hierarchy control, well accepted by Industry. An additional problem when MIMO MBPC approach is carried out, is instability due to problems in matrix as Pannochia[29] suggested. Poulser and Kouvaritakis[30] exposed some real application in chemical reactor in which MBPC offer its possibilities. Moreover, Kouvaritakis[21] presented different optimization tools in place of quadratic programming.

An special case in the MBPC is the GPC. Due to special characteristics is one of the most developed controller. It was published in 1987 by Clarke, in these papers the linear controller was proposed[7,8] . Thi first idea was to find a controller with weak computational cost to be used in adaptative applications. In fact, Clarke and Gawthrop[5,6] developed the GMV or controller of minimum variance in 1975 and 1979[5,6]. GPC inherited from GVM the prediction model CARIMA. CARIMA is based in a transfer function plus a white noise derivative model. All previous cotroller use to find some problems:

1. Minimum phase zeros process can not be used.
2. Instable process controlled by open loop architecture.
3. Robustness when delays or not expected delays.
4. Robustness when poles zeros were over stimulated.

Not in all contemporaneous controller appeared that problems, but in most of them. CARIMA model include a derivative model which made this model suitable for real applications. GPC contains the main bases of the MBPC.:

1. Prediction model: CARIMA model, what have some interesting proprieties.
2. Cost Index: with quadratic error in a time slot or prediction horizon plus a control horizon or steps of control action.
3. Optimization tool: at this point the GPC is using the standard optimization tool, roots in the first derivative and checking the second one is positive.
4. Mobil horizon: it mean applying the first control action and recalculate every step updating the new information.

During eighties the team of Clake were developing GPC. In 1988 Tsang[40] in this team explained the importance of using constraints. Once this point, two problems should be solved: include constraint in the optimization and improving stability in close loop.

In 1992 few paper about robustness in GPC were published. Soeterboek[38] tried to generalize the MBPC in one general book. The importance of T polynomial (noise polynomial poles) were widely explained, specially about robustness of the algorithm. Moreover the correct tuning of lambda parameter (or weight of control actions) in the GPC robustness. That year Kouvaritakis[21] published a paper about GPC Stability and Mosca[25] proposed a control solution about constraints in dead zone. This year Albertos[1] published a new formulation on the model based in Diophantine decomposition.

In 1994 a paper published by Clarke team is presented. Chow[4], the author of this paper, proposed some constraints applied to GPC which are: slew rate constraint, Coulomb friction, histeresys and dead zone by using two optimization algorithms; quadratic programming and Lagrange multipliers. In the University of Seville was presented by Camacho[3] an empirical approach of GPC. It consisted in using standard linear model (normalized) and setting up different model and different GPC every control area, thus a gain Schedule of GPC was proposed.

One of the most important problems available in Industrial application is the disturbance. GPC can carry out this problem due to the CARIMA model. In this field Beg in 1995, published an application of controlling a MIMO system by mean of SISO controllers and feed forward architecture.

In 1996 an approach with heuristic algorithm, genetic algorithms as a optimization tool in GPC scheme. This proposal was developed by CPOH[24], a R+D team in the UPV (Politecnical University of Valencia). In 1997 Corradini[11] proposed as hybrid solution between SMC (Sliding mode control) and GPC, in order to improve the stability.

During this century the papers in this field are focused in real application if MBPC and GPC, in particular, or Industrial applications. The team of Seville University published a high level GPC MIMO controller and PID controllers in low level, the same was done by Kennel[19]. Salcedo[34] from CPOH, published a paper in which CARIMA model is changed by state space CARIMA model. Thus the algorithm is faster, the computational effort is lower and less storage information is needed, but an state observer must be used.

In 2002 Perez de la Parte[13] from Seville, tried to improve the robustness by the same way as Corradini. Sanchis[35] used another solution for improving the robustness, he proposed the use of singular values decomposition SVD.

2.3 Planning

When MBPC is selected to be the controller for a system, some steps must be done. Firstly the system must be studied. Secondly the model system must be obtained. Thirdly the control must be designed. Fourthly the control should be implemented and tuned.

The system must be studied. We have to discover the main variables, inputs, outputs and disturbance. Moreover the main dynamics and behavior of the system should be considered.

Getting the model of the system should be done. The model is one of the key in the MBPC. Getting a good quality model is sometimes the difference between success and failure.

After that, control architecture and design should be done. Finally tuning the algorithm and checking parameters as robustness or others must be done.

3. Model based predictive controller

In this section the control algorithms will be developed. MBPC are based in 4 different beams. One of the most important is the system model. Other important tool is the optimization process.

Many time the system has a, more or less, linear behaviour. At this time a linear model uses to be a good approach. Linea model has the main advantages of the small computational cost. Specially when constraint are not considered. The optimization can be carried out off-line. Thus the control action is a combination of multiplications and additions. So the computational effort into a microcontroller is very small.

Sometimes the system has a nonlinear behaviour. Then linear approach can not be carried out by its selves. A non-linear model with constraints use to cost the highest computational effort, but with a high accurate control behaviour.

Sometimes linear approaches in gaps use to be an intelligent solution. It consists on using different linear models around linear system behaviour, and changing the model when the system is out of its gap of influence of that model. Thus an offline optimization can be carried out and the computational effort, during the control action, is very low. It looks like gain schedule but, in place of PID parameters, GPC controllers are used.

Thus a standard GPC controller is not going to be developed because it can be obtained by simplifying the GPC with disturbance measurement.

The system use to be influenced by some parameters. Those parameters many times could be considered as measurable disturbance. A measurable disturbance is an input of the system that is not controlled by the control algorithm. For instance, in a Diesel engine, the air management loop control VNT (variable nozzle turbine) and the EGR system. Mass fuel injected is controlled by the user. But the mass fuel injected in the engine affects quite a lot in the air management behaviour. Thus this important input has to be considered as measurable disturbance.

These kinds of inputs need a special management in the algorithm. So this chapter shows one way to consider them into the algorithm, for past disturbance and also for future disturbances. Two different algorithms will be developed with this approach, the one based in the linear Markov parameters and the GPC (Generalized Predictive Controller) developed by Clarke.

3.1 GPCDM, algorithm development

We now briefly describe the control algorithm which is key issues of this paper. The control algorithm was based on the model shown in the Figure 1 in which there is the CARIMA model plus an input p which is a measurable disturbance. To simplify the development one input and one output are considered.

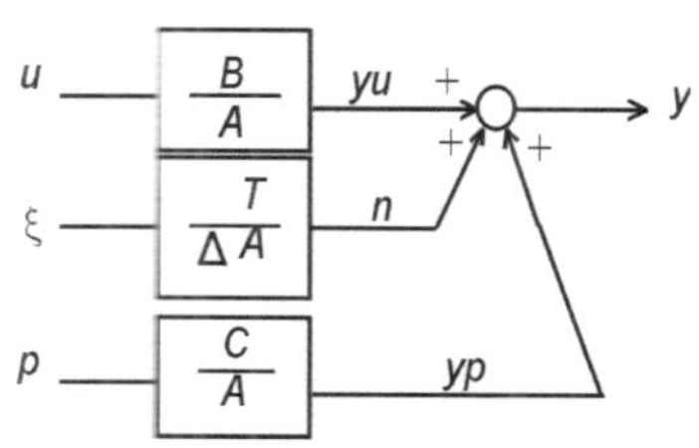

Fig. 1. CARIMA model proposed.

$$A \cdot y(z) = B \cdot u(k) + C \cdot p(k) + \frac{T}{\Delta} \cdot \xi(k)$$

$$y(k+i) = y_u(k+i) + y_P(k+i) + n(k+i) \tag{1}$$

Equation (1) is the discrete one of the model shown in figure X. The yu is the output behaviour depending on the direct input. yp is output behaviour from measurable disturbance and n is the behaviour depending on few things like non-linearity, others inputs etc.

$$n(k+i \mid k) = \frac{T}{\Delta \cdot A} \cdot \xi(k+i) \tag{2}$$

n(k+i|k) is the n prediction of (k+i) with the information in k sample time. Diophantine decomposition is applied in order to get the system response due to pass inputs and future inputs in different terms. Moreover the mean of ξ is zero, then:

$$n(k+i \mid k) = \xi(k+i) \cdot \frac{T(z^{-1})}{\Delta \cdot A} = E_i \cdot \xi(k+i) + \frac{F_i}{\Delta \cdot A} \cdot \xi(k) \tag{3}$$

and n(k+i|k) is at the same time:

$$n(k+i \mid k) = \frac{F_i}{T}[y(k) - y_u(k) - y_p(k)] \tag{4}$$

We do the same with y as a result we have:

$$y(k+i \mid k) = \Delta u(k+i-1) \cdot \frac{B \cdot E_i}{T} + \Delta p(k+i-1) \cdot \frac{C \cdot E_i}{T} + \frac{F_i}{T} \cdot y(k) \tag{5}$$

$$\begin{aligned} G_i &= B \cdot E_i \\ GP_i &= C \cdot E_i \end{aligned} \tag{6}$$

We can apply Diophantine decomposition to equation (6) and finally we can get:

$$\begin{aligned} \frac{G_i}{T} &= G_i^{|} + \frac{\Gamma_i . z^{-1}}{T} \\ \frac{GP_i}{T} &= GP_i^{|} + \frac{\Gamma P_i . z^{-1}}{T} \end{aligned} \tag{7}$$

With that we can develop the equation (5) for future prediction and sum up all terms in an equation

$$\mathrm{y} = \mathrm{G} \cdot \mathrm{u} + \mathrm{GP} \cdot \mathrm{p} + \Gamma \cdot \mathrm{u}^{\mathrm{f}} + \Gamma\mathrm{P} \cdot \mathrm{p}^{\mathrm{f}} + \mathrm{F} \cdot \mathrm{y}^{\mathrm{f}} \tag{8}$$

Where: ***y*** is the predicted system output vector; model matrices are represented by ***G, Γ, ΓP, GP***, and ***F*** (***G*** and ***GP*** represent the future response; and ***Γ, ΓP*** and ***F*** represent the past system response or natural response); Δ is the incremental operator *($1-z^{-1}$)*; ***u*** is the vector of future control action increment; ***p*** is the disturbance increment vector; the past action control

filtered by $1/T$ is u^f ; the past disturbance filtered by $1/T$ is p^f ; the past system output filtered by $1/T$ is y^f.
Equation (8) is the system model in a matrix form. More over we need the control cost which depend on error cost and control effort, represented in (9) as a matrix equation. Where y_{sp} is a vector of future set points, λ is a diagonal matrix of control effort factor and α is the diagonal matrix of error cost.

$$J=(y-y_{sp})\cdot\alpha\cdot(y-y_{sp})+u^T\cdot\lambda\cdot u \tag{9}$$

We put the model in the control cost, then using an optimization tool, like (10).

$$\frac{\partial J}{\partial\Delta u}=0 \tag{10}$$

Eventually the result is given in next equation (11)

$$\bar{u}=(G^T\cdot\alpha\cdot G+\lambda)^{-1}\cdot G^T\cdot\alpha\cdot(y_{sp}-\Gamma\cdot u^f-\Gamma P\cdot p^f-GP\cdot p-F\cdot y^f) \tag{11}$$

It is also possible to calculate the first value of equation (11) and using it like a combination of discrete time transfer functions. More details can be studied in Garcia-Ortiz[17] where the algorithm is widely explained and Camacho & Bordons[3].
There are other ways to reach the same result. In terms of model in figure 1, may be considers ad a MISO system (Multiple input single output). Getting ecuations from MIMO CARIMA developed models we can find:

$$\mathrm{y}=\mathrm{G}\cdot\mathrm{u}+\Gamma\cdot\mathrm{u}^{\mathrm{f}}+\mathrm{F}\cdot\mathrm{y}^{\mathrm{f}} \tag{12}$$

Where MISO system is:

$$y=(G_{11}\quad G_{12})\cdot\begin{pmatrix}\Delta u_1\\ \Delta u_2\end{pmatrix}+(\Gamma_{11}\quad\Gamma_{12})\cdot\begin{pmatrix}u_1^f\\ u_2^f\end{pmatrix}+F_1\cdot y^f \tag{13}$$

Thus G_{11} is the G in equation (8) and G_{12} is GP in equation (8). So is very easy to see that both ways get the same result.

3.2 DMCDM, algorithms development

A similar approach can be developed for DMC controller. The differences are that prediction model is based in Markov coefficients.

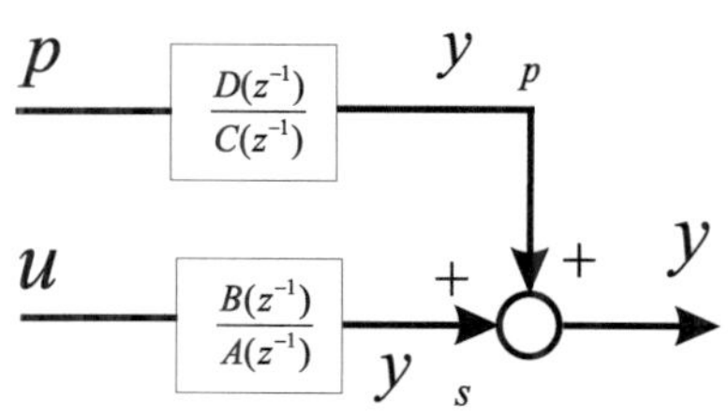

Fig. 2. Box Jenkins model. Model based for impulse response with measurable disturbances.

$$y(k)=\sum_{i=1}^{\infty} g_i \cdot u_{k-i} \tag{14}$$

Where g0 = 0 due to the delay in the discrete systems and gN = 0. From gN to gN+M=0 whatever M. from this point all systems controlled by DMC should be type 0. All systems must be stable and integer term is not allowed. Most of the industrial processes have these characteristic, which is why this controller have been developed widely in the industrial applications.

$$y(k)=\sum_{i=1}^{\infty} s_i \cdot \Delta u_{k-i} \tag{15}$$

Thus there are a equivalence between impulsion coefficients and step ones.

$$s_i=\sum_{j=1}^{i} g_i \tag{16}$$

So during the development both models will be used, depending the interest every time. Additionally some information will be defined:
P: prediction horizon. Number of time steps will be considered in the cost index and the optimization process.
M: Control Horizon. How many control actions are allowed in the optimizer to reach the target. From M to the future no movements in the control action are considered.
N: number of coefficients with information in the impulse response. From N to the future g coefficient is zero.
All that information will be used in the control cost(16).

$$J=\sum_{i=1}^{P} \psi_i^2 \cdot [y(k+i\,|\,k)-y_{SP}(k+i)]^2+\sum_{i=0}^{M-1} R_{i+1}^2 \cdot (\Delta u(k+i))^2 \tag{16}$$

Where y(k+i|k) is the value of the prediction for the step k+i, calculated in the instant k with the information available in that time k. ysp(k+i) is the set point in the step k+i (when future set point is not available, this variable is considered as the set point in k. Ψ2 is the weight of error cost and R2,i+1 is the weight of the control effort. Δu(k+i) is the increment of the control action in time=k+i.
From equation (15) we obtain:

$$y(k)=\sum_{i=0}^{\infty} s_i \cdot \Delta u(k-i)=\sum_{i=0}^{N} s_i \cdot \Delta u(k-i)+s_N \cdot \Delta u(k-N-1) \tag{17}$$

Thus we are ready to get the prediction model in the DMC. The main target is getting the future response of the system from one step time and the behaviour if the system when future control actions will be applied. The model will contain the system natural response, the one due to the past control actions applied in the system (this one is not changeable, because is produced) and the forced response, the one due to future control action (this one can be different because will be decided by the controller).
Once the bases are fixed we will start with the model development, thus we will get the system response in a generic step k+t, from the information available in k.

$$\begin{aligned}
+y(k+1) &= g_1 \cdot u(k) + g_2 \cdot u(k) + \cdots + g_N \cdot u(k-(N-1)) \\
\underline{-y(k)} &\underline{= g_1 \cdot u(k-1) + g_2 \cdot u(k-2) + \cdots + g_N \cdot u(k-N)} \\
y(k+1 \mid k) &= g_1 \cdot \Delta u(k) + g_2 \cdot \Delta u(k-1) + \cdots + g_N \cdot \Delta u(k-(N-1)) + y(k)
\end{aligned} \tag{18}$$

In the second iteration we can obtain:

$$\begin{aligned}
&y(k+2 \mid k) = g_1 \cdot u(k) + g_2 \cdot u(k) + \cdots + g_N \cdot u(k-(N-2)) \\
&y(k+1) = g_1 \cdot u(k) + g_2 \cdot u(k-1) + \cdots + g_N \cdot u(k-(N-1)) \\
&y(k+2 \mid k) = g_1 \cdot \Delta u(k+1) + g_2 \cdot \Delta u(k) + \cdots + g_N \cdot \Delta u(k-(N-2)) + y(k+1) \\
&y(k+2 \mid k) = g_1 \cdot \Delta u(k+1) + (g_1 + g_2) \cdot \Delta u(k) + (g_2 + g_3) \cdot \Delta u(k-1) + \cdots \\
&\cdots + (g_{N-1} + g_N) \cdot \Delta u(k-(N-2)) + g_N \cdot \Delta u(k-(N-1)) + y(k)
\end{aligned} \tag{19}$$

By deducting generic element:

$$\begin{aligned}
&y(k+t \mid k) = g_1 \cdot u(k+t-1) + g_2 \cdot u(k+t-2) + \cdots + g_N \cdot u(k+t-N) \\
&y(k+t-1 \mid k) = g_1 \cdot u(k+t-2) + g_2 \cdot u(k+t-3) + \cdots + g_N \cdot u(k+t-N-1) \\
&y(k+t \mid k) = g_1 \cdot \Delta u(k+t-1) + g_2 \cdot \Delta u(k+t-2) + \cdots \\
&\cdots + g_N \cdot \Delta u(k+t-N) + y(k+t-1)
\end{aligned} \tag{20}$$

Getting all the information and summarizing the data in a matrix ecuation:

$$\begin{bmatrix} y(k+1) \\ y(k+2) \\ \vdots \\ y(k+M) \\ \vdots \\ y(k+P) \end{bmatrix} = S_f \cdot \begin{bmatrix} \Delta u(k) \\ \Delta u(k+1) \\ \vdots \\ \Delta u(k+M-1) \end{bmatrix} + S_p \cdot \begin{bmatrix} \Delta u(k-1) \\ \Delta u(k+2) \\ \vdots \\ \Delta u(k-(N-1)) \end{bmatrix} + y(k) \cdot L_p \tag{21}$$

Where $L_p = (1 \quad \cdots \quad 1)^T_{Px1}$ and:

$$S_f = \begin{pmatrix} s_{\min(1,N)} & 0 & \cdots & 0 \\ s_{\min(2,N)} & s_{\min(1,N)} & \cdots & 0 \\ \vdots & \ddots & \ddots & 0 \\ s_{\min(M,N)} & s_{\min(M-1,N)} & \cdots & s_{\min(1,N)} \\ \vdots & \vdots & \ddots & \vdots \\ s_{\min(P,N)} & s_{\min(P-1,N)} & \cdots & s_{\min(N+P-1,N)} \end{pmatrix}_{Px(N-1)} \quad \text{Matrix Sf}$$

$$S_P = \begin{pmatrix} s_{\min(2,N)} - s_1 & \cdots & s_{\min(N,N)} - s_{N-1} \\ s_{\min(3,N)} - s_1 & \cdots & s_{\min(N+1,N)} - s_{N-1} \\ \vdots & \ddots & \vdots \\ s_{\min(P+1,N)} - s_1 & \cdots & s_{\min(N+P-1,N)} - s_{N-1} \end{pmatrix}_{Px(N-1)} \quad \text{Matrix Sp}$$

At this time we are able to get future response from the data available in the step k, but we have to include system information every step time so: d(k) = yreal(k) - y(k), this is the difference between real system and prediction.
Finally yf is an array of output system predictions. Δuf is an array of future increment control actions. Δup is an array of past control actions. Sf is a matrix with dynamic future system response or forced response matrix and Sp is a matrix with past dynamic system response or natural response matrix.

$$y_f = S_f \cdot \Delta u_f + S_P \cdot \Delta u_P + y_{real}(k) \cdot L_P \tag{22}$$

Applying this prediction model to the control cost:

$$J = (y_f - y_{sp})^T \cdot \psi^T \cdot \psi \cdot (y_f - y_{sp}) + \Delta u_f^T \cdot R^T \cdot R \cdot \Delta u_f \tag{23}$$

$$\begin{aligned} J = ((S_f \cdot \Delta u_f + S_P \cdot \Delta u_P + y_{real}(k) \cdot L_P) - y_{sp}) \cdot \psi^T \cdot \psi \cdot \\ ((S_f \cdot \Delta u_f + S_P \cdot \Delta u_P + y_{real}(k) \cdot L_P) - y_{sp}) + \Delta u_f^T \cdot R^T \cdot R \cdot \Delta u_f \end{aligned} \tag{24}$$

The optimization tool can be used at this point. An analytical optimization is applied by getting dJ/dΔuf and second derivative is positive defined.

$$\frac{\partial J}{\partial \Delta u_f} = 2 \cdot S_f^T \cdot \psi^T \cdot \psi \cdot (S_f \cdot \Delta u_f + S_P \cdot \Delta u_P + y_{real}(k) \cdot L_P - y_{sp}) + R^T \cdot R \cdot \Delta u_f = 0 \tag{25}$$

$$\frac{\partial J}{\partial \Delta u_f} = 2 \cdot S_f^T \cdot \psi^T \cdot \psi \cdot S_f \tag{26}$$

As a result Δuf optimal is:

$$\Delta u_f = (S_f^T \cdot \psi^T \cdot \psi \cdot S_f + R^T \cdot R)^{-1} \cdot S_f^T \cdot \psi^T \cdot \psi \cdot (y_{sp} - S_P \cdot \Delta u_P - y_{real}(k) \cdot L_P) \tag{27}$$

And simplifying, H Matrix is defined as follow:

$$H = (S_f^T \cdot \psi^T \cdot \psi \cdot S_f + R^T \cdot R)^{-1} \cdot S_f^T \cdot \psi^T \cdot \psi \qquad (\boldsymbol{H} \text{ matrix})$$

Once the standard DMC is developed and considering figure 2:

$$\begin{aligned} y_{fs} &= S_{fs} \cdot \Delta u_f + S_{Ps} \cdot \Delta u_P + y_{sreal}(k) \cdot L_P \\ y_{fp} &= S_{fp} \cdot \Delta p_f + S_{Pp} \cdot \Delta p_P + y_{preal}(k) \cdot L_P \end{aligned} \tag{28}$$

Where:
y_{fs} is the array of y_s predictions
y_{fp} is the array of predictions in disturbance output component.
S_{fs} and S_{fp} are matrixes of future system response and disturbance response respectively.
S_{ps} and S_{pp} are matrixes of past system response and disturbance response respectively.
ysreal is the system component of the output
y_{preal} is the disturbance component of the output
Δu_f is the future increment in control action.

Δp_f is the future increment in disturbance.
Δu_p is the past increment in control action.
Δp_f is the past increment in disturbance.
Thus system predictions are:

$$y_f = S_{fs} \cdot \Delta u_f + S_{Ps} \cdot \Delta u_P + S_{fp} \cdot \Delta p_f + S_{Pp} \cdot \Delta p_P + (y_{preal}(k) + y_{sreal}(k)) \cdot L_P \tag{29}$$

Using equation (23) and (29); and using the optimization tool, the result is:

$$\begin{aligned} \Delta u_f = {} & (S_{fs}^T \cdot \psi^T \cdot \psi \cdot S_{fs} + R^T \cdot R)^{-1} \cdot S_{fs}^T \cdot \psi^T \cdot \psi \cdot \\ & (y_{sp} - S_{Ps} \cdot \Delta u_P - S_{Pp} \cdot \Delta p_P - y_{real}(k) \cdot L_P) \end{aligned} \tag{30}$$

Thus ***H*** is the same as (***H*** Matrix) but, intead of S_f is S_{fs}.
Every step time the controller is applying the first control action and update the data for getting new optimal control every step time. Thus only the first line in the matrix equation should be calculated. Additionally this matrix equation, first line, can be presented as a Z transform function transfer. This is the application of the fourth basis of MBPC, receding horizon policy. It means that every step time a new control action will be calculated by means of updating data in the algorithm. Thus the control algorithm must be presented as:
In which h(z) is the first row of the ***H*** Matrix. The vector h(z) has different coefficients which will be multiplied by future set point, if the information is available. If the future set point is not available, a constant is calculated by adding all the h(z) vector coefficients.
$D(z^1)$ polynomial in which the coefficients are the result by multiplying h(z)·Sp. This array is in z^{-1}, because is using past information. All in all the algorithms structure is shown in figure 5.3.

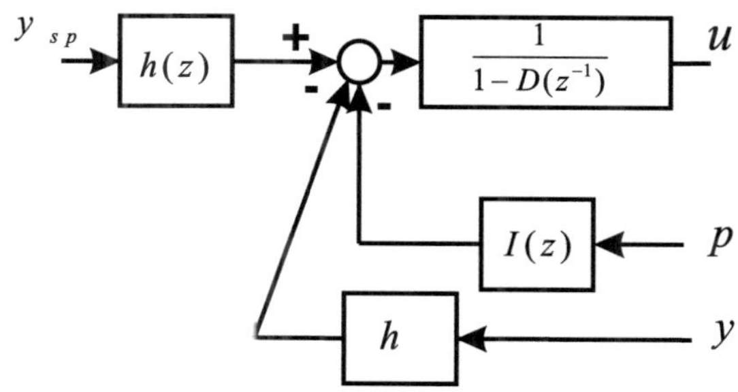

Fig. 3. DMCDM architecture with receding horizon.

When $I(z^{-1})$ is the convolution of two vectors: one array is product of h(z)·Spp, the second array is Δ operator. Spp is new matrix calculated as Sp but using Markov coefficients of disturbance of the system. More information is available at Thesis of Garcia[16].

4. Practical set up and experimental behavior

This chapter is presenting the author's experience for tuning the algorithm. Also the steps recommended for carry out a control solution. MBPC is a family on controller very suitable for controlling systems in different conditions. They can consider constraints, control cost, future information, model behavior etc... Getting the model by linearization should be considered. There are many identification algorithms, using the right ones and using the best parameters can bring you success or failure.

4.1 System identification

In control engineering ther are two important concepts that are opposite: Robustness and Performance. Usually as much Robust is a controller as less Performance it offers you. So as much Performance we demand as much easy to reach instable zones. There is one way to get both characteristic at the same time, Model accuracy. If we have a better model or a more accurate model, we will be able to demand more performance and more Robustness. So the system model is one of the most important topics in control engineering.

The most famous PID control tuning techniques are Ziegler-Nochols rules or Cohen-Coon method. Methods have a main characteristic, you can tune a PID controller with Robustness, but unfortunately performance is not good enough. The main reason for is that you have not a system model. So the quality model is very poor, thus Robustness is more important than performance. When Performance is important, model quality must be improved. As good is the model as good will be the control.

By other hand, the easiest way to get a model is by mean of physical equations. If we know the physics of the system, by using physical equation we can obtain a very good quality model. When physical equations are used model reach an important complexity, sometimes the computational effort done by the microcontroller or commuter is very high. Some models need mode than two day for calculating few second of real physical behavior.

An important approach use to be linear model. The computational cost of linear model, use to be very low (few micro seconds every step time). Additionally much real system use to have a linear behavior, so linear approach, sometimes, is the most intelligent solution. When system has a non-linear behavior other solutions can be approached. Linearization is a good solution for that.

When physical equations are hard to compute, linearization of them bring us a model easy to carry out and fast to be calculated. Another solution is linear model by identification techniques. When identification is our choice this kind of model works very well in the identified point and around of it, but we have problems when the operating point is too far from the identification point. When this problem appears a new linear model can be identified. We can do this topic as many times as we need. Thus we apply a technique similar to gain schedule but with models.

The main advantages of liner models are:

It is a simpler, so you have an easier solution for solving Differential equations.

The system behavior can be observed.

Any kind of order can be used. We should fit the order to the one system expected.

As disadvantages we can find:

The solution is exact, only in the operating pint and good enough close to it, but it is not good as far as we are.

Experimental data is needed and sometimes particular experiments are not possible to carry out.

No physical sense in poles and zeros identified can be found.

Math of the transfer functions, state space model and other can be studied in Ogata continuous , Ogata Discrete or Zhu [43] books. Additionally some identification algorithms were tested and studied widely in bibliography.

Lineal model available in bibliography:

Markov coefficients or impulse response model, there is an equivalent model, the step response.

ARX: autoregressive with exogenous variable. ARMAX: Autoregressive moving average with exogenous variable. OE: Output error. BJ: box Jenkins. CARIMA or ARIMAX:

autoregressive integral moving average with exogenous variable. State space models deterministic, advanced and stochastic ones etc. all models are descried in Garcia [17].
Finally the solution proposed is linear models by sections. When linear model is valid, then it shall be used, when linear model is not valid a new linear model must be placed.

4.1.1 Experimental data for identification

Many times data available is not a suitable data for identification. That is why system is working and particular experiments are not allowed. Thus a measurement of system behavior should be done and transient data must be used in the identification. Steady data does not include important dynamic information, so transient contains the system dynamics.
When particular experiments can be done, they must be well chosen. An easy experiment use to be a step applied to the system. The step should contain some characteristics: powerful enough, long enough, the response must be in the linear zone. Theoretically white noise is the best signal. But some authors, as Luo[22] or Zhu[43], studied physical inputs and pseudo aleatori input is the best choice.
Some times a simple step or impulse can be applied and Strejc, Csypkin methods can be applied. Simple model is obtained and some time it is good enough. Frequency identification methods are also available, based in bode diagram or graphical techniques.
Identification algorithms are proposed in this chapter. Experimental data must be pre processed and model structure should be well chosen. Parametrical and state space identification can be carried out.
When parametrical identification is proposed the following steps should be respected:

1. Experimental identification design and data acquisition.
2. Data pre processed: filtering, remove data tends, data normalization etc.
3. definition of model structure.
4. Identify the model by using an appropriate identification algorithm.
5. Studying model proprieties (zeros, poles, gain, …) and model validation.
6. If the model is the expected one, this task is finished, else come back to step 1, 2 or 3, depending on problems detected.

4.1.2 Identification algorithms

The identification algorithms proposed and tested in this work is the PEM (Prediction error method). Other where tested N4SID (Numerical Algorithm for Sub Space State System Identification) widely explained in [108,141,150]. Bud PEM is preferred by this author with very well results. PEM is an identification algorithm of state space model, so this model should be converted in a equivalent transfer function.

4.1.3 System identification

A Diesel engine is identified. Some experimental data is shown. At this studied case air path or air management system is going to be controlled. If suitable linear models cannot be found, algorithms based on the Hamertein and Wiener models [12] can be used or non linear models. The disadvantages of this latter group are: high computational cost and the difficulty of programming them in a commercial ECU (electronic control unite).
The system air-path in a Diesel engine consists on controlling VNT (variable nozzle turbine) in order to keep boost pressure in the set point. The engine controlled is a Diesel engine of heavy duty applications, see figure 4. This is a truck engine, where EGR system is not

included but VNT is available. Air-path is influenced by mass fuel injected, as much fuel injected into cylinder, more power available in the turbine and more boost pressure. Fuel injected is available information in the ECU.

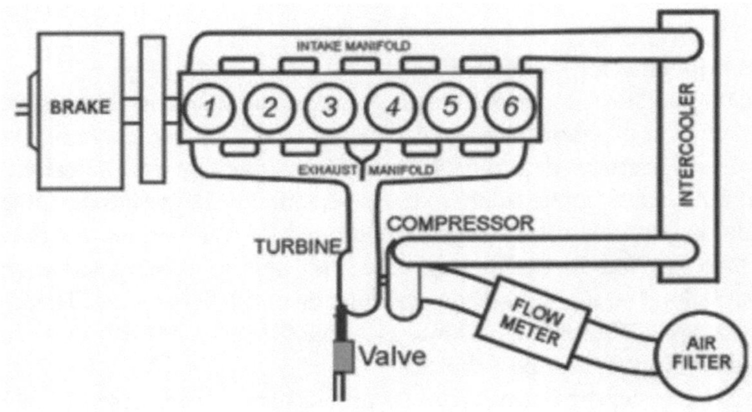

Fig. 4. System controlled.

Unfortunately the fuel injected is selected by user's needs. Thus fuel injected can not be controlled by control algorithms and it is determined as a measurable disturbance.
Experimental identification done is in figure 5.

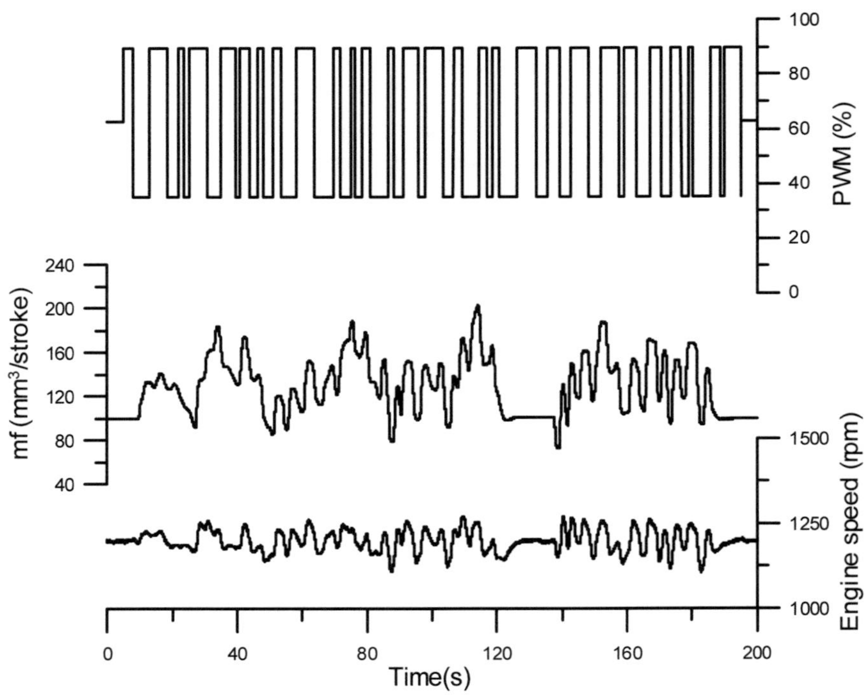

Fig. 5. Identification test performed at 1200 rpm. PWM signal is the control action on the VNT, and mf is the injected fuel. Important variations in the engine speed are due to the dynamometer response.

The VNT input applied can be seen in figure 6, where pseudo-random PWM (Pulse Width Modulation PWM) is applied in the VNT. Aleatori fuel is injected in the engine. With these inputs the boost pressure response is shown in figure below:

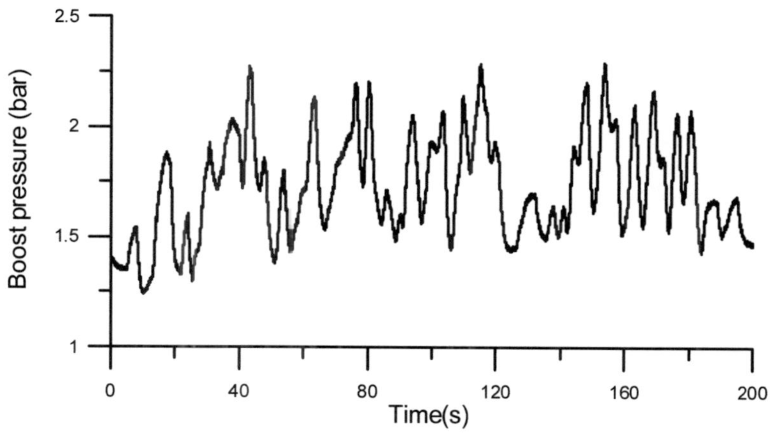

Fig. 6. Boost pressure response.

Apparently these data have not any kind of relation, but continuing with identification process we can obtain a system model. As shown the model poles are two. Physical systems don't use to have more than three poles, when model is bigger than this we could find some problems, identification algorithm tried to identify signal noise, and it has identified fast poles.

$$A=\begin{pmatrix} 0{,}983 & -0{,}022 & 0{,}008 \\ 0{,}016 & 0{,}886 & 0{,}115 \\ -0{,}002 & -0{,}027 & 0{,}907 \end{pmatrix};B=\begin{pmatrix} -1{,}7.10^{-5} & -5{,}1.10^{-5} \\ 5.8.10^{-5} & -5.10^{-4} \\ -2{,}1.10^{-4} & 5.10^{-4} \end{pmatrix}$$

$$C=\begin{pmatrix} 8{,}89 & -1{,}18 & -1{,}2 \end{pmatrix};D=0$$

Checking the results, we could realize that the system model is good enough, see figure 7.

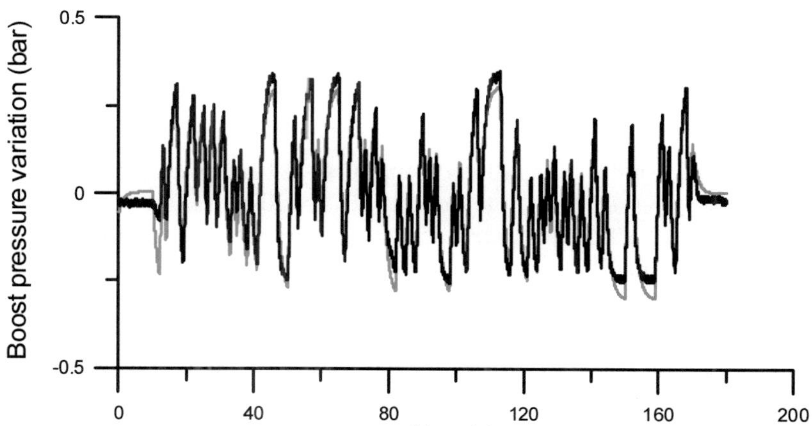

Fig. 7. Measured evolution (black) and predicted evolution (grey) of the boost pressure for the validation test at 1200 rpm.

4.2 Control system examples

Once the model is identified the controller is calculated by using control algorithms developed in section 3, it is time to control the system and testing with experimental data. A comparison is going to be shown in the following figures. The test proposed is a transient from steady conditions to full load. This is a typical situation in which a Diesel engine is requested full load to make advancement in a road. The engine is turning at 1200 rpm and no torque is need, suddenly all torque available is needed, and then engine, electronics and thermodynamics must change to the new conditions. Four different algorithms are controlling the system. The idea is testing those algorithms in order to check the engine behavior if the controller of air path is one algorithm or other one.

The algorithms tested were: PID controller with feed forward of fuel behavior, PID with Fuzzy login in parameters plus feed forward in fuel injected, GPCDM developed before and DMCDM shown in section 3.

PID controller was tuned for getting the best feasible performance. It was tuned from Ziegler Nichols parameter to updating PID until the performance was good enough. Feed forward was programmed to use measurable disturbance behavior available.

PID Fuzzy: this controller was programmed as one improvement of standard PID. The proposed algorithm is a gain schedule depending on error. Feed forward is also used here.

DMCDM: a model predictive controller based on impulse response with measurable disturbance. Control horizon and prediction horizon are chosen as long time. When performance is required long control horizon and prediction horizon must be chosen. Fro tuning the algorithm one weight is fixed to 1 and the other is changed.

GPCDM: this is an other MBPC, so many ideas explained for DMCDM are applicable here. Long time control horizon and long time prediction horizon are used for reaching good performance. The real control cost is not available, so we keep one weight fixed and we fit the second weight for tuning the system. GPC and GPCDM have a very special parameter, which is not in other kinds of MBPC. T polynomial is a parameter, which theoretically is a polynomial with zeros of the colored derivative noise from CARIMA model. In fact, it is a parameter for tuning the GPC. When noise is present in the system, this parameter must be tuned else you could put it as 1. there are many works explaining the behavior of the parameter Clarke [26]. Moreover the robustness of the system is widely improved by tuning the parameter [28,132]. Typical values of T polynomial are $T= 1-0,7*z^{-1}$; or T = convolution of ($[1-0,7.z^{-1}]$,[$1-0,7.z^{-1}$]), sometimes 0,7 are placed by 0,8 or 0,6. This is an empirical rule that can help the designer to choose the best option.

The structures are programmed as follow:

PID feed forward:

The boost pressure set point is depending on engine speed and mass fuel requested by used. Additionally the fuel injected is the measurable system disturbance. So the fuel injected is the feed forward control action. Error between boost pressure set point and measured boot pressure is the input to the PID controller. Control action calculated is the addition between feed forward and PID in figure 8. Finally control action applied is the calculated one processed by one anti-windup. Experimental result will be shown in following figures.

The controller proposed and programmed in the figure 9 is similar architecture to the PID proposed but some differences. The PID parameters are scheduled by a Fuzzy logic technique. Other sub-systems are the same, PID, feed forward and anti-windup. The behavior is similar to the PID but parameters can be tuned more accurate than a standard PID controller. The main difficult in this controller is the tuning process, because the parameters are more difficult to set up. The experience of the designer must be higher than the standard PID. For tuning the controller many experiments must be done and widely analyzed for choosing the best option.

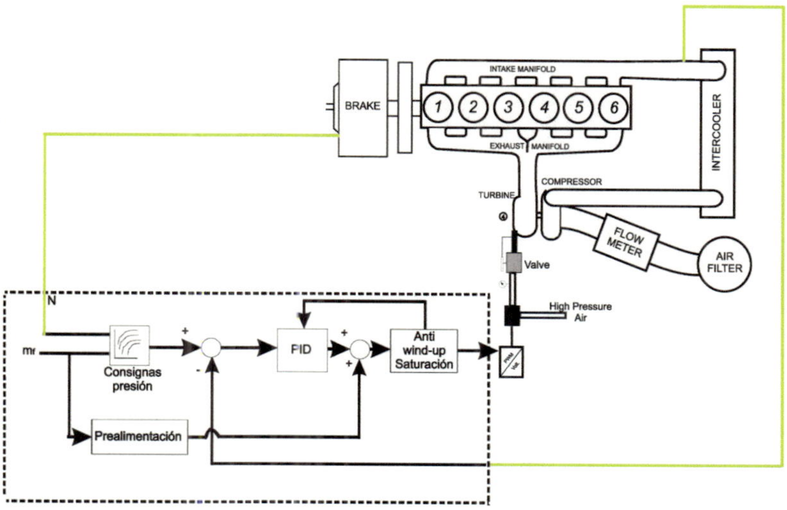

Fig. 8. PID Architecture programmed.

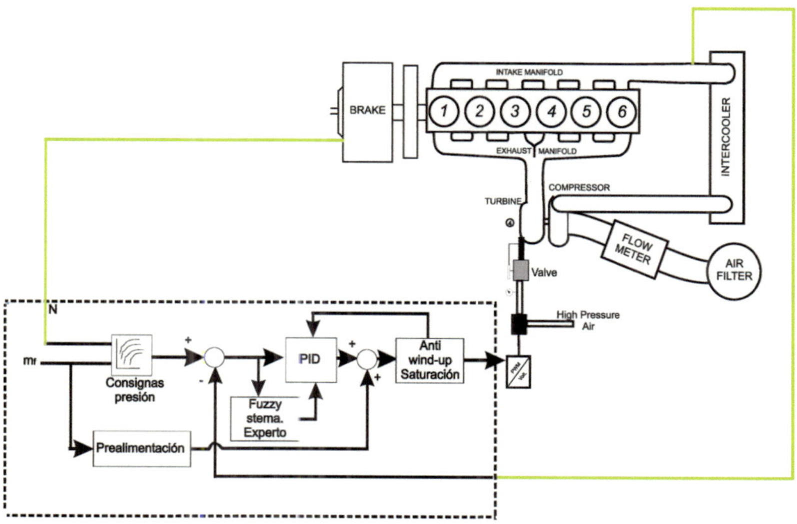

Fig. 9. PID scheduled by fuzzy contrcller and feed forward.

GPCDM architecture is the one proposed in the figure 10. Set point is processed by h(z). when future set point is not available, future ones are the actual one. Thus h(z) became as a constant term. Additionally measurable disturbance can be processed as past information

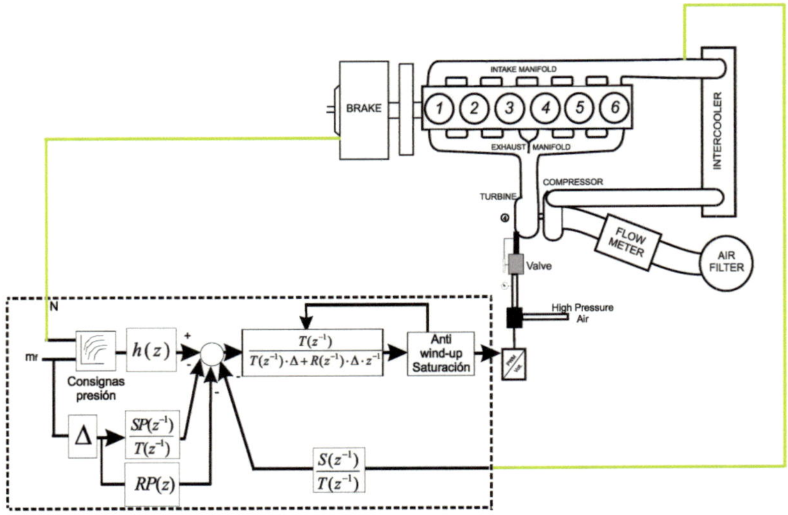

Fig. 10. GPCDM programmed.

and future information, see equation (30). In this experiment only actual and past fuel injected can be processed, but future disturbances can be considered. Anti-windup is also programmed, when control action calculated is higher than maximum, the algorithm is frozen. Thus the integral component of the algorithm can not affect to the algorithms performance.

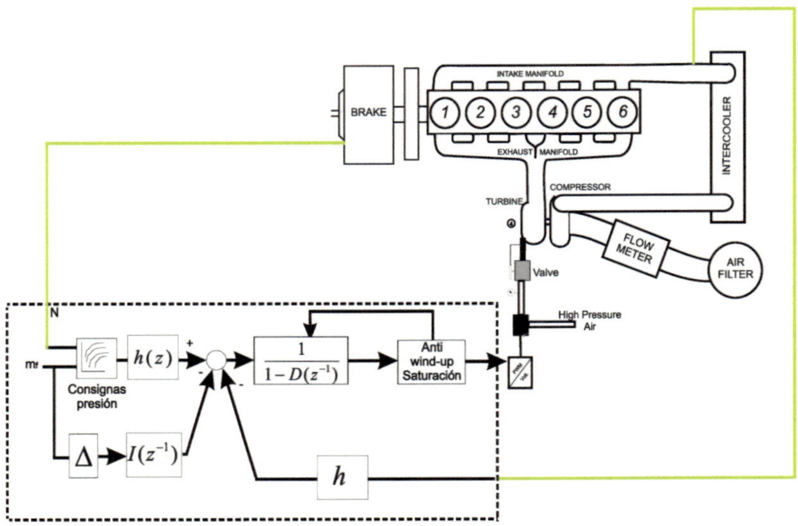

Fig. 11. DMCDM programmed.

DMCDM programmed is shown in figure 11. Measurable disturbances are considered and included in the controller. Anti-windup is programmed, a similar system to the one used in the GPCDM. When maximum or minimum control action is reached, the algorithm is frozen until control action is going to the opposite way.

5. Experimental results

As explained before full load transient test were done. The engine was in steady conditions or idle conditions at 1200 rpm. When test starts, full load is required from the engine. Thus maximum quantity of fuel is injected in order to burn it and getting the power. Additionally air mass is needed for burning the fuel, which is why VNT must be controller to give the boost pressure necessary for the new conditions. More boost pressure that necessary produces an increment in exhaust pressure and some loses of torque and power and more pollution in exhaust gases. If we have less boost pressure than set point then we will not have enough air flow for burning the fuel in the right conditions.

Many experiments were done, but only the best performance is shown in this chapter. In figure 12, boost pressure can be observed, figure 13 contains exhaust pressure behavior, and the figure 14 shows engine torque.

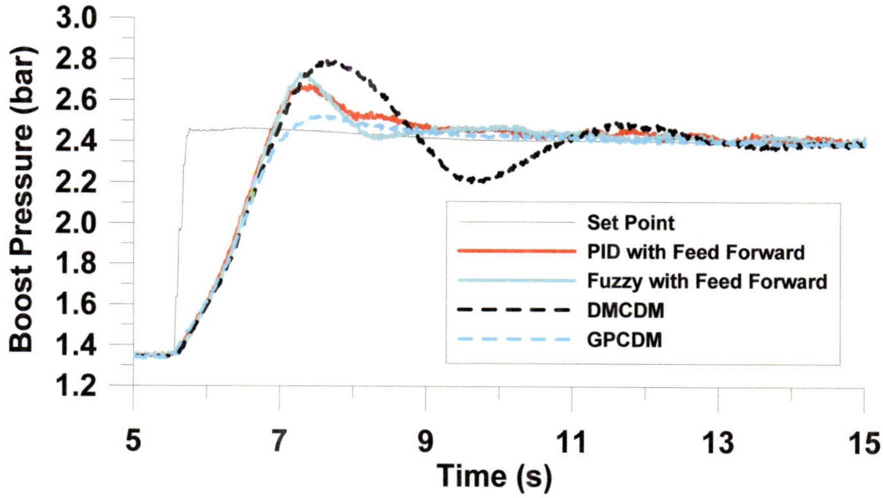

Fig. 12. Boost pressure evolution in transient conditions.

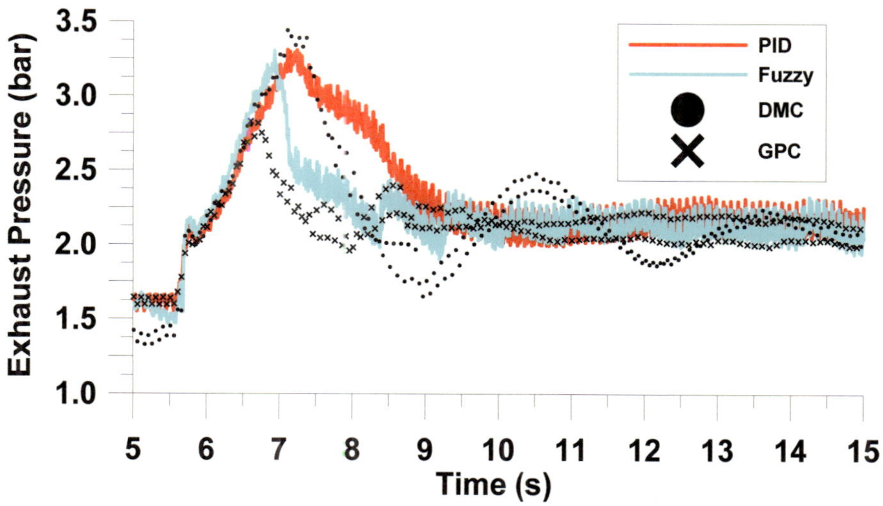

Fig. 13. Exhaust pressure evolution in transient conditions.

In figure 12, boost pressure behavior is presented. Thin line is the set point requested from idle condition to full load one. This is the best boost pressure conditions considering, efficiency, torque, power, pollution etc. so the system must be in set point as soon as possible. Continuous lines are PID controller, solid red line is the PID with feed forward and solid blue line is the PID Fuzzy logic controller. GPCDM is the dashed blue line and DMCDM is the dashed black line.

When transient test starts, the first part of the test are the same in all controllers. This happens because VNT is in bounds, completely close and no different performance can be appreciated. When VNT must be opened controller performance is different. GPCDM is opening the VNT before than others, that is why over oscillation is lower than others. The predictive model CARIMA helps the GPCDM to advance the system behavior, thus GPCDM can predict the VNT open to be before and avoiding the over oscillation. PID must be tuned with low influences of derivative term, the reason is the stability. Derivative term can produce instable conditions in the system. This controller realizes later about over oscillation and start opening the VNT later. The worst performance is done by DMCDM. Maybe the system model is very single and it is more sensible to non-linearity. Fuzzy controller improves the PID behavior because it is an improvement of PID.

Figure 12. contains the exhaust boost pressure. The steady conditions are reached in similar time to boost pressure. The over-pressure in exhaust produces over-speed in turbine or torque loses, and more pollutants than permitted.

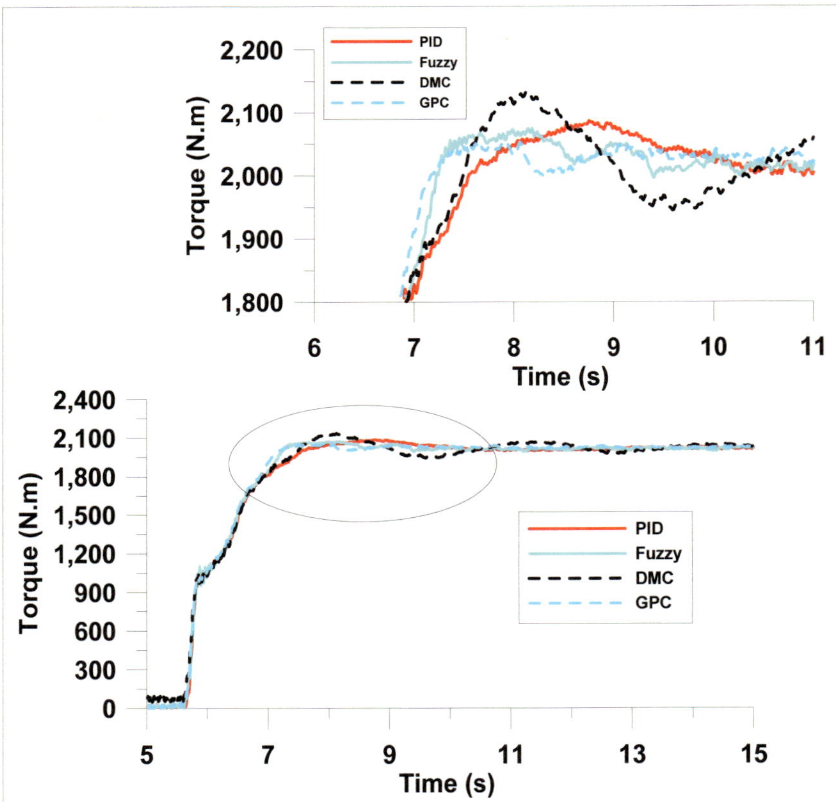

Fig. 14. Torque evolution and air-path influence during transient.

Figure 14. shows torque performance. Both GPCDM and Fuzzy reach steady conditions at the same time. So the torque is not very influenced by boost pressure. But if other controllers are analyzed engine torque is influenced and oscillations are very uncomfortable.
Finally a summary of system response are shown in table 1 and table 2. Table 1 summarizes the system response from the point of view of engine analysis. Table 2 analyzes the system from the point of view of control parameters.

Experimental results					
	Max Exhaust pressure (bar)	**Overshoot Exhaust pressure from steady conditions**	**max (Exhaust Pressure) - max(Boost Pressure)**	**Max(Torque)**	**Torque overshoot**
PID+FF	3.223	53.48	0.482	2031	85
PID+FF+Fuzzy	3.443	63.48	0.702	2100	86
DMCDM	3.220	53.33	0.428	2101	133
GPCDM	2.655	26.43	0.129	2055	36

Table 1. Experimental results of system behavior.

Experimental results				
	$\int_{5.5}^{22.5} e(t)^2 \cdot dt$	$\int_{5.5}^{22.5} \Delta u^2 \cdot dt$	J	**Boost pressure overshoot**
PID+FF	29.11	454.38	51.83	34.10
PID+FF+Fuzzy	29.13	2911.15	174.68	33.10
DMCDM	36.96	18.80	37.90	39.20
GPCDM	27.62	37.40	29.49	12.60

Table 2. Experimental results, control parameters.

The most important variable from the point of view of control is the control Cost or J. This variable summarized the control behavior from the point of view of control cost, considering control effort and control error.

5.2 Discussion

The first conclusion that can be drawn from analyzing the results is that when fuel mass is considered in the algorithm, behavior is always improved. This fact had been observed during the identification process and is logical, since the more information the algorithm has about the system, the better control it can apply. This is due to the air management process being more influenced by fuel amount than VGT control actions. The fuel amount gives energy, this energy goes to the turbine and the turbocharger increases its speed, thus the air management system changes. The delivered fuel quantity therefore has a very strong influence on the air management system. If this variable is included in control algorithms, better control will be obtained.
The analysis of the behavior of the Fuzzy controller seems to show quite a good response but at the cost of a high control effort. One of the disadvantages of this type of controller is

the high overshoot of the exhaust pressure. The control effort is one order of magnitude higher than the baseline controller. However, the greatest disadvantage is their complicated tuning process due to having a large number of independent parameters.
Another interesting aspect is that not all the MPCs work better than the existing controllers. For example, the DMC provided rather doubtful results. This was a surprise and could be explained as follows: the prediction model of the DMC is considered to be highly sensitive to nonlinearities in the system, since the system itself is highly nonlinear (in spite of having linear zone models). This algorithm finds it difficult to predict engine behavior and therefore does not take the right decisions. However, the predictive control GPC algorithm is based on the CARIMA model and is more robust to system nonlinearities. In fact, one of the most important parameters for the correct GPC tuning was the polynomial T(z).
The best all-round algorithm was the GPCDM. The most significant difference between algorithms can be seen in the exhaust pressure in Figures 8 and 9. This parameter indicates the effort made by the engine to control the air path and high pressure has a negative influence on engine torque.
The air management system, and consequently various engine parameters, was strongly influenced by the controller used. One of the most susceptible to this influence is PMEP, which also affects engine torque, turbocharger pressure, etc. If an overshoot occurs it will produce exhaust pressure peak and pumping losses. This fact produced less efficiency and torque during transient conditions.

6. Conclusions

Few linear models were obtained from the air management system of the engine and they are quiet good approximations of the nonlinear plant. These models can be used for linear controller design with low computational effort. GPCDM, DMCDM, PID and Fuzzy PID are tested in a standard test bench. In comparison to standard gain scheduled PID, setpoint tracking could be improved. The additional degrees of freedom of GPC can be used for tuning the robustness of control.

7. Acknowledgements

I would like to thank Universidad Jaume I from Castellon (Spain). I am assistant professor in the engineering department. Moreover I want strongly thank Vinci Energia. In this company Ia m working as a professional researcher and many of the improvements in control theory are applied to solar tracking system in my actual investigations. And of course, this work was impossible to do without CMT thermal engines of the "Universidad Politécnica de Valencia".

8. References

[1] Albertos P. y Ortega R. "On generalized predictive control two alternative formulations". Automatica, Vol. 25 No.5, pp. 753–755, 1989.
[2] Bordons C. y Camacho E.F. "A generalized predictive controller for a wide class of industrial process". IEEE Transactions on control system technology, Vol. 6 No. 3, May 1998.

[3] Camacho E.F., Berenguel M. y Rubio F.R. "Application of a Gain Scheduling Generalized Predictive Controller to a Solar Power Plant". Control Engineering Practice, Vol. 2(2), pp. 227–238, 1994.

[4] Chow C. y Clarke D.W. "Actuator nonlinearities in predictive control advances in model-based predictive control". Ed. Oxford university press, p´ag. 245, 1994.

[5] Clarke D.W. y Gawthrop P.J. "Self-tuning controller". Proc. IEE, Vol. 122, pp. 929–934, 1975.

[6] Clarke D.W. y Gawthrop P.J. "Self-tuning control". Proc. IEE, Vol. 126, pp. 633–640, 1979.

[7] Clarke D.W. yMothadi C. "Properties of generalized predictive control". Automatica, Vol. 25(6), pp. 859–875, 1989.

[8] Clarke D.W., Mothadi C. y Tuffs P.S. "Generalized predictive control -Part I. The basic algirithm". Automatica, Vol. 23 No.2, pp. 137–148, 1987.

[9] Clarke D.W., Mothadi C. y Tuffs P.S. "Generalized predictive control -Part II. Extension and interpretation". Automatica, Vol. 23 No.2, pp. 149–160, 1987.

[10] Clarke D.W. y Zhang L. "Long-range predictive control using weightingsecuence models". IEE Proceedings, Vol. 134 D No.3, 1987.

[11] Corradini M.L. y Orlando G. "A VSC algorithm based on generalized predictive control". Automatica, Vol. 33 ,5, pp. 927–932, 1997.

[12] Cutler C.R. y B.L. Ramaker. "Dynamic matrix control–a computer control algorithm". Proceedings of the Joint Automatic Control Conference, 1980.

[13] de la parte M. Perez, Camacho O. y Camacho E.F. "Developement of a GPC-based sliding mode controller". ISA Transaction, Vol. 41 No.1, pp. 19–30, 2002.

[14] Evans W.R. "Control system synthesis by root locus method". AIEE Transaction, Part II, Vol. 69, pp. 66–69, 1950.

[15] Galanti M., Magrini S., Mosca E. y Spicci V. "Receding horizon predictive control applied to a high performance servo-controlled pedestal". Proc. of ESPRIT (III workshop on industrial application of MBPC, Cambridge, 1992.Soeterboek R. Predictive control a unified approach. Prentice Hall int. UK Cambridge, 1992.

[16] Garcia C.E. y Morshedi A.M. "Quadratic programming solution of dynamic matrix control (QDMC)". Chemical engineering communication, Vol. 46, pp. 73–87, 1986.

[17] García-Ortiz, J.V. "Aportación a la mejora del control de la renovación de la carga en motores Diesel Turboalimentados mediante distintos algoritmos de control". Thesis in Universidad Politécnica de Valencia. CMT Motores térmicos. 9 Julio 2004.

[18] Havlena V. y F.J. Kraus. "Receding horizon MIMO LQ controller design with guaranteed stability". Automatica, Vol. 33 No.8, pp. 1567–1570, 1997.

[19] Kennel R. "Generalized predictive control GPC-ready for use in drive application?". 32nd IEEE electronics specialist conference PELS Vancouver June, 2001.

[20] Keyser R.M.C. De, de Velde G.A. Van y Dumortier F.A.G. "A comparative study of self-adaptative long-range predictive control methods". Automatica, Vol. 24(2), pp. 149–163, 1988.

[21] Kouvaritakis B., Cannon M. y Rossiter J.A. "Who needs QP for linear MPC anyway?". Automatica, Vol. 38, pp. 879–884, 2002.

[22] Kouvaritakis B., Rossiter J.A. y Chang A.O.T. "Stable gerenalized predictive control: an algorithm with guaranteed stability". Proc. IEE, Part D,, Vol. 139(4), pp. 349–362, 1992.

[23] Luo X. Analysis of automatic control system for deposition of CIGS films. Tesis Doctoral, Colorado School of Mines, Colorado Illinois, USA, 2002.

[24] Martinez M., Senent J. y Blasco X. "A comparative study of classic vs genetic algorithms optimization applied in GPC controller". IFAC world congress, San Francisco USA, 1996.
[25] Mosca E. y Zhang J. "Stable redesign of predictive control". Automatica, Vol. 28, pp. 1229-1233, 1992.
[26] Nuñez A., Cueil J.R. y Bordons C. "Modelado y control de una almazara de extraccion de aceite de oliva". XXII Jornadas de Autimatica, 2001.
[27] Nyquist H. "Regeneration theory". Bell system tech. Journal, Vol. 11, 1932.
[28] Overschee P. Van y Moor B. De. "N4SID: Subspace algorithm for the identification of combined deterministic-stochastic systems". Automatica, Vol. 30, pp. 75-93, 1994.
[29] Pannocchia G. y Rawlings J.B. "Robustness of MPC and Disturbance Models for Multivariable Ill-conditioned Processes". Technical report TWMCC No. 2001-02, 2001.
[30] Poulser N.K., Kuovaritakis B. y Cannon M. "Constrained predictive control and its application to a coupled-tanks apparatus". Int. Journal of control, Vol. 74 No. 6, pp. 552-564, 2001.
[31] Qin S.J. y Badgwell T.A. "An overview of industrial model predictive control technology". AIche symposium series, Vol. 316(93), pp. 232-256, 1996.
[32] Qin S.J. y Badgwell T.A. "A survey of industrial model predictive control technology". Control engineering practice, Vol. 11, pp. 733-764, 2003.
[33] Richalet J., Rault A., Testud J.L. y Papon J. "Model predictive heuristic control: Applications to industrial processes". Automatica, Vol. 14, pp. 413-428, 1978.
[34] Salcedo J.V, Martinez M., Sanchis J. y Blasco X. "Design of GPC's in state space". Automatika, Vol. 42(3-4), pp. 159-167, 2001.
[35] Sanchis J., Martinez M., Ramos C. y Salcedo J.V. "Principal component GPC with terminal equality constraint". In 15th IFAC World Congress , Barcelona (Spain), July 2002.
[36] Shridhar R. y Cooper D.J. "A tuning strategy for unconstrained SISO model predictive control". Ind. Eng. Chem Res., Vol. 36, pp. 729-746, 1997.
[37] Soeterboek A.R.M., Verbrugger H.B., der Bosch P.P.J Van y Butler H. "Adaptative predictive control-A unified approach". Proc. 6th Yale workshop on applied of adapt. system theory new Haver USA, 1990.
[38] Soeterboek A.R.M., Verbrugger H.B., der Bosch P.P.J Van y Butler H. "On the unification of predictive control algorithm". Proc. 29th IEEE conference on decision and control, Honolulu USA, 1990.
[39] Soeterboek R. Predictive control a unified approach. Prentice Hall int. UK Cambridge, 1992.
[40] Tsang B.A. y Clarke D.W. "Generalized predictive control with input constraints". IEE Proceedings, Vol. 135, 1988.
[41] Verhaegen M. "Identification of the deterministic part of MIMO state space models". Automatica, Vol. 30, pp. 61-74, 1994.
[42] Zafirion E. "Robust model predictive control of process with hard constraints". Comp. Chen. Eng., Vol. 14, pp. 359-371, 1990.
[43] Zheng A. y Morari M. "Stability of model predictive control with soft restraints". Proceedings 33rd CDC Florida, pp. 1018-1023, 1994.
[44] Zhu Y. y Backx T. Identification of multivariable industrial process, for simulation, diagnosis and control. Springer-verlag, ISBN 3-540-19835-0, 1993.

18

BrainWave®: Model Predictive Control for the Process Industries

W. A (Bill) Gough
Andritz Automation Ltd.
Canada

1. Introduction

This chapter describes the development and application of a model-based predictive adaptive (MPC) controller, commercially known as BrainWave®. This controller is a patented (US Patents #5,335,164 and #6,643,554) PC-based commercial software package with hundreds of installations around the world in many different process industries. The predictive control capability enables significant performance improvements compared to manual or other automatic control strategies for processes with long time delays or multi-variable interactions. Process variability reductions of 50% or more are typically achieved using this technique.

MPC technology has been used for many years in the petroleum industry but it is not yet common practice in most industries. The high cost and implementation complexity have been barriers to the wide spread use of MPC. It is therefore important that an MPC tool be designed for ease of use to reduce the cost of installation and life cycle maintenance. Model identification and controller tuning are the primary tasks involved in the installation of an MPC controller so the controller must be designed to make these functions as easy as possible. The controller should be designed to handle self-regulating systems, open loop unstable (integrating) systems, and multivariable systems, so that the one controller can be used to solve as many applications as possible, avoiding the need for the user to learn and support too many different controller designs.

BrainWave is an MPC controller that has been developed to solve the most common types of difficult regulatory control problems. It has been deployed in a wide variety of process industries including pulp & paper, mineral processing, plastics, petrochemicals, oil and gas refining, food processing, lime and cement, and glass manufacturing. BrainWave is designed for ease of use. A novel process identification and modeling method based on the Laguerre series transfer function helps to simplify the steps required to obtain a model of the process response. Internal normalization techniques simplify the setup and tuning of the controller. BrainWave is designed to control processes that are self-regulating, integrating, or multivariable, so this one MPC tool can be used for virtually all difficult regulatory control loops in a plant.

The chapter will describe the mathematics behind the Laguerre modeling method, and show the development of the predictive control law based on the Laguerre state space model used in the BrainWave controller. The user interface from the actual software implementation will be illustrated. Several application examples from the Pulp & Paper and Mineral Processing industries will be provided that highlight the different types of regulatory control problems where MPC provides clear advantages over conventional PID type loop controllers.

2. Adaptive model-based predictive controller development

Obtaining a process response model is a key part of the implementation of an MPC controller. In our design, the controller models the system response using a generic function series approximation technique based on Laguerre polynomials. This approach provides a simple and efficient method to mathematically model the process response with a minimum of *a priori* information. It also enables the controller to perform online adaptation of the process response models automatically. These factors reduce the implementation effort and contribute to quick installation times. The adaptive capabilities assist the control technician with developing the process response models, and the default configuration of the control parameters ensures excellent control performance once the process model is obtained. These features help to ensure the same good result will be achieved regardless of the expertise level of the person doing the application. For industrial customers that operate large plants with thousands of process controllers, this benefit alone is extremely valuable.

Using these models as the basis for a predictive control design, the MPC is able to control processes with long delay or response times (or fast response processes where the time delay is a significant part of the response dynamics) better than is possible using PID type controllers. This technique can also be used to automatically model and counteract the effects of measured disturbances by incorporating them into the control strategy as feed forward variables. The use of feed forward variables is particularly important for long time delay systems so that disturbances can be cancelled much sooner than is possible using feedback control alone.

The advanced process control algorithm presented here is an adaptive model-based predictive controller that has its origins in the work of (Zervos & Dumont, 1988). They proposed the use of a state-space model derived from Laguerre orthogonal basis functions so that adaptive control could be achieved without the need to know the process order or the time delay in advance. This approach reduces the *a priori* knowledge required to develop a high fidelity model of the process transient response, thus simplifying the modeling task.

An analogy to this method is the use of Cosine functions in the Fourier series method to approximate periodic signals as is common in frequency analyzers. In this case, weights for each Cosine function in the series are determined such that when the weighted Cosine functions are summed, a reasonable approximation of the original signal is obtained. In this case the signal is represented by its frequency spectrum, with each basis function weighting coefficient representing the contribution of each frequency present in the original signal. This method is efficient due to the similarity of the basis functions in the series to the signal being modeled, and also due to the special mathematical property of the basis functions called orthogonality that ensures the unique solution of the basis function weighting coefficients in the identified model.

In process control, the process transfer functions are transient in nature and are not periodic, so Cosine functions are not an appropriate choice as a basis for the model. However, the elegance of the Fourier series technique provides many advantages such as simple and efficient model structure and excellent parameter convergence when estimating the model from observed data sets due to the orthogonality property of the Fourier series. The motivation of this research was to find an equally simple and efficient method to model the transient responses common in process control applications.

The Laguerre functions are well suited to modeling the types of transient signals found in process control because they have similar behavior to the processes being modeled and are also an orthogonal function set. In addition, the Laguerre functions are able to efficiently model the dead time in the process response compared to other suitable function sets. This Laguerre model is used as a basis for the design of the predictive adaptive regulatory controller.
Laguerre basis functions are defined by:

$$L_i(t) = \sqrt{2p}\frac{\exp(pt)}{(i-1)!}\frac{d^{i-1}}{dt^{i-1}}\left(t^{i-1}\exp(-2pt)\right) \tag{1}$$

Where L_i is the ith Laguerre function, and p is a scaling parameter referred to as the Laguerre pole.
Each Laguerre basis function is a polynomial multiplied by a decaying exponential so these basis functions make an excellent choice for modeling transient behavior because they are similar to transient signals. For use in modeling processes, these basis functions are written in the form of a dynamic system. With appropriate discretization (assuming straight lines between sampling points), the orthogonality of the basis functions is preserved in discrete time.
Summing each Laguerre basis function with an appropriate weighting factor approximates a process transfer function:

$$g(t) = \sum_{i=0}^{i=\infty} c_i l_i(t) \tag{2}$$

The Laguerre function has the following Laplace domain representation:

$$L_i(s) = \sqrt{2p}\frac{(s-p)^{i-1}}{(s+p)^i} \quad , \quad i = 1,\ldots,N \tag{3}$$

where: i is the number of Laguerre filters (i = 1, N);
$p > 0$ is the time-scale;
$L_i(s)$ are the Laguerre polynomials.
The reason for using the Laplace domain is the simplicity of representing the Laguerre ladder network, as shown in Fig. 1.

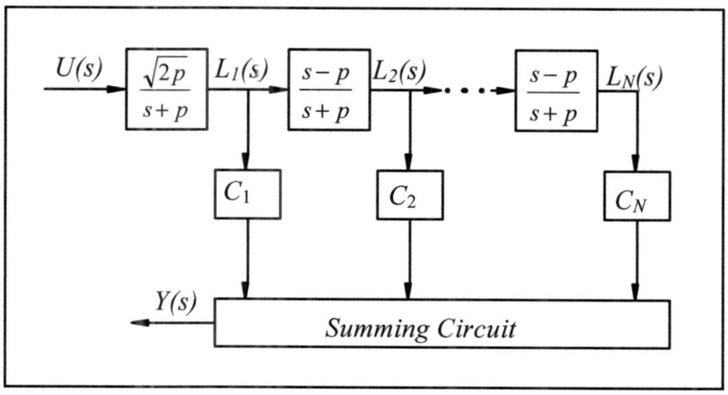

Fig. 1. Laguerre ladder network

This network can be expressed as a stable, observable and controllable state space form as:

$$l(k+1) = Al(k) + bu(k) \tag{4}$$

$$y(k) = c^T l(k) \tag{5}$$

where:

$l(k)^T = [l_1(k), \ \ldots \ ,l_N(k)]^T$ is the state of the ladder (i.e., the outputs of each block in Fig. 1);

$C_k^T(k) = [c_1(k), \ \ldots \ ,c_N(k)]$ are the Laguerre coefficients at time *k*;

A is a lower triangular square (N x N) matrix.

The Laguerre coefficients C(k) represent a projection of the plant model onto a linear space whose basis is formed by an orthonormal set of Laguerre functions.

The same concept used in the plant identification is used to identify the process response for measured feed forward variables. For unmeasured disturbances, an estimate of the load is made. This estimate is based on the observation that an external white noise feeds the disturbance model, resulting in a colored signal. The disturbance can be estimated as the difference between the plant process variable increment and the estimated plant model with the integrator removed. Using the plant and disturbance models we can then develop the Model Predictive Control (MPC) strategy.

The discrete state space model used by Zervos and Dumont for the control design takes the (velocity) form:

$$\begin{aligned} \Delta\mathbf{x}_{k+1} &= \mathbf{A}\Delta\mathbf{x}_k + \mathbf{b}\Delta u_k \\ y_k &= y_{k-1} + \mathbf{c}\Delta\mathbf{x}_k \end{aligned} \tag{6}$$

Where u is the input variable, which is adjusted to control the process variable y and $\mathbf{x}$ is the nth-order Laguerre model state. The subscript k gives the time index in terms of number of controller update steps; so the Δ operator indicates the difference from the past update. For example, $\Delta\mathbf{x}_k = \mathbf{x}_k - \mathbf{x}_{k-1}$. The n×n matrix $\mathbf{A}$ and the n×1 matrix $\mathbf{b}$ have fixed values derived from the Laguerre basis functions. Values of the 1×n matrix $\mathbf{c}$ are adjusted automatically during model adaptation to match the process model predictions to the actual plant behaviour. The dimension n gives the number of Laguerre basis functions used in the process model. The algorithm presented here does not suffer from over parameterization of the process model. The BrainWave controller uses n=15, which has proven to be enough to model almost any industrial plant behaviour.

Notice that the state equation recursively updates the process state to include the newest value of the manipulated variable. Therefore, we can view the process state as an encoded history of the input values.

Model adaptation may be executed in real time following the recursive algorithm of (Salgado et al, 1988). This algorithm updates the matrix $\mathbf{c}$ according to:

$$\mathbf{c}_{k+1}^T = \mathbf{c}_k^T + \frac{\alpha\mathbf{P}_k\Delta\mathbf{x}_k}{1+\Delta\mathbf{x}_k^T\mathbf{P}_k\Delta\mathbf{x}_k}\left(\Delta y_k - \mathbf{c}_k\Delta\mathbf{x}_k\right) \tag{7}$$

Where the covariance matrix $\mathbf{P}$ evolves according to:

$$\mathbf{P}_{k+1} = \frac{1}{\lambda}\mathbf{P}_k - \frac{\alpha \mathbf{P}_k \Delta\mathbf{x}_k \Delta\mathbf{x}_k^T \mathbf{P}_k}{1 + \Delta\mathbf{x}_k^T \mathbf{P}_k \Delta\mathbf{x}_k} + \beta\mathbf{I} - \delta\mathbf{P}_k^2 \tag{8}$$

I is the identity matrix. The constants α, λ, β and δ are essentially tuning parameters that can be adjusted to improve the convergence of the adapted Laguerre model parameters **C.**

The value of the manipulated input variable u (the controller output) that will bring the process to set point (SP) d steps in the future may be calculated using the model equations. The formula for the process variable d steps in the future is:

$$y_{k+d} = y_k + \mathbf{c}\left(\mathbf{A}^{d-1} + \ldots + \mathbf{I}\right)\mathbf{A}\Delta\mathbf{x}_k + \mathbf{c}\left(\mathbf{A}^{d-1} + \ldots + \mathbf{I}\right)\mathbf{b}\Delta u_k \tag{9}$$

Since it is desired to have the process reach set point d steps in the future, we set $y_{k+d} = SP_{k+d}$ and solve for the manipulated input variable Δu_k.

$$\Delta u_k = \frac{SP_{k+d} - y_k - \mathbf{c}\left(\mathbf{A}^{d-1} + \ldots + \mathbf{I}\right)\mathbf{A}\Delta\mathbf{x}_k}{\mathbf{c}\left(\mathbf{A}^{d-1} + \ldots + \mathbf{I}\right)\mathbf{b}} \tag{10}$$

This equation is the basic control law. As with most predictive control strategies, implementation of the strategy involves calculating the current controller output, implementing this new value, waiting one controller update period, and then repeating the calculation.

Predictive feed forward control can easily be included in this control formulation. A state equation in velocity form is created for the feed forward variable:

$$\Delta\mathbf{z}_{k+1} = \mathbf{A}_{ff}\Delta\mathbf{z}_k + \mathbf{b}_{ff}\Delta v_k \tag{11}$$

Where **z** is the feed forward state and v is the measured feed forward variable. The effect of the feed forward variable is then simply added to the output equation for the process variable to become part of the modeled process response:

$$y_k = y_{k-1} + \mathbf{c}\Delta\mathbf{x}_k + \mathbf{c}_{ff}\Delta\mathbf{z}_k \tag{12}$$

Notice that the feed forward state equation creates an encoded history of the feed forward variable in the same way that the main model state equation encodes the manipulated variable.

Re-deriving the control law from the extended set of process model equations yields:

$$\Delta u_k = \frac{SP_{k+d} - y_k - \mathbf{c}\left(\mathbf{A}^{d-1} + \ldots + \mathbf{I}\right)\mathbf{A}\Delta\mathbf{x}_k - \mathbf{c}_{ff}\left(\mathbf{A}_{ff}^{d-1} + \ldots + \mathbf{I}\right)\mathbf{A}_{ff}\Delta\mathbf{z}_k + \mathbf{c}_{ff}\left(\mathbf{A}_{ff}^{d-1} + \ldots + \mathbf{I}\right)\mathbf{b}_{ff}\Delta v_k}{\mathbf{c}\left(\mathbf{A}^{d-1} + \ldots + \mathbf{I}\right)\mathbf{b}} \tag{13}$$

The basic structure of this control law does not change as additional feed forward variables are included in the process model, so it is a simple exercise to include as many feed forwards as necessary into this control law.

Use of the Laguerre process model implicitly assumes that a self-regulating process is to be modeled. In cases where a process exhibits integrating behaviour, it is necessary to modify the Laguerre process model. The modification is two-fold. First, the Laguerre model is altered so that the Laguerre state gives the change in the slope of the process response. That is:

$$\begin{aligned} \Delta\mathbf{x}_{k+1} &= \mathbf{A}\Delta\mathbf{x}_k + \mathbf{b}\Delta u_k \\ \Delta y_k &= \Delta y_{k-1} + \mathbf{c}\Delta\mathbf{x}_k \\ y_k &= y_{k-1} + \Delta y_k \end{aligned} \tag{14}$$

Second, an unmeasured disturbance component is added to the model to estimate the steady state load on the plant. The feedback control law is then derived based on the modified Laguerre model, including the disturbance component.

The above gives the derivation of a d steps ahead adaptive model-based predictive control for a single variable. Full multivariable processes (multiple-input, multiple-output) may also be modeled using Laguerre state-space equations. However, to derive a control law for the multivariable case it is preferable to optimize for a cost function such as the one proposed in the generalized predictive control (Clarke et al, 1987), rather than back calculating the control to achieve a d steps ahead outcome. This concept of predictive control involves the repeated optimization of a performance objective over a finite horizon extending from a future time (N_1) up to a prediction horizon (N_2). Fig. 2 characterizes the way prediction is used within the MPC control strategy. Given a set point $s(k + l)$, a reference $r(k + l)$ is produced by pre-filtering and is used within the optimization of the MPC cost function. Manipulating the control variable $u(k + l)$, over the control horizon (N_u), the algorithm drives the predicted output $y(k + l)$, over the prediction horizon, towards the reference.

The control moves are determined by looking at the predicted future error, which is the difference between the predicted future output and the desired future output (reference). The user can specify the region over which these error values will be summed. The region is bounded by the initial (N1) and final (N2) prediction horizon. It is also possible to set the number of control moves that the controller will take to get to the set point by adjusting a parameter called control horizon (Nu). If we weigh the predicted squared error from set point and the total actuator movement with weighting matrices Q and R respectively, and add the two sums, we arrive at the cost function employed within the BrainWave Multivariable controller:

$$J(\Delta u) = \sum_{i=N_1}^{N_2} ||\hat{Y}(i) - \hat{S}(i)||_Q^2 + \sum_{i=1}^{Nu} ||\Delta U(i)||_R^2 \tag{15}$$

where Y (i), S(i) and U(i) are the predicted process response, the set point reference and the controller output at update i, respectively. The tuning matrices Q and R allow greater flexibility in the solution of the cost function. The above cost function is optimized with respect to U. By differentiating and solving for U, the next set of optimal control moves is obtained. Input constraints are implemented via a multivariable anti-windup scheme which was proved to be equivalent to an on-line optimization for common processes (Goodwin & Sin, 1984).

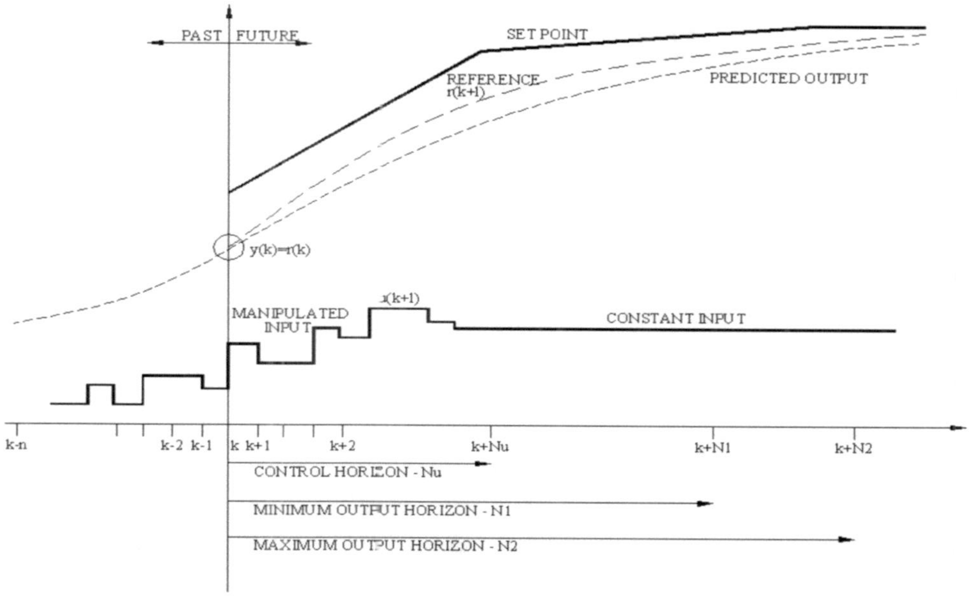

Fig. 2. MPC control concept for multivariable applications

Once the cost function is built, it remains the same until we change one of the models, Q, R, or one of the horizons. At each update, the following takes place: i) the state vector is estimated and adjusted; ii) the process response is estimated, and iii) the next control output is calculated.

Measured disturbances (feed forwards) are assumed to affect one process variable at a time. There can be up to three measured disturbances per process variable in the current controller software structure.

Aside from this change in the derivation of the control law, the multivariable algorithm is substantially the same as the single variable algorithm. That is, a deterministic control law is obtained which provides the current control move that will yield some future process response, the current move is implemented, and then a new control move is calculated based on new process data at the next control update step. For a complete mathematical development of the control law used in the multivariable case, refer to (Huzmezan, 1998).

The controller outlined above has demonstrated more than twenty years of success in industrial applications, verifying the effectiveness of this adaptive model-based predictive control algorithm.

3. Commercial implementation of the BrainWave controller

The BrainWave MPC controller is commercially available as Windows based PC software. The system connects to existing control systems using the OLE for Process Control (OPC) standard interface. A communications watchdog scheme is used to ensure that the process control automatically reverts to the existing control system in the event of any communication or hardware faults associated with the BrainWave computer.

Despite the complex mathematics use in the control and model adaptation algorithms, the software is designed to be easy to use and is suitable for control technicians to apply in an

industrial setting. The user interface for the BrainWave system is shown in Fig. 3. The interface includes a trend display of the process variables at the top right side of the interface. In this example, the set point is the yellow line, the process variable is the red line, and the controller output (the actuator) is the green line.
The identified process transfer function is shown on the lower right side. The transfer function plot is generated based on a step input at time=0 and thus shows the open loop transient response of the process to a change in the actuator or measured disturbance variable. The white line is a plot of the estimated transfer function of the process (expressed as a simple first order system using a dead time, time constant, and gain) which is used as a starting point for the model identification. The red line is a plot of the identified process transfer function open loop step response based on the Laguerre representation. The blue bars represent the relative values of the 15 Laguerre series coefficients which are the identified process model parameters.

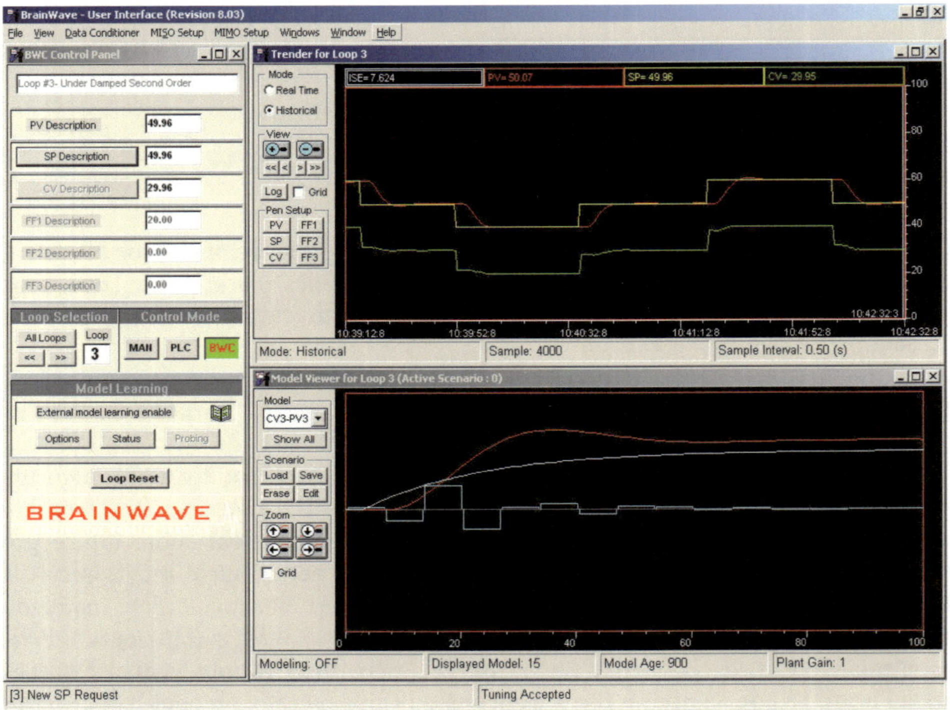

Fig. 3. BrainWave user interface

BrainWave allows the user to configure single loop controllers (Multiple Input Single Output - MISO) and multivariable controllers where more than one control loop interacts with other control loops (Multiple Input Multiple Output - MIMO). The process type is chosen as either self-regulating for stable systems or integrating for open loop unstable systems such as level controls. The primary information required is an initial estimate of the process response. This is entered using a simple first order plus dead time transfer function estimate. This estimate is used by BrainWave to determine the appropriate data sample rate for modeling and control of the process and also provides a starting point for the initial Laguerre model of the process.

Many systems are well described using the simple first order model estimate, and if the estimate is reasonably close to the actual process response, the BrainWave controller will be able to assume automatic control without any further configuration work. The default setup parameters are set to values that will provide a closed loop response time constant that is equal to the open loop response time constant of the process, thus providing effective and robust control without any additional tuning required.

Multivariable MIMO systems can be configured with a dimension of up to 48 inputs and 12 manipulated outputs. Estimates of the response of each process variable to each actuator or disturbance variable must be entered (if the cross coupling between the respective channels exists) so the complete array of process transfer functions is shown as a matrix.

For MIMO systems, it is desirable to be able to place more priority on one process variable than another as part of the controller setup. Similarly, some actuators are able to move more quickly compared to other actuators. The MIMO control setup provides simple sliders that allow the user to impose these preferences on the controller in a relative fashion between the process variables and actuators. Note that the default settings provided will result in stable and robust control of the process where the closed loop response time constant is equal to the open loop response time constant for each process variable.

3.1 Model identification

The BrainWave controller has online modeling features that help the user obtain the correct process transfer function model for the process. The model identification can be performed in open loop by observing the process response to manual changes to the actuator(s) or in closed loop during automatic control in BrainWave mode. Performing the model identification in closed loop is advantageous as it allows the user to confirm the improvement in the control performance at the same time as the convergence of the identified model is monitored in the model viewer, thus providing real-time confirmation that the identified process model is correct.

An example of the model identification procedure is shown in Fig. 4. The user starts by entering an estimate of the process response in simple first order terms as described earlier (the white line shown in the Model Viewer panel). BrainWave builds a Laguerre model that matches this estimate and uses this model as a starting point for both the predictive control operation and the model identification (the red line in the Model Viewer panel). A series of set point changes are made while the BrainWave controller is in control of the process as shown in the Trend display panel. Some overshoot is apparent during the first set point change due to the errors in the initial process response estimate provided by the user. As the series progresses, the identified process model converges to the correct value. As the model becomes more correct, the control performance during each successive set point change is improved. At the end of the sequence, the process model is completely converged to the correct value, and the resulting control performance is ideal.

Feed forward disturbance variables can also be modeled automatically while BrainWave is in control of the process. This unique capability allows the controller to identify the transfer function of the disturbance variable to the process so that the controller can calculate the correct control move with the actuator to cancel the predicted disturbance on the process. This feature enables BrainWave to provide the best disturbance cancellation possible. An example of the adaptive disturbance cancellation capability is shown in Fig. 5. The feed forward variable is the square wave signal at the bottom of the figure. In this example, the

BrainWave controller begins with no model for the feed forward disturbance variable so the process is disturbed from the set point and the actuator (the trend in the center of the figure) reacts by feedback after the disturbance has already occurred to return the process to the set point. As the feed forward transfer function is identified by BrainWave, the controller begins to take immediate action when the feed forward disturbance signal changes, thus providing improved cancellation of the disturbance. As the identified feed forward model converges to the correct value, the cancellation of the disturbance is complete.

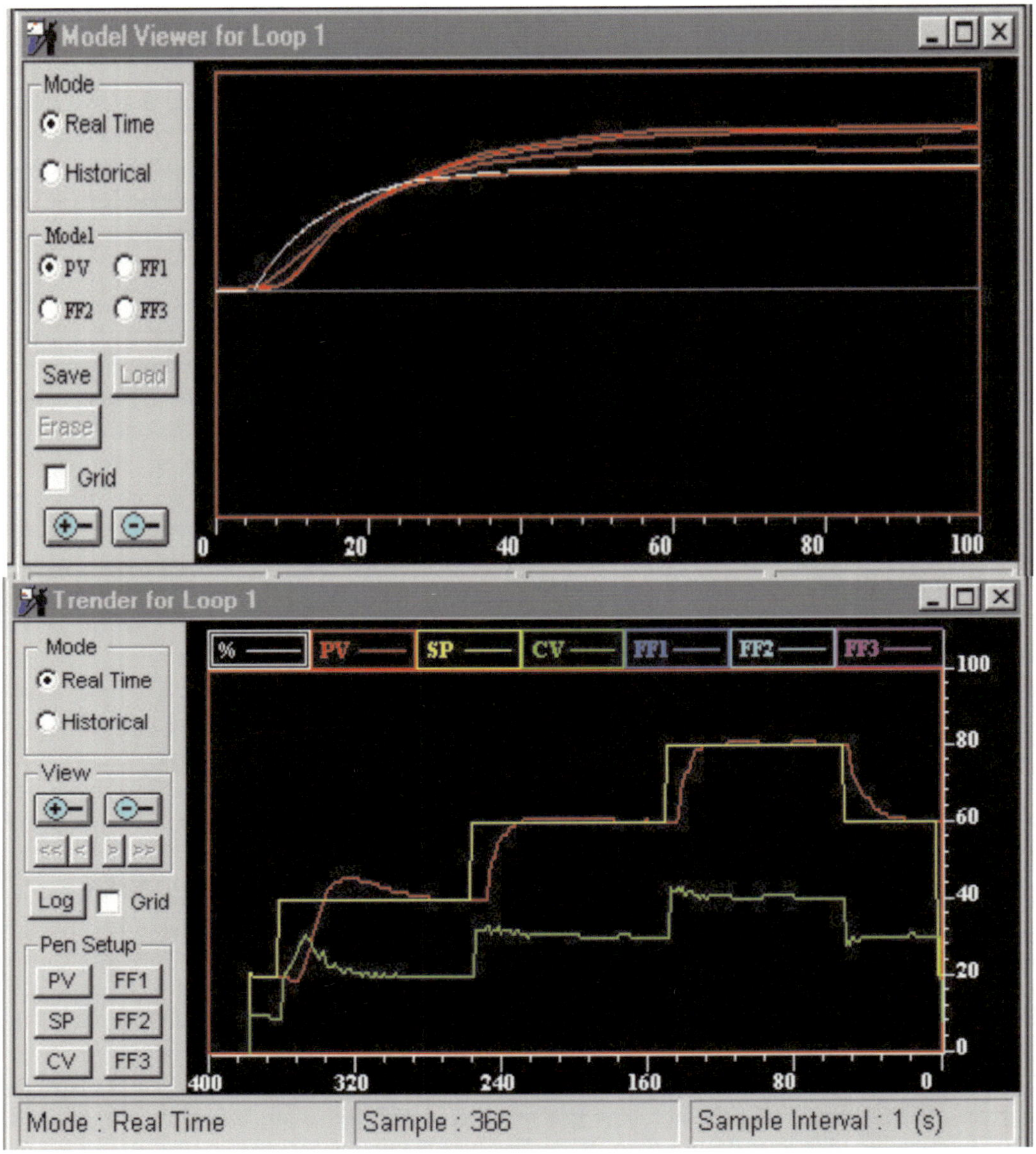

Fig. 4. Process model identification during closed loop control

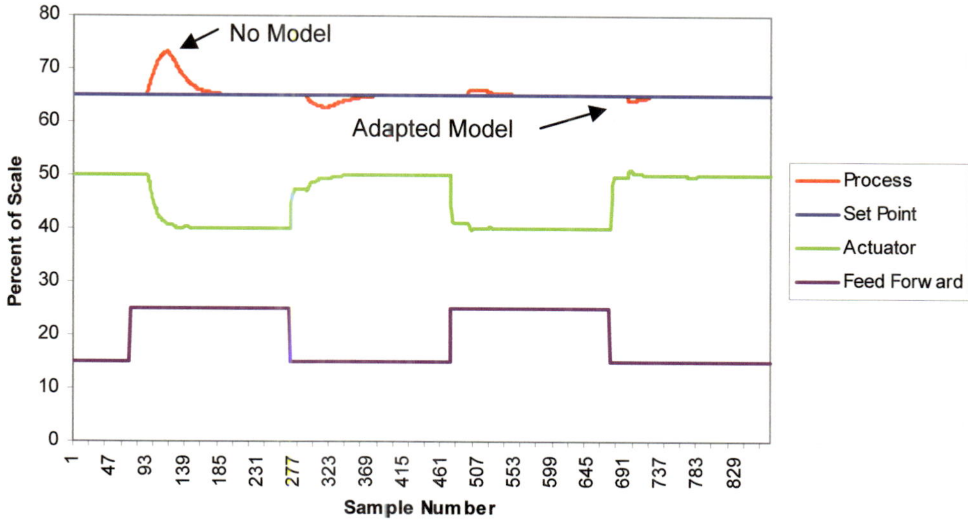

Fig. 5. Adaptive feed forward disturbance cancellation

3.2 Predictive control

The identified process model is used as the basis to calculate the required control moves with the actuator to bring the process to the set point some time in the future. For the single loop MISO controller, the primary adjustment for control performance is the prediction horizon. This is essentially the number of control steps into the future where the process variable must be on the set point. A long prediction horizon results in conservative closed loop control response; the actuator is simply positioned to the new calculated steady state value and the process reaches the set point according to its natural open loop response time. Shorter prediction horizons cause the controller to make a more aggressive transient move with the control actuator in order to bring the process to the set point in less time than the natural response. Fig. 6 shows the effect of adjusting the prediction horizon for self-regulating (open loop stable) systems

Fig. 7 shows the effect of adjusting the prediction horizon for integrating (open loop unstable) systems. In addition to adjusting the prediction horizon, the set point reference is also internally filtered to provide less aggressive actuator movement during set point changes. It should be noted that integrating systems are only stable in closed loop for a certain range of controller behavior (a marginally stable system in closed loop) so the amount of arbitrarily fast or slow control performance adjustment is limited.

In the multivariable MIMO controller, there are several parameters that can be adjusted to change the control performance. The weighting can be adjusted to change the penalty for set point tracking error for each process variable and the movement penalty for each actuator. The control horizon, which specifies the number of future control moves to be calculated, can be made longer for conservative actuator movement and slower process response or shorter for more aggressive actuator movement and faster process response. The prediction horizon is also adjusted to determine the range of future process response that will be included in the cost function minimization. Typically this range begins some time after the process with the longest time delay has elapsed and ends where the process variables should naturally arrive at steady state. As there are many parameters that have a similar effect on the control performance, typically most are set to their default values and only the set point tracking error weightings are used to adjust the relative priority of the process variables being controlled.

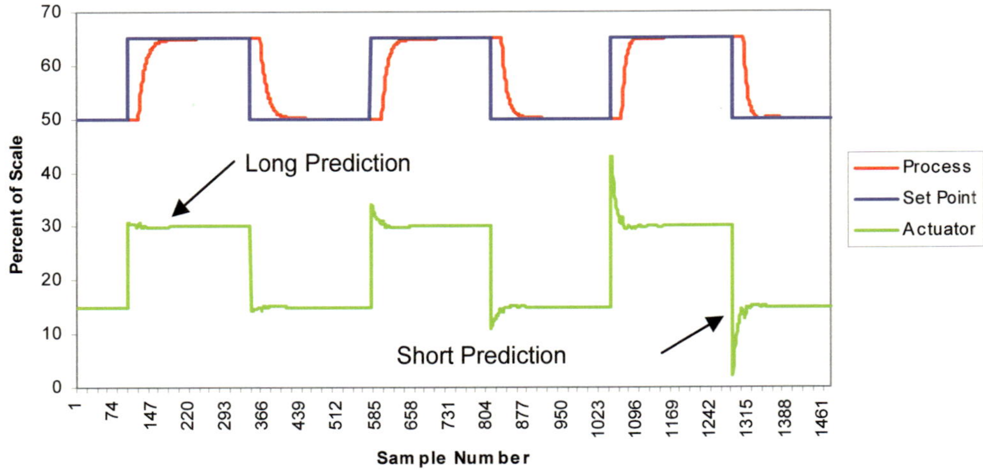

Fig. 6. Effect of Prediction Horizon Adjustment for Self-Regulating Systems

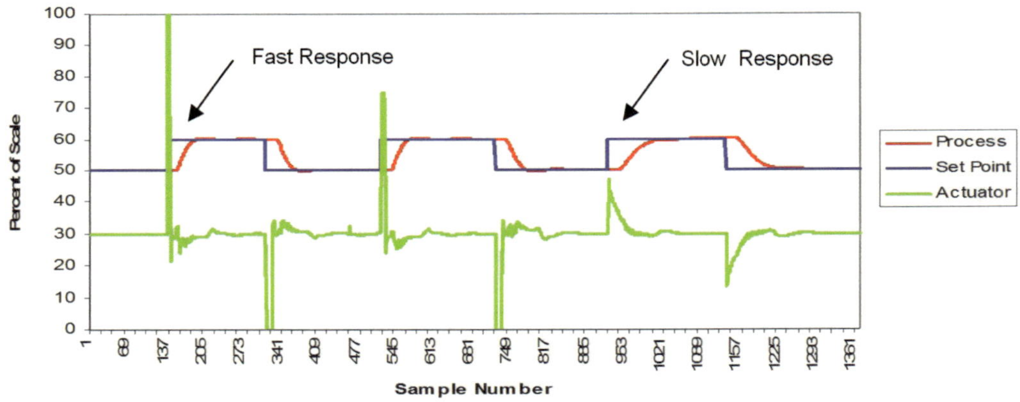

Fig. 7. Effect of prediction horizon and set point reference filtering for integrating systems

4. Mineral processing applications

Mineral recovery processes are notoriously difficult to control due to a combination of factors, including variable ore properties, complicated process dynamics, interacting process variables, and an inability to measure some key variables. Considerable time and effort has been put into the development, installation, tuning, and testing of automation systems for mineral recovery plants due to the high value of the process. A typical approach is to use PID controllers, with enhancements such as cascade control, feed forward control, ratio control, and so forth, that are available in distributed control systems (DCS). It is possible to have some degree of success in automating and bringing stability to processes with these techniques; however, PID does not perform well when there are challenging loop dynamics (such as the transport delay associated with ore feed systems, or the integrating behaviour of flotation cell levels), or multiple interacting variables (such as rotation speed, recycle flows, and ore feed rate on SAG mill weight).

A more advanced approach to automation that seems to be popular among mineral processors is the use of rule-based expert systems. A rule-based expert system that mimics the best operator actions is a logical development towards solving underlying difficult feedback control problems in the absence of other alternatives to PID control. A number of these rule-based expert systems are commercially available and they typically take the form of software running on a PC that is frequently exchanging information (receiving process data and sending control outputs) with the DCS. However, in simply trying to automate the actions of even the best human operators, there is the potential to miss an opportunity to make a full step improvement in the control of the process. For example, MPC can provide control performance improvements by managing the process dynamics and interactions in a completely different manner.

MPC is able to provide a quality of regulatory control that cannot be matched by PID control or expert system control. The superior performance of MPC in controlling challenging loops provides a clear benefit in terms of plant profitability. It is not effective to replace this advanced regulatory control layer by trying to extend traditional DCS-based controls upwards, or expert system type controls downwards; so without the advanced regulatory control layer with MPC, a comprehensive plant automation strategy cannot be properly built.

4.1 SAG mills

A semi-autogenous grinding (SAG) mill is a rotating cylinder containing steel balls that is used as an intermediate stage in the comminution (size reduction) of ore. Typically, these mills receive ore from a primary stockpile and, operating continuously, reduce the particles to a size appropriate for finer grinding in ball mills. As the name indicates, size reduction in a SAG mill occurs both from the tumbling action with the balls and from autogenous (rock on rock) grinding.

SAG mills are a critical component in mineral processing operations as they are often the bottleneck of the processing line; at the same time, they are a difficult process to control due to the transport delay inherent in the ore feed system and the integrating nature of the mill as it approaches maximum load. Depending on a number of factors, the weight of material resident in the mill will vary. It is important to maintain the weight within a specific range, as an overloaded mill quickly loses grinding efficiency, while an insufficiently loaded mill does not take proper advantage of the autogenous grinding mechanism. Grinding is also affected by rotation speed; in general, greater rotation speed allows for increased grinding. Finally, the power draw of the mill must be monitored to avoid exceeding the maximum power limit. Although brief, this description of mill operations begins to indicate the complex, multivariable nature of the SAG mill control problem.

A very direct, and popular, approach to maximizing mill throughput is to use an expert system to gradually increase the feed rate to the mill until a high limit for weight or power draw has been reached. Should a high limit be violated, feed rate to the mill is reduced until mill operations return to acceptable limits. This approach has the advantage of being direct and easy to understand, but can result in a 'saw tooth' pattern of feed rate increases and decreases, which may not be desirable. The area below this saw tooth pattern represents an opportunity to achieve higher average ore feed rate and throughput.

An alternative to this approach is to adjust mill feed rate to maintain mill weight at a set target. The idea justifying this approach is that there is a mill weight at which optimum grinding occurs. Therefore, by maintaining the correct mill weight, mill throughput is maximized. This approach has the benefit of increasing the stability of mill weight (and power draw). The drawback to this approach is that it is not always clear what weight target should be chosen, although adjusting the weight target to maximize power draw may complete this strategy.

Enhancements to these strategies may include the use of feed particle size distribution, ore hardness, and ore type as predictors for mill grinding efficiency. Mill sound may also be used as an indicator of the tumbling profile inside the mill. Some strategies have attempted to control the mill sound by adjusting mill rotation speed; however, this can be problematic due to the large and rapid effect that rotation speed can have on mill weight.

The adaptive model-based predictive control strategy based on the BrainWave MPC controller has been applied to a number of SAG mills in primary copper producing mines. The mills had a range of capacities, with the smaller mills operating at ore throughputs as low as 40,000 tons per day, and the larger mills operating at 120,000 tons per day or more.

The typical application of the BrainWave controller for SAG mills is for control of the mill load. In these cases, mill load was inferred either from a direct weight measurement or from bearing oil pressure. Fig. 8 gives a diagram showing the SAG Mill and the load control strategy. The mass feed rate of ore is measured on the feed belt and typically an enhanced PID control strategy is used to control the ore feed rate. BrainWave operates as a master controller over the feed rate controller. This controller takes the load measurement and set point, as well as key feed forward variables such as mill rotation speed, pebble recycle rate, and ore sizing information (coarse or fine fraction) and sets the set point for the feed rate controller. In some cases the BrainWave MPC may also be used for the feed rate control itself. Where there is significant transport delay between an increase in the addition of ore to the feed belt and the measurement of that feed rate increase, MPC can provide improved feed rate control.

It should be noted that in the majority of cases, this strategy was implemented on SAG mills that were already operating with expert system control. In these cases, BrainWave was integrated with the expert system, taking over the task of setting the feed rate to the mill.

The results of the MPC application demonstrated increases in mill throughput in the range of 1.5 to 2% relative to mill throughput with expert system control alone. Table 1 lists some results. To emphasize the significance of these increases, note that for a concentrator processing 120,000 tonnes per day of ore at a grade of 0.75%, with a recovery of 0.85, total annual copper production is roughly 550 million pounds. Even at a low profit margin of $0.20/pound, a production increase of 1% is worth over one million dollars of additional profit per year.

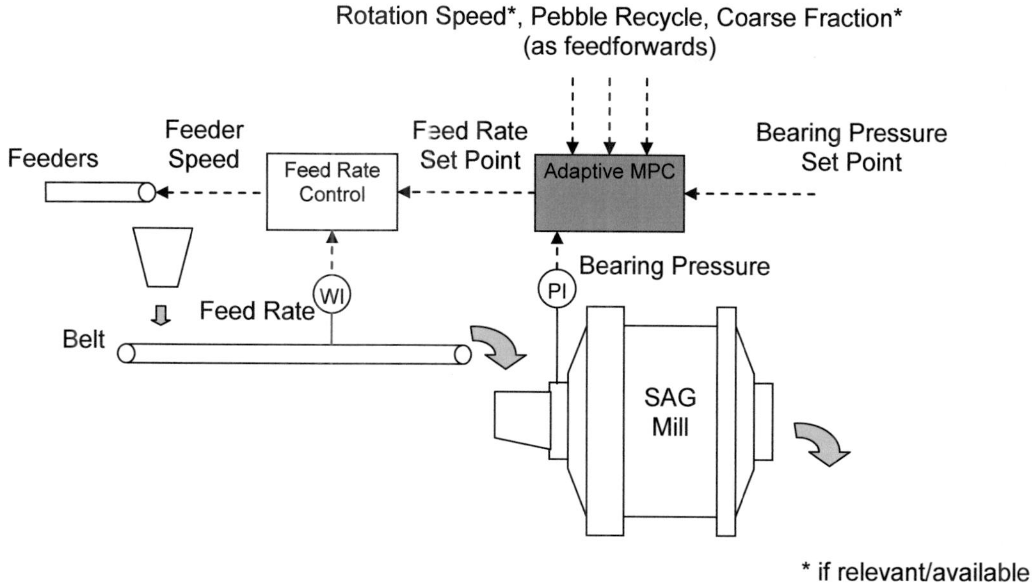

Fig. 8. BrainWave SAG mill load control

Mine Site	Number of SAG Mills	Production Increase
Minera Candelaria	2	1.6%
Minera Escondida - Laguna Seca	1	>2%
Minera Los Pelambres	2	1.5%

Table 1. SAG Mill Throughput Increases with MPC

Fig. 9 displays some process trend data from one of the applications over a period of several days. The trend also shows that when the MPC was taken off-line for some hours for maintenance the expert system was not able to maintain the mill load with the same stability (circled region in the trend plot). This particular concentrator plant is fed with a mixture of two ore types with significantly different work indexes. For this reason, grinding rates in the mill appear to change frequently and rapidly and the MPC control proved to react sooner and minimized the disturbances caused by these events. The impact of using MPC on overall mill performance was shown to go far beyond stabilization of mill load. One plant site observed reductions well above 50% in the standard deviation of ore feed rate, mill rotation speed, mill weight, and mill power draw.

4.2 Flotation cells

Mineral flotation is a process that takes advantage of differences in hydrophobicity to separate valuable minerals from tailings or one valuable mineral from a bulk concentrate. Hydrophobic ore particles become attached to air bubbles which rise to the top of the flotation cell to form a froth layer. The froth is collected as it overflows the top of the cell while the rest of the material passes through the cell outlet. Typically, a number of cells are arranged is series so that as the pulp flows through a flotation line, the hydrophobic particles have a number of opportunities to separate from the pulp.

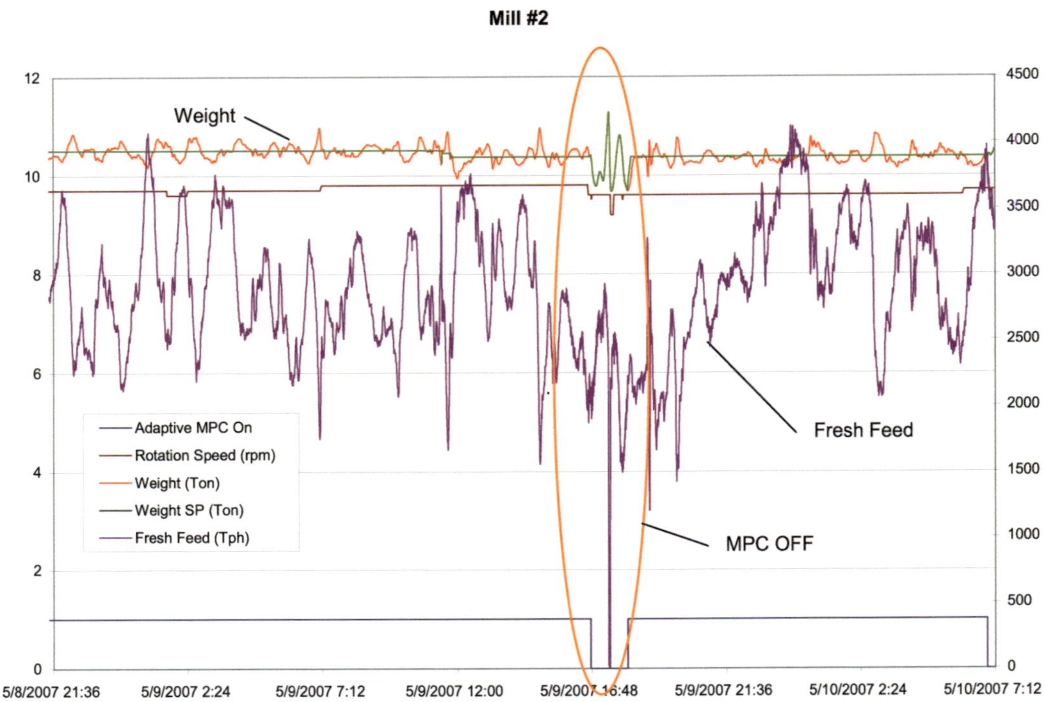

Fig. 9. BrainWave control of SAG mill load

To promote the success of flotation, it is important to have good control of the froth depth. Commonly, froth depth is inferred by measuring the location of the froth-pulp interface (with a float device) relative to the lip of the cell where the froth is able to spill into the recovery system. Under normal cell operations, froth will completely fill this space, from the top of the liquid level in the cell up to the lip of the cell. If the froth level can be maintained, a stable froth layer is formed without excessive bubble breakage, and there is a steady overflow of froth over the lip of the flotation cells. A well-chosen froth depth allows time for some 'drainage' of entrained water and tailings particles before the froth overflows the cell; however, greater froth depths can result in lower recovery and increased use of expensive frother chemicals (Wills & Napier-Munn, 2006).

Froth level control can be a challenging problem for a number of reasons. There can be large and sudden changes in the flow of pulp into the flotation line. Often this flow is unmeasured and the changes can cause large level disturbances. The arrangement of a number of flotation cells in series means that the outlet flow from one cell is the inlet flow to the next cell in the line. Therefore, as outlet valve position is adjusted to control cell level, the levels in downstream cells will be affected. The valves used in flotation applications are often dart valves or pinch valves that may not always have a fast and accurate positioning response. To summarize, control of flotation level is difficult due to the combination of significant unmeasured process disturbances, interacting process variables and imperfect actuators.

BrainWave MPC was applied to two copper rougher-scavenger flotation lines, each consisting of five 160 m^3 flotation cells. Each cell level is controlled by an individual controller which adjusts the cell's outlet valve position. To compensate for the impact of valve position changes on downstream cell levels each level controller takes the upstream cell's outlet valve position as a feed forward disturbance variable. This allows each controller to minimize disturbances by adjusting outlet valve positions at the same time as the inlet valve positions are changing. Fig. 10 provides a simple diagram illustrating the application of the control loops.

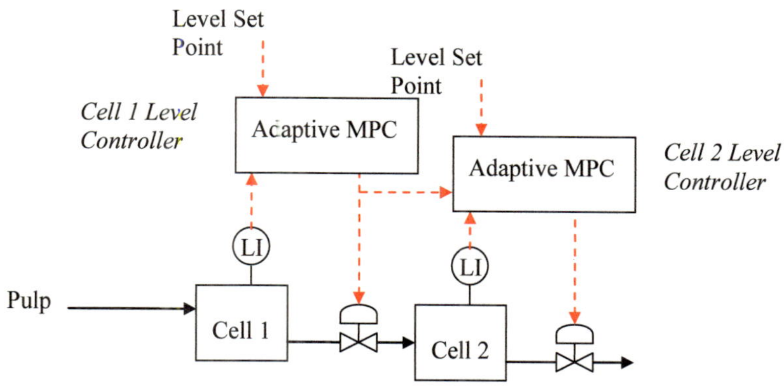

Fig. 10. Diagram of flotation control strategy

The advanced control loops replaced the existing PID loops which the plant had previously used to maintain cell level. The PID controllers had struggled to maintain tight control and had particular difficulty stabilizing the cells after large disturbances or on plant start-ups. Fig. 11 demonstrates the improvement to the cell level control that was achieved by replacing traditional PID loops with the BrainWave loop controllers. In this case there was as much as 60% reduction in level variability as measured by standard deviation.

It is suggested that improved level stability has the effect of improving mineral recovery, improving product grade, and reducing frother consumption. However, these are difficult benefits to prove, due to the problem in obtaining and collecting sufficient data, where the comparisons can be made given the same operating conditions (for example, with the same production rates and ore types). Unfortunately, to date, the plant has not been able to complete such an analysis. However, one obvious and easily measurable benefit came from the improved control performance on large production rate changes or plant start-ups. It was observed that the PID controllers had difficulty in stabilizing the cells after these events; often cell levels could swing for as much as two hours before settling into what could be considered steady-state. This settling time was greatly reduced with the BrainWave controllers. The benefits from avoiding this loss in operating time can be directly calculated. An example calculation can be made based on the business fundamentals given in Table 2.

Tonnes Milled per Day	50,000
Copper Grade	0.3%
Recovery	75%
Tonnes Copper Produced	112.5
Pounds per Metric Tonne	2,205
Price per Pound	$4.00
Production Cost	$1.50
Daily Profit	$620,156

Table 2. Example Mineral Concentrator Business Fundamentals

If the BrainWave control reaches 'steady state' in one hour less than PID control, and there is one of these large disturbances events per week, then this represents the equivalent of 2.2 more days of operating time per year. Based on daily revenue of $620,156, this gives an increase in profit of $1.36 million annually.
Note that this is only the 'easily' calculated benefit and does not include the additional benefits to be obtained through improvements in recovery. Even a modest 0.5 percentage point increase in recovery from improved froth level control yields close to an additional $1.1 million per year in profit.

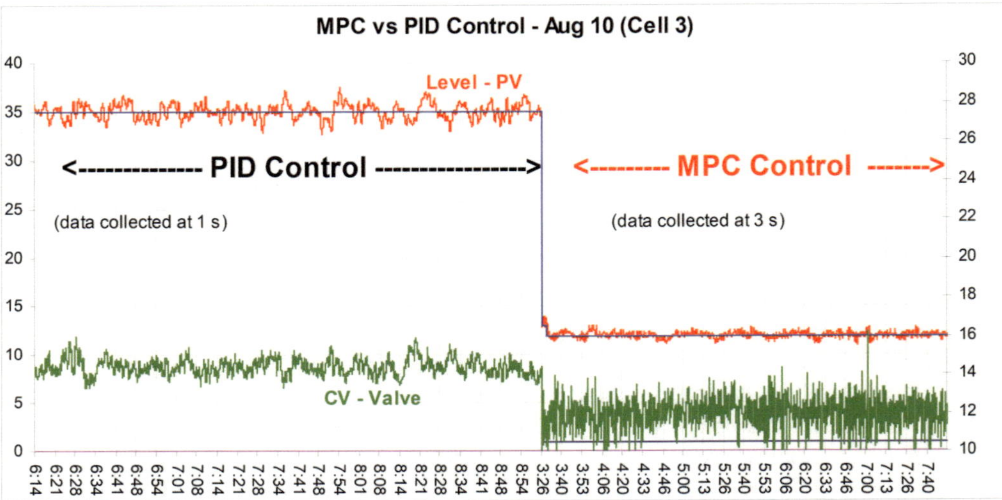

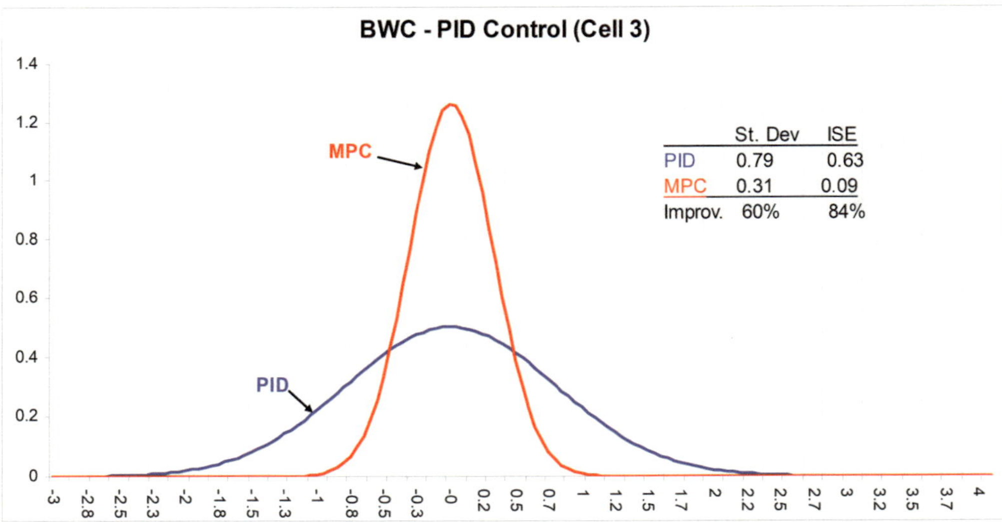

Fig. 11. Comparison of PID and BrainWave control of flotation level

5. Pulp and paper applications

Many processes involved in the manufacture of pulp and paper exhibit challenging dynamics for control due to long reaction times. The pulp digester, which cooks the wood chips with sodium hydroxide to produce pulp, has a retention time of several hours. Many properties of the pulp are affected by conditions in the digester so it is very important to maintain the correct chip level, temperature profile, and chemical concentration throughout the cooking process to produce pulp with uniform quality. The chemical recovery circuit is used to regenerate the sodium hydroxide for use in the digester as part of a closed loop sodium cycle known as the Kraft process. The chemical recovery system consists of a lime kiln and recaustification process which is also a closed circuit that uses calcium carbonate and calcium hydroxide to transform the sodium carbonate back to sodium hydroxide. This system has long response times due to the large thermal mass in the kiln and the large retention time in the causticizing reactors. The pulp bleaching stages also have long process retention times to allow complete reaction of the pulp with the expensive bleaching chemicals. Finally, the pulp dryer, which is a very large unit with a long residence time for the drying pulp, must be controlled to maintain the final moisture content of the produced pulp.

Each of these processes present unique challenges for automatic control. The BrainWave MPC controller has been successfully applied to these processes due to its ability to handle the long response times of these systems.

5.1 Digester control

Optimal digester operation requires precise control of the wood chip levels in the steaming bin, impregnation vessel, and the digester to maintain uniform residence time in the cooking process. Chemical addition, as measured by effective alkali (EA) and digester temperature profile must be constant to provide a consistent cook, or delignification, of the pulp as measured by a Kappa index. Reducing variability of these process parameters yields a corresponding reduction in Kappa variability. Precise Kappa control can reduce bleach plant costs and is certain to improve the quality of the pulp.

The BrainWave controller is used to hold these critical digester variables on target. The digester cooking control strategy is shown in Fig. 12. The level of the wood chips in the

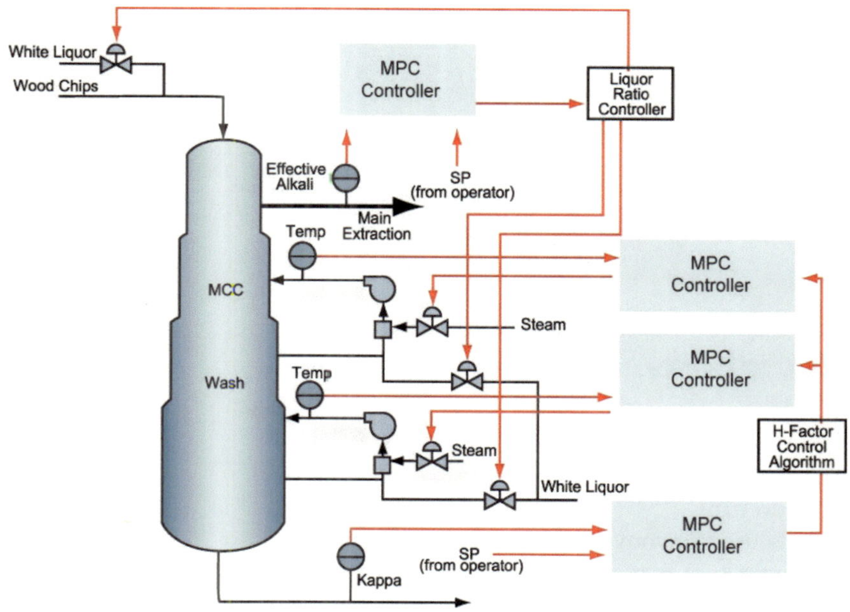

Fig. 12. Digester control scheme

digester is controlled by adjusting the flow of chips from the impregnation vessel (sluice flow) into the digester, which is fed at the top of the column. Transport delay time as well as delays in the digester level measurement, combined with the integrating level response, create a difficult control problem.

The existing digester level control was based on a PID controller in the DCS. During large disturbances, the operator assumed manual control to try and stabilize the level. The BrainWave controller was able to reduce the standard deviation of the level by more than 50% and required much less intervention from the operator. Fig. 13 provides a comparison of the control performance when using the existing DCS/Manual control approach and the improved control achieved with BrainWave.

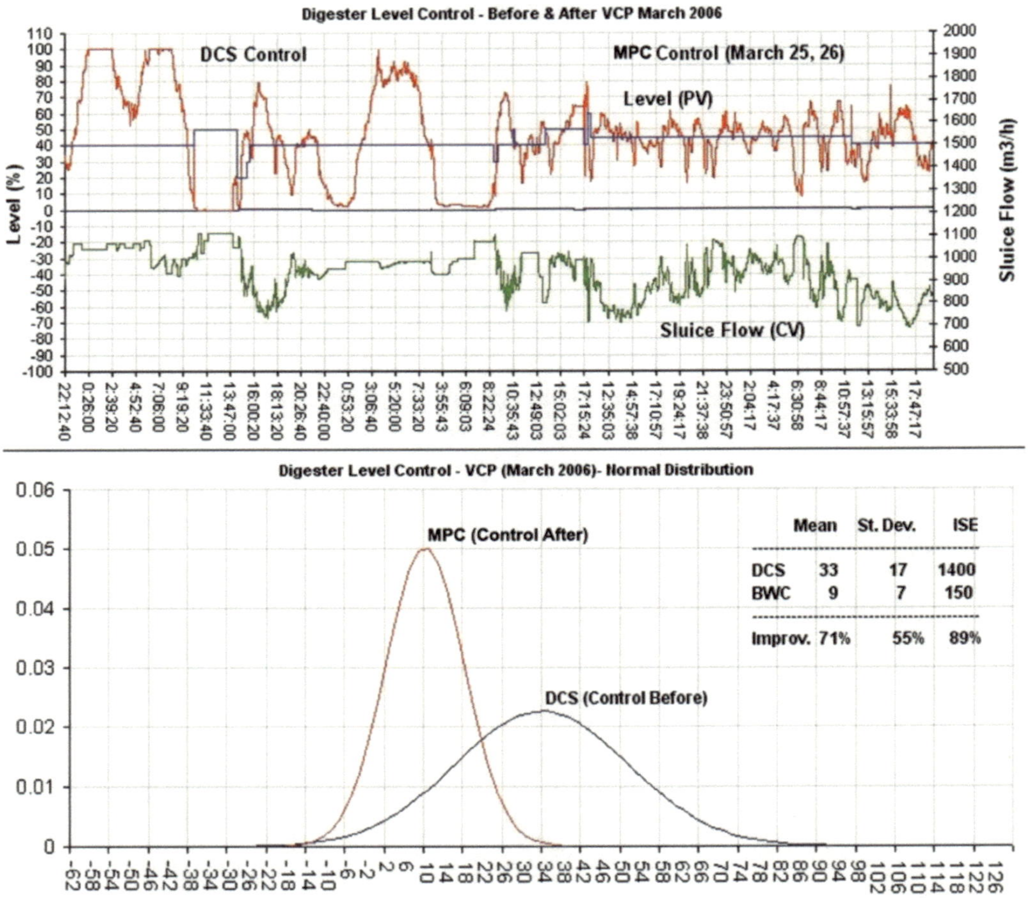

Fig. 13. Comparison of MPC vs. DCS control of digester level

5.2 Lime Kiln control

The lime kiln is used to convert calcium carbonate into calcium oxide (lime) for use in the recaustification process that converts sodium carbonate to sodium hydroxide for use in the digester. The properties of the lime are dependent on the temperature profile of the kiln. The temperature profile is typically manually controlled due to the long time delays and multivariable interactions of the draft (air flow) and fuel on the kiln temperature profile that make automatic operation with conventional PID controllers impractical. Response times of one to three hours or more are typical. Operators are often impatient with the long response time of this system and tend to make large corrections to the fuel feed rate in an attempt to recover the temperature profile quickly during process disturbances such as production rate changes. These actions result in extremes of temperature in the kiln, leading to poor lime quality, ring formation problems, and reduced refractory life. Operators also tend to control the temperature profile at a higher value than necessary for the lime burning and at a high excess oxygen level to provide a comfortable operating margin that requires less frequent attention. These practices lead to increased fuel consumption and maintenance costs.

Adjusting draft and fuel cause shifts in the flame length and excess oxygen levels. In addition to the long response times, this interaction must also be addressed by the control strategy to achieve responsive yet stable control performance. The ultimate objective of the control strategy is to maintain a constant lime discharge temperature to ensure consistent lime quality as measured by the residual calcium carbonate (un-burned lime) and reactivity with water to produce calcium hydroxide (slaking rate).

A multivariable MIMO BrainWave controller is used to control the temperature and oxygen level at the feed end of the kiln by adjusting the fuel and induced draft (ID) fan. A single loop BrainWave controller is used to control the lime discharge temperature by adjusting the target for the feed end temperature. This approach allows feed end temperature limits to be easily included in the control strategy, as low temperatures can lead to plugging of the lime mud feed and high temperatures can lead to equipment damage in the dust removal system. A schematic of the control scheme is given in Fig. 14.

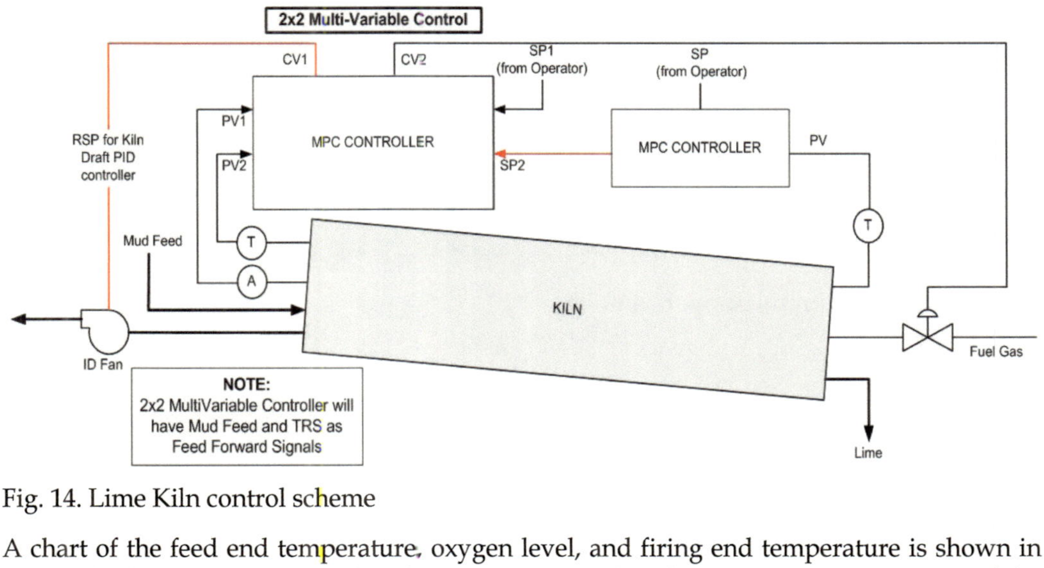

Fig. 14. Lime Kiln control scheme

A chart of the feed end temperature, oxygen level, and firing end temperature is shown in Fig. 15 for both manual control and automatic control with BrainWave. A summary of the performance improvements obtained by the BrainWave MPC controller on this application is shown in Table 3. The range of variation of each process variable was substantially reduced, allowing the kiln to operate at a lower average temperature with lower excess oxygen. These improvements resulted in reduced fuel consumption, reduced incidence of ring formation, increased production capacity, and a better quality lime (as measured by residual Calcium Carbonate $CaCO_3$) as shown in Fig. 16. Based on the results of this application, as well as experience with over 20 other similar applications, the control improvements possible with MPC provide reductions in fuel consumption of 5% or more while reducing lime quality variability by 50%.

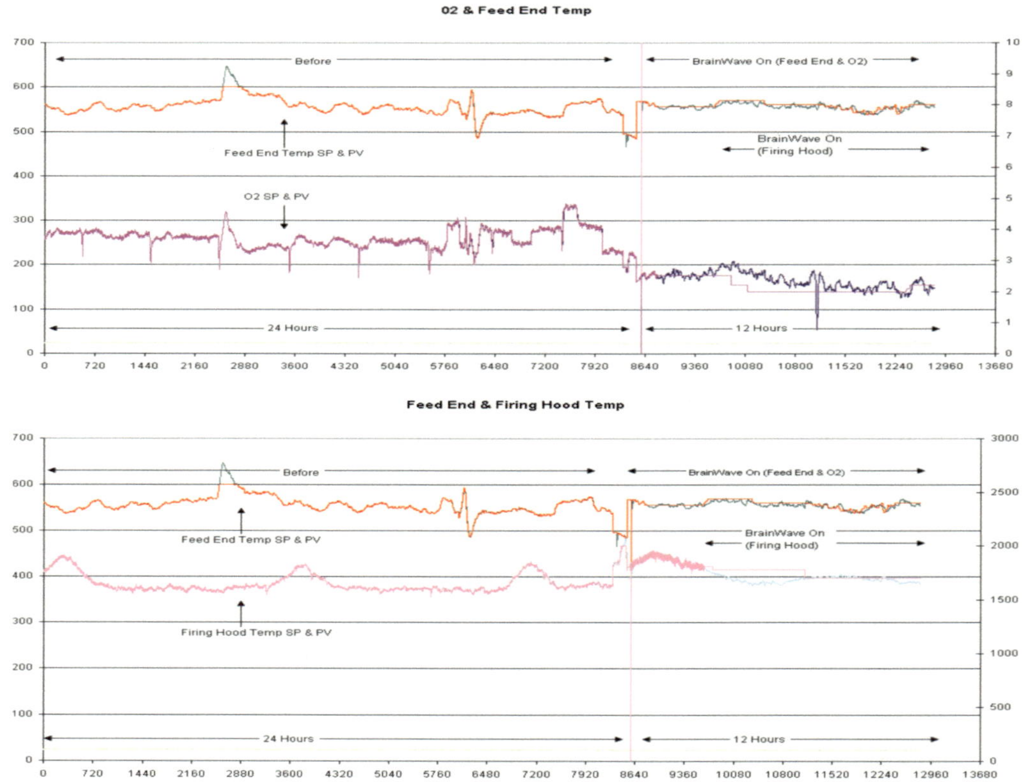

Fig. 15. Lime Kiln control comparison

Process Variable	Manual Control	MPC Control	Improvement
Excess Oxygen	1%	0.3%	70%
Feed Temperature	40°F	7°F	82%
Lime Temperature	200°F	25°F	87%

Table 3. Lime Kiln process variability comparison

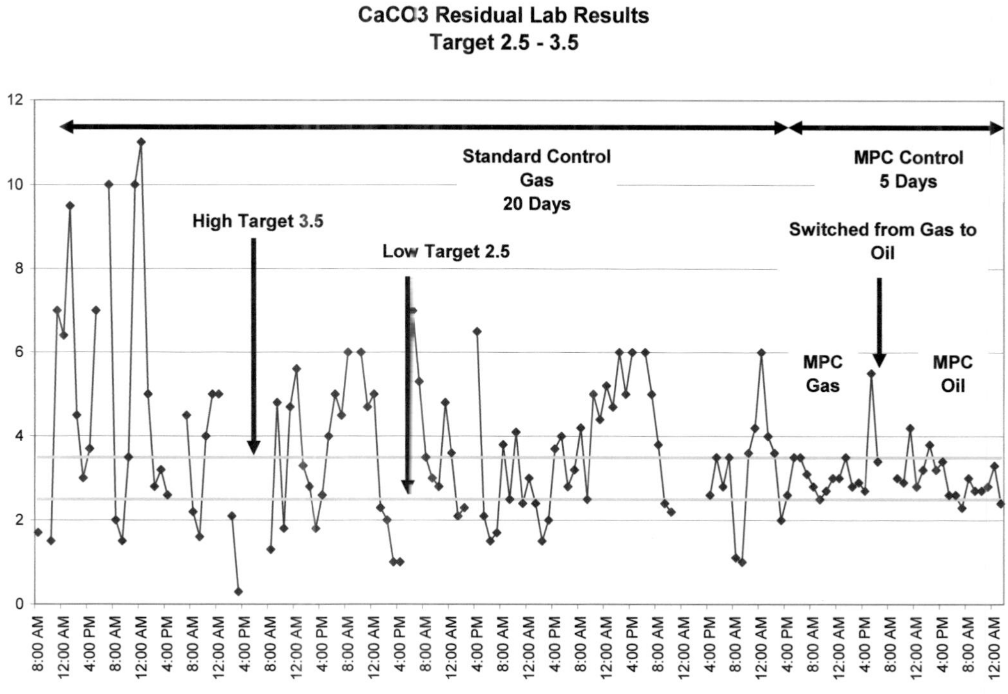

Fig. 16. Lime Kiln $CaCO_3$ residual laboratory results comparison

5.3 Pulp bleaching

The pulp bleaching process consists of several stages where bleaching chemicals are applied to the pulp to increase brightness. These reactions occur in large towers with plug flow of the pulp to allow long retention time for completion of the bleaching reaction. One of the stages in the bleaching process is known as extraction stage where sodium hydroxide is applied to the pulp to remove remaining lignin that was not removed in the digester. The addition rate of the sodium hydroxide is controlled based on the pH at the exit of the extraction stage as this measurement provides an indication if the correct amount of chemical was applied. High pH values indicate that excess chemical was applied and can result in damage to the pulp fibers and loss of pulp quality. Low pH values indicate that insufficient chemical was applied resulting in less removal of lignin. The higher lignin content will require additional expensive bleaching chemicals in the downstream stages in order to achieve the target final pulp brightness.

Control of extraction stage after tower pH is challenging due to the long and varying dead time, and that fact that the dead time and the process gain change significantly with production rate. Due to the plug flow nature of the reaction tower, the process dead time for the pH control is five times longer than the process time constant, making this control application particularly difficult. The pH response to a change in addition rate of sodium hydroxide had about two hours of dead time and a time constant of about 17 minutes.

As production rate changes affected these process dynamics significantly, the BrainWave MPC controller was configured with a set of process models to cover the entire production range. The main differences between the models were the process gain and dead time. Process gain ranged from 0.6 to 2.0 and dead time ranged from 2,000 to 6,000 seconds.

Lower production rates will require models with higher gain and longer dead time. As production rate increases, model gain and dead time will decrease. The MPC controller dynamically loaded the appropriate model according to the production rate as this provides a faster solution in this case than relying on adaptation alone to correct for the changes in the process. The mill had attempted to implement a Dahlin type controller but they had difficulty keeping the process stable. The addition rate of sodium hydroxide was applied as a ratio to the pulp production rate and the operator manually adjusted this ratio to maintain the extraction pH in the correct range.

Fig. 17 shows the control performance achieved by the operator and the MPC control. Table 4 shows the comparison between the MPC control and manual control. The improved control stability provided by the MPC control allowed operation at an average pH set point of 10.2 instead of 10.5, resulting in a reduction of sodium hydroxide addition with corresponding savings of about $100,000 per year.

Performance Index	Manual Control	MPC control	Improvement
Standard Deviation	0.089	0.032	64%

Table 4. E Stage pH control improvement summary

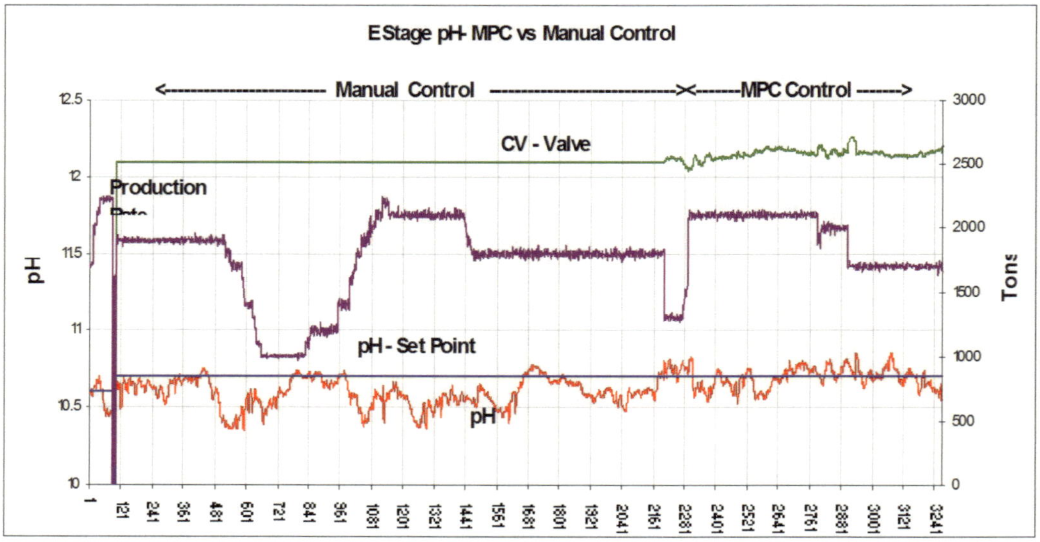

Fig. 17. Extraction stage pH control comparison

5.4 Pulp dryer control

Pulp dryers are used to control the final moisture content of the pulp before it is shipped. The moisture must not exceed a maximum specification limit so the dryer tends to be operated with slightly over-dry pulp. This energy required for pulp drying can be reduced if the pulp moisture can be controlled as close as possible to the specified limit. BrainWave MPC is ideally suited to control the drying process due to its ability to account for the long transport delay time as the pulp moves through the dryer to the moisture measurement sensor located at the dryer exit. The MPC controller also provides an effective means to incorporate measured disturbance variables such as sheet speed, broke flow, and pulp consistency as feed forward signals.

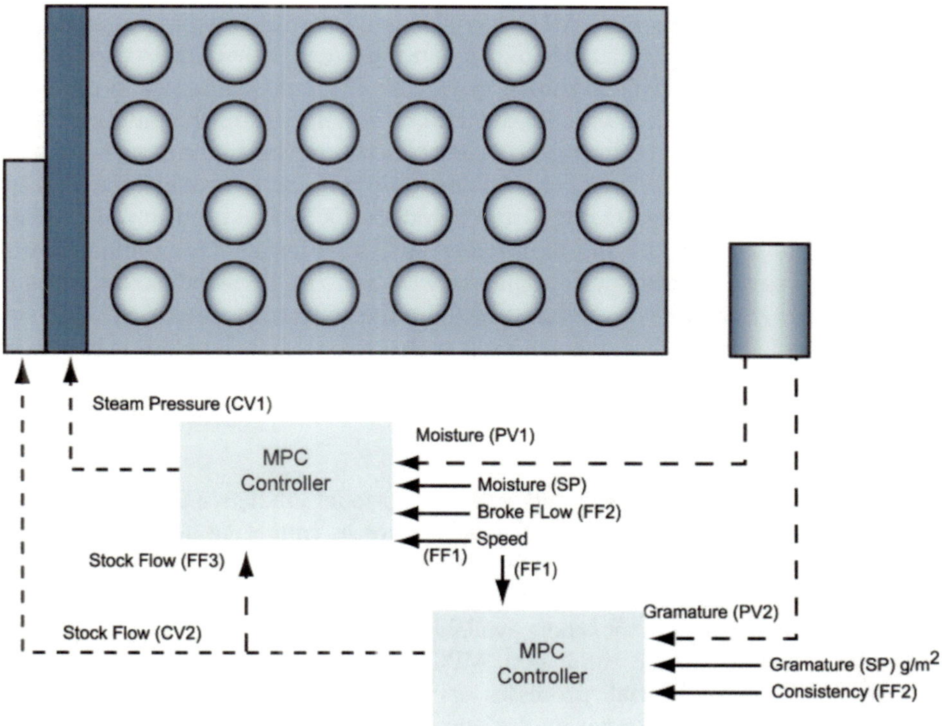

Fig. 18. Pulp dryer control scheme

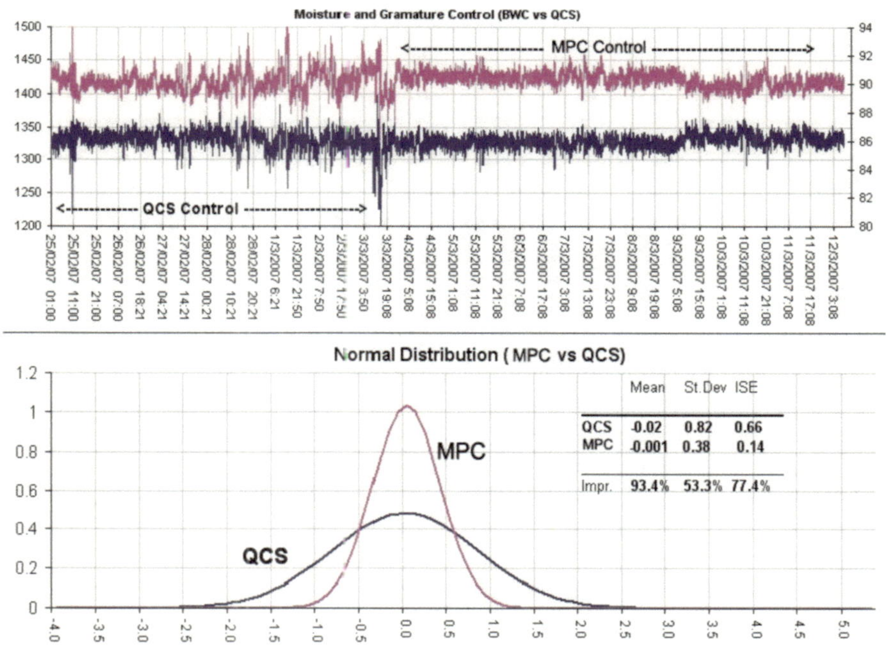

Fig. 19. Pulp dryer control performance comparison

In addition to pulp moisture control, MPC was also used to control the pulp gramature (mass of pulp per square meter) in the mat forming section at the feed end of the dryer. A diagram of the pulp dryer control schematic is shown in Fig. 18. The pulp gramature is controlled by adjusting the flow rate of pulp stock onto the wire section that forms the pulp mat. The final pulp moisture is controlled by adjusting the steam pressure applied to the dryer sections.
In this example, the pulp dryer was part of a new pulp mill and was one of the largest pulp dryers ever built. The existing Quality Control System (QCS) was used to control the pulp moisture and gramature before the BrainWave MPC was installed. A comparison of the moisture and gramature control performance is shown in Fig. 19. Standard deviation of the pulp moisture and gramature was reduced by more than 50% compared to the QCS system. With the improved stability, the average pulp moisture could be kept closer to maximum, leading to increased production and energy savings.

6. Conclusions

In this chapter, the development and application of a Model Predictive Controller (MPC) has been presented. It is clear that many industrial processes cannot be adequately controlled using conventional Proportional-Integral-Derivative (PID) control techniques due to common problems such as time delay and multivariable interactions. MPC exploits the abundance of inexpensive computing power that is now available so the limitations of the old pneumatically powered PID approach can be eliminated. MPC provides an opportunity to improve the performance of most industrial processes in order to reduce production costs and environmental impact, and improve product quality and profitability. These improvements can be achieved much faster and at less capital cost compared to modifications or upgrades of the process equipment resulting in an attractive return on investment.

7. Acknowledgements

The author would like to thank all the staff at Andritz Automation for their contributions to the development and application of the BrainWave controller. The cooperation of customers where this new technology has been installed is invaluable and the author is thankful for their willingness to share the results of their work. ISA should also be acknowledged for creating this book and making it available to the automation community.

8. References

Clarke, D.W., Mohtadi, C., & Tuffs, P.S. (1987). Generalized Predictive Control – Part I. The Basic Algorithm", *Automatica*, Vol. 23, No. 2, 1987, 137-148.

Goodwin, G.C. & K.S. Sin, K.S. (1984). *Adaptive Filtering, Prediction and Control*. Prentice-Hall Inc.

Huzmezan, M. (1998). Theory and Aerospace Applications of Constrained Model Based Predictive Control. PhD thesis, University of Cambridge.

Salgado, M.E., Goodwin, G.C., & Middleton, R.H. (1988). Modified Least Squares Algorithm Incorporating Exponential Resetting and Forgetting, *Int. J. Control*, Vol. 47, No. 2, 477-491.

Wills, B.A. & Napier-Munn, T.J. (2006). *Wills' Mineral Processing Technology*, Butterworth-Heinemann, Burlington, MA, USA.

Zervos, C.C. & Dumont, G.A. (1988). Deterministic Adaptive Control Based on Laguerre Series Representation, *Int. J. Control*, Vol. 48, No. 6, 2333-2359.